Carotenoids and Human Health

Carotenoids and Human Health

Special Issue Editor
Jaume Amengual

MDPI • Basel • Beijing • Wuhan • Barcelona • Belgrade

MDPI

Special Issue Editor
Jaume Amengual
University of Illinois Urbana-Champaign
USA

Editorial Office
MDPI
St. Alban-Anlage 66
4052 Basel, Switzerland

This is a reprint of articles from the Special Issue published online in the open access journal *Cancers* (ISSN 2072-6694) from 2018 to 2019 (available at: https://www.mdpi.com/journal/nutrients/special_issues/Carotenoids_Human_Health).

For citation purposes, cite each article independently as indicated on the article page online and as indicated below:

LastName, A.A.; LastName, B.B.; LastName, C.C. Article Title. *Journal Name* **Year**, *Article Number*, Page Range.

ISBN 978-3-03921-832-5 (Pbk)
ISBN 978-3-03921-833-2 (PDF)

Contents

About the Special Issue Editor

Jaume Amengual obtained his PhD in Biochemistry and Nutrigenomics at the University of the Balearic Islands (Spain), where he studied the role of vitamin A and β-carotene on lipid metabolism and adiposity. He continued his training at the Department of Pharmacology at Case Western Reserve University (US) as a postdoctoral fellow, where he studied the role of carotenoid cleaving enzymes and vitamin A transport and storage. Next, he moved to the Department of Cardiology at New York University where he focused on lipoprotein metabolism and atherosclerosis. Dr. Amengual is currently an Assistant Professor of Personalized Nutrition in the Department of Food Science and Human Nutrition at the University of Illinois at Urbana Champaign. He currently studies the effects of β-carotene and vitamin A in cardiometabolic diseases.

nutrients

MDPI

Editorial

Bioactive Properties of Carotenoids in Human Health

Jaume Amengual [1,2]

1   Department of Food Sciences and Human Nutrition, University of Illinois Urbana Champaign, Urbana, IL 61801, USA; jaume6@illinois.edu
2   Division of Nutritional Sciences, University of Illinois Urbana Champaign, Urbana, IL 61801, USA

Received: 7 August 2019; Accepted: 8 August 2019; Published: 6 October 2019

Keywords: retinoids; carotenoids; bioactive compounds

1. Introduction

Research shows that certain bioactive compounds in our diet have beneficial effects on human health. Among these bioactive molecules, carotenoids are some of the most chemically and functionally diverse molecules in food, which encouraged us to prepare a Special Issue on "Carotenoids and Human Health". The goal of this Special Issue is to provide a compilation of the recent advances on this important yet understudied research field. This Special Issue contains 11 original research articles, five literature reviews, one communication, and one discussion of cutting edge, peer-reviewed research papers (Figure 1).

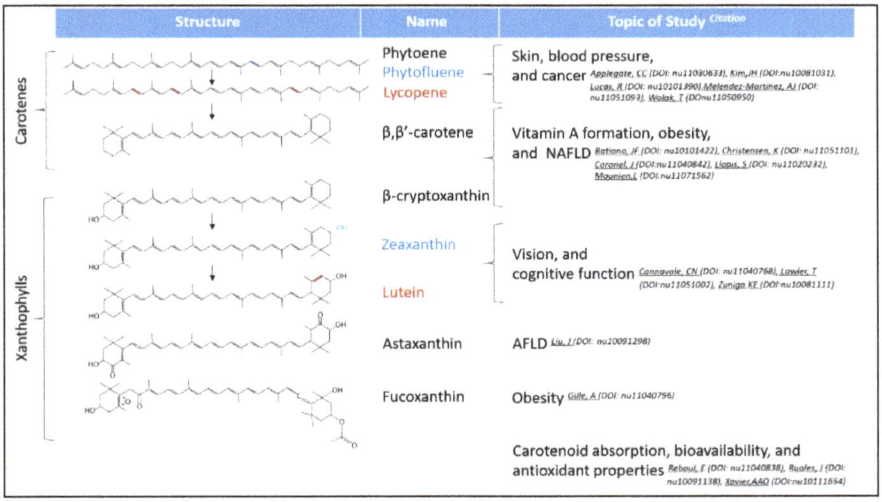

Figure 1. Topics covered by the Special Issue on carotenoids and human health. Colors present on carotenoids denote structural changes between similar carotenoids. Digital object identifier (DOI), non-alcoholic fatty liver disease (NAFLD), and alcoholic fatty liver disease (AFLD).

Carotenoids are a group of pigments produced by all photosynthetic organisms. Due to their broad distribution in nature, it is logical that heterotrophic organisms have adapted and coevolved to utilize carotenoids for their own profit, such it occurs in birds, where the plumage pigmentation is perceived as a "health signal", conditioning their reproductive success. In humans, some of these compounds serve as vitamin A precursor (pro-vitamin A carotenoids), and others are crucial for visual health (lutein), while most act as antioxidant molecules in lipid-rich environments. While there are over 650 different carotenoids described so far, the biological properties and therapeutic potential of these molecules has only been systematically studied for a handful of these molecules.

Chemically, carotenoids are 40-carbon molecules with conjugated double bonds. They are divided into carotenes, composed of only carbons and hydrogen atoms, and xanthophylls, which also contain oxygen. Humans obtain carotenoids from their diets and accumulate them in relatively large amounts in tissues and plasma, where they play a variety of biological functions. On this note, Dr. Reboul's discussion on carotenoid absorption provides a clear overview of this complex process, which will allow the reader to understand the biochemical and protein-mediated processes leading to carotenoid transport across the intestinal cell [1]. However, the first exposure to carotenoids seems to occur early during our development, as embryonic tissues express carotenoid-cleaving enzymes and transporters that allow the delivery of these compounds to the developing embryo [2,3]. Once born, breast milk will supply the developing child with the necessary carotenoids and retinoids. Xavier and colleagues studied if premature delivery affects the level of carotenoids in breast milk. They observed that in the case of premature birth, the total levels of carotenoids are decreased in the colostrum, with the exception of lutein, which in humans has a clear role in vision and possibly in brain development [4].

One of the most recognized properties of carotenoids is their antioxidant potential, as Ruales' group indicates. These authors took a modern approach focused on the beneficial effects of the discarded parts of mangos, which are rich in carotenoids. Ruales' work shows that mango skin and kennel are rich sources of bioactive compounds, and that some of these antioxidant molecules are not present in the pulp. These authors invites us to consider a better use of these fruit parts, which could also contribute to reduced waste production [5].

2. Bioactive Properties of Carotenes

Carotenes are one the first intermediates on carotenoid synthesis; therefore, these compounds are present in all photosynthetic organisms and highly abundant in our diet. Some of these early precursors such as phytoene and phytofluene are colorless, as they only contain three and five conjugated double bonds, respectively (Figure 1). As Dr. Melendez-Martinez's review points out, these carotenes have been largely ignored for many years by the carotenoid field, a topic that recently has been gaining more interest. Since these carotenes accumulate in the skin, it is not a surprise that these enigmatic carotenes play a role in skin protection against UV light and are therefore involved in skin aging and health [6].

Further saturation of phytofluene will generate lycopene, another intermediate of carotenoid and xanthophyll biosynthesis, as well as one of the most abundant carotenoids in our plasma and tissues. Unlike phytoene and phytofluene, lycopene is red, and it is predominantly found in tomato and tomato products. Several studies show that the consumption of foods rich in lycopene, and tissue levels of this acyclic carotene, are associated with a low incidence certain types of cancer and cardiovascular disease. Indeed, Applegate and colleagues prepared complete systematic review focused on the preclinical studies studying lycopene and the androgen axis that controls prostate cancer. The authors concluded that the scientific evidence published to date shows that lycopene has an inhibitory effect of this hormonal axis, which could explain the positive properties of lycopene on this devastating disease [7]. Next, Kim and colleagues studied the relationship of gastric cancer with dietary consumption of different carotenoids. In a large cohort of patients (415 patients suffering from gastric cancer and 830 control individuals), the authors showed that the consumption of lycopene-containing foods (tomatoes and their derivatives) has an inverse correlation with the development of gastric cancer [8].

As mentioned above, some studies have also shown a correlation between lycopene and the development of heart disease, at least in part by affecting blood pressure. In another exciting human study, Wolak and colleagues evaluated whether pure lycopene or a tomato-based formulation named Tomato Nutrient Complex could reduce blood pressure. The authors showed that only their tomato-based formulation was effective at reducing systolic blood pressure. Interestingly, the authors observed an increase in the plasma levels of phytoene and phytofluene, together with lycopene, suggesting that these carotenes could be the mediators of the positive effects described in their study [9].

3. Bioactive Properties of Xanthophylls

The addition of oxygen groups to carotenes, mostly hydroxyl, ketone, and epoxide groups, gives rise to xanthophylls. Due to the variety of chemical substitutions and isomers, xanthophylls are the most diverse and unknown of the two big carotenoid subfamilies. While xanthophylls are generated by plants, ingested xanthophylls can be modified in animals by isomerization or reduction of these oxidized groups as they occur systemically in rodents [10] or in the macula lutea of primates and humans [11]. Depending on the number and type of oxygen groups, xanthophylls can present different chemical properties, but overall this group is more polar than their carotene precursors, a property that facilitates their intestinal absorption.

Two of the most abundant carotenoids in our plasma are lutein and zeaxanthin. These xanthophylls are highly concentrated in our macula lutea, where they prevent photochemical damage. Many studies show that lutein supplementation, and to a lesser extent zeaxanthin supplementation, delays the development of age-related macular degeneration, a very common visual disorder in elderly. In Lawler's review, however, the authors focused on the effect of these xanthophylls, and other dietary antioxidants, on the development of glaucoma. Based on the authors' bibliographical research, a causal relationship between the risk of developing glaucoma and the consumption of antioxidants has not yet been established. However, as they point out for some preclinical studies, it is possible that lutein supplementation and zeaxanthin supplementation have some beneficial effects on this disease that deserve further study [12].

In recent years, researchers have been focusing on studying the role of lutein in cognition as this xanthophyll is preferentially accumulated in the brain. The contribution of Cannavale's work to this topic shows that lutein plasma levels correlate with elevated hippocampal function in obese individuals [13]. Additionally, Zuniga's paper establishes an association between total plasma carotenoid and inflammatory markers with cognitive function in cancer survivors. The authors observed that cancer survivors with low carotenoid plasma levels present higher rates of self-reported cognitive dysfunction [14]. Overall, these three articles contribute to understanding the positive functions of carotenoids, especially lutein, in vision and cognition. These effects could be a consequence of the specific accumulation of this pigment in neural tissues such as the brain and the eye.

Xanthophylls are the most diverse group of carotenoids, but the biological properties of most of them are still understudied. This is the example of astaxanthin, a red carotenoid present in shrimps and other seafood. Interestingly, the production for commercialization as a nutritional supplement is one of the largest growing industries in the carotenoid field. This interest is based on its antioxidant properties, together with the relatively easy production, as this carotenoid is mostly obtained from algae, avoiding chemical synthesis [15]. Liu's team examined the changes that occur in microbiota in response to astaxanthin supplementation and evaluated its effects in alcoholic fatty liver disease in mice. They observed that astaxanthin supplementation ameliorated ethanol-induced liver injury, and it is associated with the recovery of specific microbiome populations lost during ethanol supplementation [16]. Christensen's work, however, focused on non-alcoholic fatty liver disease, which is one of the most common pathologies in developed countries and is deeply associated with obesity and other lipid-related disorders. They used data from a large human cohort to compare the prevalence of the disease by the level of the main carotenoids in diets and supplements, and in plasma. The authors observed that higher levels of most carotenoids studied, including those with a provitamin A activity, were associated with lower odds of developing non-alcoholic fatty liver disease [17].

Another interesting xanthophyll covered in this Special Issue is fucoxanthin. This carotenoid is present in edible seaweeds, and is largely consumed in Asia. Gille's group evaluated the effects on obesity of a microalgae extract, rich in fucoxanthin, using mice as a preclinical model. The authors showed that this extract has beneficial effects on different adipose tissue parameters such as adipocyte size and gene expression markers. Interestingly, these effects seem to be slightly different than those observed with pure fucoxanthin in cultured adipocytes, indicating that it is possible that this carotenoid must be metabolized in the liver or intestine to exert its action, or that other components of the extract contribute to the positive effects of this carotenoid [18].

4. Carotenoids as Pro-Vitamin A Precursors

While carotenoids play various roles in human health, it is undeniable that the most important function of these compounds is the production of vitamin A. Only carotenoids with an unsubstituted β-ionone ring have pro-vitamin A activity, and the most important of these carotenoids, and the most abundant of them in our diet and tissues, is β, β'-carotene (β-carotene). Vitamin A is a potent gene regulator controlling the expression of nearly 700 genes. Additionally, vitamin A is required for vision, as retinal is the visual chromophore in mammals. One of the most interesting roles of β-carotene and vitamin A is the role they play in lipid metabolism and obesity. In this line, Coronel's group provides a bibliographical review of the effects of β-carotene in obesity. These authors divide their review into cell culture, animal, and human studies to dissect the main findings related to β-carotene and vitamin A over the past years, and use their expertise in the field to provide guidelines and considerations for future carotenoid researchers interested in joining this exciting field [19]. On the same line, Mounien summarizes the latest findings on the effect of different carotenoids on adipose tissue physiology, and the role of these compounds in the brain in relationship with obesity, proposing a provocative role of these molecules on the control of food intake, inflammation, and adipokine secretion [20].

Continuing with obesity, Llopis et al. studied the effect of β-cryptoxanthin, another provitamin A carotenoid, using *C. elegans* as a model. The authors showed that worms accumulated β-cryptoxanthin, and that this compound reduced oxidative stress. Additionally, the authors show that β-cryptoxanthin reduces lipid droplet accumulation and alters energy metabolic pathways [21], similarly to what was observed in white adipose tissue of β-carotene-fed mice [22].

As Coronel's review points out, the main transcriptionally active form of vitamin A is retinoic acid [19]. Under normal conditions, this metabolite is present in small quantities in tissues, and its quantification is challenging [23,24]. Lucas and collaborators quantified the presence and different isoforms of key carotenoids and retinoids, including all-trans retinoic acid, in plasma of patients affected with atopic dermatitis in comparison to healthy patients. This skin disorder is very common in children in Western societies, and can become chronic in adults. The exact causes of atopic dermatitis are not fully understood, but this disorder is characterized by hyper-reactivity of the immune system and skin to certain, otherwise harmless, antigens. The author shows that patients with atopic dermatitis present an overall reduction of key carotenoids such as lutein and β-carotene, accompanied by a reduction in all-trans retinoic acid. These results indicate that low levels of carotenoids/retinoids, together with an altered ratio of some of their isomers, could be associated to the maintenance or the development of atopic dermatitis [25].

Vitamin A deficiency is the single most important cause of childhood blindness in developing countries, and is responsible for millions of deaths due to immune-related disorders. Therefore, the early identification of this deficiency is crucial to prevent life-lasting consequences or even death. This exciting topic is the main focus of Bationo's paper, where the authors correlate the intake and plasma levels of pro-vitamin A carotenoids, and their association with vitamin A status among children in Burkina Faso [26].

5. Conclusions

Carotenoids and retinoids are part of our diet and present in our tissues and plasma. Understanding the implication of these bioactive molecules on human health will contribute to providing better nutritional guidance. Thanks to the work published in this Special Issue, we hope to increase the interest of this exciting field of research among researchers, and to also inform the general public of the importance of carotenoids on human health.

Funding: J.A. is funded by the American Heart Association (16SDG27550012), the U.S. Department of Agriculture (Multi-State grant project W4002), and the National Institute of Health (HL147252).

Acknowledgments: The author thanks all the authors that made this Special Issue possible through their scientific contributions.

Conflicts of Interest: The authors declare no conflicts of interest.

References

1. Reboul, E. Mechanisms of Carotenoid Intestinal Absorption: Where Do We Stand? *Nutrients* **2019**, *11*, 838. [CrossRef] [PubMed]
2. Spiegler, E.; Kim, Y.K.; Narayanasamy, H.B.S.; Jiang, H.; Savio, N.; Curley, R.W., Jr.; Harrison, E.H.; Hammerling, U.; Quadro, L. beta-apo-10'-carotenoids support normal embryonic development during vitamin A deficiency. *Sci. Rep.* **2018**, *8*, 8834. [CrossRef] [PubMed]
3. Giordano, E.; Quadro, L. Lutein, zeaxanthin and mammalian development: Metabolism, functions and implications for health. *Arch. Biochem. Biophys.* **2018**, *647*, 33–40. [CrossRef] [PubMed]
4. Xavier, A.A.O.; Diaz-Salido, E.; Arenilla-Velez, I.; Aguayo-Maldonado, J.; Garrido-Fernández, J.; Fontecha, J.; Sánchez-García, A.; Pérez-Gálvez, A. Carotenoid Content in Human Colostrum is Associated to Preterm/Full-Term Birth Condition. *Nutrients* **2018**, *10*, 1654. [CrossRef]
5. Ruales, J.; Baenas, N.; Moreno, D.A.; Stinco, C.M.; Melendez-Martinez, A.J.; Garcia-Ruiz, A. Biological Active Ecuadorian Mango 'Tommy Atkins' Ingredients—An Opportunity to Reduce Agrowaste. *Nutrients* **2018**, *10*, 1138. [CrossRef]
6. Melendez-Martinez, A.J.; Stinco, C.M.; Mapelli-Brahm, P. Skin Carotenoids in Public Health and Nutricosmetics: The Emerging Roles and Applications of the UV Radiation-Absorbing Colourless Carotenoids Phytoene and Phytofluene. *Nutrients* **2019**, *11*, 1093. [CrossRef]
7. Applegate, C.C.; Rowles, J.L., 3rd; Erdman, J.W., Jr. Can Lycopene Impact the Androgen Axis in Prostate Cancer? A Systematic Review of Cell Culture and Animal Studies. *Nutrients* **2019**, *11*, 633. [CrossRef]
8. Kim, J.H.; Lee, J.; Choi, I.J.; Kim, Y.I.; Kwon, O.; Kim, H.; Kim, J. Dietary Carotenoids Intake and the Risk of Gastric Cancer: A Case-Control Study in Korea. *Nutrients* **2018**, *10*, 1031. [CrossRef]
9. Wolak, T.; Sharoni, Y.; Levy, J.; Linnewiel-Hermoni, K.; Stepensky, D.; Paran, E. Effect of Tomato Nutrient Complex on Blood Pressure: A Double Blind, Randomized Dose(-)Response Study. *Nutrients* **2019**, *11*, 950. [CrossRef]
10. Amengual, J.; Lobo, G.P.; Golczak, M.; Li, H.N.; Klimova, T.; Hoppel, C.L.; Wyss, A.; Palczewski, K.; von Lintig, J. A mitochondrial enzyme degrades carotenoids and protects against oxidative stress. *FASEB J.* **2011**, *25*, 948–959. [CrossRef]
11. Johnson, E.J.; Neuringer, M.; Russell, R.M.; Schalch, W.; Snodderly, D.M. Nutritional manipulation of primate retinas, III: Effects of lutein or zeaxanthin supplementation on adipose tissue and retina of xanthophyll-free monkeys. *Investig. Ophthalmol. Vis. Sci.* **2005**, *46*, 692–702. [CrossRef] [PubMed]
12. Lawler, T.; Liu, Y.; Christensen, K.; Vajaranant, T.S.; Mares, J. Dietary Antioxidants, Macular Pigment, and Glaucomatous Neurodegeneration: A Review of the Evidence. *Nutrients* **2019**, *11*, 1002. [CrossRef] [PubMed]
13. Cannavale, C.N.; Hassevoort, K.M.; Edwards, C.G.; Thompson, S.V.; Burd, N.A.; Holscher, H.D.; Erdman, J.W., Jr.; Cohen, N.J.; Khan, N.A. Serum Lutein is related to Relational Memory Performance. *Nutrients* **2019**, *11*, 768. [CrossRef] [PubMed]
14. Zuniga, K.E.; Moran, N.E. Low Serum Carotenoids Are Associated with Self-Reported Cognitive Dysfunction and Inflammatory Markers in Breast Cancer Survivors. *Nutrients* **2018**, *10*, 1111. [CrossRef] [PubMed]
15. Barreiro, C.; Barredo, J.L. Carotenoids Production: A Healthy and Profitable Industry. *Methods Mol. Biol.* **2018**, *1852*, 45–55.
16. Liu, H.; Liu, M.; Fu, X.; Zhang, Z.; Zhu, L.; Zheng, X.; Liu, J. Astaxanthin Prevents Alcoholic Fatty Liver Disease by Modulating Mouse Gut Microbiota. *Nutrients* **2018**, *10*, 1298. [CrossRef]
17. Christensen, K.; Lawler, T.; Mares, J. Dietary Carotenoids and Non-Alcoholic Fatty Liver Disease among US Adults, NHANES 2003(-)2014. *Nutrients* **2019**, *11*, 1101. [CrossRef]
18. Gille, A.; Stojnic, B.; Derwenskus, F.; Trautmann, A.; Schmid-Staiger, U.; Posten, C.; Briviba, K.; Palou, A.; Bonet, M.L.; Ribot, J. A Lipophilic Fucoxanthin-Rich Phaeodactylum tricornutum Extract Ameliorates Effects of Diet-Induced Obesity in C57BL/6J Mice. *Nutrients* **2019**, *11*, 796. [CrossRef]

19. Coronel, J.; Pinos, I.; Amengual, J. Beta-carotene in Obesity Research: Technical Considerations and Current Status of the Field. *Nutrients* **2019**, *11*, 842. [CrossRef]

20. Mounien, L.; Tourniaire, F.; Landrier, J.F. Anti-Obesity Effect of Carotenoids: Direct Impact on Adipose Tissue and Adipose Tissue-Driven Indirect Effects. *Nutrients* **2019**, *11*, 1562. [CrossRef]

21. Llopis, S.; Rodrigo, M.J.; Gonzalez, N.; Genovés, S.; Zacarías, L.; Ramón, D.; Martorell, P. Beta-Cryptoxanthin Reduces Body Fat and Increases Oxidative Stress Response in Caenorhabditis elegans Model. *Nutrients* **2019**, *11*, 232. [CrossRef] [PubMed]

22. Amengual, J.; Gouranton, E.; van Helden, Y.G.; Hessel, S.; Ribot, J.; Kramer, E.; Kiec-Wilk, B.; Razny, U.; Lietz, G.; Wyss, A.; et al. Beta-carotene reduces body adiposity of mice via BCMO1. *PLoS ONE* **2011**, *6*, e20644. [CrossRef] [PubMed]

23. Kane, M.A.; Napoli, J.L. Quantification of endogenous retinoids. *Methods Mol. Biol.* **2010**, *652*, 1–54. [PubMed]

24. Ruhl, R. Method to determine 4-oxo-retinoic acids, retinoic acids and retinol in serum and cell extracts by liquid chromatography/diode-array detection atmospheric pressure chemical ionisation tandem mass spectrometry. *Rapid Commun. Mass Spectrom.* **2006**, *20*, 2497–2504. [CrossRef]

25. Lucas, R.; Mihaly, J.; Lowe, G.M.; Graham, D.L.; Szklenar, M.; Szegedi, A.; Töröcsik, D.; Rühl, R. Reduced Carotenoid and Retinoid Concentrations and Altered Lycopene Isomer Ratio in Plasma of Atopic Dermatitis Patients. *Nutrients* **2018**, *10*, 1390. [CrossRef]

26. Bationo, J.F.; Zeba, A.N.; Abbeddou, S.; Coulibaly, N.D.; Sombier, O.O.; Sheftel, J.; Bassole, I.H.N.; Barro, N.; Ouedraogo, J.B.; Tanumihardjo, S.A. Serum Carotenoids Reveal Poor Fruit and Vegetable Intake among Schoolchildren in Burkina Faso. *Nutrients* **2018**, *10*, 1422. [CrossRef]

nutrients

MDPI

Discussion

Mechanisms of Carotenoid Intestinal Absorption: Where Do We Stand?

Emmanuelle Reboul

Aix-Marseille University, INRA, INSERM, C2VN, 13005 Marseille, France; Emmanuelle.Reboul@univ-amu.fr; Tel.: +33-4-91-324-278

Received: 26 February 2019; Accepted: 6 April 2019; Published: 13 April 2019

Abstract: A growing literature is dedicated to the understanding of carotenoid beneficial health effects. However, the absorption process of this broad family of molecules is still poorly understood. These highly lipophilic plant metabolites are usually weakly absorbed. It was long believed that β-carotene absorption (the principal provitamin A carotenoid in the human diet), and thus all other carotenoid absorption, was driven by passive diffusion through the brush border of the enterocytes. The identification of transporters able to facilitate carotenoid uptake by the enterocytes has challenged established statements. After a brief overview of carotenoid metabolism in the human upper gastrointestinal tract, a focus will be put on the identified proteins participating in the transport and the metabolism of carotenoids in intestinal cells and the regulation of these processes. Further progress in the understanding of the molecular mechanisms regulating carotenoid intestinal absorption is still required to optimize their bioavailability and, thus, their health effects.

Keywords: carotenes; xanthophylls; bioavailability; intestine; membrane transporters

1. Introduction

Carotenoids are hydrophobic molecules synthesized by plants and by some microorganisms (bacteria, algae, or fungi). Carotenoid physicochemical properties determine their distribution in the cellular environment: carotenoids are associated with membrane lipid bilayers and cytosolic lipid droplets. More than 600 different carotenoids have been found in nature, but only 40 in the human diet and about 20 have been clearly identified in human blood and tissues [1,2].

Carotenoids are polymers of isopentenyl diphosphate with 40 carbon atoms. They derive chemically from a basic structure formed by the linear sequence of 8 isoprenic units, associated in two groups of four units head to tail. The first molecule of this biosynthesis pathway is phytoene, which presents 3 conjugated double bonds in the center of the molecule and 6 other unconjugated double bonds through the length of the molecule. Phytoene is then enzymatically desaturated to produce phytofluene and eventually lycopene, a linear basic structure ($C_{40}H_{56}$) with many conjugated double bonds showing a characteristic red color [3]. Other carotenes derive from lycopene by cyclization and dehydrogenation, and xanthophylls derive from carotenes by oxidation [4] (Table 1). Some xanthophyll carotenoids such as β-cryptoxanthin [5] or lutein [6] can be found in esterified forms. Additionally, each carotenoid double bond can take a *trans* or *cis* configuration. Most of natural carotenoids are *all-trans* molecules, but *cis*-isomers can be produced during heat treatments [7]. Finally, it is worth mentioning that small amounts of apocarotenoids, i.e., cleavage products of parent 40-C carotenoids, can be naturally found in foods and/or produced during food processing, but usually represent less than 5% of the parent carotenoid levels [8,9].

The main dietary sources of carotenoids are colored fruits and vegetables (Table 1).

Table 1. Main dietary carotenoids.

Carotenoids	Molecular Structure	Examples of Food Sources (mg/100 g) [10,11]
Phytoene		Tomato juice: 2.24 Carrot juice: 0.94
Phytofluene		Tomato juice: 0.86 Carrot juice: 0.59
Lycopene		Tomato sauce: 15.92 Tomatoes: 3.03 Watermelon: 4.87
β-carotene		Raw carrot: 8.84 Canned carrot: 5.78 Cooked spinach: 5.24
α-carotene		Carrot juice: 1.70
β-cryptoxanthin		Sanguinello juice: 0.02
Lutein		Cooked spinach: 7.04 Lettuce: 2.64

Carotenoid health properties were initially mainly accredited to their antioxidant properties as carotenoid are, at least in vitro, powerful radical quenchers [12]. Later, new investigations have highlighted the carotenoid ability to regulate intracellular signalling cascades, thus influencing both gene expression and protein translation in a broad number of metabolic pathways related to inflammatory and oxidative stress modulation [13]. However, the physiological relevance of these observations still needs to be fully established in humans. Interestingly, randomized placebo-controlled clinical trials have evidenced that supplementation with the xanthophylls lutein and zeaxanthin, which specifically accumulate into the human macula, was associated with improved visual function and decreased risk of progression to late age macular degeneration. These xanthophylls also display encouraging preventive and therapeutic effects on cataracts and retinopathies [14]. Finally, some carotenoids, such as α-carotene, β-carotene and β-cryptoxanthin, are vitamin A precursors. Indeed, they can be cleaved, mainly at the intestinal level, and metabolized into retinol (cf. Section 3.2).

After a short description of carotenoid fate in the human upper gastrointestinal lumen, a focus will be put on the identified proteins participating in carotenoid transport and metabolism in intestinal cells and on the regulation of these processes.

2. Digestion Process of Carotenoids

Fat-soluble micronutrients, including carotenoids, follow the fate of lipids in the human upper digestive tract. The first step of their digestion is thus their dissolution in the fat phase of the meal [15,16]. This phase is emulsified into lipid droplets in the stomach and duodenum.

The hypothesis that a carotenoid *cis*-isomerization could take place during gastric digestion was emitted [17], but finally refuted by a study in humans [16]. Recently, in vitro digestion experiments mimicking duodenal conditions showed that no significant isomerization of lycopene, β-carotene, or lutein occurred either [18].

It has been suggested that xanthophyll ester hydrolysis by lipases is indispensable prior to absorption. The cholesterol ester hydrolase (CEH) from pancreatic juice is likely responsible for the release of free xanthophyll from xanthophyll esters [5]. The remaining xanthophyll esters, if any, may either be cleaved at the brush border level or enter the enterocyte to be hydrolyzed in the cytosol [19].

During duodenal digestion, carotenoids are incorporated with other lipids (i.e., cholesterol, phospholipids) and lipid digestion products (i.e., free fatty acids, monoacylglycerols, lysophospholipids) into mixed micelles [20]. A fraction of carotenoids may also associate with proteins. For instance, the milk lipocalin β-lactoglobulin is able to bind β-carotene and does not alter its absorption compared to mixed micelles [21]. However, the mechanisms of carotenoid absorption may depend on carotenoid binding vehicles. Mixed micelles are likely isolated from the rest of the bolus in the unstirred water layer of the glycocalyx area and approach the brush border membrane [22] where carotenoids can be absorbed by passive diffusion and/or via a transporter-dependent process (see Section 3).

Carotenoid bioaccessibility (i.e., the fraction of carotenoids released form their food matrix and included in mixed micelles—which represents the fraction of carotenoids potentially able to be absorbed by the intestine) is highly variable. An in vitro digestion study highlighted that lycopene bioaccessibility was very limited (from 0.1% in raw tomatoes to 1.5% in tomato puree), β-carotene bioaccessibility was fairly low (from about 4% in carrot puree to 14% in carrot juice), while lutein bioaccessibility was the highest (from 37% in raw spinach leaves to 48% in boiled spinach). These values correlate with in vivo data and highlight the fact that the disruption of the food matrix by thermal treatment or processing can increase carotenoid bioaccessibility [10]. Xanthophylls were consistently shown to display a higher bioaccessibility than carotenes in different studies [10,11], probably because the presence of one or two hydroxylated group(s) increases their solubility into the micellar structures. Interestingly, phytoene and phytofluene also displayed a very high bioaccessibility. This may be linked to their more flexible molecular structure, compared to other carotenoids, which likely increase their incorporation into mixed micelles as well [11].

3. Carotenoid Absorption through the Enterocyte

Absorption efficiency of labelled β-carotene is widely variable among clinical studies, fluctuating from ≈3% to 80%, but usually ranging from 10% to 30% [23,24]. This can partly be due to the variable bioaccessibility of β-carotene (see above), but it may also reflect its moderate uptake and transport through the enterocyte. It should be mentioned that β-carotene absorption efficiency was usually measured following a single meal. However, the intestine can store β-carotene from a first meal to release it during subsequent postprandial phases in humans [25]. β-carotene absorption efficiency may, thus, be underestimated in some trials.

Studies using differentiated Caco-2 cell monolayers showed that phytofluene, β-carotene, and lutein uptakes were similar and significantly higher than that of phytoene, while lycopene uptake was the lowest [26,27]. In the same way as bioaccessibility, uptake efficiency thus seems to correlate with carotenoid polarity and flexibility. This may be explained by the fact that polar and flexible carotenoids present a better affinity for lipid transporters and/or for plasma membranes, which would lead to an increased absorption.

3.1. Apical Transport Across the Brush Border Membrane of the Enterocyte

Carotenoid uptake by the enterocytes has been considered to occur by passive diffusion for four decades, which was inconsistent with the high inter-individual variability in absorption observed in humans, as well as with the isomer selectivity and the competition for absorption between carotenoids and other fat-soluble micronutrients observed at the intestinal level (see [20] for review). Different teams started to re-explore carotenoid absorption mechanisms in the 2000s and several lipid transporters playing a role in carotenoid uptake by the intestinal cell have since been identified.

A first critical result was the identification of the gene ninaD encoding a class B scavenger receptor, essential for xanthophyll cellular distribution in *Drosophila* [28]. In 2005, we then identified the Scavenger Receptor class B type I: SR-BI as a key transporter of lutein in human intestinal Caco2 TC7 cells. This ubiquitous transmembrane glycoprotein found at the apical membrane of the enterocytes is expressed following a decreasing gradient from the duodenum to the colon [29]. Intestinal SR-BI was shown to facilitate the uptake of free cholesterol, but also of other lipids such as cholesterol esters, phospholipids, and triacylglycerol hydrolysis products, thus presenting a low substrate specificity [30,31]. The effective role of SR-BI in terms of cholesterol transport is still subject to debate [32] and SR-BI was recently presented as a cholesterol sensor [33], regulating chylomicron secretion [34]. Its involvement in the intestinal uptake of carotenoids has been extended to lycopene [35], provitamin A carotenoids [36], as well as to phytoene and phytofluene [27]. As SR-BI is also involved in the uptake of vitamin D [37], E [38], and K [39], in cultured cells and in mice, we suggest that another primary role of SR-BI in the gut is the transport of minor molecules, such as fat-soluble vitamins and carotenoids. However, we specifically showed, using both Caco2 cells and transfected HEK cells, that SR-BI was not involved in the uptake of micellar preformed vitamin A (retinol) [36].

Another pervasive scavenger receptor of interest is CD36 (CD 36 molecule). This membrane protein is highly expressed at the brush border level of the duodenum and the jejunum [40]. It is supposed to play a key role in the intestinal uptake of long-chain fatty acids [41], but also displays a broad substrate specificity [42,43]. Recently, CD36 has been described as a lipid sensor and its impact on chylomicron secretion has been established in many studies [44]. Besides, CD36 facilitates, directly or indirectly, fat-soluble vitamin uptake in the intestine [37,39,45]. CD36 was also shown to facilitate the uptake of lycopene, β-carotene, α-carotene, β-cryptoxanthine, and lutein, but not that of phytoene and phytofluene, in transfected Griptite cells and/or cultured adipocytes [27,36,46]. This result was confirmed ex vivo for β-carotene using brush-border membrane vesicles from CD36-deficient and wild-type mouse intestines [47].

A last candidate for carotenoid uptake is the NPC1-like transporter 1 (NPC1L1), which is a major sterol transporter in the intestine [37,48]. NPC1L1 was suggested to be involved in α-carotene, β-carotene, β-cryptoxanthin, and lutein intestinal uptake [49,50], but not in that of lycopene, phytoene, and phytofluene [27,35].

It is still possible that a fraction of carotenoid is absorption via a passive diffusion process, depending on the carotenoid concentration in the lumen. We previously showed in Caco-2 cells that vitamin D absorption is carrier-mediated at physiological concentrations and occurs by passive diffusion at pharmacological concentrations [37]. We suggest that a similar phenomenon occurs for carotenoids.

Recently, we showed that a fraction of phytoene and phytofluene taken up by the intestinal cells could be effluxed back to the lumen [27]. This phenomenon was previously acknowledged for fat-soluble vitamins such as vitamin D, E, and K and was shown to be, at least partly, SR-B-dependent [37–39]. This efflux may contribute to the limited absorption efficiency of carotenoids. Further research is needed to clearly identify the membrane transporters participating in this pathway. Besides SR-BI, ABCB1 (ATP binding cassette B1, also known as P-glycoprotein) and ABCG transporters, such as ABCG5, appear as good candidates. Indeed, a recent study combining in silico, cell culture, animal, and genetic approaches showed that ABCB1 was involved in vitamin D intestinal efflux [51]. Additionally, polymorphisms in the *ABCG5* gene tended to contribute to individual response to lutein supplementation in humans [52].

3.2. Cytosolic Transport and Intracellular Metabolism

No carotenoid carrier protein has clearly been identified in the human gut so far. However, the lutein-binding protein HR-LBP (Human Retinal Lutein-Binding Protein) present in the human retina cross-reacts with antibodies raised against a carotenoid-binding protein present in the *Bombyx mori* midgut [53], suggesting that it could be an intestinal intracellular transporter of xanthophylls [54]. As carotenoid membrane transporters SR-BI, CD36, and NPC1L1 can traffic in the enterocyte, especially after a fat load, we previously suggested that they may act as cytosolic carotenoid transporters [20]. However, this hypothesis still remains to be verified. Similarly, the association found between a genetic variant in the Intestinal Fatty-Acid Binding Protein (IFABP) and the fasting plasma lycopene concentrations in humans [55] still need to be challenged to assess whether IFABP is actually a carotenoid-carrier.

Up to 40% of absorbed carotenoids remain unmetabolized [56]. β-carotene can be cleaved into retinal by a cytosolic enzyme, BCO1 (β-carotene oxygenase 1), via a one-step process in the enterocyte [57]. β-Carotene "low-converter" phenotypes, which have been reported in several clinical studies [58], are likely due to genetic variation in *BCO1* gene. Other provitamin A carotenoids, such as β-cryptoxanthin, can be cleaved into retinal though a multi-step process involving both mitochondrial BCO2 (β-carotene oxygenase 2) and cytosolic BCO1 [59]. The produced retinal is subsequently converted into retinol and esterified into retinyl esters by the lecithin:retinol acyltransferase (LRAT) and probably by the diacylglycerol acyltransferase 1 (DGAT1) that displays an acyl-CoA:retinol acyltransferase activity [60,61]. Both provitamin A and nonprovitamin A carotenoids can also be cleaved asymmetrically in apocarotenoids by BCO2 [62]. However, a recent study showed that only traces of asymmetric [^{13}C]-β-apo-carotenals were found in plasma after [^{13}C]-β-carotene ingestion, suggesting a lack of significant postprandial intestinal BCO2 activity in healthy humans [63].

No *cis-trans* isomerization of β-carotene was measured in intestinal cultured cells [26]. As *cis*-isomerization does not occur in the gastrointestinal lumen (see above), the site of the 9-*cis* isomerization of β-carotene reported in vivo [64] remain undetermined. Conversely, lycopene isomerization in *cis*-isomers was identified to occur at the enterocyte level [65].

A summary of carotenoid transport pathways across the enterocyte is depicted in Figure 1.

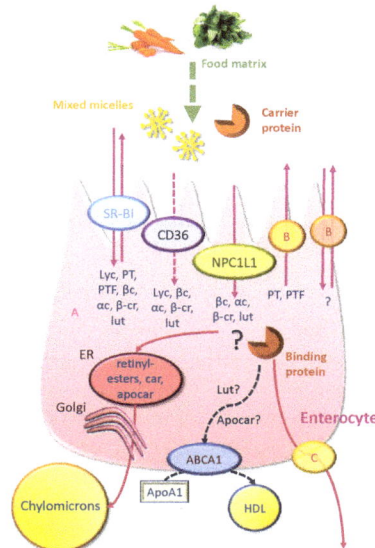

Figure 1. Uptake, transport, and secretion pathways of carotenoids across the enterocyte. PT = phytoene;

PTF = phytofluene; Lyc = lycopene; βC = β-carotene; αC = α-carotene; βCr = β-cryptoxanthine; Lut = lutein; Car = carotenoids; Apocar = apocarotenoids; A = passive diffusion; B = unidentified apical transporter; C = unidentified basolateral efflux transporter; ? = putative pathway, and ER = endoplasmic reticulum. Carotenoids are captured from mixed micelles and possibly from carrier proteins by apical membrane transporters SR-BI, CD36, and NPC1L1. A fraction of PT and PTF can then be effluxed back to the intestinal lumen via apical membrane transporters (likely SR-BI and possibly other transporters). Another fraction is transported to the site where they are incorporated into chylomicrons. Some proteins may be involved in intracellular transport of carotenoids, but none has been clearly identified. Provitamin A carotenoids are partly metabolized into retinyl-esters. Retinyl-esters and carotenoids are secreted in the lymph into chylomicrons, while a part of xanthophylls and a part of the more polar metabolites, such as some apocarotenoids, may be secreted via an HDL pathway.

3.3. Secretion Through the Basolateral Membrane of the Enterocyte

During the postprandial period, the major fraction of free carotenoids and retinyl esters originating from provitamin A carotenoid cleavage are packaged into chylomicrons (apoB-dependent pathway) that are secreted into the lymph to further join the bloodstream [20]. A non-apoB-dependent pathway (via high-density lipoproteins, HDL), mediated by the ABCA1 transporter, has been involved in vitamin E absorption [66] and possibly allows a part of free retinol absorption [20]. This HDL pathway may also exist for xanthophylls, such as lutein and zeaxanthin [67], but has not been proven to occur for other carotenoids. However, recent studies showed that several genetic variants in *ABCA1* gene were associated with lycopene [68], β-carotene [69], and lutein [70] postprandial responses in healthy subjects. Thus, further research is needed to fully understand the contribution of the intestinal HDL pathway to carotenoid absorption in humans.

4. Regulation of Carotenoid Transporter Expressions in the Enterocyte

Crucial factors modulating the expression and/or the activity of intestinal proteins involved in carotenoid absorption are provitamin A carotenoids, through a feedback regulation. Indeed, studies have pointed out that SR-BI activity is partly controlled by retinoids. Using both mouse models and human cell lines, it was specifically shown that retinoic acid produced from dietary precursors by BCO1 induced the expression of the intestinal transcription factor ISX that repressed the expression of both BCO1 [71] and SR-B1 [72], thus impacting both carotenoid conversion and uptake [73].

Additionally, many dietary factors other than retinoids were shown to regulate transporter expression in the intestine and may, thus, indirectly impact on carotenoid absorption.

Among these dietary factors, fat and fatty acids seem to play major roles. For instance, SR-BI expression in Caco-2 cells is increased by micellar oleic and ecosapentaenoic acids [74]. Conversely, CD36, NPC1L1, or ABCA1 expressions in rodent intestines are downregulated by dietary fat, including oleic acid [75,76] and cholesterol [77]. Such downregulation in NPC1L1 and ABCA1 expressions was also found after exposure of cultured intestinal FHs 74 or Caco-2 cells to phytosterols [78,79] and to long-chain polyunsaturated fatty acids [74,80].

In addition, dietary glucose increases SR-BI expression in both Caco-2 cells and mouse intestines [81] and decreases Caco2 cell ABCA1 expression [82].

Finally, some polyphenols were shown to decrease both SR-BI, NPCL1, and ABCA1 expressions in Caco-2 cells [83,84] and so did the cholesterol-lowering drug ezetimibe [49].

Host factors can also regulate carotenoid transporters. Among them, insulin resistance increases SR-BI intestinal expression in hamsters [85]. SR-BI post-transcriptional regulation also seems be dependent on bile secretion, with bile salts leading to a rise of intestinal SR-BI expression in rodents [86]. Finally, NPC1L1 expression was increased by estrogen [87] and cholecystokinin [88] in mouse intestines, while its expression was decreased by peptide YY in Caco-2 cells [89] (see [90] for review).

As the above results were exclusively obtained in cultured cells and animal studies, further investigations are deeply needed to address their relevance in humans.

5. Conclusions

To conclude, the understanding of carotenoid intestinal absorption by the intestine is far from being fully understood. Proteins including lipid membrane transporters (i.e., SR-BI, CD36, NPC1L1), the cleavage enzyme BCO1, and the transcription factor ISX have been showed to play important roles in carotenoid intestinal uptake and metabolism, but other proteins likely remain to be identified.

Genome-wide association studies (GWAS) and candidate gene association studies have identified correlations between single nucleotide polymorphisms in *SCARB1* (encoding SR-BI), *CD36*, *NPC1L1*, *BCO1*, and *ISX* and carotenoid blood concentrations. Interestingly, these studies also highlighted the impact of polymorphisms in genes encoding proteins likely indirectly linked to carotenoid metabolism (i.e., ELOVL fatty acid elongase 2). The involvement of such proteins in carotenoid intestinal metabolism still needs to be defined [91].

Carotenoid "low responder" or "high responder" phenotypes presumably correspond to individuals bearing associations of several "disadvantageous" or "advantageous" polymorphisms, respectively. In the future, it would thus be of major interest to take into account the carotenoid "low responder" or "high responder" phenotypes that are due to different transport and/or conversion efficiency, to propose tailored dietary recommendations to individuals and to thus optimize carotenoid health benefits.

Funding: This research received no external funding.

Conflicts of Interest: The author declares no conflict of interest.

Abbreviations

ABCA1	ATP binding cassette A1
ABCB1	ATB binding cassette B1
ABCG5	ATP binding cassette G5
BCO1	β-carotene-oxygenase 1
BCO2	β-carotene-oxygenase 2
CD36	CD36 molecule
CEH	cholesterol ester hydrolase
DGAT1	diacylglycerol acyltransferase 1
HR-LBP	human retinal lutein-binding protein
ISX	intestine-specific homebox
FABP	fatty-acid-binding protein
HDL	high-density lipoproteins
LRAT	lecithin retinol acyltransferase
NPC1L1	NPC1-like transporter 1
SR-BI	scavenger receptor class B type 1

References

1. Khachik, F.; Spangler, C.J.; Smith, J.C., Jr.; Canfield, L.M.; Steck, A.; Pfander, H. Identification, quantification, and relative concentrations of carotenoids and their metabolites in human milk and serum. *Anal. Chem.* **1997**, *69*, 1873–1881. [PubMed]
2. Tapiero, H.; Townsend, D.M.; Tew, K.D. The role of carotenoids in the prevention of human pathologies. *Biomed. Pharmacother.* **2004**, *58*, 100–110. [CrossRef]
3. Engelmann, N.J.; Clinton, S.K.; Erdman, J.W., Jr. Nutritional aspects of phytoene and phytofluene, carotenoid precursors to lycopene. *Adv. Nutr.* **2011**, *2*, 51–61. [PubMed]
4. Moise, A.R.; Al-Babili, S.; Wurtzel, E.T. Mechanistic aspects of carotenoid biosynthesis. *Chem. Rev.* **2014**, *114*, 164–193. [CrossRef] [PubMed]
5. Breithaupt, D.E.; Bamedi, A.; Wirt, U. Carotenol fatty acid esters: Easy substrates for digestive enzymes? *Comp. Biochem. Physiol. B Biochem. Mol. Biol.* **2002**, *132*, 721–728. [CrossRef]

6. Bowen, P.E.; Herbst-Espinosa, S.M.; Hussain, E.A.; Stacewicz-Sapuntzakis, M. Esterification does not impair lutein bioavailability in humans. *J. Nutr.* **2002**, *132*, 3668–3673. [CrossRef]

7. Shi, J.; Le Maguer, M. Lycopene in tomatoes: Chemical and physical properties affected by food processing. *Crit. Rev. Biotechnol.* **2000**, *20*, 293–334. [CrossRef]

8. Rodriguez, E.B.; Rodriguez-Amaya, D.B. Formation of apocarotenals and epoxycarotenoids from beta-carotene by chemical reactions and by autoxidation in model systems and processed foods. *Food Chem.* **2007**, *101*, 563–572. [CrossRef]

9. Kopec, R.E.; Riedl, K.M.; Harrison, E.H.; Curley, R.W., Jr.; Hruszkewycz, D.P.; Clinton, S.K.; Schwartz, S.J. Identification and quantification of apo-lycopenals in fruits, vegetables, and human plasma. *J. Agric. Food Chem.* **2010**, *58*, 3290–3296. [CrossRef]

10. Reboul, E.; Richelle, M.; Perrot, E.; Desmoulins-Malezet, C.; Pirisi, V.; Borel, P. Bioaccessibility of carotenoids and vitamin e from their main dietary sources. *J. Agric. Food Chem.* **2006**, *54*, 8749–8755. [CrossRef] [PubMed]

11. Mapelli-Brahm, P.; Corte-Real, J.; Melendez-Martinez, A.J.; Bohn, T. Bioaccessibility of phytoene and phytofluene is superior to other carotenoids from selected fruit and vegetable juices. *Food Chem.* **2017**, *229*, 304–311. [CrossRef] [PubMed]

12. Krinsky, N.I.; Yeum, K.J. Carotenoid-radical interactions. *Biochem. Biophys. Res. Commun.* **2003**, *305*, 754–760. [CrossRef]

13. Kaulmann, A.; Bohn, T. Carotenoids, inflammation, and oxidative stress—Implications of cellular signaling pathways and relation to chronic disease prevention. *Nutr. Res.* **2014**, *34*, 907–929. [CrossRef] [PubMed]

14. Scripsema, N.K.; Hu, D.N.; Rosen, R.B. Lutein, zeaxanthin, and meso-zeaxanthin in the clinical management of eye disease. *J. Ophthalmol.* **2015**, *2015*, 865179. [CrossRef]

15. Borel, P.; Grolier, P.; Armand, M.; Partier, A.; Lafont, H.; Lairon, D.; Azais-Braesco, V. Carotenoids in biological emulsions: Solubility, surface-to-core distribution, and release from lipid droplets. *J. Lipid Res.* **1996**, *37*, 250–261. [PubMed]

16. Tyssandier, V.; Reboul, E.; Dumas, J.F.; Bouteloup-Demange, C.; Armand, M.; Marcand, J.; Sallas, M.; Borel, P. Processing of vegetable-borne carotenoids in the human stomach and duodenum. *Am. J. Physiol. Gastrointest. Liver Physiol.* **2003**, *284*, G913–G923. [CrossRef] [PubMed]

17. Re, R.; Fraser, P.D.; Long, M.; Bramley, P.M.; Rice-Evans, C. Isomerization of lycopene in the gastric milieu. *Biochem. Biophys. Res. Commun.* **2001**, *281*, 576–581. [CrossRef]

18. Kopec, R.E.; Gleize, B.; Borel, P.; Desmarchelier, C.; Caris-Veyrat, C. Are lutein, lycopene, and beta-carotene lost through the digestive process? *Food Funct.* **2017**, *8*, 1494–1503. [CrossRef]

19. Dhuique-Mayer, C.; Borel, P.; Reboul, E.; Caporiccio, B.; Besancon, P.; Amiot, M.J. Beta-cryptoxanthin from citrus juices: Assessment of bioaccessibility using an in vitro digestion/caco-2 cell culture model. *Br. J. Nutr.* **2007**, *97*, 883–890. [CrossRef]

20. Reboul, E. Absorption of vitamin a and carotenoids by the enterocyte: Focus on transport proteins. *Nutrients* **2013**, *5*, 3563–3581. [CrossRef]

21. Mensi, A.; Borel, P.; Goncalves, A.; Nowicki, M.; Gleize, B.; Roi, S.; Chobert, J.M.; Haertle, T.; Reboul, E. Beta-lactoglobulin as a vector for beta-carotene food fortification. *J. Agric. Food Chem.* **2014**, *62*, 5916–5924. [CrossRef]

22. Phan, C.T.; Tso, P. Intestinal lipid absorption and transport. *Front. Biosci.* **2001**, *6*, D299–D319. [CrossRef] [PubMed]

23. van Lieshout, M.; West, C.E.; van Breemen, R.B. Isotopic tracer techniques for studying the bioavailability and bioefficacy of dietary carotenoids, particularly beta-carotene, in humans: A review. *Am. J. Clin. Nutr.* **2003**, *77*, 12–28. [CrossRef]

24. Van Loo-Bouwman, C.A.; Naber, T.H.; van Breemen, R.B.; Zhu, D.; Dicke, H.; Siebelink, E.; Hulshof, P.J.; Russel, F.G.; Schaafsma, G.; West, C.E. Vitamin a equivalency and apparent absorption of beta-carotene in ileostomy subjects using a dual-isotope dilution technique. *Br. J. Nutr.* **2010**, *103*, 1836–1843. [CrossRef]

25. Borel, P.; Grolier, P.; Mekki, N.; Boirie, Y.; Rochette, Y.; Le Roy, B.; Alexandre-Gouabau, M.C.; Lairon, D.; Azais-Braesco, V. Low and high responders to pharmacological doses of beta-carotene: Proportion in the population, mechanisms involved and consequences on beta-carotene metabolism. *J. Lipid Res.* **1998**, *39*, 2250–2260.

26. During, A.; Hussain, M.M.; Morel, D.W.; Harrison, E.H. Carotenoid uptake and secretion by caco-2 cells: Beta-carotene isomer selectivity and carotenoid interactions. *J. Lipid Res.* **2002**, *43*, 1086–1095. [CrossRef]

27. Mapelli-Brahm, P.; Desmarchelier, C.; Margier, M.; Reboul, E.; Melendez Martinez, A.J.; Borel, P. Phytoene and phytofluene isolated from a tomato extract are readily incorporated in mixed micelles and absorbed by caco-2 cells, as compared to lycopene, and sr-bi is involved in their cellular uptake. *Mol. Nutr. Food Res.* **2018**, *62*, e1800703. [CrossRef] [PubMed]

28. Kiefer, C.; Sumser, E.; Wernet, M.F.; Von Lintig, J. A class b scavenger receptor mediates the cellular uptake of carotenoids in drosophila. *Proc. Natl. Acad. Sci. USA* **2002**, *99*, 10581–10586. [CrossRef] [PubMed]

29. Reboul, E.; Soayfane, Z.; Goncalves, A.; Cantiello, M.; Bott, R.; Nauze, M.; Terce, F.; Collet, X.; Comera, C. Respective contributions of intestinal niemann-pick c1-like 1 and scavenger receptor class b type i to cholesterol and tocopherol uptake: In vivo v. In vitro studies. *Br. J. Nutr.* **2012**, *107*, 1296–1304. [CrossRef] [PubMed]

30. Hauser, H.; Dyer, J.H.; Nandy, A.; Vega, M.A.; Werder, M.; Bieliauskaite, E.; Weber, F.E.; Compassi, S.; Gemperli, A.; Boffelli, D.; et al. Identification of a receptor mediating absorption of dietary cholesterol in the intestine. *Biochemistry* **1998**, *37*, 17843–17850. [CrossRef] [PubMed]

31. Bietrix, F.; Yan, D.; Nauze, M.; Rolland, C.; Bertrand-Michel, J.; Comera, C.; Schaak, S.; Barbaras, R.; Groen, A.K.; Perret, B.; et al. Accelerated lipid absorption in mice overexpressing intestinal sr-bi. *J. Biol. Chem.* **2006**, *281*, 7214–7219. [CrossRef]

32. Nguyen, D.V.; Drover, V.A.; Knopfel, M.; Dhanasekaran, P.; Hauser, H.; Phillips, M.C. Influence of class b scavenger receptors on cholesterol flux across the brush border membrane and intestinal absorption. *J. Lipid Res.* **2009**, *50*, 2235–2244. [CrossRef]

33. Saddar, S.; Carriere, V.; Lee, W.R.; Tanigaki, K.; Yuhanna, I.S.; Parathath, S.; Morel, E.; Warrier, M.; Sawyer, J.K.; Gerard, R.D.; et al. Scavenger receptor class b type i is a plasma membrane cholesterol sensor. *Circ. Res.* **2013**, *112*, 140–151. [CrossRef] [PubMed]

34. Lino, M.; Farr, S.; Baker, C.; Fuller, M.; Trigatti, B.; Adeli, K. Intestinal scavenger receptor class b type i as a novel regulator of chylomicron production in healthy and diet-induced obese states. *Am. J. Physiol. Gastrointest. Liver Physiol.* **2015**, *309*, G350–G359. [CrossRef]

35. Moussa, M.; Landrier, J.F.; Reboul, E.; Ghiringhelli, O.; Comera, C.; Collet, X.; Frohlich, K.; Bohm, V.; Borel, P. Lycopene absorption in human intestinal cells and in mice involves scavenger receptor class b type i but not niemann-pick c1-like 1. *J. Nutr.* **2008**, *138*, 1432–1436. [CrossRef] [PubMed]

36. Borel, P.; Lietz, G.; Goncalves, A.; Szabo de Edelenyi, F.; Lecompte, S.; Curtis, P.; Goumidi, L.; Caslake, M.J.; Miles, E.A.; Packard, C.; et al. Cd36 and sr-bi are involved in cellular uptake of provitamin a carotenoids by caco-2 and hek cells, and some of their genetic variants are associated with plasma concentrations of these micronutrients in humans. *J. Nutr.* **2013**, *143*, 448–456. [CrossRef]

37. Reboul, E.; Goncalves, A.; Comera, C.; Bott, R.; Nowicki, M.; Landrier, J.F.; Jourdheuil-Rahmani, D.; Dufour, C.; Collet, X.; Borel, P. Vitamin d intestinal absorption is not a simple passive diffusion: Evidences for involvement of cholesterol transporters. *Mol. Nutr. Food Res.* **2011**, *55*, 691–702. [CrossRef]

38. Reboul, E.; Klein, A.; Bietrix, F.; Gleize, B.; Malezet-Desmoulins, C.; Schneider, M.; Margotat, A.; Lagrost, L.; Collet, X.; Borel, P. Scavenger receptor class b type i (sr-bi) is involved in vitamin e transport across the enterocyte. *J. Biol. Chem.* **2006**, *281*, 4739–4745. [CrossRef]

39. Goncalves, A.; Margier, M.; Roi, S.; Collet, X.; Niot, I.; Goupy, P.; Caris-Veyrat, C.; Reboul, E. Intestinal scavenger receptors are involved in vitamin k1 absorption. *J. Biol. Chem.* **2014**, *289*, 30743–30752. [CrossRef]

40. Terpstra, V.; van Amersfoort, E.S.; van Velzen, A.G.; Kuiper, J.; van Berkel, T.J. Hepatic and extrahepatic scavenger receptors: Function in relation to disease. *Arterioscler. Thromb. Vasc. Biol.* **2000**, *20*, 1860–1872. [CrossRef]

41. Drover, V.A.; Nguyen, D.V.; Bastie, C.C.; Darlington, Y.F.; Abumrad, N.A.; Pessin, J.E.; London, E.; Sahoo, D.; Phillips, M.C. Cd36 mediates both cellular uptake of very long chain fatty acids and their intestinal absorption in mice. *J. Biol. Chem.* **2008**, *283*, 13108–13115. [CrossRef] [PubMed]

42. Rigotti, A.; Acton, S.L.; Krieger, M. The class b scavenger receptors sr-bi and cd36 are receptors for anionic phospholipids. *J. Biol. Chem.* **1995**, *270*, 16221–16224. [CrossRef] [PubMed]

43. Endemann, G.; Stanton, L.W.; Madden, K.S.; Bryant, C.M.; White, R.T.; Protter, A.A. Cd36 is a receptor for oxidized low density lipoprotein. *J. Biol. Chem.* **1993**, *268*, 11811–11816. [PubMed]

44. Buttet, M.; Traynard, V.; Tran, T.T.; Besnard, P.; Poirier, H.; Niot, I. From fatty-acid sensing to chylomicron synthesis: Role of intestinal lipid-binding proteins. *Biochimie* **2014**, *96*, 37–47. [CrossRef] [PubMed]

45. Goncalves, A.; Roi, S.; Nowicki, M.; Niot, I.; Reboul, E. Cluster-determinant 36 (cd36) impacts on vitamin e postprandial response. *Mol. Nutr. Food Res.* **2014**, *58*, 2297–2306. [CrossRef] [PubMed]

46. Moussa, M.; Gouranton, E.; Gleize, B.; El Yazidi, C.; Niot, I.; Besnard, P.; Borel, P.; Landrier, J.F. Cd36 is involved in lycopene and lutein uptake by adipocytes and adipose tissue cultures. *Mol. Nutr. Food Res.* **2011**, *55*, 578–584. [CrossRef]

47. van Bennekum, A.; Werder, M.; Thuahnai, S.T.; Han, C.H.; Duong, P.; Williams, D.L.; Wettstein, P.; Schulthess, G.; Phillips, M.C.; Hauser, H. Class b scavenger receptor-mediated intestinal absorption of dietary beta-carotene and cholesterol. *Biochemistry* **2005**, *44*, 4517–4525. [CrossRef]

48. Davis, H.R., Jr.; Altmann, S.W. Niemann-pick c1 like 1 (npc1l1) an intestinal sterol transporter. *Biochim. Biophys. Acta* **2009**, *1791*, 679–683. [CrossRef]

49. During, A.; Dawson, H.D.; Harrison, E.H. Carotenoid transport is decreased and expression of the lipid transporters sr-bi, npc1l1, and abca1 is downregulated in caco-2 cells treated with ezetimibe. *J. Nutr.* **2005**, *135*, 2305–2312. [CrossRef] [PubMed]

50. Sato, Y.; Suzuki, R.; Kobayashi, M.; Itagaki, S.; Hirano, T.; Noda, T.; Mizuno, S.; Sugawara, M.; Iseki, K. Involvement of cholesterol membrane transporter niemann-pick c1-like 1 in the intestinal absorption of lutein. *J. Pharm. Pharm. Sci.* **2012**, *15*, 256–264. [CrossRef]

51. Margier, M.; Collet, X.; le May, C.; Desmarchelier, C.; Andre, F.; Lebrun, C.; Defoort, C.; Bluteau, A.; Borel, P.; Lespine, A.; et al. Abcb1 (p-glycoprotein) regulates vitamin d absorption and contributes to its transintestinal efflux. *FASEB J.* **2019**, *33*, 2084–2094. [CrossRef] [PubMed]

52. Herron, K.L.; McGrane, M.M.; Waters, D.; Lofgren, I.E.; Clark, R.M.; Ordovas, J.M.; Fernandez, M.L. The abcg5 polymorphism contributes to individual responses to dietary cholesterol and carotenoids in eggs. *J. Nutr.* **2006**, *136*, 1161–1165. [CrossRef] [PubMed]

53. Tabunoki, H.; Sugiyama, H.; Tanaka, Y.; Fujii, H.; Banno, Y.; Jouni, Z.E.; Kobayashi, M.; Sato, R.; Maekawa, H.; Tsuchida, K. Isolation, characterization, and cdna sequence of a carotenoid binding protein from the silk gland of bombyx mori larvae. *J. Biol. Chem.* **2002**, *277*, 32133–32140. [CrossRef] [PubMed]

54. Bhosale, P.; Li, B.; Sharifzadeh, M.; Gellermann, W.; Frederick, J.M.; Tsuchida, K.; Bernstein, P.S. Purification and partial characterization of a lutein-binding protein from human retina. *Biochemistry* **2009**, *48*, 4798–4807. [CrossRef] [PubMed]

55. Borel, P.; Moussa, M.; Reboul, E.; Lyan, B.; Defoort, C.; Vincent-Baudry, S.; Maillot, M.; Gastaldi, M.; Darmon, M.; Portugal, H.; et al. Human fasting plasma concentrations of vitamin e and carotenoids, and their association with genetic variants in apo c-iii, cholesteryl ester transfer protein, hepatic lipase, intestinal fatty acid binding protein and microsomal triacylglycerol transfer protein. *Br. J. Nutr.* **2009**, *101*, 680–687. [PubMed]

56. Castenmiller, J.J.M.; West, C.E. Bioavailability and bioconversion of carotenoids. *Annu. Rev. Nutr.* **1998**, *18*, 19–38. [CrossRef] [PubMed]

57. dela Sena, C.; Riedl, K.M.; Narayanasamy, S.; Curley, R.W., Jr.; Schwartz, S.J.; Harrison, E.H. The human enzyme that converts dietary provitamin a carotenoids to vitamin a is a dioxygenase. *J. Biol. Chem.* **2014**, *289*, 13661–13666. [CrossRef] [PubMed]

58. Lobo, G.P.; Amengual, J.; Palczewski, G.; Babino, D.; von Lintig, J. Mammalian carotenoid-oxygenases: Key players for carotenoid function and homeostasis. *Biochim. Biophys. Acta* **2012**, *1821*, 78–87. [CrossRef] [PubMed]

59. Amengual, J.; Widjaja-Adhi, M.A.; Rodriguez-Santiago, S.; Hessel, S.; Golczak, M.; Palczewski, K.; von Lintig, J. Two carotenoid oxygenases contribute to mammalian provitamin a metabolism. *J. Biol. Chem.* **2013**, *288*, 34081–34096. [CrossRef] [PubMed]

60. O'Byrne, S.M.; Wongsiriroj, N.; Libien, J.; Vogel, S.; Goldberg, I.J.; Baehr, W.; Palczewski, K.; Blaner, W.S. Retinoid absorption and storage is impaired in mice lacking lecithin:Retinol acyltransferase (lrat). *J. Biol. Chem.* **2005**, *280*, 35647–35657. [CrossRef]

61. Wongsiriroj, N.; Piantedosi, R.; Palczewski, K.; Goldberg, I.J.; Johnston, T.P.; Li, E.; Blaner, W.S. The molecular basis of retinoid absorption: A genetic dissection. *J. Biol. Chem.* **2008**, *283*, 13510–13519. [CrossRef]

62. Palczewski, G.; Amengual, J.; Hoppel, C.L.; von Lintig, J. Evidence for compartmentalization of mammalian carotenoid metabolism. *FASEB J.* **2014**, *28*, 4457–4469. [CrossRef]

63. Kopec, R.E.; Caris-Veyrat, C.; Nowicki, M.; Gleize, B.; Carail, M.; Borel, P. Production of asymmetric oxidative metabolites of [13c]-beta-carotene during digestion in the gastrointestinal lumen of healthy men. *Am. J. Clin. Nutr.* **2018**, *108*, 803–813. [CrossRef]

64. You, C.S.; Parker, R.S.; Goodman, K.J.; Swanson, J.E.; Corso, T.N. Evidence of cis-trans isomerization of 9-cis-beta-carotene during absorption in humans. *Am. J. Clin. Nutr.* **1996**, *64*, 177–183. [CrossRef]

65. Richelle, M.; Sanchez, B.; Tavazzi, I.; Lambelet, P.; Bortlik, K.; Williamson, G. Lycopene isomerisation takes place within enterocytes during absorption in human subjects. *Br. J. Nutr.* **2010**, *103*, 1800–1807. [CrossRef]

66. Reboul, E.; Trompier, D.; Moussa, M.; Klein, A.; Landrier, J.F.; Chimini, G.; Borel, P. Atp-binding cassette transporter a1 is significantly involved in the intestinal absorption of alpha- and gamma-tocopherol but not in that of retinyl palmitate in mice. *Am. J. Clin. Nutr.* **2009**, *89*, 177–184. [CrossRef]

67. Niesor, E.J.; Chaput, E.; Mary, J.L.; Staempfli, A.; Topp, A.; Stauffer, A.; Wang, H.; Durrwell, A. Effect of compounds affecting abca1 expression and cetp activity on the hdl pathway involved in intestinal absorption of lutein and zeaxanthin. *Lipids* **2014**, *49*, 1233–1243. [CrossRef]

68. Borel, P.; Desmarchelier, C.; Nowicki, M.; Bott, R. Lycopene bioavailability is associated with a combination of genetic variants. *Free Radic. Biol. Med.* **2015**, *83*, 238–244. [CrossRef]

69. Borel, P.; Desmarchelier, C.; Nowicki, M.; Bott, R. A combination of single-nucleotide polymorphisms is associated with interindividual variability in dietary beta-carotene bioavailability in healthy men. *J. Nutr.* **2015**, *145*, 1740–1747. [CrossRef]

70. Borel, P.; Desmarchelier, C.; Nowicki, M.; Bott, R.; Morange, S.; Lesavre, N. Interindividual variability of lutein bioavailability in healthy men: Characterization, genetic variants involved, and relation with fasting plasma lutein concentration. *Am. J. Clin. Nutr.* **2014**, *100*, 168–175. [CrossRef]

71. Seino, Y.; Miki, T.; Kiyonari, H.; Abe, T.; Fujimoto, W.; Kimura, K.; Takeuchi, A.; Takahashi, Y.; Oiso, Y.; Iwanaga, T.; et al. Isx participates in the maintenance of vitamin a metabolism by regulation of beta-carotene 15,15′-monooxygenase (bcmo1) expression. *J. Biol. Chem.* **2008**, *283*, 4905–4911. [CrossRef] [PubMed]

72. Choi, M.Y.; Romer, A.I.; Hu, M.; Lepourcelet, M.; Mechoor, A.; Yesilaltay, A.; Krieger, M.; Gray, P.A.; Shivdasani, R.A. A dynamic expression survey identifies transcription factors relevant in mouse digestive tract development. *Development* **2006**, *133*, 4119–4129. [CrossRef]

73. Lobo, G.P.; Hessel, S.; Eichinger, A.; Noy, N.; Moise, A.R.; Wyss, A.; Palczewski, K.; von Lintig, J. Isx is a retinoic acid-sensitive gatekeeper that controls intestinal beta,beta-carotene absorption and vitamin a production. *FASEB J.* **2010**, *24*, 1656–1666. [CrossRef] [PubMed]

74. Goncalves, A.; Gleize, B.; Roi, S.; Nowicki, M.; Dhaussy, A.; Huertas, A.; Amiot, M.J.; Reboul, E. Fatty acids affect micellar properties and modulate vitamin d uptake and basolateral efflux in caco-2 cells. *J. Nutr. Biochem.* **2013**, *24*, 1751–1757. [CrossRef] [PubMed]

75. de Vogel-van den Bosch, H.M.; de Wit, N.J.; Hooiveld, G.J.; Vermeulen, H.; van der Veen, J.N.; Houten, S.M.; Kuipers, F.; Muller, M.; van der Meer, R. A cholesterol-free, high-fat diet suppresses gene expression of cholesterol transporters in murine small intestine. *Am. J. Physiol. Gastrointest. Liver Physiol.* **2008**, *294*, G1171–G1180. [CrossRef] [PubMed]

76. Chen, M.; Yang, Y.; Braunstein, E.; Georgeson, K.E.; Harmon, C.M. Gut expression and regulation of fat/cd36: Possible role in fatty acid transport in rat enterocytes. *Am. J. Physiol. Endocrinol. Metab.* **2001**, *281*, E916–E923. [CrossRef] [PubMed]

77. Davis, H.R., Jr.; Zhu, L.J.; Hoos, L.M.; Tetzloff, G.; Maguire, M.; Liu, J.; Yao, X.; Iyer, S.P.; Lam, M.H.; Lund, E.G.; et al. Niemann-pick c1 like 1 (npc1l1) is the intestinal phytosterol and cholesterol transporter and a key modulator of whole-body cholesterol homeostasis. *J. Biol. Chem.* **2004**, *279*, 33586–33592. [CrossRef]

78. Jesch, E.D.; Seo, J.M.; Carr, T.P.; Lee, J.Y. Sitosterol reduces messenger rna and protein expression levels of niemann-pick c1-like 1 in fhs 74 int cells. *Nutr. Res.* **2009**, *29*, 859–866. [CrossRef]

79. Brauner, R.; Johannes, C.; Ploessl, F.; Bracher, F.; Lorenz, R.L. Phytosterols reduce cholesterol absorption by inhibition of 27-hydroxycholesterol generation, liver x receptor alpha activation, and expression of the basolateral sterol exporter atp-binding cassette a1 in caco-2 enterocytes. *J. Nutr.* **2012**, *142*, 981–989. [CrossRef] [PubMed]

80. Alvaro, A.; Rosales, R.; Masana, L.; Vallve, J.C. Polyunsaturated fatty acids down-regulate in vitro expression of the key intestinal cholesterol absorption protein npc1l1: No effect of monounsaturated nor saturated fatty acids. *J. Nutr. Biochem.* **2010**, *21*, 518–525. [CrossRef] [PubMed]

81. Malhotra, P.; Boddy, C.S.; Soni, V.; Saksena, S.; Dudeja, P.K.; Gill, R.K.; Alrefai, W.A. D-glucose modulates intestinal niemann-pick c1-like 1 (npc1l1) gene expression via transcriptional regulation. *Am. J. Physiol. Gastrointest. Liver Physiol.* **2013**, *304*, G203–G210. [CrossRef] [PubMed]

82. Boztepe, T.; Gulec, S. Investigation of the influence of high glucose on molecular and genetic responses: An in vitro study using a human intestine model. *Genes Nutr.* **2018**, *13*, 11. [CrossRef]

83. Kim, B.; Park, Y.; Wegner, C.J.; Bolling, B.W.; Lee, J. Polyphenol-rich black chokeberry (aronia melanocarpa) extract regulates the expression of genes critical for intestinal cholesterol flux in caco-2 cells. *J. Nutr. Biochem.* **2013**, *24*, 1564–1570. [CrossRef] [PubMed]

84. Feng, D.; Zou, J.; Zhang, S.; Li, X.; Lu, M. Hypocholesterolemic activity of curcumin is mediated by down-regulating the expression of niemann-pick c1-like 1 in hamsters. *J. Agric. Food Chem.* **2017**, *65*, 276–280. [CrossRef]

85. Hayashi, A.A.; Webb, J.; Choi, J.; Baker, C.; Lino, M.; Trigatti, B.; Trajcevski, K.E.; Hawke, T.J.; Adeli, K. Intestinal sr-bi is upregulated in insulin-resistant states and is associated with overproduction of intestinal apob48-containing lipoproteins. *Am. J. Physiol. Gastrointest. Liver Physiol.* **2011**, *301*, G326–G337. [CrossRef] [PubMed]

86. Voshol, P.J.; Schwarz, M.; Rigotti, A.; Krieger, M.; Groen, A.K.; Kuipers, F. Down-regulation of intestinal scavenger receptor class b, type i (sr-bi) expression in rodents under conditions of deficient bile delivery to the intestine. *Biochem. J.* **2001**, *356*, 317–325. [CrossRef]

87. Duan, L.P.; Wang, H.H.; Ohashi, A.; Wang, D.Q. Role of intestinal sterol transporters abcg5, abcg8, and npc1l1 in cholesterol absorption in mice: Gender and age effects. *Am. J. Physiol. Gastrointest. Liver Physiol.* **2006**, *290*, G269–G276. [CrossRef]

88. Zhou, L.; Yang, H.; Okoro, E.U.; Guo, Z. Up-regulation of cholesterol absorption is a mechanism for cholecystokinin-induced hypercholesterolemia. *J. Biol. Chem.* **2014**, *289*, 12989–12999. [CrossRef]

89. Grenier, E.; Garofalo, C.; Delvin, E.; Levy, E. Modulatory role of pyy in transport and metabolism of cholesterol in intestinal epithelial cells. *PLoS ONE* **2012**, *7*, e40992. [CrossRef]

90. Reboul, E. Vitamin e intestinal absorption: Regulation of membrane transport across the enterocyte. *IUBMB Life* **2019**, *71*, 416–423. [CrossRef]

91. Desmarchelier, C.; Landrier, J.F.; Borel, P. Genetic factors involved in the bioavailability of tomato carotenoids. *Curr. Opin. Clin. Nutr. Metab. Care* **2018**, *21*, 489–497. [CrossRef] [PubMed]

nutrients

MDPI

Article

Carotenoid Content in Human Colostrum is Associated to Preterm/Full-Term Birth Condition

Ana A. O. Xavier [1], Elena Díaz-Salido [2], Isabel Arenilla-Vélez [2], Josefa Aguayo-Maldonado [2], Juan Garrido-Fernández [1], Javier Fontecha [3], Alicia Sánchez-García [4] and Antonio Pérez-Gálvez [1,*]

[1] Department of Food Phytochemistry, Instituto de la Grasa (CSIC), Campus Universitario, Building 46, 41013 Sevilla, Spain; anaaugustax@gmail.com (A.A.O.X.); jgarrido@ig.csic.es (J.G.-F.)

[2] Unidad de Neonatología, Hospital Virgen del Rocío, 41013 Sevilla, Spain; elenam.diaz.sspa@juntadeandalucia.es (E.D.-S.); Isaareni@yahoo.es (I.A.-V.); pepaaguayo@gmail.com (J.A.-M.)

[3] Institute of Food Science Research (CSIC-UAM), 28049 Madrid, Spain; j.fontecha@csic.es

[4] Laboratorio de Espectrometría de Masas, Instituto de la Grasa (CSIC), 41013 Sevilla, Spain; asangar@ig.csic.es

* Correspondence: aperez@ig.csic.es; Tel.: +34-954-611-550

Received: 13 October 2018; Accepted: 31 October 2018; Published: 3 November 2018

Abstract: Factors such as lactation stage and premature and small-for-gestational conditions could lead to great inter-individual variability in the carotenoid content of human milk. The aim was to analyze the carotenoid content in colostrum and mature milk of preterm (PT) and full-term (FT) mothers to establish whether they are significantly different and, if so, the stage of lactation when the differences are established. Samples of blood, colostrum, and mature milk were collected from Spanish donating mothers who gave birth to PT or FT infants. Carotenoids from serum and milk samples were analyzed by HPLC-atmospheric pressure chemical ionization (APCI)-MS. Quantitatively, colostrum from PT mothers presented lower total carotenoid content when compared to that from FT mothers. The only exception was lutein, where levels were not different. The transition from colostrum to mature milk makes observed differences in the carotenoid content disappear, since there were no variances between PT and FT groups for both individual and total carotenoid content. The premature birth condition affects the quantitative carotenoid composition of the colostrum but has no effect on the lutein content. This fact could be related to the significant role of this xanthophyll in the development of infant retina and feasibly to cognitive function.

Keywords: breastfeeding; new-born; full-term mothers; preterm mothers; xanthophylls; carotenes; lutein; colostrum; mature milk

---

1. Introduction

Breastfeeding provides health advantages for the immediate growth and development of the new-born, benefits that will extend into adulthood [1], and decreases the risk of child morbidity and mortality [2,3]. International policy agencies recommend exclusive breastfeeding for six months, followed by continued breastfeeding with appropriate complementary foods for up to two years or beyond [4]. Breast milk contains the macro and micronutrients necessary to fulfil nutritional requirements of the infant, as well as immune and nonimmune properties and growth factors [5,6]. Many of these constituents in the milk are derived from maternal blood, which is enriched through diet. This is the case of carotenoid pigments that are exclusively synthesized de novo by photosynthetic organisms, bacteria, yeasts, and some fungi, and incorporated into human tissues and blood through food sources, fruits, vegetables, algae, eggs, and fish [7]. The human milk carotenoids with provitamin

A activity contribute to the vitamin A needs of the infant, a key issue in those developing countries where the dietary supply of vitamin A sources is often limited [8]. Independently of the provitamin A activity, carotenoids exert important functions in immunity, participate in the antioxidant defense system, and are related to a reduced risk of developing chronic diseases [9,10]. The specific accumulation of lutein and zeaxanthin in the fovea of the retina is of remarkable interest, given that this tissue continues to develop during infancy [11]. These carotenoids, known as macular pigments, prevent retinal damage through their blue light filtration and antioxidant properties [12–14], and contribute to the proper structural development of the retina [15]. Furthermore, it has been shown that enriching the macular pigment density by lutein supplementation positively impacts on contrast sensitivity and glare disability [16]. Recent research correlates macular pigment content with cognitive function through different evidence, as it has been shown that the increase in visual performance reached with lutein supplementation is associated with better cognition [17,18], in a similar fashion as the levels of macular lutein and zeaxanthin are related to brain lutein and zeaxanthin content [19]. Also, it would be important to point out that these levels were significantly related to pre-mortem measures of cognitive function [20,21].

These statements may apply in the prenatal and postnatal developing infant. Vishwanathan et al. [22] have shown that lutein is the predominant carotenoid in the infant brain, with much lower levels of zeaxanthin, β-cryptoxanthin, and β-carotene. This accumulation is the starting point to support that carotenoids contribute to early neurodevelopment [23,24]. It could be assumed that the accumulation of carotenoids in the infant brain behave in a similar fashion to the macular pigments in the infant retina, where the carotenoids tend to accumulate in the final months of gestation [25,26]. Indeed, the content of macular pigments is variable in the developing retina [27] and such variability would extend to their brain content if both quantities are correlated as noted above.

The macular and brain carotenoid levels after delivery are a consequence of the infant intake of carotenoids from colostrum and mature milk or infant formula, because these are the only dietary sources containing carotenoids that the new-born receives [23]. Also, it is well established that there is a high-amount of inter-individual and intra-individual variability in milk carotenoid content, which it is associated to maternal dietary habits [28,29], lactation stage [30–33], and health status of the mother, including alcohol intake and smoking [34,35]. With this knowledge, we are beginning to understand the factors contributing to the efficiency of the carotenoid accumulation in infant tissues during the initial months of age. In addition, some inputs are now identified regarding the similarities and differences associated with the carotenoid transfer from maternal blood with the transport of bulk lipids during lactation. Even the existence of transport mechanism of xanthophylls [36] and an active acylation pathway in the mammary glands [37] have been observed. Other features influencing the carotenoid status of the infant at birth, which are now under analysis, are the premature and small-for-gestational conditions. The highest utero accretion of energy and nutrients takes place during the last weeks of pregnancy and prematurity impairs the accumulation of carotenoids into the tissues of the child as shown by Vishwanathan et al. [22], who reported lower carotenoid content in brain of preterm (PT) infants compared to those born at full-term (FT) condition.

The aim of this work was to analyze the carotenoid content in the colostrum and mature milk of preterm and full-term mothers to determine whether they are significantly different and, if so, the stage of lactation when the differences are established and whether they are transitory or not. Prematurity is not a condition only affecting the developmental and nutritional stages of the infant, but also the maturity of the mammary gland and its ability to secrete milk with the appropriate composition for the situation of the new-born. An incomplete gestational period affects the deposition of nutrients in the mammary gland—as it has been described for the accumulation of fatty acids [38]—and the deficiencies may happen at birth or appear during the lactation transition to mature milk [39]. Thus, the incomplete maturity of the mammary gland could be a significant factor that may affect the potential of the breastfeeding to compensate for the nutritional deficits of the PT child. Neither the carotenoid content of the colostrum nor the dynamics of carotenoid content changes occurring in mothers who gave birth prematurely and the comparison with mothers whose children were born at full-term, have been analyzed so far.

2. Materials and Methods

2.1. Subjects

The study population comprised 144 healthy women, which were recruited within a 12 month period between 2015 and 2016 and classified into two groups: mothers (n = 72) who gave birth to full-term neonates (37–40 weeks) and mothers (n = 72) who gave birth to preterm neonates (28–35 weeks). Eligible participants in this study were non-smoking mothers with no chronic diseases. Mothers following any special diets, vegetarians, or those taking supplements were not included. Exclusion criteria applied were pathologies and/or infections during the gestation, developmental anomalies in the fetus, or death of the child. They provided informed consent to participate in the research program and none of the initially enrolled mothers dropped out of the study.

Our previous study [37] indicated that 70 volunteers per group would be necessary to detect significant differences of the individual carotenoid profile between preterm and full-term delivery conditions and between lactation stages at a significance level of p = 0.05.

2.2. Measures

2.2.1. Milk and Blood Sample Collection

The samples were collected at the Unidad de Neonatología of the Hospital Universitario Virgen del Rocío (Seville, Spain). Preterm and full-term mothers donated colostrum at 3–5 days postpartum and the same mothers donated mature milk at 30 days postpartum. Milk samples were obtained by collection of the total milk volume of one breast during one milk expression session into a polypropylene bottle. Fasting blood samples (10 mL) were collected at the same time of milk samples. Blood was centrifuged after clotting for 10 min at 2000× g at 4 °C to obtain serum. The samples were transported directly to the laboratory and stored at −80 °C until analysis.

2.2.2. Extraction of the Carotenoid Fraction

The experimental conditions previously described by Ríos et al. [37] were used for extraction of carotenoids from human milk samples. Human milk (3 mL) was mixed with 3 mL of KOH:methanol (20% w/v), and the mixture was incubated for 1 h. After hydrolysis, 6 mL of methanol were added, and the mixture was vortex-mixed for 2 min and cooled at −20 °C for 20 min. Subsequently, the cooled mixture was centrifuged at 10,000× g and 4 °C for 5 min, discarding the upper layer. Diethyl ether (5 mL) and hexane (2 mL) were added to the pellet and vortex-mixed for 2 min. Then, 5 mL of NaCl 10% (w/v) was added, and the sample was vortex-mixed again for 2 min. After centrifugation (10,000× g at 4 °C for 5 min), the organic layer was washed with water until neutral pH was reached. The organic extract was evaporated to dryness in a rotatory evaporator at 25 °C, and the residue was dissolved in 1 mL of methanol:methyl tert-butyl ether (8:2). This solution was filtered through a 0.22-µm filter and stored at −20 °C until analysis. Carotenoid extraction from serum samples was carried out according to the method described by Pérez-Gálvez et al. [40]. Serum (0.1 mL) was extracted with 6 mL hexane:dichloromethane (5:1). The mixture was vortex-mixed for 1 min and centrifuged at 2000× g at 20 °C for 10 min. A portion of the upper phase (5 mL) was withdrawn and the solvent evaporated under N_2. The extract was dissolved in 100 µL of acetone and stored at −20 °C until analysis by HPLC-MS, which was performed within 1 week.

2.2.3. Identification and Quantification of Carotenoids in the Carotenoid Extracts

The carotenoids from human milk and serum samples were separated using a Dionex Ultimate 3000RS U-HPLC (Thermo Fisher Scientific, Waltham, MA, USA) using the method developed by

Breithaupt et al. [41] with slight modifications [37]. High-resolution mass spectrometry measurements were completed on the basis of mass accuracy and in combination with the isotopic pattern in the SigmaFit algorithm [37]. The characteristics of experimental mass spectrum and the MS^2 data were compared with the data available in the literature for carotenoid identification in human milk and serum samples [42–48]. For carotenoid quantification, stock solutions of β-carotene, β-cryptoxanthin, lutein, and lycopene were prepared at a concentration of 25 mg/L. Once the exact concentration was determined (2% maximum total error), working stock solutions for external calibration curves were prepared at 5 concentration levels ranging from 0.15 to 10.0 mg/L. The content of xanthophylls and carotenes was determined in the unsaponified extract. Zeaxanthin was quantified with the calibration curve of lutein, while xanthophyll esters were quantified as their corresponding free xanthophyll [47,48].

2.2.4. Determination of the Lipid Content

The lipid content of human milk samples was determined according to the solvent extraction procedure and then by gravimetry [49].

2.3. Ethics Approval

All subjects gave their informed consent for inclusion before they participated in the study, which was conducted in accordance with the Declaration of Helsinki. The study protocol was approved by the Ethics Committee of the Hospital Universitario Virgen del Rocío and the Bioethics Subcommittee of the Spanish National Research Council (AGL2013-42757R).

2.4. Statistical Analysis

Data are reported as the median, including 25th and 75th percentiles. Due to the non-normality of content of individual carotenoids (Kolmogorov–Smirnov test), the data were analyzed using a non-parametric statistical procedure in the SPSS software (IBM® SPSS® Statistics version 24, IBM, New York, NY, USA). The Mann–Whitney test was applied to analyze the differences in carotenoid content of colostrum and mature milk of PT and FT mothers, and to compare the carotenoid content of colostrum and mature milk within each group of lactating mothers. The same test was applied for carotenoid content in serum samples of both groups and at both collection stages (colostrum and mature milk). The significance was set at $p < 0.05$.

3. Results

The carotenoid composition of the colostrum and mature milk samples from PT and FT mothers is shown in Table 1. Colostrum samples from both groups were qualitatively equal regarding the carotenoid profile, but significant differences were observed when the groups were compared quantitatively. Specifically, the colostrum of FT mothers showed higher carotenoid content, both for individual carotenoids and for the total amount (median of total carotenoid content of 4961.1 nM in the colostrum of full-term mothers vs. 2641.1 nM in the colostrum from preterm mothers, $p < 0.05$), except for the lutein content that was not different between groups (486.31 nM vs. 432.83 nM, $p = 0.238$). Figure 1 shows the percentage distribution of the carotenoids in colostrum and mature milk of both groups of mothers. Xanthophylls, which are carotenoids containing oxygenated functions in their structure, and carotenes, which are strictly hydrocarbons, were similarly accumulated in colostrum samples of PT and FT mothers, as the ratio between both groups of pigments was close to 1 (0.97 and 0.90 for the preterm and the full-term group, respectively). The ranking of individual carotenoids in the colostrum samples was recurrent for both groups, with zeaxanthin presenting the lowest content value followed by the pool of xanthophyll esters (sum of lutein, zeaxanthin, and β-cryptoxanthin esters), the free xanthophylls lutein and β-cryptoxanthin, and finally by the carotenes α- and β-carotene and lycopene. The total fat content of colostrum was not related with the PT or FT condition, with values in the range of 35–45 mg/mL showing no significant differences. Regarding the carotenoid content of the mature milk, both individual and total amounts of carotenoids from PT mothers did not differ from that of FT mothers, either qualitatively or quantitatively. Thus, the total carotenoid content was not statistically different (median 768.0 nM vs. 693.54 nM, $p = 0.278$). The ratio of xanthophylls/carotenes in the mature milk samples was higher than 1 (1.61 and 1.70 for the FT and the PT groups, respectively), unlike the distribution pattern of carotenoids found in the colostrum samples. Zeaxanthin was the minor carotenoid present in the mature milk samples, likewise in the colostrum samples, but the xanthophyll esters were undetectable in both groups. Then, α- and β-carotene and lycopene followed, and finally β-cryptoxanthin and lutein. Again, the total milk fat content was similar in both groups, similar to the values reached in colostrum samples (30–40 mg/mL) and unrelated with the PT or FT condition, with no significant differences between groups.

Table 1. Carotenoid content in colostrum and mature milk of preterm (PT, $n = 70$) and full-term (FT, $n = 70$) mothers. Data are expressed in nM concentration.

	PT Mothers [1]			FT Mothers [1]		
	25th Percentile	Median	75th Percentile	25th Percentile	Median	75th Percentile
Colostrum [2]						
Zeaxanthin (X) [3]	35.9	63.2	112.6	73.7	106.4	141.4
Lutein (X)	231.6	432.8 [a]	667.9	322.9	486.3 [a]	745.8
β-Cryptoxanthin (X)	231.4	406.7	853.0	429.6	754.6	1486
Lycopene (C) [4]	388.1	669.9	931.6	483.0	1065	1846
α+β-Carotene (C)	331.0	594.6	1364	602.3	1103	2238
Xanthophyll esters (X)	144.5	334.6 [b]	733.4	251.5	598.7 [b]	1211
Total	1810	2641	4498	2854	4961	7209
Mature milk [5]						
Zeaxanthin	36.7	46.6	65.9	38.4	59.9	91.9
Lutein	170.0	217.3	283.0	150.3	195.9	270.8
β-Cryptoxanthin	53.0	135.1	224.6	87.4	190.6	353.9
Lycopene	89.4	125.8	173.6	118.8	192.7	221.2
α+β-Carotene	50.0	106.4	202.2	59.4	84.8	315.0
Xanthophyll esters	0	0	0	0	0	0
Total	508.7	693.5	934.0	512.3	768.0	1174

[1]: Data are significantly different when colostrum and mature milk values are compared within each group of lactating mothers, except for zeaxanthin of the preterm group. [2]: Data of colostrum samples were significantly different between both groups of lactating mothers except for values marked with superscript letters ($p < 0.05$). [3]: X means xanthophyll. [4]: C means carotene. [5]: No significant differences were observed for data of both groups of lactating mothers.

Evolution of the carotenoid content from colostrum to mature milk followed a similar sharp decline but it was lower for the carotenoid content in mature milk of PT mothers (74% of the total carotenoid content) than for the FT mothers (85% of the total carotenoid content). Significant differences were observed when individual and total carotenoid contents of colostrum and mature milk samples were compared within each group of lactating mothers, except for zeaxanthin content of the group of PT mothers.

No significant differences were observed in both groups both for individual and total carotenoid content, as shown in Table 2 and Figure 1, in the serum from lactating mothers, either at the initial lactation stage (3–5 days after delivery) or once the milk reached the mature state (30 days after delivery). Within each group and during the progress of lactation, individual and total carotenoid content did not differ also, as shown in Table 2 and Figure 1. Xanthophylls and carotenes were similarly distributed in both groups of lactating mothers and this ratio was the same at both stages of lactation. Zeaxanthin was the minor carotenoid found in the serum, and the close similarity in the data of the rest of carotenoids makes it tough to establish a clear ranking.

Table 2. Carotenoid content in serum of preterm (PT, *n* = 70) and full-term (FT, *n* = 70) mothers at two lactation stages. Data are expressed in nM concentration.

	PT Mothers			FT Mothers		
	25th Percentile	Median	75th Percentile	25th percentile	Median	75th Percentile
Serum samples collected at 3–5 day postpartum [1]						
Zeaxanthin	108.2	142.9	171.8	94.4	122.1	152.4
Lutein	658.3	744.8	911.9	552.4	656.0	834.7
β-Cryptoxanthin	419.8	814.1	964.7	617.8	737.3	1084
Lycopene	481.7	712.9	980.2	450.3	673.5	887.3
α+β-Carotene	519.6	816.8	1089	494.6	671.2	1547
Xanthophyll esters	0	0	0	0	0	0
Total	2984	3288	3751	2869	3222	3811
Serum samples collected at 15–30 day postpartum [1]						
Zeaxanthin	105.3	136.1	187.4	96.0	135.5	158.6
Lutein	622.6	768.4	1010	579.0	673.9	809.1
β-Cryptoxanthin	329.9	729.3	1152	669.2	822.4	925.2
Lycopene	555.3	824.0	1007	473.9	730.4	905.7
α+β-Carotene	583.3	967.4	1594	549.2	743.8	1087
Xanthophyll esters	0	0	0	0	0	0
Total	3279	3636	3898	2938	3295	3711

[1]: No significant differences were observed for data of both groups of lactating mothers or at each collection time within a group.

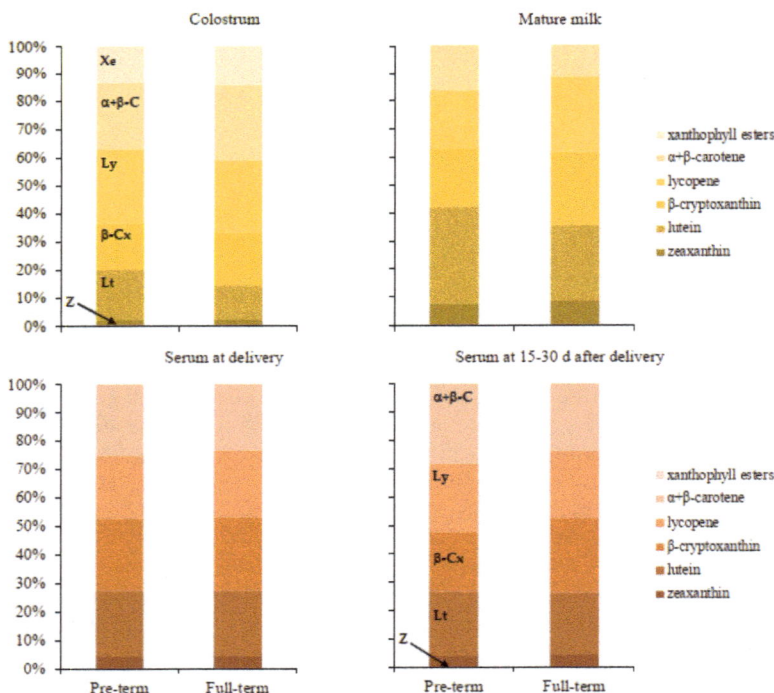

Figure 1. Percentage distribution of the individual carotenoids and xanthophyll esters in colostrum and mature milk samples, and in serum collected at delivery or at 15–30 days after delivery, from preterm and full-term mothers. Xe: xanthophyll esters; α+β-C: α+β-carotene; Ly: lycopene; β-Cx: β-cryptoxanthin; Lt: lutein; Z: zeaxanthin.

4. Discussion

Individual and total content of the major carotenoids in human milk samples have been determined in previous studies [28,29,32,33,36], and our data are in the same order of magnitude and follow similar changes and trends regarding the transition from colostrum to mature milk and the distribution pattern of individual carotenoids. We also noted the high variability of data observed before. Giuliano et al. [28] showed that although the carotenoid content might fluctuate daily and even during feeding, the large variability in carotenoid content is due to inter-individual differences in dietary intake of food sources of carotenoids, the carotenoid bioavailability, and the bioconversion of pro-vitamin A carotenoids to retinoid. Indeed, the bioconversion to retinoid in addition to the nutritional status of the mother may explain the extreme inter-individual differences in the case of α- and β-carotene.

Studies focused on carotenoid content in human colostrum are less frequent in the literature [30,36,37,50], and the characterization of the full carotenoid profile in some of them is incomplete, in comparison with the extensive surveys done with mature human milk samples. Colostrum is a secretion of the mammary gland only available during the few days immediately after delivery, and its composition does not represent the stabilized nutrient contents of mature milk. These are reasonable causes to explain the less intensive attention paid to carotenoids in human colostrum. Most longitudinal studies do not include colostrum collection in the sampling, while some others do. In the latter cases, data should be considered carefully as the carotenoid content in colostrum is significantly higher than in mature milk and even the distribution of carotenoids (xanthophylls, carotenes, and presence of xanthophyll esters) may be different as well [37]. It is well established that carotenoid levels decrease significantly from the colostrum to the mature milk samples as observed in preceding longitudinal and comparative studies [32]. We found a drop of carotenoid levels similar (4-fold for PT mothers and 6-fold for FT mothers) to those reported by Sommerburg et al. [51], Macias and Schweigert [52], and Schweigert et al. [36].

The total carotenoid content observed for mature milk in this study are in the highest range of data reported for North American mothers [28], close to the mean values reported for Irish mothers [53], and higher than those found in a study on milk of German mothers [36]. Finally, the data presented in this study are higher when compared with the largest multinational study of breast milk carotenoids published by Canfield et al. [29]. The main reasons for the differences and similarities are dietary habits including individual preferences, seasonal/regional changes in fruit and vegetable availability, and cooking methods. Qualitatively, the carotenoids were distributed in colostrum samples in a different fashion in comparison to mature milk, with an almost equal presence of carotenes and xanthophylls, while this equilibrium is significantly displaced to the preferential accumulation of xanthophylls during the progress of lactation in agreement with previous reports [32,36]. Mature milk showed a distribution pattern with the prevalence of the xanthophylls lutein and β-cryptoxanthin, followed by the apolar carotenes lycopene and α+β-carotene. Zeaxanthin was a minor carotenoid. This displacement is not correlated with the carotenoid distribution in plasma samples, which was the same at both collection stages, with similar distribution of carotenes and xanthophylls, as shown in Figure 1. This constancy of carotenoid distribution and quantitative content in plasma could indicate that the volunteers had similar food sources of carotenoids throughout the study and that they did not introduce significant changes in their dietary habits after delivery. Circulating carotenoid content is positively associated with the daily intake of fruits and vegetables. It has been shown that qualitative carotenoid composition of serum correlates with short-term carotenoid intake [54,55] and particularly, β-cryptoxanthin and lutein are robust biomarkers of fruit and vegetable consumption [56]. Therefore, the shift of carotenoid distribution towards a higher xanthophyll presence in mature milk would not likely be due to a change in dietary habits. It seems that during the transition of the carotenoid content from colostrum to mature milk, xanthophylls are preferentially accumulated in the mammary epithelium. Transport mechanisms operative at the intestinal epithelium that are involved in the uptake of carotenoids from the lumen, may be active at the mammary epithelium. In fact, it has been demonstrated that SR-BI, CD36, and NPC1L1 facilitate the uptake of carotenoids in the small intestine [57–59], and these proteins are also expressed in the mammary glands [60,61]. Even the expression of some of these proteins has been shown to increase in the mammary tissue of lactating animals. This is the case of CD36 in lactating cows [62] which increases its expression levels after delivery, reaching a maximum at 6 weeks postpartum. Whether such an increase in the expression levels of any of these proteins take place coordinately after delivery in the human mammary gland, and whether they are specific for some protein kind remains to be elucidated, though this would help to explain differences in the xanthophyll to carotene proportion in colostrum and mature milk.

Regarding the influence of the gestational age on the carotenoid content, we observed that significant differences appeared only in colostrum samples. The group of PT mothers produced colostrum with lower carotenoid levels, except for lutein that reached a similar content to the one observed in the group of FT mothers. The carotenoid distribution in colostrum was similar in both groups (equal xanthophyll:carotene ratio), so that the differences were only quantitative. Jewell et al. [53] reported differences in lutein and zeaxanthin content in colostrum between PT and FT mothers, but the statistical power was limited in their study and differences were not significant. The differences observed here should not be attributed to diverse dietary habits of the mothers of the PT group, since the plasma carotenoid contents were not different from those observed in the mothers of the FT group. Several changes take place in the mammary glands during pregnancy to prepare for lactation, including gland maturation and alveologenesis [63], with the start of several signaling pathways that develop the structural and functional changes required to generate and deliver milk. The maternal metabolism is remodeled to flow nutrients to the placenta and the mammary glands for support fetal and infant growth. In this scenario, the premature delivery breaks the regular process of utero accretion and transfer of nutrients to the developing fetus, as well as the accumulation of nutrients and phytochemicals in the mammary glands from the mother stores. It has been shown that carotenoid content in PT infants' tissues is abnormally low or undetectable [26]. Our data show the differences in the carotenoid contents are also extended to the colostrum secretion in the case of premature delivery, with a trend to lower carotenoid content. However, it does not affect the lutein content in colostrum of PT mothers, which reached the same value as in the group of FT mothers. This fact could be related with the significance of lutein to the health of the new born. Lutein content is associated with proper condition of the infant retina [64] and its accumulation in the brain points to a role in cognitive function [21]. Such impact at the initial stage of life implies a mechanism assuring the lutein supply from human milk. Thus, it would not be impaired by factors like the premature condition. Again, the presence of a facilitated transport mechanism, principally active for lutein accumulation, is a reasonable hypothesis to explain this result. As it has been noted above, the measurement of the levels of those proteins involved in carotenoid transport in mammary glands would provide evidence to support this hypothesis. The subsequent regularization process of the carotenoid content from colostrum to mature milk is not affected by the premature condition since the mature milk of both groups of mothers showed carotenoid levels without significant differences both qualitatively and quantitatively.

To our knowledge, this is the first study to examine and compare the qualitative and quantitative carotenoid profile in colostrum and mature milk samples from lactating mothers that gave birth either prematurely or at full-term. The homogeneous sample, regarding the dietary habits, controlled one of the main factors that determine carotenoid levels in human tissues, including colostrum and milk samples. The qualitative findings replicated those reported in previous studies, but through the inclusion of an adequate number of volunteers, we had statistical power to find the differences in carotenoid content of the colostrum samples. Thus, our data give support to the hypothesis of the targeted accumulation of lutein and that this process is not limited by the premature condition at birth. Indeed, this condition reduced the amount of the rest of carotenoids in colostrum but the subsequent transition stage to mature milk made differences disappear. However, we are not able to predict whether our findings will reproduce in other populations with different dietary habits.

5. Conclusions

In our study, we have determined the qualitative and quantitative features of the carotenoid composition of colostrum and mature milk of mothers who gave birth to babies born between 28 and 35 weeks of gestation and to infants at full-term. Our results have shown that the qualitative profile does not vary between groups, but there is a shift from a balanced composition in carotenes and xanthophylls in colostrum to a higher presence of the latter in mature milk. Quantitatively, we found significant differences between the groups in the colostrum samples, with lower content in the preterm group, except for lutein, which behaves in a different fashion. The transition from colostrum to mature milk results in no difference in the carotenoid content between the two groups. These results lead to the conclusion that the premature condition affects the quantitative carotenoid composition of the colostrum, but that condition does not impair the lutein content, which is related to the significant role of the xanthophyll in development of the infant retina and brain. Our study points to the need of research regarding the presence of protein-type transporters in mammary epithelium as one key issue for the existing dynamics on carotenoid content in human milk.

Author Contributions: E.D.-S., I.A.-V., and J.A.-M. carried out all human studies and sample collection. J.G.-F., A.A.O.X., and A.S.-G. conducted research. J.A.-M., J.G.-F., and A.P.-G. analyzed data. A.A.O.X., J.G.-F., J.F., and A.P.-G. wrote the paper. All authors read and approved the final manuscript.

Acknowledgments: Source of financial support: Ministry of Economy and Competitiveness of the Spanish Government, grants AGL2013-42757-R and AGL2017-87884-R. A.A.O.X. is a fellow of the Science without Borders Program of the Brazilian National Council for Scientific and Technological Development, fellowship 238163/2012-1.

Conflicts of Interest: The authors declare no conflicts of interest.

Abbreviations

PT	preterm
FT	full-term
HPLC	high-performance liquid chromatography
APCI	atmospheric pressure chemical ionization
UHR	ultra-high resolution
TOF	time of flight

References

1. Leon-Cava, N.; Lutter, C.; Ross, J.; Martin, L. *Quantifying the Benefits of Breastfeeding: A Summary of the Evidence*; Pan American Health Organization: Washington, DC, USA, 2002; pp. 1–168, ISBN 9275123977.
2. Betran, A.P.; de Onis, M.; Lauer, J.A.; Villar, J. Ecological study of effect of breast feeding on infant mortality in Latin America. *BMJ* **2001**, *323*, 303–306. [CrossRef] [PubMed]
3. WHO Collaborative Study Team on the Role of Breastfeeding on the Prevention of Infant Mortality. Effect of breastfeeding on infant and child mortality due to infectious diseases in less developed countries: A pooled analysis. *Lancet* **2000**, *355*, 451–455. [CrossRef]
4. *WHO Global Strategy for Infant and Young Child Feeding*; World Health Organization: Geneva, Switzerland, 2003; pp. 1–30, ISBN 9241562218.
5. Picciano, M.F. Nutrient composition of human milk. *Pediatr. Clin. North Am.* **2001**, *48*, 53–67. [CrossRef]
6. Ballard, O.; Morrow, A.L. Human milk composition: Nutrients and bioactive factors. *Pediatr. Clin. North Am.* **2013**, *60*, 49–74. [CrossRef] [PubMed]
7. Liaaen-Jensen, S. Basic carotenoid chemistry. In *Carotenoids in Health and Disease*; Krinsky, N.I., Mayne, S.T., Sies, H., Eds.; Marcel Dekker: New York, NY, USA, 2004; pp. 1–30, ISBN 0203026640.
8. Stoltzfus, R.J.; Hakima, M.; Miller, K.W.; Rasmussen, K.M.; Dawiesah, S.I.; Habicht, J.P.; Dibley, M.J. High dose vitamin A supplementation of breast-feeding Indonesian mothers: Effects on the vitamin A status of mother and infant. *J. Nutr.* **1993**, *123*, 666–675. [CrossRef] [PubMed]
9. Bendich, A. Non-provitamin A activity of carotenoids: Immunoenhancement. *Trends Food Sci. Tech.* **1991**, *2*, 127–130. [CrossRef]

10. Stahl, W.; Sies, H. Bioactivity and protective effects of natural carotenoids. *Biochim. Biophys. Acta* **2005**, *1740*, 101–107. [CrossRef] [PubMed]

11. Provis, J.; Diaz, C.; Dreher, B. Ontogeny of the primate fovea: A central issue in retinal development. *Prog. Neurobiol.* **1998**, *54*, 549–581. [CrossRef]

12. Landrum, J.; Bone, R. Lutein, zeaxanthin, and the macular pigment. *Arch. Biochem. Biophys.* **2001**, *385*, 28–40. [CrossRef] [PubMed]

13. Pintea, A.; Socaciu, C.; Rugina, D.O.; Pop, R.; Bunea, A. Xanthophylls protect against induced oxidation in cultured human retinal pigment epithelial cells. *J. Food Compost. Anal.* **2011**, *26*, 830–836. [CrossRef]

14. Junghans, A.; Sies, H.; Stahl, W. Macular pigments lutein and zeaxanthin as blue light filters studied in liposomes. *Arch. Biochem. Biophys.* **2001**, *391*, 160–164. [CrossRef] [PubMed]

15. Leung, I.; Sandstrom, M.M.; Zucker, C.; Neuringer, M.; Snodderly, D. Nutritional manipulation of primate retinas, II: Effects of age, n-3 fatty acids, lutein, and zeaxanthin on retinal pigment epithelium. *Invest. Ophthalmol. Vis. Sci.* **2004**, *45*, 3244–3256. [CrossRef] [PubMed]

16. Loughman, J.; Nolan, J.M.; Howard, A.N.; Connolly, E.; Meagher, K.; Beatty, S. The impact of macular pigment augmentation on visual performance using different carotenoid formulations. *Invest. Ophthalmol. Vis. Sci.* **2012**, *53*, 7871–7880. [CrossRef] [PubMed]

17. Kelly, D.; Coen, R.F.; Akuffo, K.O.; Beatty, S.; Dennison, J.; Moran, R.; Stack, J.; Howard, A.N.; Mulcahy, R.; Nolan, J.M. Cognitive function and its relationship with macular pigment optical density and serum concentrations of its constituent carotenoids. *J. Alzheimers. Dis.* **2015**, *48*, 261–277. [CrossRef] [PubMed]

18. Vishwanathan, R.; Iannaccone, A.; Scott, T.M.; Kritchevsky, S.B.; Jennings, B.J.; Carboni, G.; Forma, G.; Satterfield, S.; Harris, T.; Johnson, K.C.; et al. Macular pigment optical density is related to cognitive function in older people. *Age Ageing* **2014**, *43*, 271–275. [CrossRef] [PubMed]

19. Vishwanathan, R.; Neuringer, M.; Snodderly, D.M.; Schalch, W.; Johnson, E.J. Macular lutein and zeaxanthin are related to brain lutein and zeaxanthin in primates. *Nutr. Neurosci.* **2013**, *16*, 21–29. [CrossRef] [PubMed]

20. Craft, N.E.; Haitema, T.B.; Garnett, K.M.; Fitch, K.A.; Dorey, C.K. Carotenoid, tocopherol, and retinol concentrations in elderly human brain. *J. Nutr. Health Aging* **2004**, *8*, 156–162. [PubMed]

21. Johnson, E.J.; Vishwanathan, R.; Johnson, M.A.; Hausman, D.B.; Davey, A.; Scott, T.M.; Green, R.C.; Miller, S.; Gearing, M.; Woodard, J.; et al. Relationship between serum and brain carotenoids, α-tocopherol, and retinol concentrations and cognitive performance in the oldest old from the Georgia Centenarian Study. *J. Aging Res.* **2013**, *2013*, 951786. [CrossRef] [PubMed]

22. Vishwanathan, R.; Kuchan, M.; Sen, S.; Johnson, E.J. Lutein and preterm infants with decreased concentration of brain carotenoids. *J. Pediatr. Gastroenterol. Nutr.* **2014**, *59*, 659–665. [CrossRef] [PubMed]

23. Hammond, B.R. Jr. Possible role for dietary lutein and zeaxanthin in visual development. *Nutr. Rev.* **2008**, *66*, 695–702. [CrossRef] [PubMed]

24. Johnson, E.J. Role of lutein and zeaxanthin in visual and cognitive function throughout the lifespan. *Nutr. Rev.* **2014**, *72*, 605–612. [CrossRef] [PubMed]

25. Henriksen, B.S.; Chan, G.; Hoffman, R.O.; Sharifzadeh, M.; Ermakov, I.V.; Gellermann, W.; Bernstein, P.S. Interrelationships between maternal carotenoid status and newborn infant macular pigment optical density and carotenoid status. *Invest. Ophthalmol. Vis. Sci.* **2013**, *54*, 5568–5578. [CrossRef] [PubMed]

26. Bernstein, P.S.; Sharifzadeh, M.; Liu, A.; Ermakiv, I.; Nelson, K.; Sheng, X.; Panish, C.; Carlstrom, B.; Hoffman, R.O.; Gellermann, W. Blue-light reflectance imaging of macular pigment in infants and children. *Invest. Ophthalmol. Vis. Sci.* **2013**, *54*, 4034–4040. [CrossRef] [PubMed]

27. Bone, R.A.; Landrum, J.T.; Friedes, L.M.; Gomez, C.M.; Kilburn, M.D.; Menendez, E.; Vidal, I.; Wang, W. Distribution of lutein and zeaxanthin stereoisomers in the human retina. *Exp. Eye Res.* **1997**, *64*, 211–218. [CrossRef] [PubMed]

28. Giuliano, A.R.; Neilson, E.M.; Yap, H.; Baier, M.; Canfield, L.M. Carotenoids of mature human milk: Inter/intraindividual variability. *J. Nutr. Biochem.* **1994**, *5*, 551–556. [CrossRef]

29. Canfield, L.M.; Clandinin, M.T.; Davies, D.P.; Fernandez, M.C.; Jackson, J.; Hawkes, J.; Goldman, W.J.; Pramuk, K.; Reyes, H.; Sablan, B.; et al. Multinational study of major breast milk carotenoids of healthy mothers. *Eur. J. Nutr.* **2003**, *42*, 133–141. [PubMed]

30. Patton, S.; Canfield, L.M.; Huston, G.E.; Ferris, A.M.; Jensen, R.G. Carotenoids of human colostrum. *Lipids* **1990**, *25*, 159–165. [CrossRef] [PubMed]

31. Khachik, F.; Spangler, C.J.; Smith, J.C.; Canfield, L.M.; Pfander, H.; Speck, A. Identification, quantification, and relative concentrations of carotenoids, and their metabolites in human milk and serum. *Anal. Chem.* **1997**, *69*, 1873–1881. [CrossRef] [PubMed]

32. Gossage, C.P.; Deyhim, M.; Yamini, S.; Douglas, L.W.; Moser-Veillon, P.B. Carotenoid composition of human milk during the first month postpartum and the response to β-carotene supplementation. *Am. J. Clin. Nutr.* **2002**, *76*, 193–197. [CrossRef] [PubMed]

33. Lipkie, T.E.; Morrow, A.L.; Jouni, Z.E.; McMahon, R.J.; Ferruzzi, M.G. Longitudinal survey of carotenoids in human milk from urban cohorts in China, Mexico, and the USA. *PLoS ONE* **2015**, *10*. [CrossRef] [PubMed]

34. Galan, P.; Viteri, F.E.; Bertrais, S.; Czernichow, S.; Faure, H.; Arnaud, J.; Ruffieux, D.; Chenal, S.; Arnault, N.; Favier, A.; et al. Serum concentrations of beta-carotene, vitamins C and E, zinc and selenium are influenced by sex, age, diet, smoking status, alcohol consumption and corpulence in a general French adult population. *Eur. J. Clin. Nutr.* **2005**, *59*, 1181–1190. [CrossRef] [PubMed]

35. Forman, M.R.; Beecher, G.R.; Lanza, E.; Reichman, M.E.; Graubard, B.I.; Campbell, W.S.; Marr, T.; Yong, L.C.; Judd, J.T.; Taylor, P.R. Effect of alcohol consumption on plasma carotenoid concentrations in premenopausal women, a controlled dietary study. *Am. J. Clin. Nutr.* **1995**, *62*, 131–135. [CrossRef] [PubMed]

36. Schweigert, F.J.; Bathe, K.; Chen, F.; Büscher, U.; Dudenhausen, J.W. Effect of the stage of lactation in humans on carotenoid levels in milk, blood plasma and plasma lipoprotein fractions. *Eur. J. Nutr.* **2004**, *43*, 39–44. [CrossRef] [PubMed]

37. Ríos, J.J.; Xavier, A.A.O.; Díaz-Salido, E.; Arenilla-Vélez, I.; Jarén-Galán, M.; Garrido-Fernández, J.; Aguayo-Maldonado, J.; Pérez-Gálvez, A. Xanthophyll esters are found in human colostrum. *Mol. Nutr. Food Res.* **2017**, *61*. [CrossRef] [PubMed]

38. Bokor, S.; Koletzko, B.; Decsi, T. Systematic review of fatty acid composition of human milk from mothers of preterm compared to full-term infants. *Ann. Nutr. Metab.* **2007**, *51*, 550–556. [CrossRef] [PubMed]

39. Bobiński, R.; Mikulska, M.; Mojska, H.; Simon, M. Comparison of the fatty acid composition of transitional and mature milk of mothers who delivered healthy full-term babies, preterm babies and full-term small for gestational age infants. *Eur. J. Clin. Nutr.* **2013**, *67*, 966–971. [CrossRef] [PubMed]

40. Pérez-Gálvez, A.; Martin, H.D.; Sies, H.; Stahl, W. Incorporation of carotenoids from paprika oleoresin into human chylomicrons. *Br. J. Nutr.* **2003**, *89*, 787–793. [CrossRef] [PubMed]

41. Breithaupt, D.E.; Wirt, U.; Bamedi, A. Differentiation between lutein monoester regioisomers and detection of lutein diesters from marigold flowers (*Tagetes erecta* L.) and several fruits by liquid chromatography-mass spectrometry. *J. Agric. Food Chem.* **2002**, *50*, 66–70. [CrossRef] [PubMed]

42. Liaaen-Jensen, S. Combined approach, identification and structure elucidation of carotenoids. In *Carotenoids, Volume 1B, Spectroscopy*; Britton, G., Liaaen-Jensen, S., Pfander, H., Eds.; Birkhäuser Verlag: Basel, Switzerland, 1995; pp. 343–354, ISBN 0817629092.

43. Enzell, C.R.; Back, S. Mass spectrometry. In *Carotenoids, Volume 1B, Spectroscopy*; Britton, G., Liaaen-Jensen, S., Pfander, H., Eds.; Birkhäuser Verlag: Basel, Switzerland, 1995; pp. 261–320, ISBN 0817629092.

44. Britton, G.; Liaaen-Jensen, S.; Pfander, H. *Carotenoids Handbook*; Birkhäuser Verlag: Basel, Switzerland, 2004; ISBN 3034878362.

45. Rodriguez-Amaya, D.B. *Food Carotenoids, Chemistry, Biology, and Technology*; Wiley Blackwell: Singapore, 2016.

46. Rivera, S.M.; Christou, P.; Canela-Garayoa, R. Identification of carotenoids using mass spectrometry. *Mass Spectrom. Rev.* **2014**, *33*, 353–372. [CrossRef] [PubMed]

47. De Rosso, V.V.; Mercadante, A.Z. Identification and quantification of carotenoids, by HPLC-PDA-MS/MS, from Amazonian fruits. *J. Agric. Food Chem.* **2007**, *55*, 5062–5072. [CrossRef] [PubMed]

48. Weller, P.; Breithaupt, D.E. Identification and quantification of zeaxanthin esters in plants using liquid chromatography-mass spectrometry. *J. Agric. Food Chem.* **2003**, *51*, 7044–7049. [CrossRef] [PubMed]

49. Jensen, R.G.; Lammi-Keefe, C.J.; Koletzko, B. Representative sampling of human milk and the extraction of fat for analysis of environmental lipophilic contaminants. *Toxicol. Environ. Chem.* **1997**, *62*, 229–247. [CrossRef]

50. Gebre-Medhin, M.; Vahlquist, A.; Hofvander, Y.; Uppsall, L.; Vahlquist, B. Breast milk composition in Ethiopean and Swedish mothers. I. Vitamin A and beta-carotene. *Am. J. Clin. Nutr.* **1976**, *29*, 441–451. [CrossRef] [PubMed]

51. Sommerburg, O.; Meissner, K.; Nelle, M.; Lenhartz, H.; Leichsenring, M. Carotenoid supply in breast-fed neonates. *Eur. J. Pediatr.* **2000**, *159*, 86–90. [CrossRef] [PubMed]

52. Macias, C.; Schweigert, F.J. Changes in the concentration of carotenoids, vitamin A, alpha-tocopherol and total lipids in human milk throughout early lactation. *Ann. Nutr. Metab.* **2001**, *45*, 82–85. [CrossRef] [PubMed]

53. Jewell, V.C.; Northrop-Clewes, C.A.; Tubman, R.; Thurnham, D.I. Nutritional factors and visual function in premature infants. *Proc. Nutr. Soc.* **2001**, *60*, 171–178. [CrossRef] [PubMed]

54. Olmedilla, B.; Granado, F.; Southon, S.; Wright, A.J.A.; Blanco, I.; Gil-Martínez, E.; van den Berg, H.; Corridan, B.; Roussel, A.M.; Chopra, M.; et al. Serum concentrations of carotenoids and vitamins A, E, and C in control subjects from five European countries. *Br. J. Nutr.* **2001**, *85*, 227–238. [CrossRef] [PubMed]

55. Thurnham, D.I.; Northrop-Clewes, C.A.; Chopra, M. Biomarkers of vegetable and fruit intakes. *Am. J. Clin. Nutr.* **1998**, *68*, 756–757. [CrossRef] [PubMed]

56. Couillard, C.; Lemieux, S.; Vohl, M.C.; Couture, P.; Lamarche, B. Carotenoids as biomarkers of fruit and vegetable intake in men and women. *Br. J. Nutr.* **2016**, *116*, 1206–1215. [CrossRef] [PubMed]

57. Borel, P.; Lietz, G.; Goncalves, A.; Edelenyi, F.S.; Lecompte, S.; Curtis, P.; Goumidi, L.; Caslake, M.J.; Miles, E.A.; Packard, C.; et al. CD36 and SR-BI are involved in cellular uptake of provitamin A carotenoids by Caco-2 and HEK cells, and some of their genetic variants are associated with plasma concentrations of these micronutrients in humans. *J. Nutr.* **2013**, *143*, 448–456. [CrossRef] [PubMed]

58. Moussa, M.; Gouranton, E.; Gleize, B.; Yazidi, C.E.; Niot, I.; Besnard, P.; Borel, P.; Landrier, J.F. CD36 is involved in lycopene and lutein uptake by adipocytes and adipose tissue cultures. *Mol. Nutr. Food Res.* **2011**, *55*, 578–584. [CrossRef] [PubMed]

59. Sato, Y.; Suzuki, R.; Kobayashi, M.; Itagaki, S.; Hirano, T.; Noda, T.; Mizuno, S.; Sugawara, M.; Iseki, K. Involvement of cholesterol membrane transporter Niemann-Pick C1-like 1 in the intestinal absorption of lutein. *J. Pharm. Pharm. Sci.* **2012**, *15*, 256–264. [CrossRef] [PubMed]

60. Landschulz, K.; Pathak, R.; Rigotti, A.; Krieger, M.; Hobbs, H. Regulation of scavenger receptor, class B, type I, a high density lipoprotein receptor, in liver and steroidogenic tissues of the rat. *J. Clin. Invest.* **1996**, *98*, 984–995. [CrossRef] [PubMed]

61. Pussinen, P.J.; Karten, B.; Wintersperger, A.; Reicher, H.; McLean, M.; Malle, E.; Sattler, W. The human breast carcinoma cell line HBL-100 acquires exogenous cholesterol from high-density lipoprotein via CLA-1 (CD-36 and LIMPII analogous 1)-mediated selective cholesteryl ester uptake. *Biochem. J.* **2000**, *349*, 559–566. [CrossRef] [PubMed]

62. Bionaz, M.; Loor, J.J. Gene networks driving bovine milk fat synthesis during the lactation cycle. *BMC Genomics* **2008**, *9*, 366. [CrossRef] [PubMed]

63. Macías, H.; Hinck, L. Mammary gland development. *Wiley Interdiscip. Rev. Dev. Biol.* **2012**, *1*, 533–557. [CrossRef] [PubMed]

64. Zimmer, J.P.; Hammond, B.R.Jr. Possible influences of lutein and zeaxanthin on the developing retina. *Clin. Ophthalmol.* **2007**, *1*, 25–35. [PubMed]

nutrients

MDPI

Article

Biological Active Ecuadorian Mango 'Tommy Atkins' Ingredients—An Opportunity to Reduce Agrowaste

Jenny Ruales [1], Nieves Baenas [2], Diego A. Moreno [2], Carla M. Stinco [3], Antonio J. Meléndez-Martínez [3] and Almudena García-Ruiz [1,4,*]

[1] Department of Food Science and Biotechnology, Escuela Politécnica National, Quito 17-01-2759, Ecuador; jenny.ruales@epn.edu.ec
[2] Phytochemistry and Healthy Foods Lab., Department of Food Science and Technology, CEBAS-CSIC, Campus de Espinardo-Edificio 25, E-30100 Murcia, Spain; nbaenas@cebas.csic.es (N.B.); dmoreno@cebas.csic.es (D.A.M.)
[3] Food Colour & Quality Lab., Department of Nutrition & Food Science, Universidad de Sevilla, Facultad de Farmacia, 41012 Sevilla, Spain; cstinco@us.es (C.M.S.); ajmelendez@us.es (A.J.M.-M.)
[4] Laboratory of Epigenetics of Lipid Metabolism, Madrid Institute for Advanced Studies (IMDEA)-Food, CEI UAM + CSIC, 28049 Madrid, Spain
*   Correspondence: almudena.garcia@epn.edu.ec or almudena.garcia@imdea.org; Tel.: +34-91-72-78-100

Received: 17 July 2018; Accepted: 19 August 2018; Published: 21 August 2018

Abstract: Mango is a commercially important tropical fruit. During its processing, peel and seed kernel are discarded as waste but they could be recovered as an excellent and cost-effective source of health-promoting ingredients. This study aimed to characterize some of them, including carotenoids like the provitamin A β-carotene and lutein, with an interest beyond its role in eye health. Other health-promoting compounds like tocopherols and polyphenols were also evaluated, as well as the in vitro antioxidant capacity of mango by-products. Regarding isoprenoids, α-tocopherol was mainly found in the peels and carotenoids concentration was higher in the pulps. β-carotene was the most abundant carotene in pulp and seed kernel, whereas peel was the only source of lutein, with violaxanthin the most abundant xanthophyll in the different mango organs tested. With regard to polyphenols, peels exhibited greater variability in its phenolic composition, being the total content up to 85 and 10 times higher than the pulp and seed kernels, respectively. On the other hand, peels also stood out for being a very rich source of mangiferin. Seed kernels and peels showed higher antioxidant capacity values than the pulps. These results contribute to the valorization of mango by-products as new natural ingredients for the pharma and food industries.

Keywords: mango by-products; lutein; β-carotene; α-tocopherol; mangiferin; food ingredients

1. Introduction

Mango (*Mangifera indica*) is considered one of the most consumed fresh fruits in the world, with extensive marketing and production taking place in 115 countries [1]. The global area of mangoes harvested is approximately 5.41 million hectares and its global production is 42.66 million metric tons (MMT) [2]. India, with a production of 18 MMT, is the world's largest mango producer, whereas Mexico and the United States are the main mango exporter and importer, respectively [2]. There are several hundreds of cultivars of mango; but by its long shelf life, excellent ratings in handling and transport tolerance the cultivar Tommy Atkins is the most commercialized [1,3]. Ecuador, with a global area of mangoes harvested of 13,300 hectares and a production of 61,300 metric tons, is the second and sixth mango exporter to USA and to worldwide, respectively, being an important fruit in the Ecuadorian economy [2].

Apart of being consumed fresh, about 20% of mango are processed for products such as juices, desserts, mango jam, among others [4]. During processing, 33% of the fruit is removed in the form of waste, generating, as a result, several million tons per year of mango waste from factories [4,5]. The mango fractions discarded, peel and seed kernel (35–60% total weight of the fruit) are a source of pollution, among other reasons because they are prone to microbial spoilage causing objectionable odors and environmental problems [1,3]. However, the mango peel and seed kernel may be interesting because their high levels of health-enhancing substances, such as carotenoids, polyphenols, vitamins C and E and dietary fiber, among others [6]. The benefit effect of these phytochemicals may be associated with their antioxidant capacity, since in the pathogenesis of many chronic disease is involved the overproduction of oxidants [7]. On the other hand, various studies have described that efficient, inexpensive and environmentally friendly use of agri-food industry waste is highly cost-effective and minimizes environmental impact [8,9]. In this line, the characterization, recovery and utilization of valuable compounds from mango by-products is an important challenge, whose result would have a significant positive impact both at the environmental level (reduction of pollution of mango industry) and economically (contribution to more sustainable production in the food and pharmaceutical industries). In addition, the revaluation of mango peels and seed kernels as a natural bioactive ingredient for the industry, would have a positive socio-economical effect on Ecuadorian mango and tree fruit producing areas, contributing to a reduction of nutritional deficiencies and promoting health benefits, as well as reducing the environmental implications associated with the mango processing.

In this context, the main goal of the present work was to characterize and evaluate Ecuadorian mango by-products in their phytochemical composition: two types of isoprenoids, specifically carotenoids and α-tocopherol and polyphenols by RRLC (Rapid Resolución Liquid Chromatographic) and HPLC–DAD–ESI/MS^n (High-performance Liquid Chromatographic Diode-array detector Electrospray ionization Mass Spectometry), respectively and their total antioxidant capacity by ORAC (Oxygen Radical Absorbance Capacity) and DPPH (2,2,-diphenyl-2-picrylhydrazyl) methods, as well as by Folin–Ciocalteu assay. These evaluations were carried out to establish the potential applications of these mango discards or non-commercial products (peels and seed kernels) compared with the pulp (the main fraction consumed of the mango fruits) as a valuable source of natural ingredients and/or additives for the pharmaceutical and food industry.

2. Materials and Methods

2.1. Fruit Samples

Mangoes (*Mangifera indica* L. cv. Tommy Atkins), were obtained in local markets in Quito (Ecuador). The fruits were selected free of damage. Two kilograms of samples were separated into peels, pulps and seed kernels and stored at −20 °C until freeze drying in an Alpha 2–4 LD drying manifold (Martin Christ Gefriertrocknungsanlagen GmbH, Osterode am Harz, Germany). Then, samples were ground as a fine powder and stored at −20 °C until analyses.

2.2. Standards, Chemicals and Solvents

The commercially available standards (+)-catechin and rutin (Quercetin-3-rutinoside) were acquired from Phytoplan GmbH (Heidelberg, Germany). Cyanidin 3-O-glucoside was purchased from Polyphenols (Sandnes, Norway). The β-carotene and β-cryptoxanthin were obtained from Sigma-Aldrich Chemie GmbH (Steinheim, Germany) and α-tocopherol was purchased from Calbiochem (Merck, Darmstadt, Germany). Violaxanthin and phytoene were isolated from natural sources by classical chromatographic techniques [10]. Luteoxanthin, neoxanthin and lutein were obtained as described by Meléndez-Martínez et al. [11].

Trolox (6-hydroxy-2,5,7,8-tetramethylchroman-2-carboxylic acid) was obtained from Fluka Chemika (Neu-Ulm, Switzerland). The reagents 2,2-diphenyl-1-picrylhidracyl radical (DPPH˙), monobasic and dibasic sodium phosphate, Folin Ciocalteu's reagent and fluorescein (free acid) were purchased from Sigma–Aldrich (Steinheim, Germany). Finally, formic acid and solvents (ethanol, methanol, hexane, acetone, diclhoromethane and acetonitrile) were all of analytical grade and were obtained from Merck (Darmstadt, Germany).

2.3. Identification and Quantification of Isoprenoids (Carotenoids and α-Tocopherol) by Rapid Resolution Liquid Chromatography (RRLC)

The extraction and analyses of carotenoids were carried out according to the method described by Stinco et al. [12]. Mango samples (200 mg) were extracted with 1mL of hexane/acetone (1:1 v/v) using a vortex and an ultrasonic bath for 2 min. Then, samples were centrifuged at 18,000× g for 5 min and the colored fractions were recovered. The extraction was performed twice more until color extinction. Finally, the carotenoid extracts were concentrated to dryness in a rotary evaporator at temperature below 30 °C. To obtain saponified carotenoids, the extracts were treated with 1000 μL of dichloromethane and 1000 μL of methanolic KOH (30% w/v) for 1 h under dim light and at room temperature, after which they were washed with water to remove any trace of base. The extracts obtained were concentrated to dryness in a rotary evaporator and redissolved in ethyl acetate prior to their injection in the RRLC system. Samples were extracted and analyzed in triplicate.

The RRLC acquisitions were made by using an Agilent 1260 system equipped with a diode-array detector, which was set to scan from 200 to 770 nm and a Poroshell 120 C18 column (2.7 μm, 5 cm × 4.6 mm) (Agilent, Palo Alto, CA, USA) kept at 28 °C, according to Stinco et al. [12]. The injection volume was set at 10–20 μL. The mobile phase was pumped at 1 mL/min and consisted of three solvents: solvent A, acetonitrile, solvent B, methanol and solvent C, ethyl acetate. The linear gradient elution was 0 min, 85% A + 15% B; 5 min, 60% A + 20% B + 20% C; 7 min, 60% A + 20% B + 20% C; 9 min, 85% A + 15% B; 12 min, 85% A + 15% B. Chromatograms were monitored at 450 nm. The identification and quantification of isoprenoids were performed by comparison of their chromatographic UV–vis spectroscopic characteristics with the standards, as well as by comparison with the external calibration line calculated. Results were expressed as μg/g dry weight (D.W.).

2.4. Folin-Ciocalteu Assay

Folin assay was performed following the method described by Slinkard and Singleton [13]. Briefly, 500 μL of the extracts, blank or standards were placed in a 15 mL tube, where 2.5 mL of the Folin–Ciocalteu reagent was added, allowing to react for 2 min while shaking. Then, 2 mL of a solution of sodium carbonate (75 g/L) was added and properly mixed. The solution was thus incubated 15 min at 50 °C. After that, the absorbance was measured at 750 nm in a spectrophotometer (Shimadzu UV-160A, Kyoto, Japan). Gallic acid was used as a standard (10–90 mg/L) and the results were expressed as mg of gallic acid equivalents (GAE) per gram.

2.5. Identification and Quantification of Phenolic Compounds by HPLC–DAD–ESI-MS^n

Phenolic compounds were extracted and analyzed following the protocol and method of Gironés-Vilaplana et al. [14]. Briefly, samples were extracted with MeOH 70% using the ultrasound technology and kept at 4 °C overnight. Then, samples were filtered and the identification of phenolic compounds was carried out following their MS2 fragmentations by an HPLC–DAD–ESI-MS^n, constituted by an Agilent series model HPLC (High-performance Liquid Chromatographic) 1100 with a photodiode array detector and a mass spectrometer detector in series (model G2445A) equipped with an electrospray ionization interface (Agilent Technologies, Waldbronn, Germany). The ionization conditions were selected according to those described in the method, covering an m/z range from 100 to 1200. The acquisition of the mass spectrometry data (MS^n) was performed in the negative ionization mode for flavonoids, except for anthocyanins, where the positive ionization mode was used. The quantification equipped with a Luna C18 column (25 cm × 0.46 cm, 5 μm particle size) (Phenomenex, Macclesfield, UK) using the acquisition conditions described before. Flavan-3-ols were quantified using the external standard (+)-catechin at 280 nm, flavonols at 360 nm using the standard rutin (quercetin-3-rutinoside) and the anthocyanins by using cyanidin 3-O-glucoside at 520 nm. Samples were extracted and analyzed in triplicate. Results were expressed as μg/g D.W.

2.6. Antioxidant Capacity

The antioxidant capacity was evaluated using the methods DPPH· and ORAC, both adapted to a microscale and performed using 96-well micro plates (Nunc, Roskilde, Denmark), which were measured using an Infinite® M200 microplate reader (Tecan, Grödig, Austria). The power of scavenge DPPH radicals were determined according to Mena et al. [15], briefly, 2 μL of the corresponding diluted sample was added to the wells containing 250 μL of DPPH· dissolved in methanol up to absorbance ~1. Then, the plate was shaken and left for 50 min at 37 °C, thus, the variation in absorbance was measured at 515 nm. Regarding the ORAC method and according to Ou et al. [16], 25 μL of the properly diluted sample was added to 150 μL of fluorescein (1 μM) and, after 30 min of incubation, 25 μL of the radical AAPH (2,2′-azobis(2-methyl-propionamidine)-dihydrochloride) (250 mM) was added to the wells. Results were studied by measuring the variation in fluorescence each 2 min during 120 min of reaction with the radical. Trolox was used as a standard in both methods, following the same procedure as with the samples. Results were expressed as mmol Trolox/100 g D.W.

2.7. Statistical Analysis

All assays were conducted in triplicate. The data were processed using the software Statgraphics Centurion version 16.1.18 (Statgraphics.Net, Madrid, Spain). All values were subjected to analysis of variance (ANOVA) with a 95% confidence level. Pearson's correlation coefficients were also calculated to corroborate relationships among the selected parameters.

3. Results and Discussion

The first step to establish the potential applications of the mango by-products, peels and seed kernels, as a valuable source of natural ingredients and/or additives for the pharmaceutical and food industry is crucial to characterize the phytochemical composition with efficient techniques for the identification and quantitation of the nutrients and compounds of interest. Bearing this in mind, different chromatographic methods were used to characterize the isoprenoids, especially carotenoids and α-tocopherol and the phenolic composition of mango by-products.

3.1. Isoprenoids: Carotenoids and α-Tocopherol

Many studies have evaluated the carotenoid fraction of mango pulps but not peels and seed kernels. In this sense, this study provides information on the characterization of carotenoid in non-edible parts of the mango. In particular, six carotenoids (4 xanthophylls and 2 carotenes) were identified and quantified in the different mango organs tested (peels, pulps and seed kernels) (Table 1). The carotenoid composition, both qualitative and quantitative, showed differences between the different mango organs tested, which is in line with data previously described in the literature [17]. Violaxanthin and β-carotene stood out as the only two carotenoids present in the different mango fractions tested (Table 1). Regarding pulps and seed kernels, the most abundant carotene was β-carotene, whereas the most important xanthophyll was violaxanthin (Table 1). This is comparable with the results reported by Ornelas-Paz et al. [18] in seven Mexican mango cultivars. Moreover, several authors have described β-carotene as the most predominant carotenoid in Australian and Taiwanese mango pulps [19,20]. In terms of biological effects, β-carotene is considered, theoretically, the carotenoid with the highest provitamin A activity [21] while the health benefits of violaxanthin have yet to be established [17]. With regard to mango peels, lutein highlighted as major carotenoid (Table 1). This agrees with the result reported by Ajila et al. [22] in peels Bandami mango variety. Lutein is an essential nutrient with health promoting effects, especially for eye health. In this line, the use of lutein in the formulation of nutritional supplements have gained increasing popularity for the prevention of age-related macular degeneration, as well as for its antioxidant properties, after the public awareness of its potential to prevent the disease [23]. Thus, the particular interest of the industry, especially pharmaceutical, in the search for new cost-effective sources of lutein as could be mango peels. In addition to its effects on the retina, it has recently been reported the accumulation of lutein in brain, being its content in neural tissue positively correlated with cognitive function, which has intensified interest in identifying functions of lutein in this organ [24]. Furthermore, lutein is a natural colorant, so the mango peels could be employed as additive in the industry such as food, cosmetic and nutraceutical.

Table 1. Isoprenoid, carotenoids and α-tocopherol, composition in mango organs (peels, pulps and seed kernels).

	Concentration (μg/g D.W.)		
	Peel	**Pulp**	**Seed Kernel**
Carotenoids			
Violaxanthin *	1.58 [b] ± 0.13	3.97 [a] ± 0.19	0.18 [c] ± 0.02
Lutein	3.26 ± 0.19	-	-
Luteoxanthin	-	1.69 [a] ± 0.08	0.16 [b] ± 0.04
β-cryptoxanthin	-	2.72 ± 0.04	-
β-carotene	2.78 [b] ± 0.05	4.86 [a] ± 0.01	0.50 [c] ± 0.01
Phytoene	-	1.23 [a] ± 0.01	0.23 [b] ± 0.03
∑*carotenoids*	7.62 [b] ± 0.37	14.47 [a] ± 0.33	1.07 [c] ± 0.10
α-tocopherol	10.20 [a] ± 1.13	0.39 [b] ± 0.21	-

* In peel, the concentration of violaxanthin corresponds to violaxanthin + neoxanthin; [a–c] Mean values with different letter on the right in the same row indicate statistically significant differences among the three treatments ($p < 0.05$).

On the other hand, the results obtained on the qualitative and quantitative characterization of carotenoid showed differences with the results reported in other mango cultivars (Keitt, Ataulfo, Haden and Kent) from Mexico, Brazil, Taiwan [18,20,25]. These differences could be due to factors such as genetics, agricultural and industrial practices, temperature, harvest, maturity, among others, which can modify the composition of carotenoids [3,17].

The total content of carotenoids (TCC), evaluated as the sum of the content of individual pigments, showed significant differences between the different fractions of mango. The mango pulps showed the highest content (Table 1). TCC obtained in the pulps (14.47 μg/g) is within the range described in other several mango cultivars (9.0 to 92 μg/g) [26]. Although the pulps showed the highest TCC, the result also reflected that mango by-products could be a valuable source of carotenoid, especially mango peels. TCC obtained in mango peels (7.62 μg/g) is higher or comparable than TCC values reported in tropical fruits (Table 2), which reflects that mango peels are not only a disposable waste but an extraordinary source natural of carotenoids. The use of mango peels offers a window of opportunity for configuring alternative food supply chains. The raw material is valuable in terms of nutritional and functional properties and besides, the use of this side stream is of great interest from the point of view of environmental concerns and for food and nutrition purposes.

Table 2. Carotenoid concentration in tropical fruits.

	Carotenoid Concentration (μg/g D.W.)
Eugenia stipitata	8.06 [27]
Solanum quitoense	7.94 [28]
Ananas comosus	4.97 [29]
Psidium guajava	6.04 [29]
Carica papaya	7.93−51.34 [30]

In relation to α-tocopherol, it is an antioxidant with an effective chemoprotectant agent against lipid oxidation. In the present study, the α-tocopherol was detected and quantified in peels and pulps but not in seed kernels. Concretely, its content was 26 times greater in peels than pulps (Table 1). Abbasi et al. [31] also detected higher content of α-tocopherol in peels than pulps in nine mango cultivars from China. α-tocopherol amount in mango pulps was higher than that described by Burns et al. [32] (0.05 µg/g) in mango from Costa Rica, in agreement with Ornelas-Paz et al. [18] (0.2–0.5 µg/g) in seven Mexican mango cultivars but lower than the content described by Vilela et al. [33] (12–94 µg/g) and Gong et al. [34] (2.0 µg/g) in twelve Portuguese mango cultivars and in mango from China, respectively. As polyphenols and carotenoids, the α-tocopherol amount depends on the genotype, the environmental factors and analytical methods, among other factors [35], which explains the differences observed among the results obtained in this study and others. On the other hand, α-tocopherol concentration obtained in peels was greater than observed in some exotic fruits such as dragon fruit (4.5 µg/g), durian (3.6 µg/g) and papaya (2.6 µg/g) [35]. This result reflects that mango peels could be exploited as natural antioxidant in cosmetic (i.e., anti-aging products), pharmaceutical and agro-industry.

3.2. Characterization of the Phenolic Composition

3.2.1. Identification of Phenolic Compounds

Seventeen phenolic compounds were separated and tentatively identified as procyanidins (1–3), anthocyanins (4, 6), xanthones (5, 7, 9, 16, 17) and flavonols (8, 10–15) by HPLC–DAD–ESI/MS^n, which are shown in Table 3. In addition, the separation of polyphenolics in mango peel is shown in Figure 1.

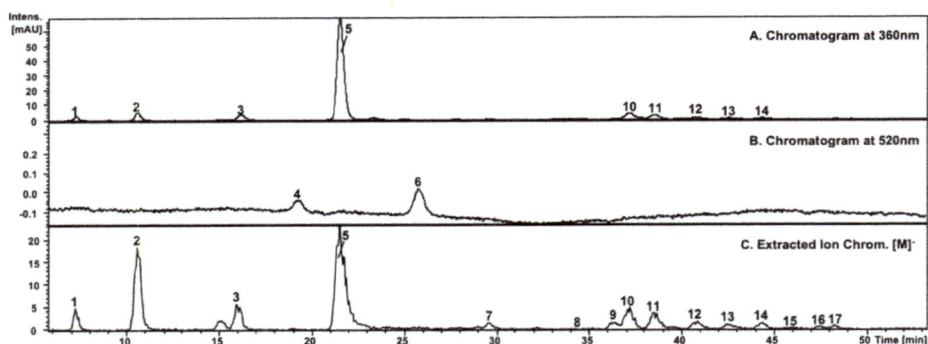

Figure 1. Typical chromatogram of mango organs (e.g., peels), registered at 360 nm (**A**) and 520 nm (**B**) for the identification; Extracted Ion Chromatogram of (M)$^-$ of parental ions (**C**) of the phenolic compounds in the mango samples is also included. For the compound assignment numbers, please see Table 3.

Table 3. Tentative identification of phenolic compounds in mango organs (peels, pulps and seed kernels) by HPLC–DAD–ESI-MS[a].

Peak Number	Rt (min)	DAD (Max. Abs.) λnm	(M)⁻	Fragment Ions (MS[a])	Phenolic Compounds (Tentative Identification)	Peels	Pulp	Seed Kernels
1	7.2	280	423	303, 289	Procyanidin (catechin derivative)	√	-	-
2	10.5	280	575	423, 289	Procyanidin dimer	√	-	-
3	15.9	280	559	407, 289	(Epi)afzelechin-(epi)catechin dimer	√	-	-
4	19.2	280, 520	-	-	Unidentified anthocyanin	√	-	-
5	21.5	330,360	421	403, 331, 301, 258–259	Mangiferin	√	√	√
6	25.7	280, 520	-	-	Unidentified anthocyanin	√	-	-
7	29.5	360	573	421, 403, 331, 301	Mangiferin gallate	√	-	-
8	34.4	360	599	285	Kaempferol derivative	√	-	-
9	36.3	360	573	421, 403, 331, 301	Mangiferin gallate	√	-	-
10	37.1	360	463	301	Quercetin galactoside	√	-	-
11	38.5	360	463	301	Quercetin glucoside	√	-	-
12	40.8	360	433	301	Quercetin xyloside	√	-	-
13	42.5	360	433	301	Quercetin arabinopyranoside	√	-	-
14	44.4	360	433	301	Quercetin arabinofuranoside	√	-	-
15	45.8	360	447	301	Quercetin rhamnoside	√	-	-
16	47.4	330,360	421	403, 373, 331, 301	Mangiferin (isomer)	√	-	-
17	48.2	360	573	421, 403, 331, 301	Mangiferin gallate (isomer)	√	-	-

Rt: retention time; DAD: dyode-array detrector.

Procyanidins

Peaks 1 and 2 showed the MS spectra of procyanidin dimers with molecular intact ions at m/z of 423 and 575, respectively and a characteristic deprotonated molecular ion in MS2 of m/z 289 that corresponds to a (epi)catechin. It is important to highlight that both phenolic compounds had never been determined in mango peels before. On the other hand, peak 3 was identified as a propelargonidin dimer due to its molecular ion (M–H) at m/z at 559–560 and its MS2/MS3 fragment ions m/z 407 and 289, corresponding probably to (epi)afzelechin-(epi)catechin. These procyanidins were only characterized in mango peels.

The compounds 4 and 6 exhibited peak absorption maximum at ~280 nm and ~520 nm wavelengths, characteristic of anthocyanins but in extremely low concentration that did not allow the tentative identification of the compounds or the isolation of their aglycones in the (M)$^-$ and its corresponding MS/MS fragmentation experiments. These compounds were detected in mango peels but not in pulps and seed kernels.

Xanthones

Peak 5 showed a molecular anion at m/z 421, being therefore tentatively assigned as mangiferin. Peaks 7 and 9, with a corresponding m/z ion of 573 (M)$^-$ and characteristic MS/MS fragmentation, were tentatively identified as mangiferin gallates [36]. In the last part of the chromatogram, tiny peaks of possible isomers of mangiferin (peak 16) and mangiferin gallate (peak 17) were also detected in the MS/MS experiments. These compounds, with exception of mangiferin, were only detected in peels. Mangiferin was also the unique xanthone detected in the pulps of the cultivar Haden, and in the seed kernels of the cultivar Ubá, but the pulps of the cultivar Ubá were constituted by a greater number of xanthones [37]. This result reflects the variability of the phenolic composition in the mango.

Flavonols

Peak 8 was identified as kaempferol derivative according to their UV spectra and MS fragmentation leading to the kaempferol aglycone at m/z 285 in negative mode [38]. With regard to compounds 10–14, they were identified as quercetin glycosides based on their UV-Vis data and characteristic mass spectra and elution order [39] (Table 3). In the quercetin glycosides, the most abundant fragment ion in MS2/MS3 was m/z 301 that corresponds with the radical anion of the aglycone quercetin. Peaks 10 and 11 displayed identical (M)$^-$ ion, m/z 463 and could be assigned as quercetin-galactoside and quercetin glucoside, respectively. The formation of the ion at m/z 433 [M]- as the main fragment ions in the peaks 12, 13 and 14 revealed the presence of three different quercetin pentosides and recognized as quercetin xyloside (12), arabinopyranoside (13) and arabinofuranoside (14), respectively. Finally, the occurrence of an ion (M–H)$^-$ at m/z 447 in the compound 15 indicated the existence of a quercetin rhamnoside. These flavonoids were observed in mango peels but not in pulps and seed kernels. This agrees with previously reported data by Berardini et al. [40] in cultivar Tommy Atkins and Gómez-Caravaca et al. [41] in Keitt mango.

These results reflected that the phenolic profile of peels was different to the profiles obtained in pulps and seed kernels (Table 3). This agrees with information previously reported in the literature, where it is indicated that of different mango phenolics differ in the different plant parts [42]. In particular, in this study the peels showed a higher number of phenolic compounds than pulps and seed kernels. This is in accordance with results previously described in mango [37,43].

3.2.2. Quantification of Phenolic Compounds

The results of the quantification of phenolic compounds in mango are shown in Table 4. Mangiferin was the predominant phenolic compound in the three mango fractions but its quantity was different in each organ. In particular, mango peels presented the highest concentration (2500 µg/g D.W.) followed by seed kernels and pulp (Table 4). This result reflected that non-edible parts of the mango fruit are good sources of mangiferin. Similar results were obtained by Luo et al. [43], Gómez-Caravaca et al. [41] and Ribeiro et al. [37] in 11 Chinese cultivars, cultivar Keitt and cultivar Ubá of mango, respectively. However, the mangiferin concentrations obtained were different to the described by other authors in cultivar Tommy Atkins (peels: 1190.9–1690.4 µg/g; pulps: 2.2 µg/g) [36,43] and cultivars Ataulfo, Keitt, Van Dyke and Ubá, among others (peels: 62.3–21530 µg/g; pulps: not detected-200 µg/g; seed kernels: traces-2340 µg/g) [37,41,43–45]. This agrees with data previously reported in the literature, where it has been reported that factors such as cultivar, environment, harvest stage, maturity as well as the method extraction, among other factors, have an effect on the phenolic composition [3,19,44,45]. Regarding the mangiferin biological effects, this phenol possesses an antioxidant capacity higher than other natural antioxidants like vitamin C and E. In this line, this phenolic compound could be used as a food preservative. In addition, mangiferin has a special particular interest for the pharmacological industry by its cancer chemopreventive potential [41].

Table 4. Concentration of phenolic composition in mango peel, pulp and seed kernel.

Peak	Phenolic Compounds	Concentration (µg/g D.W.)		
		Peel	Pulp	Seed Kernel
1	Procyanidin (catechin derivative)	560 ± 60	-	-
2	Procyanidin dimer	<LOQ	-	-
3	Epiafzelechin-epicatechin dimer	600 ± 60	-	-
4	Unidentified anthocyanin	<LOQ	-	-
5	Mangiferin	2500 [a] ± 320	50 [c] ± 20	430 [b] ± 90
6	Unidentified anthocyanin	30 ± 0	-	-
7	Mangiferin gallate	<LOQ	-	-
8	Kaempferol derivative	<LOQ	-	-
9	Mangiferin gallate	<LOQ	-	-
10	Quercetin-galactoside	220 ± 20	-	-
11	Quercetin glucoside	180 ± 10	-	-
12	Quercetin xyloside	<LOQ	-	-
13	Quercetin arabinopyranoside	80 ± 10	-	-
14	Quercetin arabinofuranoside	50 ± 0	-	-
15	Quercetin rhamnoside	50 ± 0	-	-
16	Mangiferin (isomer)	<LOQ	-	-
17	Mangiferin gallate (isomer)	<LOQ	-	-
	$\sum$phenolic compounds	4270 [a] ± 480	50 [c] ± 20	430 [b] ± 90

LOQ: limit of quantification; [a-c] Mean values with different letter on the right in the same row indicate statistically significant differences among the three treatments ($p < 0.05$).

Regarding the presence of procyanidins, these compounds were found only in the peel and two of them, a catechin derivative (560 µg/g D.W.) and epiafzelechin-epicatechin (600 µg/g D.W.) dimers, constituted the most abundant phenolic compounds after mangiferin (Table 4). These compounds, accounting for the 25% of total phenolic compounds, have been described in mango peels for the first time. According to other authors, these low-molecular weight procyanidins have been described as interesting because of their potent antioxidant capacity and possible protective effects on human health [42,46], especially because dimers can be absorbed intact in the intestinal tract [47] and have recently shown to promote the growth of *Bifidobacterium* in vitro [48]. As natural antioxidants and antimicrobials, proanthocyanidins can be also used in the industry as a preservative, to stabilize food colors, to prevent rancidity due to oxidation of unsaturated fats and to avoid the growth of bacteria and molds [49].

On the other hand, flavonol glycosides obtained in mango peel were identified as quercetin glycosides, being quercetin galactoside and quercetin glucoside the most abundant (Table 4), according to the literature [36,41,50]. Quercetin glycosides biological effects are mainly associated to their antioxidant capacity which could be exploited as a food preservative and stabilizer in the agro-industry, as well as in the development of new drugs in the pharmaceutical industry, among others [51].

The total phenolic content (TPC), evaluated as the sum of the content of individual phenolic, showed significant differences between the different fractions of mango. The order observed was peels > seed kernels > pulps (Table 4). This is in line as described the literature, where it has been reported that phenolic compounds are preferentially located in non-edible parts (peels and seed kernels) and in a lesser extent in the edible part (pulps) [3,42,52].

3.3. Antioxidant Capacity and Folin Assay

The health-promoting effects of both bioactive compounds characterized, polyphenols and isoprenoids, have been strongly associated with their antioxidant capacity. Thus, in this present study has been also evaluated the antioxidant capacity of the mango by-products. As shown Table 5, antioxidant capacity of different mango organs (pulp, peel and seed kernel) was evaluated by DPPH radical and ORAC method. Mango by-products highlighted by showing a higher antioxidant capacity than pulps, being seed kernels the mango fraction that exhibited the highest values of antioxidant capacity. The order described is comparable with that obtained by Abbasi et al. [30], Guo et al. [53] and Sogi et al. [54] in nine Chinese mango cultivars, in Chinese mango cultivar and in Tommy Atkins mango from USA, respectively. However, the values obtained in mango peels and seed kernels were lower than that described by Sogi et al. [54] (seed kernels = 154.7–181.9 mmol Trolox/100 g; peels = 41.8–77.6 mmol Trolox/100 g), while mango pulp was higher than that reported by Noratto et al. [55] in five Mexican mango cultivars (15.0–32.7 mmol Trolox/100 g). In this line, it has been described that mango antioxidant capacity can be affected by different factors such as variety, maturity, agricultural and industrial practices [36]. Moreover, it is important to indicate that ORAC value of seed kernels was greater than the ORAC values reported in other tropical fruits with demonstrated high polyphenol content such as banana passion fruit and Andean blackberry [56,57]. These results suggested that mango by-products, especially seed kernels, could represent a valuable ingredient with high antioxidant capacity in the agro-industry. In spite of this, it is important to note that the results were obtained by in vitro methods, which do not take into account the (among others) the metabolic transformations and interactions that clearly affect the bioavailability and biological action of phytochemicals [58,59]. In particular, the predominant forms of polyphenols in plasma are conjugates (glucuronates or sulfates, with or without methylation), which are chemically distinct from their parent compounds and therefore their properties are also different [59]. Thus, prior to the use of mango by-product as a new ingredient in the industry is necessary to go deeper into the analysis of their antioxidant activity and carry out assays of antioxidant capacity in vivo that corroborate the results obtained in vitro; as well as analyze the synergism or antagonism between the bioactive compounds present in the mango by-product matrix both at the level of bioavailability and biological action.

Table 5. Antioxidant capacity in mango organs (peels, pulps and seed kernels).

	Antioxidant Capacity (mmol Trolox/100 g D.W.)		
	Peel	**Pulp**	**Seed Kernel**
DPPH	11.06 [b] ± 0.42	2.11 [c] ± 0.12	44.34 [a] ± 0.89
ORAC	29.87 [b] ± 2.69	1.83 [c] ± 0.45	126.08 [a] ± 2.44

[a–c] Mean values with different letter on the right in the same row indicate statistically significant differences among the three treatments ($p < 0.05$).

On the other hand, the order observed in Folin-Ciocalteu assay by mango organs, seed kernels (23,759.13 mg/100 g GAE) > peels (2,874.97 mg/100 g GAE) > pulps (563.01 mg/100 g GAE), coincided with that described in the antioxidant activity. This result makes sense, because Folin-Ciocalteu assay determines the content of polyphenols and other reducing bioactive compounds (sugars, amino acids, etc.) that have antioxidant properties.

Finally, the correlation phytochemical composition (polyphenol, carotenoid and α-tocopherol) versus antioxidant capacity (DPPH and ORAC) was positive and direct (Table 6).

Table 6. Pearson's correlation coefficients (*r*) between bioactive compounds (isoprenoids polyphenols and) in mango organs (peels, pulps and seed kernels) and its antioxidant capacity (DPPH and ORAC).

	Bioactive Compounds Mango		
Assay	Peel	Pulp	Seed Kernel
DPPH	0.97	0.89	0.92
ORAC	1.00	0.90	0.97

4. Conclusions

In summary, the results obtained demonstrate that the non-edible parts of mango, the peels and seed kernels commonly managed as a waste in the industry, are a good source of bioactive compounds like isoprenoids, especially carotenoids and α-tocopherol and polyphenols with high nutritional value and health benefit effects. In particular, mango peels constituted the great part of the fruit in terms of bioactive compounds contents, following by the seed kernels and both compared to the pulp, being some of these components essential nutrients in the human diet, such as lutein and α-tocopherol. In addition, mango by-products, especially seed kernels, exhibited higher antioxidant activity than pulps. Therefore, mango by-products could be exploited as a natural preservative by the pharmaceutical and agro-food industry, as well as being used as a natural ingredient contributing to different health benefits to consumers. Besides, the amount of waste generated during the processing of the mango could be reduced. Finally, future studies will be carried out deeply to elucidate the bioavailability and safety of bioactive compounds of mango by-products.

Author Contributions: Author N.B. carried out the assays, collecting data and helped to draft the article. Author C.M.S. carried out the assays, collecting data and helped to draft the manuscript. Authors J.R., D.A.M. and A.J.M.-M. participated in the experimental design and helped to draft the manuscript. Author A.G.-R. carried out the studies participated in collecting data, experimental design and drafted the manuscript. All authors read and approved the final manuscript.

Funding: The authors would like to express gratitude for the funding received from the Spanish Ministry of Economy and Competitiveness (MINECO) and CSIC–VITRI with the Project I-COOP+2014 (Ref. COOPB20125) and Escuela Politécnica Nacional (PIS 12-21 and PIMI 14-14). Part of this work was carried out as a collaboration within the framework of the CYTED Program (Refs. 112RT0460-CORNUCOPIA and 112RT0445-IBERCAROT, Thematic Networks).

Acknowledgments: Quality technical work from Ms. Ana Benítez González (Universidad de Sevilla) is acknowledged. AGR also thanks the Ecuadorian SENECYT for the Prometeo Postdoctoral Grants (Ref. PROMETEO-CEB-018-2015).

Conflicts of Interest: The authors declare that they have no current or potential conflicts of interest.

References

1. Torres-León, C.; Rojas, R.; Contreras-Esquivel, J.C.; Serna-Cock, L.; Belmares-Cerda, R.E.; Aguilar, C.N. Mango seed: Functional and nutritional properties. *Trends Food Sci. Technol.* **2016**, *55*, 109–117. [CrossRef]
2. Evans, E.A.; Ballen, F.H.; Siddiq, M. Mango production, global trade, consumption trends and postharvest processing and nutrition. In *Handbook of Mango Fruit*; John Wiley & Sons: Chichester, UK, 2017; pp. 1–16.
3. Burton-Freeman, B.M.; Sandhu, A.K.; Edirisinghe, I.; Trevisan, M.T.; Viana, D.d.A.; Rao, V.S.; Santos, F.A.; Gali-Muhtasib, H.; Chae, S.; Saido, T.C.; et al. Mangos and their bioactive components: Adding variety to the fruit plate for health. *Food Funct.* **2017**, *8*, 3010–3032. [CrossRef] [PubMed]
4. Ravani, A.; Joshi, D.C. Mango and it's by product utilization—A review. *Energy* **2013**, *1*, 55–67.
5. Purnachandra, M.R.; Saritha, K.V. Bio-catalysis of mango industrial waste by newly isolated *Fusarium* sp. (PSTF1) for pectinase production. *Biotech* **2015**, *5*, 893–900. [CrossRef] [PubMed]

6. Jahurul, M.H.A.; Zaidul, I.S.M.; Ghafoor, K.; Al-Juhaimi, F.Y.; Nyam, K.-L.; Norulaini, N.A.N.; Sahena, F.; Mohd Omar, A.K. Mango (*Mangifera indica* L.) by-products and their valuable components: A Review. *Food Chem.* **2015**, *183*, 173–180. [CrossRef] [PubMed]

7. Zhang, Y.J.; Gan, R.Y.; Li, S.; Zhou, Y.; Li, A.N.; Xu, D.P.; Li, H.B. Antioxidant Phytochemicals for the Prevention and Treatment of Chronic Diseases. *Molecules* **2015**, *20*, 21138–21156. [CrossRef] [PubMed]

8. De Oliveira, A.C.; Valentim, I.B.; Silva, C.A.; Bechara, E.J.H.; de Barros, M.P.; Mano, C.M.; Goulart, M.O.F. Total phenolic content and free radical scavenging activities of methanolic extract powders of tropical fruit residues. *Food Chem.* **2009**, *115*, 469–475. [CrossRef]

9. Salas, S.; Anton, A.; McLaren, S.J.; Notarnicola, B.; Saouter, E.; Sonesson, U. In quest of reducing the environmental impacts of food production and consumption. *J. Clean Prod.* **2017**, *140*, 387–398. [CrossRef]

10. Meléndez-Martínez, A.J.; Vicario, I.M.; Heredia, F.J. Carotenoids, color and ascorbic acid content of a novel frozen-marketed orange juice. *J. Agric. Food Chem.* **2007**, *55*, 1347–1355. [CrossRef] [PubMed]

11. Meléndez-Martínez, A.J.; Britton, G.; Vicario, I.M.; Heredia, F.J. The complex carotenoid pattern of orange juices from concentrate. *Food Chem.* **2008**, *109*, 546–553. [CrossRef]

12. Stinco, C.M.; Benítez-González, A.M.; Hernanz, D.; Vicario, I.M.; Meléndez-Martínez, A.J. Development and validation of a rapid resolution liquid Chromatography method for the screening of dietary plant isoprenoids: Carotenoids, tocopherols and chlorophylls. *J. Chromatogr. A* **2014**, *1370*, 162–170. [CrossRef] [PubMed]

13. Slinkard, K.; Singleton, V.L. Total Phenol Analysis: Automation and comparison with Manual Methods. *Am. J. Enol. Vitic.* **1977**, *28*, 49–55.

14. Gironés-Vilaplana, A.; Baenas, N.; Villaño, D.; Speisky, H.; García-Viguera, C.; Moreno, D.A. Evaluation of Latin-American fruits rich in phytochemicals with biological effects. *J. Funct. Foods* **2014**, *7*, 599–608. [CrossRef]

15. Mena, P.; García-Viguera, C.; Navarro-Rico, J.; Moreno, D.A.; Bartual, J.; Saura, D.; Martí, N. Phytochemical characterisation for industrial use of pomegranate (*Punica granatum* L.) cultivars grown in Spain. *J. Sci. Food Agric.* **2011**, *91*, 1893–1906. [CrossRef] [PubMed]

16. Ou, B.; Hampsch-Woodill, M.; Prior, R.L. Development and validation of an improved oxygen radical absorbance capacity assay using fluorescein as the fluorescent probe. *J. Agric. Food Chem.* **2001**, *49*, 4619–4626. [CrossRef] [PubMed]

17. Rodriguez-Amaya, D.B.; Kimura, M.; Godoy, H.T.; Amaya-Farfan, J. Updated Brazilian database on food carotenoids: Factors affecting carotenoid composition. *J. Food Compos. Anal.* **2008**, *21*, 445–463. [CrossRef]

18. Ornelas-Paz, J.D.J.; Yahia, E.M.; Gardea-Bejar, A. Identification and quantification of xanthophyll esters, carotenes and tocopherols in the fruit of seven mexican mango cultivars by liquid chromatography-atmospheric pressure chemical ionization-time-of-flight mass spectrometry [LC-(APcI+)-MS]. *J. Agric. Food Chem.* **2007**, *55*, 6628–6635. [CrossRef] [PubMed]

19. Hewavitharana, A.K.; Tan, Z.W.; Shimada, R.; Shaw, P.N.; Flanagan, B.M. Between fruit variability of the bioactive compounds, β-carotene and mangiferin, in mango (*Mangifera indica*). *Nutr. Diet.* **2013**, *70*, 158–163. [CrossRef]

20. Chen, J.P.; Tai, C.Y.; Chen, B.H. Improved liquid chromatographic method for determination of carotenoids in Taiwanese mango (*Mangifera indica* L.). *J. Chromatogr. A* **2004**, *1054*, 261–268. [CrossRef]

21. Watson, R.R.; Preedy, V.R. *Bioactive Foods in Promoting Health: Fruits and Vegetables*; Academic Press: London, UK, 2010.

22. Ajila, C.M.; Jaganmohan Rao, L.; Prasada Rao, U.J.S. Characterization of bioactive compounds from raw and ripe *Mangifera indica* L. peel extracts. *Food Chem. Toxicol.* **2010**, *48*, 3406–3411. [CrossRef] [PubMed]

23. Fernández-Sevilla, J.M.; Acién Fernández, F.G.; Molina Grima, E. Biotechnological production of lutein and its applications. *Appl. Microbiol. Biotechnol.* **2010**, *86*, 27–40. [CrossRef] [PubMed]

24. Erdman, J.W.; Smith, J.W.; Kuchan, M.J.; Mohn, E.S.; Johnson, E.J.; Rubakhin, S.S.; Wang, L.; Sweedler, J.V.; Neuringer, M.; Neuringer, M. Lutein and brain function. *Foods* **2015**, *4*, 547–564. [CrossRef] [PubMed]

25. Mercadante, A.Z.; Rodriguez-Amaya, D.B.; Britton, G. HPLC and Mass Spectrometric analysis of carotenoids from mango. *J. Agric. Food Chem.* **1997**, *45*, 120–123. [CrossRef]

26. Litz, R.E. *The Mango: Botany, Production and Uses*; CABI: Wallingford, UK, 2009.

27. Garzón, G.A.; Narváez-Cuenca, C.-E.; Kopec, R.E.; Barry, A.M.; Riedl, K.M.; Schwartz, S.J. Determination of carotenoids, total phenolic content and antioxidant activity of arazá (*Eugenia stipitata* McVaugh), an Amazonian fruit. *J. Agric. Food Chem.* **2012**, *60*, 4709–4717. [CrossRef] [PubMed]

28. Gancel, A.-L.; Alter, P.; Dhuique-Mayer, C.; Ruales, J.; Vaillant, F. Identifying Carotenoids and phenolic compounds in naranjilla (*Solanum quitoense* Lam. Var. puyo hybrid), an Andean fruit. *J. Agric. Food Chem.* **2008**, *56*, 11890–11899. [CrossRef] [PubMed]

29. Ellong, E.N.; Billard, C.; Adenet, S.; Rochefort, K. Polyphenols, carotenoids, vitamin C content in tropical fruits and vegetables and impact of processing methods. *Food Nutr. Sci.* **2015**, *6*, 299–313. [CrossRef]

30. Wall, M.M. Ascorbic acid, vitamin A and mineral composition of banana (*Musa* sp.) and papaya (*Carica papaya*) cultivars grown in Hawaii. *J. Food Compos. Anal.* **2006**, *19*, 434–445. [CrossRef]

31. Abbasi, A.M.; Liu, F.; Guo, X.; Fu, X.; Li, T.; Liu, R.H. Phytochemical composition, cellular antioxidant capacity and antiproliferative activity in mango (*Mangifera indica* L.) pulp and peel. *Int. J. Food Sci. Technol.* **2017**, *52*, 817–826. [CrossRef]

32. Burns, J.; Fraser, P.D.; Bramley, P.M. Identification and quantification of carotenoids, tocopherols and chlorophylls in commonly consumed fruits and vegetables. *Phytochemistry* **2003**, *62*, 939–947. [CrossRef]

33. Vilela, C.; Santos, S.A.O.; Oliveira, L.; Camacho, J.F.; Cordeiro, N.; Freire, C.S.R.; Silvestre, A.J.D. The ripe pulp of *Mangifera indica* L.: A rich source of phytosterols and other lipophilic phytochemicals. *Food Res. Int.* **2013**, *54*, 1535–1540. [CrossRef]

34. Gong, X.; Qi, N.; Wang, X.; Li, J.; Lin, L. A new method for determination of α-tocopherol in tropical fruits by Ultra Performance Convergence Chromatography with Diode Array Detector. *Food Anal. Methods* **2014**, *7*, 1572–1576. [CrossRef]

35. Kodad, O.; Company, R.S.i.; Alonso, J.M. Genotypic and environmental effects on tocopherol content in almond. *Antioxidants* **2018**, *7*, 6. [CrossRef] [PubMed]

36. Schieber, A.; Berardini, N.; Carle, R. Identification of flavonol and xanthone glycosides from mango (*Mangifera indica* L. cv. "Tommy Atkins") peels by High-Performance Liquid Chromatography-Electrospray Ionization Mass Spectrometry. *J. Agric. Food Chem.* **2003**, *51*, 5006–5011. [CrossRef] [PubMed]

37. Ribeiro, S.M.R.; Barbosa, L.C.A.; Queiroz, J.H.; Knödler, M.; Schieber, A. phenolic compounds and antioxidant capacity of Brazilian mango (*Mangifera indica* L.) Varieties. *Food Chem.* **2008**, *110*, 620–626. [CrossRef]

38. KajdŽanoska, M.; Gjamovski, V.; Stefova, M. HPLC-DAD-ESI-MSn Identification of phenolic compounds in cultivated strawberries from Macedonia. *Maced. J. Chem. Chem. Eng.* **2010**, *29*, 181–194.

39. Schieber, A.; Hilt, P.; Conrad, J.; Beifuss, U.; Carle, R. Elution order of quercetin glycosides from apple pomace extracts on a new HPLC stationary phase with hydrophilic endcapping. *J. Sep. Sci.* **2002**, *25*, 361–364. [CrossRef]

40. Berardini, N.; Carle, R.; Schieber, A. Characterization of gallotannins and benzophenone derivatives from mango (*Mangifera indica* L. cv. 'Tommy Atkins') peels, pulp and kernels by High-Performance Liquid Chromatography/Electrospray Ionization Mass Spectrometry. *Rapid Commun. Mass Spectrom.* **2004**, *18*, 2208–2213. [CrossRef] [PubMed]

41. Gómez-Caravaca, A.M.; López-Cobo, A.; Verardo, V.; Segura-Carretero, A.; Fernández-Gutiérrez, A. HPLC-DAD-q-TOF-MS as a powerful platform for the determination of phenolic and other polar compounds in the edible part of mango and its by-products (peel, seed and seed husk). *Electrophoresis* **2016**, *37*, 1072–1084. [CrossRef] [PubMed]

42. Masibo, M.; Qian, H. Major mango polyphenols and their potential significance to human health. *Compr. Rev. Food Sci. Food Saf.* **2008**, *7*, 309–319. [CrossRef]

43. Luo, F.; Lv, Q.; Zhao, Y.; Hu, G.; Huang, G.; Zhang, J.; Sun, C.; Li, X.; Chen, K. Quantification and purification of mangiferin from Chinese mango (*Mangifera indica* L.) cultivars and its protective effect on human umbilical vein endothelial cells under H_2O_2-induced stress. *Int. J. Mol. Sci.* **2012**, *13*, 11260–11274. [CrossRef] [PubMed]

44. Berardini, N.; Fezer, R.; Conrad, J.; Beifuss, U.; Carle, R.; Schieber, A. Screening of mango (*Mangifera indica* L.) cultivars for their contents of flavonol *O*- and xanthone *C*-glycosides, anthocyanins and pectin. *J. Agric. Food Chem.* **2005**, *53*, 1563–1570. [CrossRef] [PubMed]

45. Barreto, J.C.; Trevisan, M.T.S.; Hull, W.E.; Erben, G.; De Brito, E.S.; Pfundstein, B.; Würtele, G.; Spiegelhalder, B.; Owen, R.W. Characterization and quantitation of polyphenolic compounds in bark, kernel, leaves and peel of mango (*Mangifera indica* L.). *J. Agric. Food Chem.* **2008**, *56*, 5599–5610. [CrossRef] [PubMed]

46. Ramirez, J.E.; Zambrano, R.; Sepúlveda, B.; Simirgiotis, M.J. Antioxidant properties and hyphenated HPLC-PDA-MS Profiling of Chilean Pica mango fruits (*Mangifera indica* L. cv. Piqueño). *Molecules* **2014**, *19*, 438–458. [CrossRef] [PubMed]

47. Gu, L.; Kelm, M.A.; Hammerstone, J.F.; Beecher, G.; Holden, J.; Haytowitz, D.; Gebhardt, S.; Prior, R.L. Concentrations of proanthocyanidins in common foods and estimations of normal consumption. *J. Nutr.* **2004**, *134*, 613–617. [CrossRef] [PubMed]

48. Sáyago-Ayerdi, S.G.; Zamora-Gasga, V.M.; Venema, K. Prebiotic Effect of predigested mango peel on gut microbiota assessed in a dynamic in vitro model of the human colon (TIM-2). *Food Res. Int.* **2017**. [CrossRef]

49. Vélez-Rivera, N.; Cárdenas Perez, S.; Chanona-Pérez, J.J.; Perea-Flores, M.J.; Claderón-Domínguez, G.; Blasco Ivars, J.; Farrera-Rebollo, R.R. Industrial applications and potential pharmaceutical uses of mango (Mangifera indica) kernel. In *Seeds as Functional Foods & Nutraceuticals: New Frontiers in Food Science*; Nova Science Publishers Inc.: New York, NY, USA, 2014.

50. Ribeiro, S.M.R.; Schieber, A. Bioactive compounds in mango (*Mangifera indica* L.). In *Bioactive Foods in Promoting Health*; Elsevier: New York, NY, USA, 2010; pp. 507–523.

51. Baenas, N.; Abellán, Á.; Rivera, S.; Moreno, D.A.; García-Viguera, C.; Domínguez-Perles, R. Foods and supplements. In *Polyphenols: Properties, Recovery and Applications*; Elsevier: New York, NY, USA, 2018; pp. 327–362.

52. Ayala-Zavala, J.F.; Vega-Vega, V.; Rosas-Domínguez, C.; Palafox-Carlos, H.; Villa-Rodriguez, J.A.; Siddiqui, M.W.; Dávila-Aviña, J.E.; González-Aguilar, G.A. Agro-industrial potential of exotic fruit byproducts as a source of food additives. *Food Res. Int.* **2011**, *44*, 1866–1874. [CrossRef]

53. Guo, C.; Yang, J.; Wei, J.; Li, Y.; Xu, J.; Jiang, Y. Antioxidant activities of peel, pulp and seed fractions of common fruits as determined by FRAP assay. *Nutr. Res.* **2003**, *23*, 1719–1726. [CrossRef]

54. Sogi, D.S.; Siddiq, M.; Greiby, I.; Dolan, K.D. Total phenolics, antioxidant activity and functional properties of "Tommy Atkins" mango peel and kernel as affected by drying methods. *Food Chem.* **2013**, *141*, 2649–2655. [CrossRef] [PubMed]

55. Noratto, G.D.; Bertoldi, M.C.; Krenek, K.; Talcott, S.T.; Stringheta, P.C.; Mertens-Talcott, S.U. Anticarcinogenic effects of polyphenolics from mango (*Mangifera indica*) varieties. *J. Agric. Food Chem.* **2010**, *58*, 4104–4112. [CrossRef] [PubMed]

56. García-Ruiz, A.; Girones-Vilaplana, A.; León, P.; Moreno, D.A.; Stinco, C.M.; Meléndez-Martínez, A.J.; Ruales, J. Banana passion fruit (*Passiflora mollissima* (Kunth) L.H. Bailey): Microencapsulation, phytochemical composition and antioxidant capacity. *Molecules* **2017**, *22*, 85. [CrossRef] [PubMed]

57. Zapata, S.; Piedrahita, A.M.; Rojano, B. Capacidad Atrapadora de Radicales Oxígeno (ORAC) y fenoles totales de frutas y hortalizas de Colombia. *Perspect. Nutr. Humana* **2014**, *16*, 25–36.

58. Halliwell, B.; Rafter, J.; Jenner, A. Health promotion by flavonoids, tocopherols, tocotrienols and other phenols: Direct or indirect effects? Antioxidant or not? *Am. J. Clin. Nutr.* **2005**, *81*. [CrossRef] [PubMed]

59. Kroon, P.A.; Clifford, M.N.; Crozier, A.; Day, A.J.; Donovan, J.L.; Manach, C.; Williamson, G. How should we assess the effects of exposure to dietary polyphenols in vitro? *Am. J. Clin. Nutr.* **2005**, *80*, 15–21. [CrossRef] [PubMed]

nutrients

MDPI

Review

Skin Carotenoids in Public Health and Nutricosmetics: The Emerging Roles and Applications of the UV Radiation-Absorbing Colourless Carotenoids Phytoene and Phytofluene

Antonio J. Meléndez-Martínez *, Carla M. Stinco and Paula Mapelli-Brahm

Food Colour & Quality Laboratory, Area of Nutrition & Food Science, Universidad de Sevilla, 41012 Seville, Spain; cstinco@us.es (C.M.S.); pmapelli@us.es (P.M.-B.)
* Correspondence: ajmelendez@us.es; Tel.: +34-95455-7017

Received: 21 February 2019; Accepted: 9 May 2019; Published: 16 May 2019

Abstract: In this work, the importance of dietary carotenoids in skin health and appearance is comprehensively reviewed and discussed. References are made to their applications in health-promoting and nutricosmetic products and the important public health implications that can be derived. Attention is focused on the colourless UV radiation (UVR)-absorbing dietary carotenoids phytoene and phytofluene, which are attracting increased interest in food science and technology, nutrition, health and cosmetics. These compounds are major dietary carotenoids, readily bioavailable, and have been shown to be involved in several health-promoting actions, as pinpointed in recent reviews. The growing evidence that these unique UVR-absorbing carotenoids with distinctive structures, properties (light absorption, susceptibility to oxidation, rigidity, tendency to aggregation, or even fluorescence, in the case of phytofluene) and activities can be beneficial in these contexts is highlighted. Additionally, the recommendation that the levels of these carotenoids are considered in properly assessing skin carotenoid status is made.

Keywords: colourless carotenoids; cosmeceuticals; functional foods; nutraceuticals; nutricosmetics; photoprotection; phytoene; phytofluene; public health

---

1. Introduction

Epithelial tissues like those that form the skin are essential to protect the human body from diverse aggressions. The skin is the largest organ of the human body and consists of several layers and structures. It is an essential physical barrier that protects us from the external environment (radiation, xenobiotics, microorganisms, etc.) but also intervenes in essential processes. Examples are thermoregulation, metabolism, the homeostasis of fluids, sensing, or the production of vitamin D and other important compounds [1–5].

Given its importance at different levels, damage to the skin lead to disorders of different natures that can eventually lead to diseases, for instance those derived from infections or even skin cancers, with consequent negative effects in terms of well-being and associated healthcare costs. Skin appearance is also important for individuals at various levels, since some attributes (pigmentation, colour uniformity, wrinkling, elasticity, etc.) have a significant influence on attractiveness judgements, and skin signalling has been associated to aspects ranging from mating choices to socioeconomic status [6–10].

Within this context, protection of the skin against sunlight and other sources of radiation (for example tanning lamps) is of great relevance, as excessive exposure in particular to UV radiation (UVR) can lead to photosensitivity, sunburn, photoaging, immunosuppresive effects, or even development of skin cancer, disorders with associated negative health and aesthetic outcomes [11].

The role of nutrition in skin health and appearance is undeniable and long-known. In this context, several terms are commonly used and sometimes overlap, such as cosmetics, cosmeceuticals, nutricosmetics, functional foods, or nutraceuticals. Although there is no broad standard consensus definition, the term "nutricosmetics" is associated with the oral consumption of products containing food components (e.g., vitamins, peptides, polysaccharides, polyphenols, coenzyme Q10, polyunsaturated fatty acids, and carotenoids) for cosmetic purposes. Thus, the term is clearly linked to concepts like 'beauty pills', 'beauty from within', 'beauty foods', 'nutraceuticals for skin care', or 'oral cosmetics' [5,12–14].

Carotenoids are natural dietary products ingested from foods, as additives, or from supplements, among other products [15,16]. Some carotenoids can be consistently found in human plasma, milk, and various tissues, including the skin, with major carotenoids usually being lutein, zeaxanthin, β-cryptoxanthin, α-carotene, β-carotene, lycopene, phytoene, and phytofluene [17]. Apart from their role as natural colourants, carotenoids have been important in nutrition because some of them function as provitamin A, and evidence has accumulated over the last 30 years that carotenoids may contribute to decreasing the risk of developing various non-communicable diseases including several types of cancers, cardiovascular disease, and skin or eye disorders, among others [18–21]. This is the main source of their indisputable value in the context of functional foods, nutraceuticals and related products.

The UVR-absorbing colourless carotenoids phytoene and phytofluene are rarities in the carotenoid family and have been largely ignored in studies dealing with food science and technology, nutrition, public health, and cosmetics. This is surprising since recent comprehensive reviews have indicated that they are major dietary carotenoids (found in widely consumed products, including tomatoes, carrots, citrus, and derivatives), readily bioavailable (they are present as major carotenoids in plasma, human milk, skin and other tissues) and involved in several health-promoting actions, as revealed by various studies. Notably, evidence is accumulating that they could be involved in the health benefits traditionally associated to lycopene [22–24]. Being the unique major dietary carotenoids absorbing maximally in the UV region and possessing other distinctive characteristics within the carotenoid family, research, and applications in the use of these carotenoids for the promotion of health and cosmetics is a live and expanding area recently featuring in the carotenoid field.

2. Structure and Functions of the Skin

Epithelial tissues, including the skin, are essential for survival since they protect from physical, chemical, and microbial damage and intervening in homeostasis [3].

2.1. Skin Structure

The skin is the largest organ of the human body with a surface area of 1.5–2 m^2 and accounting for about 15% of the total body weight of an adult. It is composed of epidermis, dermis, subcutaneous fat and structures like hair follicles, as well as sweat and sebaceous glands. The different layers that are distinguished in the epidermis are characterized by the state of differentiation of keratinocytes, the most numerous cells in this skin layer. The outermost epidermis layer is the stratum corneum, with a thickness of 10 μm to 30 μm. The cells in this external layer are called corneocytes and lack nuclei and cytoplasmic organelles. Below the stratum corneum there are keratinocytes (~95% of epidermal cells), melanocytes, Langerhans cells and Merkel cells [1,4,5].

The dermis is made up of ~60% water and has a thickness ranging from 0.5–5 mm depending on the location. In it, two main layers are distinguished: the papillary dermis (rich in blood vessels and nerve endings); and the reticular dermis, the main part of the dermis that is in contact with the subcutis. The main dermal cells are fibroblasts, mast cells, plasma cells, lymphocytes, dermal dendritic cells and histiocytes. The dermis also contains interstitial materials, including collagen fibres (~70–90% of the dermis dry weight), elastic tissue or ground substance (that together form part of the so-called extracellular matrix proteins complex) as well as sweat pores and hair follicles. It is estimated that in the dermis there are 1.5–6 million sweat glands, which are of two types, namely eccrine (from which most thermoregulatory sweating and sweat fluid comes from) and apocrine. Within the dermis, different types of mechanoreceptors can also be found [1,4,5].

The hypodermis or subcutis is the most internal layer of the skin and contains cells known as lipocytes. It is estimated that almost 80% of all body fat is deposited in the subcutis (in non-obese individuals) [1,4,5].

2.2. Skin Functions

The skin is an essential physical barrier that protects the body from the external environment (radiation, xenobiotics, microorganisms, etc.), the epidermis being fundamental for this function. The dermis confers mechanical strength and elasticity to the skin and the subcutis acts as insulation and mechanical protection and contributes to thermoregulation [1,4,5]

In relation to the protection against radiation, it is to be noted that the degree of light penetration in the skin is wavelength-dependent. Thus, the longer the wavelength (and the lower the energy of the radiation), the deeper is the penetration. Long wave UVA (320–400 nm) can penetrate the whole dermis but the more energetic medium wave UVB (280–320 nm) mainly reaches only as deep as the epidermis. The shortwave UVC (200–280 nm) is absorbed by the ozone layer and does not reach the Earth's surface. Visible radiation (approximately 400–770 nm) penetrates into subcutaneous tissue. [5,11,25,26] (Figure 1).

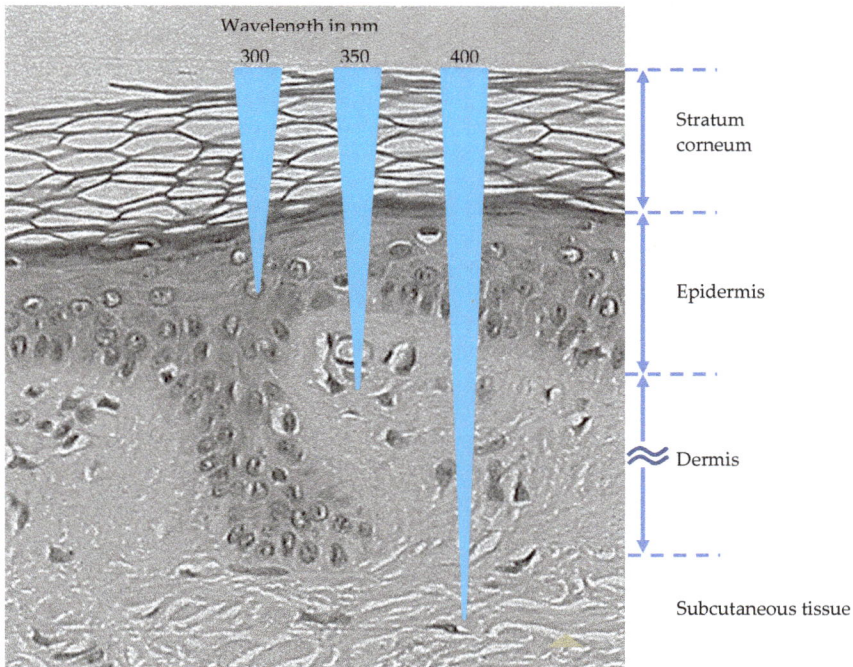

Figure 1. Skin penetration depth of different wavelenghts.

Protection against UV light is of great importance to prevent lesions of different gravity, some of which may lead to an undesirable appearance. For instance, excessive exposure to sunlight or artificial sources of UV light, a situation commonly associated with tanning, lead to adverse effects. These can be categorized into different disorders, including sunburn, immunosuppression, photocarcinogenesis, and photoaging. The latter is characterized by collagen proteolysis that results in signs such as wrinkles. Other signs of photoaging are teleangiectasia (dilated and deformed microvasculature) or hyper-pigmentation [11], all of them undesirable signs from an aesthetic point of view.

Concerning the barrier function of the skin, the stratum corneum has a prominent role as a water-proof, relatively impermeable barrier that is key in preventing the entry of diverse xenobiotics [2] However, many compounds can absorb through the stratum corneum, from which derives the existence of a wide variety of pharmaceutical and cosmetic products intended for topical application [27–30].

The skin also play key roles in vital processes such as thermoregulation, metabolism and the homeostasis of fluids. In relation to the latter, the skin, particularly the water-impermeable stratum corneum, helps reduce water loss and, therefore, dehydration. The skin also contributes to innate and adaptive immunity. For instance, substances produced in keratinocytes or derived from the sweat can exhibit antimicrobial activity against a wide variety of bacteria, viruses, and fungi. Additionally, the skin's Langerhans cells act in the immune response against threats. Regarding the role of skin in the regulation of body temperature, the regulation of body temperature and heat loss is effected by sweating and the vasodilatation and vasoconstriction of vascular plexi. Dissipation of excess heat occurs mostly through the skin, by means of the production of sweat from plasma and the subsequent heat transfer to the environment (perspiration) [1,2].

The skin is also important for sensory perception due to the presence of abundant free nerve endings and end-corpuscles. Thus, different kinds of mechanoreceptors (including Merkel's disks, Meissner's corpuscles, Ruffini's endings, Pacinian corpuscles and, in some locations, Krause's end bulbs) are present [31–33].

Skin is important for the synthesis of vitamin D and, therefore, has a role in bone synthesis, calcium metabolism and other processes associated with the vitamin. More specifically, the epidermis is considered the major source of vitamin D for humans, as ultraviolet radiation (specifically UVB) can lead to the conversion of 7-dehydrocholesterol to vitamin D. Furthermore, the keratinocytes contain the enzymes required to further convert the vitamin into 1,25 dihydroxyvitamin D, its active form. The layer of subcutaneous fat is important to provide insulation and for mechanical protection, acting as a cushion. Finally, this layer has a role as an energy reserve and in the secretion of hormones [1,4,5,34].

The properties of the skin, related to its functions and individual quality of life, are dependent on manifold factors of varying nature, such as the ones summarized in Figure 2 [4]. Most of the factors related to the individual are congenital. In this regard, it is considered that some factors, notably ethnicity and gender, influence the skin's properties to a lesser degree. The Fitzpatrick–Pathak skin type, a classification of six different types of skin based on their response to sun exposure (Table 1), has a marked impact on protection against radiation and photoaging, which features reduced hydration and elasticity. Factors dependent on the skin characteristics can vary markedly depending on the body site considered. Additionally, the presence or penetration of different substances may have a marked effect on the skin properties. Indeed, the cosmetic and pharmaceutical industries have made use of the skin's capacity to absorb such substances [4].

Table 1. Characteristics of the different types of skin according to the Fitzpatrick–Pathak classification.

		Characteristics [a]	
Skin type [a]	Colour [b]	Sunburn	Tan
I	White	Yes	No
II	White	Yes	Minimal
III	White	Yes	Yes
IV	White	No	Yes
V	Brown	No	Yes
VI	Black	No	Yes

[a] Based on the responses to a verbal questionnaire related to the response of the skin to initial sun exposure, i.e., three minimum erythema doses (MEDs) or about 45–60 min of noon exposure in northern latitudes in the early summer. [b] Colour of the unexposed skin. From Fitzpatrick [35].

Dependent on the individual

Age
Anatomical region
Body mass index (BMI)
Ethnicity
Fitzpatrick-Pathak skin type
Gender
Lifestyle

Skin characteristics

Elasticity
Flexibility
Health status
Hydration status
pH
Roughness
Sebum secretion
Sweating rate
Thickness

Associated to substances and external agents

Active compounds
Dietary components
Cosmetics
Drugs
Allergens
Fluids (liquids, gases)
Microbes and other organisms
Particles
Radiation

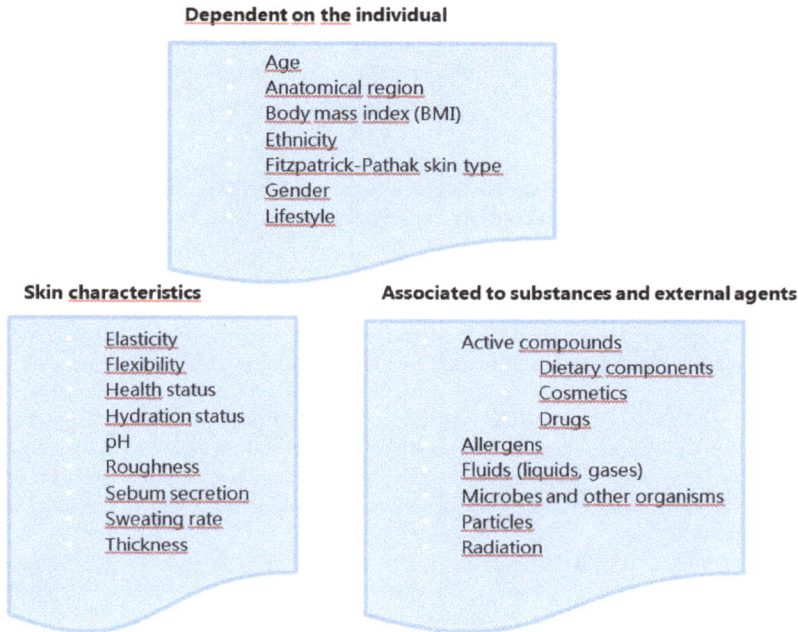

Figure 2. Factors affecting skin properties. Adapted from Dąbrowska et al. [4].

3. Skin Disorders

It is estimated that over one thousand disorders involving the skin have been described (infections, drug reactions, psoriasis, eczema, urticaria, acne vulgaris, pityriasis rubra pilaris, Darier's disease, ichthyosis, lichen ruber, and different skin cancers, including basal cell carcinoma, squamous cell carcinoma, or melanoma, etc.) and that 1/5 of all patient referrals to general practitioners involve skin pathologies. From this it can be inferred that skin disorders involve an important economic cost to both the citizens and the healthcare services [1,28].

3.1. UV Radiation Types and Consequences of Exposure

UV radiation (UVR) can be categorized into three types as a function of wavelengths, namely UVA (320–400 nm, accounting for ~95% of the UV radiation reaching the earth), UVB (280–320 nm, accounting for ~5%), and UVC (200–280 nm), the latter not reaching the surface of our planet [26].

Exposure to UV radiation may lead to the production of reactive oxygen species (ROS), a generic term used to refer to both oxygen radicals and also some non-radical derivatives of oxygen, including the deleterious singlet oxygen, $^1O_2$. This is generated from O_2 by an input of energy rearranging the electrons. Due to continuous exposure to light, this non-radical ROS may be easily formed in the skin or even the eye, where they can directly damage lipids, proteins and even DNA [36]. This compound and/or UVA are thought to be able to regulate the expression of a wide variety of genes, including genes intervening in the cell cycle, apoptosis or photoaging (such as matrix metalloproteases or heme oxygenase). UVB can also cause sunburn and lead to the appearance of mutations and skin cancer through direct interaction with DNA [11]. Interestingly, UVA light is not filtered by window glass (unlike UVB light) and it is estimated that half of UVA exposure takes place in the shade [26].

Some of the skin's mechanisms for protection against UVR are increasing the thickness of epidermis; DNA repair mechanisms; programmed cell death; antioxidant enzymes; and skin pigmentation [26]. There is also ample evidence that dietary components, among which carotenoids, may have a protective role in some cases [11,37,38]. UVR can also increase sensitivity to some drugs, e.g., corticoids, and have beneficial effects beyond the induction of vitamin D synthesis. Thus, UVR may be positive for the treatment of skin pathologies, including psoriasis, morphea, scleroderma, vitiligo, or atopic dermatitis [5].

3.2. Skin Disorders Associated to UVR

Depending on the intensity and the continuity, exposure to UVR may lead to disorders of different nature and severity, such as photosensitivity disorders, photocarcinogenesis, sunburn, photoaging, and even photoimmune modulation.

3.2.1. Photosensitivity Disorders

One example of photosensitivity disorder is erythropoietic protoporphyria. This is an uncommon inborn haematological disease that leads to increased levels of protoporphyrin in plasma, red globules, skin and faeces. Protoporphyrin is an endogenous photosensitizer. Upon exposure to light can eventually favour the formation of ROS that cause cellular damage and leads to photosensitivity clinical symptoms. Specifically, the patients experience itching, burning, and pain in the skin exposed to sunlight after even only minutes of exposure, followed by edema, erythema, and purpura [11,39–41]. Apart from porphyrins, other endogenous compounds (for instance, flavins or amino acids) act as sensitizing molecules [38].

The first classification of skin based on the skin response to the sun was based exclusively on the phenotype (hair and eye colour) [42]. However, it was soon realized that it was not fully valid since people with the same hair or eye phenotype had very different responses to the sun. Therefore, Fitzpatrick developed a new classification based on the responses of people to a brief and simple questionnaire, about the response of the skin to sun exposure. Based on this questionnaire six types of skin were eventually defined. The first four types correspond to white skins and were defined by Fitzpatrick [42]. The last two types are made up of skins that do not suffer sunburn and get suntanned after the initial sun exposure and whose colours before the sun exposure are brown (Type V) or black (Type VI) (Table 1). Types V and VI were defined by Fitzpatrick [43] and Pathak [44]. It is considered that the higher the phototype, the higher the melanin levels and the lower the risk of skin cancer [5].

3.2.2. Sunburn

Sunburn, or erythema solare, is the acute inflammatory reaction of the skin, originated by an excessive exposure to natural of artificial UV radiation. The effectiveness of UV in causing sunburn is wavelength-dependent, decreasing with longer wavelengths. In this sense it has been estimated that it takes 1000 times more UVA relative to UVB to produce the same erythemal response. Therefore, it is commonly assumed that sunburn is caused by UVB, which induces cytokine-mediated processes and neuroactive and vasoactive mediators in the skin, resulting in inflammatory responses [5].

The classical symptom of this disorder is the reddening of the skin and vasodilation of cutaneous blood vessels, although blisters and ablation of the epidermis can appear in more severe episodes. The sunburn episode occurs approximately 4 h after exposure and reaches the peak in the interval 8–24 h. Normally, it disappears in approximately one day, although the persistence of the erythema depends on age and skin type. Thus, it is considered that sunburn lasts longer (up to several days) in older individuals and those with fairer skin. Sunburn cells are keratinocytes undergoing apoptosis [11,25,26].

Minimal erythema dose (MED) refers to the minimal dose of UVB required to cause sunburn and is markedly dependent on the skin type. This parameter is used to evaluate the protection factor of sun screens [11].

3.2.3. Photoaging

The structural and functional decline of skin is categorized into two main categories, namely aging due to the passage of time (intrinsic or chronological aging) or to external factors (extrinsic aging). The latter includes skin aging due to chronic exposure to high UVR (or photoaging) and other agents, namely smoking, pollution, sleep deprivation and poor nutrition [45]. Intrinsic and extrinsic aging can be superimposed. However, chronological aging and photoaging display some differences histologically and are sometimes clinically distinguishable. Chronological aging is clinically associated with increased fragility and loss of elasticity, and photoaging with elasticity loss, irregular pigmentation, dryness and wrinkling, signs mostly derived from the degradation of proteins of the extracellular matrix and aberrations in the production of melanin (leading to the appearance of spots or irregular skin tone). Skin aging in general has an impact on manifold characteristics, processes and functions (regenerative capability, pigmentation, thermoregulation, the strength and resiliency conferred by the collagenous extracellular matrix, etc.) converging into increased skin fragility and susceptibility to diseases [2,5].

UVA has a prominent role in photoaging. This radiation favours the production of ROS that may eventually cause mutations in the mitochondrial DNA, leading to defects in the cell energy metabolism that result in further inductions of ROS. Additionally, singlet oxygen can also upregulate matrix metalloproteases directly, leading to further damage to the extracellular matrix and the formation of wrinkles and skin sagging [11,46]. Other features of photoaging include pigmented lesions (actinic lentigines or "age spots", ephelides or "freckles", and pigmented solar and seborrhoeic keratosis), teleangiectasia (a widening and deformation of the fine capillaries in skin), dryness and inelasticity. Photoaging signs are more common in locations that are more frequently exposed to UVR like the face, neck, forearms, or the back of the hands [2,11].

Besides UVR, both near-infrared (NIR, IRA: 760–1440 nm and IRB: 1440–3000 nm) and visible light (400–760 nm) may also be involved in the development of photoaging. More specifically, it is estimated that the solar radiation energy received at the surface of the Earth is divided in this way: 6.8% UV (0.5% UVB, 6.3% UVA), 38.9% VIS, and 54.3% NIR [47]. NIR can generate ROS in the skin and contribute to photoaging [48]. Analogously, visible light may also cause the generation of ROS in the skin [49].

3.2.4. Photoimmune Modulation

There is consistent evidence that excessive UVR induces a number of immunological changes to the immune system [11]. Both UVA and UVB have been shown to be able to cause local and systemic immunosuppressive effects. For example, after UVR exposure, the Langerhans cells (epidermal antigen-presenting dendritic cells produced in the bone marrow) undergo functional and morphological changes leading to their depletion. Exposure to UVR may also result in T cell tolerance [1,5,26].

3.2.5. Photocarcinogenesis

Photocarcinogenesis refers to events initiated by the exposure to solar or artificial light that eventually lead to the development of skin cancer [50]. UVB is thought to be absorbed directly by DNA, causing DNA base structural damage, for instance formation of cyclobutane pyrimidine dimers and pyramidine-pyrimidone photodimers, which may eventually cause mutations and cancer. UVA is considered to provoke indirect DNA damage through the generation of ROS resulting in single-strand breaks and in DNA–protein crosslinks. Indeed, DNA absorbs maximally in the range of UV wavelengths from 245–290 nm, implicating UVB as a primary mutagen. Overall, UVR can cause alterations in DNA integrity and homeostasis, and affect genes, including oncogenes and tumour suppressor genes [5,26].

4. Skin Beauty

From evidence referred to in different sections of this and other works, it is arguable that the skin reflects health-status as well as intrinsic and extrinsic aging. Also, that apart from serving diverse essential functions, the skin is very important in terms of beauty and appeal. Some key attributes of skin appearance (for instance colour, texture, or elasticity) have been long known to be affected by nutritional status, and even by micronutrients or other minor food components of nutritional interest. Thus, vitamin C deficiency can cause various alterations in epithelial tissues, including impaired wound healing or skin fragility, since this vitamin is involved in collagen synthesis and antioxidant protection. Furthermore, it is well-known that retinoic acid (a form of vitamin A) and derivatives modulate the expression of genes involved in cellular differentiation and proliferation at the skin level, hence their many therapeutic and cosmetic applications. Furthermore, increases of dietary intakes of fruits and vegetables may cause perceptible changes in the yellowness of the skin, whilst excessive provitamin A (usually β-carotene) or lycopene may lead to increased skin pigmentation [5,30,51–55].

4.1. Colour and Other Parameters Associated to Skin Beauty

4.1.1. Colour

The main skin pigment is melanin. However, there also exist other chromophore-containing molecules that contribute to skin colouration, including oxyhemoglobin, deoxyhemoglobin, and carotenoids [26]. The distribution of carotenoids across the different skin layers and anatomical regions is not uniform. The highest levels are usually found in the stratum corneum especially close to the skin surface [56,57]. Oxyhemoglobin and deoxyhemoglobin are especially important in the papillary dermis, which is rich in blood vessels [1]. Melanin is a term used to refer to a group of natural pigments that also confer colour to the hair. There are two types of melanin, namely, eumelanin (dark brown-black insoluble polymer) and pheomelanin (light red-yellow sulphur-containing polymer). Melanin is synthesized in melanosomes, special organelles within the melanocytes in the basal layer of the epidermis, and then distributed to surrounding keratinocytes. The biosynthesis of melanin (melanogenesis) can be induced by diverse factors (α-melanocyte-stimulating hormone, stem cell factor, endothelin-1, nitric oxide, adrenocorticotropic hormone, prostaglandins, thymidine dinucleotide, or histamine) that eventually lead to the expression of genes codifying for microphthalmia-associated transcription factor (MITF). This is a master regulator of melanogenesis that, among other actions related to melanogenesis, upregulates the melanogenesis enzymes tyrosinase (TYR), tyrosine-related protein-1 (TRP-1), and tyrosine-related protein-2 (TRP-2). Of these three enzymes, TYR is thought to be the only one exclusively necessary for melanogenesis. The amino acid L-tyrosine acts as a substrate for the enzyme and is generally transported into the melanosome by facilitated diffusion. Skin pigmentation is regarded to be one important photoprotective factor as melanin, besides functioning as a broadband UV absorbent, seems to protect against oxidation. Evidence is accumulating that UVR-induced damage and its repair are also signals for the induction of the synthesis of melanin. Many epidemiological studies point to the fact that there is a lower incidence of skin cancer in individuals with darker skin relative to those with fair skin [1,2,10,26,58].

The number of melanocytes across different ethnicities is very similar. Indeed, the differences in skin melanin-related colour are mainly attributed to the size and distribution pattern of melanosomes, their contents in eumelanin and phaeomelanin, and to metabolic and tyrosinase activity within the melanocytes. For example, there is evidence that East Asians have a yellowish skin colour due to a higher proportion of phaeomelanin to eumelanin, and a spherical rather than ellipsoidal arrangement of the clustered pigments. Likewise, other studies indicated that the most fair-skinned ethnical groups (including European, Chinese, and Mexican) have about half the amount of melanin in their epidermis in comparison to the most darkly pigmented groups (including sub-Saharan African and Indian ethnicities) and that the size of melanosomes seems to vary across these groups, so that the largest melanosomes are found in sub-Saharan African individuals, followed in turn by Indian, Mexican, Chinese, and Europeans [9].

Interestingly, independently of the ethnicity group, there is evidence that the melanin volume fraction in the epidermis is positively correlated with the mean dose of surface solar UV-radiation received at the geographical location of the group in question [9].

Skin pigmentation and colour homogeneity are well-known to have a significant influence on attractiveness judgements, above all in women, since they can convey information about age, health status and social status, with some cultural differences [6–9].

4.1.2. Other Skin Aesthetic Parameters

Physical appeal in general and facial attractiveness in particular, are well-known to be largely associated with health, mate choices, and social status, such that youthful appearance confers advantages, including socioeconomic ones. Thus, there is constant interest in skin rejuvenation, mostly of the face, which is permanently exposed and perceived by others. In this context, there is evidence in the scientific and technical literature that skin structure conducive to a youthful appearance is a very important attribute of attractiveness, along with such others as facial symmetry and proportions [8,10]. Thus, the cosmetic industry aims to fight signs impairing youthful appearance like loss of elasticity, wrinkles and folds, dryness, or pigmentation disorders (spots, uneven colour, sallowness), which are associated with aging in general [2,5,10,59].

4.2. Geographic and Ethnic Differences

Regarding aesthetic preferences, the cosmetic industry must take into account aspects as diverse as differences in preferences across ethnicities or even environmental factors. Thus, it is well-known that many Asians (above all of the east) have preference for fair skin, whereas Caucasians prefer tanned skin, which in both cases is often associated to better health or economic status. Skin tone is an attribute of beauty of current great relevance for some population groups like African American individuals, since there is evidence that skin colouration in individuals of colour has a deep impact in the way they are perceived. Thus, some studies indicate that individuals of colour with lighter skin are perceived more favourably compared to others with darker skins. Additionally, that among African Americans, African American women with lighter skin tones are perceived as more attractive than their darker skinned counterparts [60–63]. Additionally, the environmental differences across geographical regions (for instance tropical vs. temperate) have an important impact on skin characteristics (colour, photosensitivity, hydration, etc.), so that cosmetics must be conceived, designed, and customized accordingly [64].

5. Functional Foods, Nutraceuticals, Cosmetics, Cosmeceuticals, and Nutricosmetics: Definitions and Concepts

The interest of scientists, physicians, pharmacists and other professionals in the impact of nutrition on skin is not new. However, the growing interest of citizens in health-promoting products and the consequent expansion of functional foods and nutraceuticals in different industries experienced in the last 30–40 years have also had an impact on the cosmetic sector. In this context, there is a series of terms, interrelated to a greater or lesser extent, which are commonly used in those industries.

Although there is no consensus definition for the term "functional foods", at least in Europe, there is a widely accepted working definition stemming from the concerted action "Functional Food Science in Europe" (FUFOSE), as follows:

(1) "A food that beneficially affects one or more target functions in the body beyond adequate nutritional effects in a way that is relevant to either an improved state of health and well-being and/or reduction of risk of disease"; (2) "not a pill, a capsule or any form of dietary supplement"; (3) "consumed as part of a normal food pattern" [65].

Stephen DeFelice (founder of the Foundation for Innovation in Medicine, FIM) is usually credited with the coining of the term "nutraceutical" some 30 years ago. According to him "a nutraceutical is any substance that is a food or part of a food and provides medical or health benefits, including the prevention and treatment of disease. Such products may range from isolated nutrients, dietary supplements and specific diets to genetically engineered designer foods, herbal products, and processed foods such as cereals, soups and beverages" [66]. Although it refers to products at the intersection between foods and drugs (sometimes they are referred to as "pharma-foods") there is no consensus definition or legal frame for "nutraceuticals" either [67].

According to Regulation (EC) No 1223/2009 of the European Parliament and of the Council of 30 November 2009 on cosmetic products, "'cosmetic product' means any substance or mixture intended to be placed in contact with the external parts of the human body (epidermis, hair system, nails, lips and external genital organs) or with the teeth and the mucous membranes of the oral cavity with a view exclusively or mainly to cleaning them, perfuming them, changing their appearance, protecting them, keeping them in good condition or correcting body odours." [68]. The US Federal Food, Drug, and Cosmetic Act (FD&C Act) defines "cosmetics" as "(1) articles intended to be rubbed, poured, sprinkled, or sprayed on, introduced into, or otherwise applied to the human body or any part thereof for cleansing, beautifying, promoting attractiveness, or altering the appearance, and (2) articles intended for use as a component of any such articles; except that such term shall not include soap" [69].

According to Saint-Leger [70], the composite term "cosmeceuticals", which also lacks consensus definition and legal identity to date, was coined by Mr. R. E. Reed (as President of the Society of Cosmetic Chemists) in 1962. According to this allegedly first definition, a cosmeceutical was a scientifically designed product meeting rigid chemical, physical and medical standards, intended for external application to the human body, which produces useful, desired results and has desirable aesthetic properties [70].

There is no consensus definition of "nutricosmetics" either, although this term is commonly used to refer to products containing food components that are intended for cosmetic purposes and are administered orally. Thus, the term is also associated to concepts like "inside-out approach" "beauty pills", "beauty from within", "beauty foods", "nutraceuticals for skin care", or "oral cosmetics" [12–14]. Common food components used in these products are vitamins, peptides, polysaccharides, polyphenols, coenzyme Q10, polyunsaturated fatty acids, and carotenoids [5].

6. Dietary Carotenoids

Carotenoids are widespread isoprenoid compounds that appeared very early in the history of life on earth. Their structures, physicochemical properties and activities have evolved so that they play key roles in photosynthesis, communication between species through colour signalling, nutrition, and health and other processes [21,71].

6.1. Basics on Dietary Carotenoids

6.1.1. Sources and Intakes

The main dietary sources of carotenoids are fruits and vegetables [15]. However, carotenoids are also present in other components of the diet, like herbs, legumes or cereals [72], algae [73,74], foods of animal origin (egg yolk, mammals' milk and tissues, seafood) [75–79] additives (colourants) [80,81] and in the form of supplements [16,82].

The diets of humans usually contain ~50 carotenoids, although not all of them are found at detectable levels in humans [17,21]. The normal dietary intakes of major carotenoids in various countries, as recently reviewed, are within the range 0–10 mg/day [21].

6.1.2. Presence in Plasma, Other Biological Fluids and Tissues

Traditionally, it has been considered that the main dietary carotenoids are lutein, zeaxanthin, β-cryptoxanthin, α-carotene, β-carotene and lycopene. These, along with phytoene and phytofluene are usually the major carotenoids in human fluids and tissues [15,17,83–85], at typical levels in the range of 0–2 μmol/L (plasma) and 0–1 nmol/g (tissues) [84,86].

6.1.3. Health-Promoting Biological Actions

Provitamin A carotenoids are key in the fight against vitamin A deficiency, a global health problem leading to different manifestations (xerophthalmia, negative effects on growth, compromised immunity, etc.) and that is responsible for large numbers of child mortality [87,88]. In addition, evidence has accumulated over the last 30 years that carotenoids contribute to decreasing the risk of developing non-communicable diseases like several kind of cancers, cardiovascular disease, bone, skin, neurological, metabolic or eye disorders [18,19,54,89–96]. The importance of carotenoids in the context of functional foods, nutraceuticals and related products for health promotion is therefore undeniable.

6.2. The "Undercover" Colourless Carotenoids Phytoene and Phytofluene

Phytoene and phytofluene are carotenoid rarities since they are colourless. They are very important from a biosynthetic point of view because they are precursors of the rest of the carotenoids; hence, they have been thoroughly studied in relation to their biosynthesis. The typical committed step of carotenoid biosynthesis within the isoprenoid route is the condensation of two C_{20} molecules to form (15Z)-phytoene, the major geometrical isomer of this carotenoid in most carotenogenic organisms [21,97].

Surprisingly, phytoene and phytofluene have been largely ignored in food science and technology, nutrition, health or cosmetics, relative to the other major carotenoids found in human fluids and tissues (lutein, zeaxanthin, β-cryptoxanthin, α-carotene, β-carotene, lycopene). This is probably largely due to a considerable lack of analytical data compared to these latter carotenoids. This in turn is attributable to their lack of colour. Indeed, before the advent of modern detectors attached to liquid chromatography (like diode-arrays allowing the simultaneous monitoring of different wavelengths, or mass spectrometry detectors) their lack of colour made their analysis more difficult, since they absorb mainly UV radiation rather than visible light like virtually all the rest of carotenoids [23]. Currently, there is a fast-expanding interest in these colourless carotenoids because the critical analysis of the literature in recent reviews [22–24] has revealed that:

- They are present in widely consumed foods;
- Their intakes and levels in plasma and tissues are comparable (or even superior in some cases) to those of other major dietary carotenoids; and
- They are involved in biological actions that result in health and cosmetic benefits.

6.2.1. Distinctive Chemical Features among Carotenoids

The system of conjugated double bonds (c.d.b.), that is, of alternate single and double bonds, is the main structural characteristic of carotenoids. It is the main feature responsible for their physicochemical characteristics (light-absorbing properties, shape or reactivity) and activities [98]. Phytoene and phytofluene are linear hydrocarbons with three and five c.d.b., respectively (Figure 3). Since their polyene chain is markedly shorter than that of other dietary carotenoids (for instance lutein contains 10 c.d.b., zeaxanthin, lycopene and β-carotene 11 c.d.b., and astaxanthin and canthaxanthin 13 c.d.b.) (Figure 3), both colourless carotenoids are expected to exhibit important differences in some physico-chemical properties and biological actions, compared to other carotenoids [23].

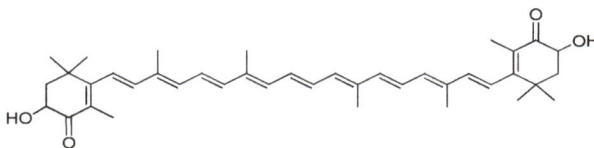

Astaxanthin (13 c.d.b.)

Figure 3. *Cont.*

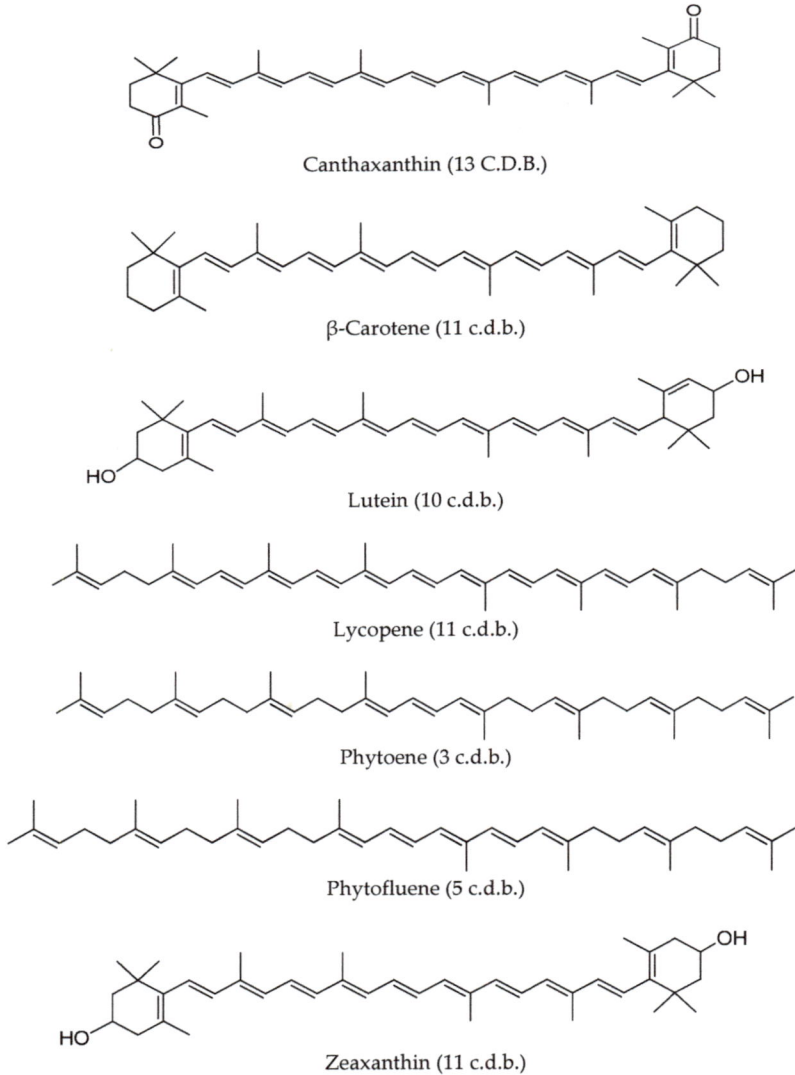

Figure 3. Chemical structures of some carotenoids of interest for skin health and appearance.

UV Light Absorption

Due to their shorter polyene chain, phytoene and phytofluene absorb maximally in the UV region and are colourless (Table 2). Indeed, it is accepted that 7 c.d.b. are needed for a carotenoid to exhibit appreciable colour [99]. More specifically, phytoene absorbs maximally in the UVB region (280–320) and phytofluene in the UVA region (320–400 nm), unlike virtually the rest of carotenoids in general, including the main carotenoids being studied in relation to skin health and appearance in particular (Table 2) [23]. Their UV–VIS spectra are shown in Figure 4.

Table 2. Absorption maxima in acetone or petroleum ether, number of conjugated double bonds (c.d.b.) and colour of carotenoids important for skin health and appearance [99,100].

Carotenoid	Colour	c.d.b. (in rings)	Absorption maxima (nm)		
Canthaxanthin	Red	13 (4)		472	
Astaxanthin *	Red	13 (4)		468	
Lycopene	Red	11 (0)	446	474	504
β-Carotene	Orange	11 (2)		454	480
Zeaxanthin	Orange	11 (2)		454	480
Lutein	Yellow	10 (1)	424	448	476
Phytofluene *	Colourless	5 (0)	331	348	367
Phytoene *	Colourless	3 (0)		286	

* Absorption maxima in petroleum ether.

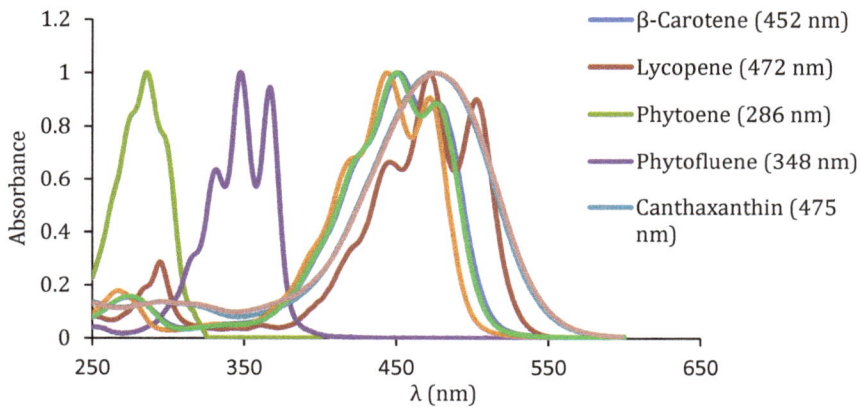

Figure 4. UV–VIS spectra of diverse carotenoids important for photoprotection and cosmetics.

The detection of phytofluene can be enhanced by its intense greenish-white fluorescence. Indeed, it is known to fluorescence at around 510 nm when it is excited with near-UV light [101,102].

Susceptibility to Oxidation

The characteristic long chromophore of c.d.b. of coloured carotenoids is thought to be closely related to their susceptibility to oxidation and antioxidant/prooxidant properties [98]. Being significantly less unsaturated, the colourless carotenoids could be expected to be more stable towards oxidation under certain conditions. In a recent study, the interaction of phytoene, phytofluene and lycopene with the synthetic oxidizing 2,29-azinobis-(3-ethylbenzothiazoline-6-sulfonic acid) (ABTS) radical cation was assessed by Density Functional Theory and ABTS radical cation decolouration assay strategies. The results suggested that the colourless carotenoids (with 3 and 5 c.d.b., respectively, Figure 3) are not as effective antiradicals as lycopene (11 c.d.b., Figure 3) [103]. Although data on the oxidative stability of phytoene and phytofluene in comparison to other carotenoids in tissues are lacking, there is evidence that they could be more stable in food matrices, for example thermally-treated tomato products [104]

Rigidity and Tendency to Aggregation

Their characteristic system of c.d.b. confer carotenoid molecules with rigidity. Since the number of c.d.b. is much lower in phytoene and phytofluene as compared to lycopene for example (another linear carotene), they are expected to adopt shapes notably less rigid relative to most carotenoids in general and lycopene in particular [23,105]. This is expected to have an important impact in terms of their release from the food matrix and incorporation into micelles ("bioaccessibility"). Bioaccessibility is one of the key factors explaining carotenoid bioavailability, since carotenoids must be incorporated into these structures to be taken up by the enterocytes prior to their transfer into the bloodstream [17,106]. In this sense, evidence is accumulating that the bioaccessibility of the colourless carotenoids is markedly higher relative to other carotenoids, probably partly due to their less rigid shape (due to their lower number of c.d.b., but also to their geometrical configuration) and they are less prone to aggregate and form crystals, which are regarded to impair carotenoid release from the food matrix [105,107,108].

Carotenoids undergo geometrical isomerization, so all-*trans* (all-E) and *cis* (Z) isomers can exist. The shapes and sizes of these isomers vary considerably (Figure 5). Thus, the former are linear and rigid and the latter have a bent shape. The geometrical isomerism in carotenoids is thought to have an impact in their solubility, tendency to aggregation, reactivity, bioavailability, or interaction with enzymes, hence the importance of discerning between geometrical isomers [85,105]. Regarding phytoene and phytofluene, it is thought that the (15Z)-isomer is usually the major isomer [102].

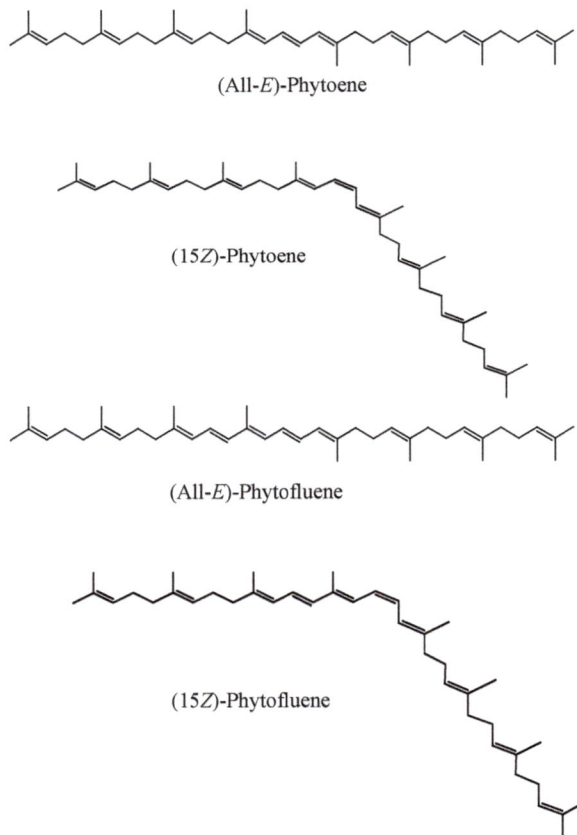

(All-*E*)-Phytoene

(15*Z*)-Phytoene

(All-*E*)-Phytofluene

(15*Z*)-Phytofluene

Figure 5. Chemical structures of (all-E)- and (15Z)-isomers of phytoene and phytofluene.

It has been recently shown that both carotenoids are incorporated into mixed micelles much more efficiently than other carotenes, notably lycopene. Unexpectedly, it has also been observed that their micellization efficiency was similar to that of the xanthophyll lutein, observation that challenges the paradigm that xanthophylls have a higher micellization efficiency compared to carotenes. These results have been attributed to the fact that phytoene and phytofluene were mainly present in the form of Z-isomers and/or to these carotenes' higher molecular flexibility, due to their shorter polyene chain [109].

6.2.2. Sources and Intakes

High contents of the colourless carotenoids are found in tomatoes, red grapefruits, watermelon (typically along with lycopene in these three cases), apricot, carrots, and some peppers. Other sources are cantaloupe, banana, melon, oranges, lemon, clementines, avocado, mandarin, nectarine, peach, and exotic fruits (caja, buriti, mamey, marimari, physalis, or gac) [23,24]. The daily dietary intakes of colourless carotenoids was recently assessed in Luxembourg, The percentage intake of phytoene+phytofluene was estimated to be 16% (2.7 mg, especially 2.0 mg for phytoene and 0.7 mg for phytofluene) of total carotenoid intake. Noticeably, the estimated daily intake of lycopene was lower (1.8 mg) than that of phytoene [110].

6.2.3. Presence in Plasma, Other Biological Fluids, and Tissues

Plasma levels of phytoene and phytofluene in the range of 0.04–0.33 μM have been reported [111–113] (Table 3). These compounds are also present in breast milk at comparable levels to the other major carotenoids (in the μg/dL range) [114]. Their presence in several human tissues (lung, breast, liver, prostate, cervix, colon, and skin) has been reported, at ng/g levels [115] (Table 3).

Table 3. Concentration of phytoene and phytofluene in body fluids (μM) and tissues (ng/g).

Body Fluid/ Tissue	Phytoene	Phytofluene	Reference
Blood	0.11 ± 0.01	0.30 ± 0.02	[111]
Blood	0.14 ± 0.08	0.14 ± 0.08	[112]
Blood	0.06 ± 0.04	0.33 ± 0.15	[113]
Blood	0.04	0.17	[114]
Breast	69	416	[115]
Cervix	-	106	[115]
Colon	70	116	[115]
Liver	168	261	[115]
Lung	1275	195	[115]
Milk	0.002	0.016	[114]
Prostate	45	201	[115]

6.2.4. Health-Promoting Biological Actions

Evidence is accumulating that phytoene and phytofluene could be involved in the health benefits traditionally attributed to lycopene, since some reviews suggest that such benefits have usually been observed when tomato products (also containing the colourless carotenoids and other compounds) rather than pure lycopene were used [116,117]. Aside from this, there are studies of different nature (in humans, animals, cell cultures, isolated lipoproteins) indicating that phytoene and phytofluene may be involved, either on their own or together with other compounds, in health-promoting biological actions such as protection against oxidation [118–120], inflammation [121–123], or anticarcinogenic activity [124–128].

6.2.5. Safety of Phytoene and Phytofluene

The safe use of phytoene and phytofluene in topical applications is backed by various evidence, namely in vitro cytotoxicity and genotoxicity studies or human in vivo 48-h patch tests and a longer-term Human Repeat Insult Patch Test [29]. The safety of oral formulations can be presumed. On one hand, these compounds are naturally present in many fruits and vegetables. Indeed, estimated daily intakes of phytoene + phytofluene of ~2.70 mg have been proposed for Luxembourg citizens [110]. Furthermore, several phytoene and phytofluene-rich products, notably tomato-derived, have been regarded as safe for human consumption by different competent bodies, including the US FDA or the European Food Safety Authority (EFSA). More especifically, several tomato products have been categorized as GRAS by the FDA, including tomato pulp powder or concentrated tomato lycopene extract [129,130]. Similarly, EFSA considers colourless carotenoid-containing tomato oleoresins safe for human consumption [131].

7. Carotenoids in the Skin

7.1. Deposition Mechanism

Regarding the delivery and distribution of carotenoids into the skin layers, there appear to be two main pathways, namely 1) diffusion from the adipose tissue, blood and lymph; and 2) secretion through sweat and/or sebaceous glands onto the skin surface and subsequent penetration [56].

Based on the observation that in untreated skin the larger carotenoid fraction is detected in the external part of the stratum corneum, it has been hypothesized that carotenoids are delivered to that location through secretions by eccrine sweat glands and/or sebaceous glands, similarly to what occurs with vitamin E [57]. In relation to this, it has been observed that, upon consumption of carotenoid-rich products, the increase in the dermal level of carotenoids occurs within 1–3 days, whereas the process of stratum corneum renewal takes 2–3 weeks [132]. According to these authors, the deposition of carotenoids in the skin surface requires their diffusion from the blood, hypodermis and the dermis to the epidermis. Then, the carotenoids would be transported from the blood, the hypodermis and the dermis into the sweat glands, and subsequently to the skin surface with sweat. This explanation is consistent with the fact that the highest levels of skin carotenoids are detected in locations with high numbers of sweat glands [132].

It is hypothesized that the subcutaneous tissue is a storage site for carotenoids and that these are loaded into the keratinocytes which are continuously formed at the basal layer and then migrate to the skin surface, transporting along the carotenoids. In this regard, the time the keratinocytes take to travel from the basal layer to the skin surface (ca. six weeks) matches very well with the time supplementation studies with carotenoids take in producing perceived photoprotection [38].

Finally, it is known that the application of carotenoids topically, results in increased levels in the stratum corneum [56].

7.2. Assessment of Skin Carotenoid Levels

To date, the two major strategies to assess dermal carotenoid levels are classical HPLC analysis and non-invasive spectroscopic methods.

7.2.1. HPLC Analysis

The HPLC analysis of skin requires the obtaining of biopsies and a suitable extraction method. Apart from the volunteer's discomfort, this is a time-consuming approach that may also lead to the oxidative loss of carotenoids. Therefore, it is not suitable for high throughput analyses or the analysis of kinetics of carotenoid levels changes in human skin. Some methodologies are found in recent studies for validation of resonance Raman spectroscopy approaches [133,134].

7.2.2. Non-Invasive Spectroscopic Methods

Resonance Raman spectroscopy (RRS) and light reflection spectroscopy (LRS), including the assessment of colour parameters from reflectance measurement are optical methods used for the rapid, almost instantaneous evaluation of carotenoid levels in skin.

They are non-invasive, so that the performance of biopsies is not required, with obvious advantages. Given that skin carotenoids are a biomarker of fruits and vegetables intake, the use of these techniques is very useful to screen large population sets for nutritional and epidemiological purposes in the context of public health. To avoid interferences from other sources of colour in skin, the measurements should be made in areas with minimal melanin content, for example the heel of the palm, the tip of a finger, or the heel of the foot. RRS is a more widely used and thoroughly validated method. LRS is an emerging alternative with some advantages, including higher simplicity, portability and the possibility of quantifying the major tissue chromophores and taking them into account to estimate skin carotenoid levels, made possible through detecting reflection over a wider spectral range, from the near UV to the near IR regions [135].

A major drawback of spectroscopic approaches relative to HPLC analyses is that the former do not take into account the levels of the colourless carotenoids. Thus, in RRS carotenoids are usually excited in the typical maximal absorption band of coloured carotenoids in the blue region of the visible spectrum. LRS results in the calculation of a score for coloured carotenoid compounds absorbing in the 460–520 nm range [135].

Light Reflection Spectroscopy (LRS)

The LRS methodology, which was first applied to the assessment of dermal carotenoids ~20 years ago [136,137] is more simple and requires much smaller devices, some of which are portable. Thus, even a low-power white lamp can be used as a light source and the carotenoid levels are derived from the reflection spectra. Currently, the LRS leads to the calculation of a composite score for all carotenoids absorbing in the 460–520 nm range. Some limitations that this methodology needs to overcome are the interference of the optical properties of the skin (related to autofluorescence, skin hydration, and sweat and sebum production), including those derived from other coloured molecules, namely melanin, haemoglobin, or oxyhemoglobin [138]. Although the technique is not as widely validated as RRS [56,135,139] its use is becoming more popular [140,141].

Another approach based on reflectance measurement consists in measuring colour parameters [135] Indeed, correlations between skin colour parameters and carotenoid levels have been recently reported [142,143]. The assessment of carotenoids from colour parameters derived from reflection spectra has also been widely used in food science and technology to estimate concentrations of a wide diversity of carotenoids [144–147], notably such as those occurring in industrially-processed orange juices, with different chromophores and functional groups [148].

Resonance Raman Spectroscopy (RRS)

Details about the basics of the technique are provided elsewhere [135,139]. RRS methods are based on the use of laser or LED light sources of narrow spectrum' and detects the characteristic vibrational/rotational energy levels of a molecule. The characteristic polyene chain of carotenoids makes them well suited for detection by this method. They are characterized by three intense Stokes lines at 1005 cm^{-1} (due to the rocking motion of the methyl group), 1156 cm^{-1} (owed to a carbon-carbon single-bond stretch vibration of the polyene chain) and 1523 cm^{-1} (carbon–carbon double-bond stretch vibration of the conjugated polyene) [56]. The Raman line intensities have been shown to have good linear correlation with the physiological skin carotenoid levels [134,135,139].

RSS has become a valuable tool for different types of studies, for instance the use of dermal carotenoids as nutritional biomarkers in diverse groups [133,149–151] or the study of associations between carotenoid status and several conditions [152].

7.2.3. Major Skin Carotenoids

It is accepted that, in general, carotenoid levels in the skin reflect those present in plasma [153]. Major human circulating carotenoids (lutein, zeaxanthin, β-cryptoxanthin, β-carotene, lycopene, phytoene, and phytofluene) have been detected in the skin, usually in the range of 0–10 nmol/g wet tissue (Table 4), although higher concentrations are detected in case of dietary supplementation or in carotenodermia [11,113,154]. The presence of other less common carotenoids (α-cryptoxanthin and anhydrolutein) has also been reported [155]. Interestingly, carotenoid esters have been found in the skin, unlike in other tissues. Indeed, it is well-known that esterified xanthophylls are hydrolyzed during digestion and absorbed as free xanthophylls [17,156,157]. Specifically, up to eighteen esters (including both mono- and di-esters) of lutein, zeaxanthin, 2′,3′-anhydrolutein, α-cryptoxanthin, and β-cryptoxanthin have been described in human skin, although at levels considerably lower (pmol/g) than those of free carotenoids [155].

Table 4. Carotenoid concentrations reported in human skin (nmol/g).

Carotenoid	Tissue	Concentration	Reference
α-Carotene	Abdominal skin	0.01	[154]
β-Carotene	Epidermis	0.39	[11]
β-Carotene	Dermis	0.01	[11]
β-Carotene	Epidermis	4.1	[11]
β-Carotene	Dermis	1.3	[11]
β-Carotene	Subcutis	3.5	[11]
β-Carotene	Surface lipid	10.0	[11]
β-Carotene	Comedones	14.5	[11]
β-Carotene	Whole skin	0.09	[11]
β-Carotene	Whole skin	1.41	[11]
β-Carotene	Punch biopsy	8.3	[11]
β-Carotene	Abdominal skin	0.05	[154]
γ-Carotene	Abdominal skin	0.04	[154]
ζ-Carotene	Abdominal skin	0.02	[154]
Lycopene	Abdominal skin	0.13	[154]
Phytoene	Abdominal skin	0.12	[154]
Phytoene	Whole skin	0.12	[115]
Phytofluene	Abdominal skin	0.03	[154]
Phytofluene	Whole skin	0.03	[115]
Total carotenoids	Whole skin	0.17	[113]

7.2.4. Factors Affecting Skin Carotenoid Levels

Obviously, there are important individual differences in the skin carotenoid levels, as a result of the manifold factors affecting carotenoid bioavailability, from dietary patterns to lifestyle and genotypic factors [17,158,159]. The skin levels of antioxidants are known to depend not only on the diet, but also on skin location or type, individual characteristics (gender, age), health status, or stress factors, both environmental and derived from lifestyle [56,132].

Diet

Dermal carotenoid levels can be increased in a relatively short time by dietary means. There is ample evidence that not only supplements, but also fruits and vegetables increase carotenoid skin levels [160–162] and that this may have a noticeable impact in terms of skin colour [55,163]. Occasionally, excessive carotenoid intake leads to carotenodermia, an apparently innocuous phenomenon featuring noticeable orange pigmentation of the skin as a result of elevated carotenoid deposition mainly in the stratum corneum, sweat, and sebum. The condition can be due to high carotenoid intake (normally of β-carotene), metabolic disorders, or familial carotenemia. It is believed that it appears when the serum carotenoid levels are in the range of 2.5 mg/L. Following the cessation of excessive carotenoid ingestion, the possible altered biochemical parameters normalize within several weeks. A similar condition due to very high intakes of lycopene has been sometimes termed lycopenemia [52,53,164].

Skin Location

Concerning carotenoids' distribution in different skin areas, there are marked differences, such that it has been reported that the highest levels can be detected in the forehead and the palm of the hands and lower concentrations are found in the back of the hands, inside the arm or in the dorsal area [57,136]. Additionally, the distribution of carotenoids across skin layers is not homogeneous.

It appears that the maximum levels are found close to the skin surface (depth of ~4–8 μm), and levels decrease at least up to the depth of 30 μm [56,57].

Individual Characteristics

There is evidence that gender and body mass index (BMI) could have an impact on dermal carotenoid status. Thus, a recent study observed higher values in women and volunteers with BMI below 30. Age did not seem to have an effect, except in the case of lycopene [165].

Stress Factors

External stress factors (e.g., radiation) or derived from lifestyle (smoking) or disease (cold, psoriasis or cancer) have been shown to have a negative impact on carotenoid levels in the skin.

Thus, in a classical study, controlled exposure to sunlight was shown to lead to noticeable decreases of circulating and skin carotenoid levels in human volunteers [166]. Exposure to visible blue-violet and infra-red light has also been shown to lead to the decrease of skin carotenoids, likely through the production of ROS [167–169]. In an interesting study the topical administration of β-carotene resulted in a protective effect [169].

As an example of lifestyle factors, it has been observed that, overall, smokers had lower levels of skin carotenoids relative to non-smokers [165].

Psoriasis is an inflammatory disease that affects the skin but can also have impacts at the nutritional level. In a cross-sectional study, it was concluded that patients with psoriasis had lower dermal carotenoid levels as assessed in the palm of the hand. However, the carotenoid levels were not significantly associated with the severity of the disorder [170].

A recent study involving 102 breast cancer patients evaluated possible associations between anxiety and skin carotenoid levels (assessed non-invasively as skin carotenoid score, SCS) as a measure of oxidative stress, since chronic stress has been associated with tumour progression, higher recurrence rates and increased risk of metastasis of this malignancy. The results indicated that higher levels of skin carotenoids were associated with decreased severity of anxiety and other miscellaneous parameters (lower body max index (BMI), higher intakes of fruits and vegetables, Hispanic race, lower educational status, and non-smoking) [171].

7.2.5. Kinetic Aspects

In a very interesting work Darvin et al. [132] evaluated the changes in the levels of β-carotene and lycopene (assessed by resonance Raman spectroscopy) in human skin over one year in relation to diet and exposure to stress factors as assessed via questionnaires. The skin carotenoid levels in all volunteers were noticed to be higher in summer and autumn, which was mainly attributed to the increased intake of fruits and vegetables and decrease occurrence of illnesses. Overall, they concluded that large intakes of fruits and vegetables clearly enhanced the levels of these two carotenoids in the skin, while stress factors (radiation, fatigue, illness, smoking, or alcohol intake) decreased them, probably in relation to the production of reactive oxygen species. The kinetics of the observed rises in carotenoid levels owed to fruit and vegetable intakes, and the corresponding decreases seemed due to stress factors behave differently. The decreases took place relatively quickly (over the course of 2h), while the subsequent recovery usually took up to three days to level. Both increases and decreases in carotenoid levels took place much faster than the subsequent return to basal levels [132].

Another study showed that skin carotenoid levels rose quickly with increased carotenoid intakes, and that such rises were useful to evaluate compliance in intervention studies as soon as two weeks after start of the dietary intervention [161]. Circulating carotenoid levels returned to baseline after three weeks, which is in good agreement with depletion studies showing decreases in blood carotenoids within 2–3 weeks. In the latter, the dermal carotenoid levels did not return to baseline until volunteers returned to their usual diets for ca. one month. The authors argued that this could be due to the fact that skin acts as a storage tissue for carotenoids. This argument agrees with the findings of other authors after a supplementation trial suggesting that the delayed drop in the skin carotenoid levels relative to those in blood may indicate a peripheral buffer function of the skin for carotenoids [172]. The study by Jahns et al. [161] concluded that skin carotenoid levels assessed by RRS was a valid biomarker of change in skin carotenoid status through fruit and vegetable consumption at recommended intake levels in the US, but with a longer half-life relative to blood carotenoids and the additional advantage that the methodology is non-invasive [161].

8. Carotenoids and Photoprotection

8.1. Carotenoids and Protection Against Light in Diverse Organisms and Locations

Carotenoids are used in photosynthesis to aid in collecting light of certain wavelengths and protecting from light-derived damage in photosynthetic organisms, including cyanobacteria, considered to be the most ancient oxygenic photosynthetic organisms and the origin of plant chloroplasts [173]. Other organisms that appeared on earth later, also used them in liaison with light: for instance, they play essential roles (quenching of excited chlorophyll or singlet oxygen, light harvesting, assembly of protein-pigment complexes, among others) in photosynthetic tissues in other organisms. Additionally, they are important for vision, including in functions beyond the role of some carotenoids as precursors of retinal for the visual cycle. For instance, some birds have carotenoid-containing oil droplets, through which light passes before reaching the photoreceptors in the retina. This seems to be a means of filtering shorter (more energetic) visible wavelengths and enhancing the sensitivity and perception of colours [174].

Strikingly, the human macula lutea, the yellow spot in the central part of the retina (with a diameter of 5–6 mm) has the highest local concentration of carotenoids in humans. This area contains the fovea (with a diameter of ~1.5 mm), which is the area with the highest visual acuity. More specifically, only lutein and zeaxanthin (in the form of two stereoisomers (3R,3'R)-zeaxanthin and (3R,3'S)-zeaxanthin, the latter of non-dietary origin and also referred to as *meso*-zeaxanthin) of the circulating human carotenoids are found in the macula lutea [175]. *Meso*-zeaxanthin is thought to be formed in vivo from lutein [176]. This selective accumulation is due to the existence of specific transporters for these two carotenoids, such as a Pi isoform of glutathione S-transferase (GSTP1) (which has been shown to be a zeaxanthin-binding protein in the human macula) [177] or a member of the steroidogenic acute regulatory domain (StARD) (which has been demonstrated to bind lutein in the primate retina) [178]. The presence of such high amounts of just two specific carotenoids in this area of the retina, where light reaches with high intensity is unlikely to be fortuitous. Indeed, the macular carotenoids are thought to be important to reduce the risk of developing age-related macular degeneration, maybe, at least in part, by absorbing blue light and, thus, protecting the photoreceptor cell layer from light-induced damage that could be initiated by the formation of reactive oxygen species during a photosensitized reaction [179]. Thus, carotenoids are important in the protection of different organisms against damage caused by light. Consequently, it is reasonable to expect that they can also have such a role in the skin, which is exposed to sunlight on a daily basis.

8.2. Sunscreens vs. Dietary Approaches

When it comes to providing protection from sunlight, sunscreens are usually the method of choice. However, dietary approaches may also be used, which entails additional benefits.

Photoprotection based on dietary components in terms of sun protection factor is regarded to be markedly lower than that achievable by using topical sunscreens. Additionally, whilst the protection conferred by the latter is virtually instantaneous, the intervention studies carried out with coloured carotenoids indicate that it takes 7–10 weeks until the protection against sunburn becomes significant. However, it is argued that the presence of dietary antioxidants and other compounds from the diet is a good and natural strategy to endow the skin with basal defences against photodamage and other aggressions that can also affect appearance [54]. β-carotene has been shown to protect against photodamage caused by visible and infra-red radiations and may be an effective antioxidant in sunscreens [49,169]. However, carotenoids are very unstable compounds, and strategies to overcome this fact are needed for the meaningful formulation of carotenoid-containing products. In this context, as a result of a study in which the topical photoprotection of β-carotene was assessed it was recommended that products intended for topical use should consist of a mixture of antioxidants reflecting those present in the skin rather than single antioxidants [169].

8.3. Mechanisms of Skin Protection by Carotenoids

Apart from possible cosmetic benefits, another great advantage of increasing the intake of carotenoids and other dietary components for photoprotection is that their levels in plasma and other tissues can also be enhanced, with consequent health-promoting benefits. There is ample evidence indicating that carotenoids contribute to decreasing the risk of developing diseases, including cancer, cardiovascular, and metabolic diseases [19,21–24].

Goralczyk and Wertz [11] identified some major mechanisms, among which inhibition of lipid peroxidation, inhibition of UVA-induced expression of heme oxygenase 1, prevention of mitochondrial DNA mutations, inhibition of metalloproteases and photoimmune modulation. Recent experiments in hairless mice subjected to UVB radiation indicated that tangerine tomato carotenoids may exert beneficial effects by attenuating DNA-damage and inflammation, with interesting differences between males and females [180]. Noticeably, new possible mechanisms are hypothesized thanks to the use of omics technologies for the study of changes in gene expressions associated to exposure to carotenoids. In any case, attribution of the effect to a single mechanism does not seem possible, and there may be connections among them in some cases [11].

Further, provitamin A carotenoids are beneficial for the skin through the production of retinoic acid, which plays important roles at this level, as do other retinoids. They are thought to intervene in processes including keratinocyte proliferation, epidermal differentiation and keratinisation, reduction of inflammation or oxidation, or even the enhancement of the penetration of agents administered topically, among many others. Thus, retinoids are applied for different purposes, for example improving wound healing, preventing skin aging or the treatment of acne, psoriasis, or other skin conditions [28,181]. However, this topic will not be discussed in detail in this review.

8.3.1. Inhibition of Lipid Peroxidation

The peroxidation of diverse lipids (importantly free and esterified cholesterol, and polyunsaturated fatty acids) by enzymatic or non-enzymatic means leads to the formation of various products that act as redox signalling molecules. They may have negative effects at the level of structures like membranes or molecules (proteins or DNA bases) that eventually lead to disease states [182].

It is well-known that carotenoids intervene by quenching singlet oxygen or scavenging free radicals, both in solution and in other systems such as membranes or cells, although they can also interact with other antioxidant and non-antioxidant compounds or even act as pro-oxidants under certain conditions [19,46,183,184]. Discussion about the protection by carotenoid-containing products against lipid oxidation is found elsewhere [185].

Interestingly, there is evidence that carotenoids may also be prooxidants depending on factors including their concentration, oxygen tension, exposure to radiation, or interaction with other compounds. The resulting prooxidant effects could be harmful or beneficial [186–188].

8.3.2. Inhibition of UVA-Induced Expression of Heme Oxygenase 1

Heme oxygenase-1 (HO1) is a ubiquitous enzyme that catalyses the first reaction in heme degradation, which eventually leads to the formation of CO, biliverdin, and Fe^{2+}. The enzyme is involved many processes related to the regulation of cell proliferation, differentiation and apoptosis [189]. The expression of the gene codifying the enzyme is inducible. As an example, it is activated via singlet oxygen within the first hours after the exposure of skin fibroblasts to UVA. Some studies in cell cultures involving mainly β-carotene in different formulations suggest that this could be a possible mechanism of carotenoid protection [11].

8.3.3. Prevention of Mitochondrial DNA Mutations

Mutations (typically a ~5000 base pair deletion usually termed "common mutation") at the level of the mitochondrial DNA, are thought to be involved in negative effects including skin chronological and UV-induced aging and carcinogenesis. Indeed, these mutations have been shown to be increased in aged skin. There is evidence from cell culture studies that β-carotene (and/or some oxidative metabolites) could protect fibroblasts by reducing the occurrence of such mutations [11].

8.3.4. Metalloprotease Inhibition

Matrix metalloprotease (MMPs) encoding genes are regarded to be among the most important genes involved in photoaging, whose expression is induced by singlet oxygen. MMPs are members of a family of enzymes (collagenases, gelatinases, stromelysins, some elastases, and aggrecanases) that catalyze the normal degradation of extracellular matrix (ECM) macromolecules including collagens, proteoglycans (aggrecan, decorin, biglycan, fibromodulin, and versican) and accessory ECM proteins like fibronectin [190].

UVR can activate cell-surface growth factors and cytokine receptors by a ligand-independent mechanism. This induces several signalling pathways that result in the stimulation of the transcription factor AP-1, which upregulates genes of several members of the matrix metalloprotease (MMP) family. This increases the proteolysis of proteins of the extracellular matrix, mostly collagen, but also fibronectin, proteoglycans, and elastin, and results in skin elastosis and wrinkling. MMPs are also thought to be important in photocarcinogenesis since they intervene in cell growth, angiogenesis and metastasis. If the high exposure to UVR is chronic, other signs, such as dilated and twisted microvasculature (teleangiectasia) or hyper-pigmentation may appear, these being clinical features of photoaging [5,11].

The use of omics technologies for the overall assessment of changes in the expression of genes in whole genomes is expanding the knowledge of possible biological actions of carotenoids and mechanistic aspects. In this sense, the results of a classical study about overall changes of gene expressions due to treatment of keratinocytes with β-carotene (at physiological dose levels, namely 0.5, 1.5, and 3.0 μmol/L) prior to UVA exposure indicated that carotenoids can act in these cells by modulating the expression of genes related to multiple pathways. More specifically, it was concluded that of the 568 genes whose expression was regulated by UV, the carotene reduced the effect of radiation for 143 and enhanced it for 180 [191]. In this sense, it would not be surprising that carotenoids protect the skin by several of the major mechanisms pinpointed. Thus, a recent study has shown that supplementation with two products containing different carotenoids (lycopene and lutein) results in the modulation of the expression of not only heme-oxygenase 1 and matrix metallopeptidase 1 genes, but also of the intercellular adhesion molecule 1 [192].

8.4. Visible Light-Absorbing Coloured Carotenoids and Photoprotection in Humans

8.4.1. Astaxanthin

Astaxanthin is a xanthophyll biosynthesized by microalgae (*Haematococcus pluvialis*, *Chlorella zofingiensis*, and *Chlorococcum* sp), the yeast *Phaffia rhodozyma* and the bacterium *Agrobacterium aurantiacum*. Animals (zooplankton, crustaceans, fish) incorporate it through the diet [193]. In general, the major dietary source for humans is salmon. Commercially, this carotenoid, which can be used as feed additive or human dietary supplement, is largely obtained by synthesis or biotechnologically from *H. pluvialis* [194]. There is evidence that astaxanthin could provide benefits at the level of skin, such as protection against erythema or reduced wrinkling [194].

8.4.2. Canthaxanthin

Canthaxanthin, which is not one of the major dietary carotenoids but is approved as a food additive in many countries, has also proven useful to treat erythropoietic protoporphyria, although its accumulation in the retina over extended use raised concern about use of this carotenoid in general [175,195].

8.4.3. Beta-Carotene

Beta-carotene is a widely distributed carotenoid found in many foods. Some important sources are carrots, palm oil, mango, sweet potato, apricot, and green vegetables [15,72]. Beta-carotene has been long known to be beneficial for the treatment of erythropoietic protoporphyria, a rare inborn haematological disease that leads to increased levels of protoporphyrins in plasma, red blood cells, skin, and faeces. Protoporphyrin is an endogenous photosensitizer. This compound, upon exposure to UVR, becomes excited, and may eventually pass on the excitation energy to O_2 in the ground state, thus generating singlet oxygen, which is a ROS that can interact with different molecules (DNA, proteins, lipids), causing cellular damage and leading to clinical symptoms of photosensitivity. Specifically, the patients experience itching, burning and pain in the skin exposed to sunlight after even only minutes of exposure, followed by edema, erythema and purpura [11,39–41,54]. The photoprotective effect of β-carotene has been the subject of several original research articles and reviews. Altogether, they indicate that it protects against erythema caused by UVR and that such protection requires daily dosages of about 10 mg for ~10 weeks [38,54,196–199]. On the other hand, a (9Z)-β-carotene-rich algal powder from *Dunaliella bardawil* has been shown to reduce the severity of psoriasis in adult patients with mild, chronic, plaque-type psoriasis as assessed by changes in Psoriasis Area and Severity Index (PASI) [200].

8.4.4. Lutein

Lutein is a widely distributed dietary carotenoid present in all green vegetables, a wide variety of fruits, and egg yolk [72,75,201]. A recent placebo-controlled, double-blinded, randomized, crossover study has shown that an intervention with lutein capsules containing free lutein stabilized by 10% of the antioxidant carnosic acid, can protect against photodamage by decreasing the expression of UVR-modulated genes, including heme-oxygenase 1, intercellular adhesion molecule 1, and matrix metallopeptidase 1 genes [192].

8.4.5. Lycopene

Lycopene is an acyclic carotenoid found in foods including some varieties of tomatoes, watermelons, guava, papaya, apricots, and grapefruits, together with other carotenoids, including the colourless phytoene and phytofluene [23,72,110,202].

Having been shown to be a quencher of singlet oxygen in vitro [203], several studies indicate that lycopene may provide photoprotection, although most of the in vivo studies have used products containing lycopene extracts (e.g., tomato paste in olive oil, diverse tomato extracts, or even a tomato-containing carrot juice) that also include other accompanying compounds (e.g., vitamin E or the colourless carotenoids phytoene and phytofluene) [204]. Being precursors of carotenoids and very close to lycopene in the biosynthetic path of carotenoids, phytoene and phytofluene are always found together with lycopene in the main sources of the latter, and also occur in other foods where lycopene is not found, such as carrots, citrus and others [23]. In this sense, the possible contribution of such components must be considered, in addition to the possibilities of interactions such as synergisms [50]. Indeed, some reviews indicate that there is little evidence of the health benefits of lycopene alone, since in most cases tomato extracts also containing other carotenoids are used [116,117]. Concerning photoprotection, this is well illustrated in an elegant study in which human volunteers received synthetic lycopene, a tomato extract or a tomato-based drink (all of them supplying virtually the same lycopene amount) for 12 weeks. The results indicated that the intervention resulted in the prevention of the UV-induced erythema formation in all groups, although the effect was more intense in the volunteers receiving tomato based products than in the group receiving synthetic lycopene alone, suggesting that phytofluene and phytoene may have contributed to the effect [113]. Recently, evidence became available that a lycopene-rich tomato product also containing phytoene and phytofluene, tocopherols, and phytosterols can provide photoprotection by inhibiting UVR-induced upregulation of heme-oxygenase 1, intercellular adhesion molecule 1 and matrix metallopeptidase 1 genes [192]

Lastly it has been recently shown in SKH-1 hairless and immunocompetent mice that male mice that received diets containing lycopene, phytoene and phytofluene and other constituents from tomatoes developed fewer UVB-induced skin tumours relative to controls [205].

8.4.6. Other Carotenoid-Containing Products

Recently, paprika oleoresin has been shown to be effective in humans in increasing MED and reducing UV-induced skin darkening. Paprika is a carotenoid-rich product obtained from red peppers containing red xanthophylls (capsanthin or capsorubin), as well as other xanthophylls (zeaxanthin and β-cryptoxanthin, among others) and carotenes (β-carotene, phytoene, and phytofluene) [72,110]. In a randomized, placebo-controlled, parallel-group comparative clinical study, daily oral supplementation with a red paprika product led to a significant enhancement of the MED and reduction of skin tanning in skin exposed to UV on the back, allegedly through antioxidant and anti-inflammatory mechanisms. At the end of the four-week study the treatment did not lead to statistically significant changes in a* (redness), transepidermal water loss (TEWL) or stratum corneum hydration (SCH) in the back area exposed to UVR, although the change in skin lightness was statistically significant, indicating that the intervention effects a change in skin colour after UV irradiation. To evaluate the effect of paprika xanthophylls on non-UV-irradiated skin, facial skin colour (L*, a*, b*), TEWL and SCH values were determined, although no significant differences were observed [206].

8.5. UV Light-Absorbing Colourless-Carotenoids (Phytoene and Phytofluene) and Photoprotection in Humans

One of the first studies pointing to the beneficial role of carotenoids in skin health concluded that injected doses of phytoene provided protection against UVR-induced erythema in guinea pigs [207]. The pioneer Mathews-Roth [208] described that the prolonged administration of phytoene to mice reduced the appearance of UV-B light-induced skin tumours and their multiplicity. Another major finding of this study was that no anticarcinogenic activity was noticed when the tumours were induced chemically.

Regarding humans, aside from studies in which products containing lycopene plus colourless carotenoids were tested, there is further evidence that they exert photoprotection. Thus, a proprietary oral food supplement in the form of a tomato powder rich in PT and PTF (Israeli Biotechnology Research Ltd., Yavne, Israel) provided to women resulted in an mean increase of 10% in the minimum erythemal dose (MED), with 2/3 of the volunteers having a 20% increase at the end of the 12-week study [209].

Possible Mechanisms

Evidence is accumulating in relation to the beneficial effects of colourless carotenoids in terms of skin photoprotection, although the mechanistic aspects are still unclear. However, there is evidence of varying nature about actions that merit further investigation.

As already commented, phytoene absorbs maximally in the UVB region (280–320) and phytofluene in the UVA region (320–400 nm), unlike virtually all other carotenoids (Table 2). Taking this into consideration, it is expected that these colourless carotenoids could provide photoprotection through the absorption of UVR [23].

Additionally, it has been reported that phytoene and phytofluene could protect against erythema and DNA damage caused by UVR and hydroxyl radicals and that they could have anti-inflammatory effects, as observed in human peripheral blood lymphocytes and in vivo in a mouse ear edema model [210]. Finally, in an in vitro study, human neonatal dermal fibroblast cell cultures were either subjected to UVR or exposed to interleukin-1 and then treated with coenzyme Q10, phytoene+phytofluene or combinations of the coenzyme and the carotenoids. The latter treatment was shown to lead to an enhanced anti-inflamatory response [122].

9. Carotenoids and Cosmetic Benefits

9.1. Carotenoids and Colour Signalling in Animals

The importance of the colour of the skin of animals or associated structures, e.g., feathers in birds, from several points of view (biological, social) is beyond doubt. Thus, carotenoid-based signals in animals advertise information about health/disease or nutritional status, genetic quality, aggressiveness, fertility, and so on [211]. For example, it is thought that male birds devoting higher levels of carotenoids to sexual-related colouration are communicating their better health [212]. Thus, according to the carotenoid trade-off hypothesis, using carotenoids for colouration prevents their use as antioxidants, such that the trade-off leads to a correlation between external colouration and health as healthy individuals can afford to use more carotenoids for colour. Contrastingly, according to the carotenoid protection hypothesis, carotenoid-based external colouration in animals could be a sign of the occurrence of other antioxidants that prevent carotenoids from being degraded by oxidation, with concomitant colour loss [59].

9.2. Carotenoids as Cosmetics

9.2.1. From Cleopatra to Tanning Pills

Curiously, humans already used carotenoids topically with cosmetic purposes long before these compounds were "officially" discovered. Thus, it is reported that Cleopatra, the famous Pharaoh of Egypt, used saffron profusely for cosmetic purposes. Saffron is a product derived from the stigmas of *Crocus sativus* that exhibits a vivid orangish colour due to the presence of high levels of the apocarotenoid crocetin associated to sugar moieties. Similarly, indigenous people from South and Central America used to paint their faces with carotenoid-rich products, like annatto. This is obtained from the seeds of *Bixa orellana*, a small tree that bears in its scientific name the surname of the Spanish adventurer Francisco de Orellana, who is thought to have "discovered" it during his travels in the Amazonian region [213].

In modern times, carotenoids in supplements have been used since the 1970s as tanning agents, especially in Northern Europe. However, the use of canthaxanthin supplements raised concerns as it was observed that, in some cases, their continuous intake at high doses (from 30 mg/day) could lead to the formation of crystals of this carotenoid in the eye—though these could disappear when the intake of the product was discontinued [16,175]. Currently, EFSA recommends lower doses, specifically the Panel on Food Additives and Nutrient Sources added to Food (ANS) established an acceptable daily intake (ADI) of 0.03 mg/kg bw/day, which is in agreement with earlier recommendations made by the Joint FAO/WHO Expert Committee on Food Additives (JECFA) and the Scientific Committee on Food (SCF) [195].

9.2.2. New Trends

Evidence that the intake of carotenoids from fruits and vegetables leads to perceived cosmetic benefits is accumulating in the last years. More specifically, the association between skin carotenoid-based colour and facial appeal has drawn attention only in the last years [59].

Dietary interventions with carotenoid-containing products leads to perceived changes in skin colour. Whitehead et al. [163] informed that self-reported increases in the dietary intakes of fruits and vegetables for 6 weeks led to increases of the CIELAB colour parameters a* and b* (regarded as estimations of redness and yellowness contributions to overall colour, respectively) in the skin. More recently, Pezdirc et al. [55] clearly demonstrated that the consumption of a diet high in fruits and vegetables by young women led to significant increases in the readings of the b* parameter in skin, consistent with increased yellowness.

In a very interesting paper, data on the contribution of melanin and carotenoids to the colouration of skin and the healthy appearance of human faces were evaluated [214]. It was concluded that the skin colour attributed to carotenoids is associated to the quality of the diet and health status and that such "carotenoid skin colouration" can be a valid cue of importance for the choice of mate [214], a situation similar to that described for other animals [211,212] Furthermore, the effect of facial skin carotenoid and melanin colouration on the perceived health was investigated. For this purpose, Caucasian participants were allowed to manipulate the skin colour of computer-generated facial images from the same racial group along carotenoid and melanin colour axes that were empirically measured. They were asked to "make the face as healthy as possible". In this survey, it was concluded that the participants chose to enhance derived "carotenoid colouration" more than "melanin colouration" to maximize apparent facial health. This, according to the authors, indicated that the colouration imparted by carotenoids had more impact on the perceived human facial health than that attributable to melanin [214].

Evidence that skin colour linked to carotenoids is preferred over that attributable to melanin in certain populations is accumulating. Thus, in a recent report Lefevre et al. [215], provides more interesting information in this respect. The authors used controlled facial images manipulated to be high or low on carotenoid or melanin colouration to evaluate not only that aspect, but also whether both kinds of pigments had an impact on the perceived appeal and if the results were dependent on the gender of the face. The results of the study indicated that, under the experimental conditions, carotenoid colouration was consistently preferred over melanin with more marked preferences for carotenoids in female compared to male faces, regardless of the sex of the judging observer [215]. More recently, it has been concluded that young Australian adults see the skin colour of the face linked with carotenoids (derived from the consumption of fruit and vegetables) and melanin (owed to sunlight exposure) as conveying a healthy appearance in young adults, although carotenoid colouration was more important to health perception [216].

9.3. The Colourless UV-Absorbing Carotenoids Phytoene and Phytofluene in Cosmetics

Given their unique characteristics as colourless UV-absorbing carotenoids, phytoene and phytofluene offer distinctive possibilities relative to other carotenoids to provide cosmetic benefits at different levels.

9.3.1. Skin Whitening by Phytoene and Phytofluene-Rich Products

An increase of the deposition of melanin in certain areas of the skin to produce dark spots and freckles is related to exposure to sun, skin disorders, aging or hormonal disorders. The importance of dark spots goes beyond aesthetic preferences since they may lead to cancer in some cases, hence the interest of using whitening agents [210]. The use of whitening agents for cosmetic and other purposes (psychological sociological, political or economic reasons) has been common in African and Asian societies, having been traced back to the 16th century in some Asian countries (India, China, Japan, or Korea). Currently, some commonly used whitening agents (corticosteroids, hydroquinone, monobenzyl hydroquinone, tretinoin, or mercury salts) raise concerns and are even forbidden in some countries due to associated side effects [217]. For example, hydroquinone at low levels (2%) has been the whitening treatment of choice for dyspigmentation, for decades, due to its ability of inhibit the activity of tyrosinase, the key enzyme for melanogenesis. However, the safety of its use raises controversies as it may cause irritation and has been associated to the possible development of malignancies [10]. In this context, the use of phytoene and phytofluene-containing products for skin whitening offers obvious advantages, including their safety, which may be assumed from their constant consumption in the human diet worldwide [29]. The use of these carotenoids for this purpose may also result in additional cosmetic and health benefits, due to their distinctive strong UV radiation absorbing properties and various attributed effects (antioxidant, antiinflammatory, anticarcinogenic [22–24].

Previous experiments with phytoene and phytofluene-rich topical formulations (IBR, Israeli Biotechnology Research Ltd., Yavne, Israel) result in skin lightening effects and beneficial effects like anti-ageing and anti-wrinkling effects [218,219]. More specifically, in a 12-week clinical study the skin whitening effect of a dietary supplementation with capsules providing 5 mg of phytoene and phytofluene per day) was evaluated in women with Fitzpatrick skin phototype IV (Table 1). In this study it was reported that a lightening effect, as assessed from increases in L* values and ITA, was observed in up to 82% of the volunteers [219]. L* is a measurement of the relative lightness of the skin, and places a given skin colour on the grey scale, between black (L* = 0) and white (L* = 100) [220], whereas ITA° is a parameter that helps assess the degree of skin pigmentation and is inversely related to skin lightness [209].

The study by von Oppen-Bezalel et al. [209] showed that, contrarily to the perceived change of skin colouration produced by other carotenoid-rich products [55,163], a tomato powder rich in colourless carotenoids (Israeli Biotechnology Research Ltd., Yavne, Israel) administered orally to women for 12 weeks at dietary achievable doses (~5 mg of PT plus PTF per day) increased MED but did not lead to statistically significant changes in the colour parameters L*, a*, and b* nor the individual typological angle (ITA°) Indeed, although the aim of the study was not to measure skin whitening (for example, women qualifying as skin phototype II were enrolled; see Table 1 for characteristics of this light phototype), the results obtained from the instrumental measurements carried out with a spectrocolourimeter revealed clear, albeit non-statistically significant, changes in L* [209]

The results of these studies using phytoene and phytofluene-rich products open the doors to the oral use of these carotenoids in nutricosmetics as whitening agents [192], for the treatment of dark spots or to cater for the preferences of certain populations (like East Asians or Afro-Americans) for lighter skin colours.

9.3.2. Effect on Other Skin Aesthetical Parameters

The photoprotective effects of the intervention described in the study by von Oppen-Bezalel et al. [209] was accompanied by perceptible improvements in skin radiance, suppleness, evenness, smoothness, moisturization, elasticity, visible skin health, visible skin youthfulness, and overall skin beauty at the end of the 12-week study. Interestingly some enhancements were already noticed after 6 weeks. Notably, the improvements in the parameters were perceived both clinically and by the human volunteers [192,209]. Evidence of skin evening and anti-wrinkling effects derived from the use of topical formulations containing phytoene and phytofluene have also been reported [218,219].

Finally, the fact that phytofluene is fluorescent may have cosmetic implications and be harnessed for the development of innovative products.

9.4. Carotenoids and Aesthetic Benefits: Public Health Implications

It is well-known that the skin tone of certain ethnicities (for instance, Caucasians) is linked to the healthiness of the diet [8,214,221]. Overall, the results of some of the studies commented in previous sections, point to the fact that individuals, including young adults, from different geographical locations show preference for skin colouration linked to carotenoids (that is, to higher intakes of fruit and vegetables, their main dietary sources) over skin colouration associated with melanin (derived from sunlight exposure), which may be harnessed to increase fruits and vegetables intakes, especially in sectors with low intakes, like young adults [214–216]. Beyond skin tone, the intake of colourless carotenoid-containing products seem to have other aesthetical benefits at the level of the skin, resulting in the self-perception of the skin as more beautiful, healthier and younger [192,209]. On the other hand, a study involving young college women indicated that nutrition educators and health practitioners should be aware that such population groups (and probably many more similar groups) are less aware of and probably less interested in the health benefits derived from increasing fruits and vegetables intakes compared with older adults. Within this context, it seems sensible that such professionals revise their strategies and focus on aspects like showing to young men and women the relationship of fruit and vegetable intake with satiety, weight and appearance [222]. In this scenario, it is not surprising that the relationships between perceived attractiveness, skin pigmentation and the intake of carotenoid-rich fruits and vegetables could be used to boost the consumption of such products as a measure to tackle unhealthy dietary patterns leading to higher risks of several diseases, as envisaged by Whitehead et al. [223,224] who argued that this new paradigm based on an "appeal to vanity" in relation to carotenoid pigmentation can be an alternative to health-based messages. One example of this would be the recommendation of increasing the consumption of fruit and vegetables to reduce the risk of developing chronic diseases [225] which, according to Whitehead et al. needs alternatives, like interventions appealing to vanity, some of which have already proven promising [224].

Whitehead et al. [223] concluded that the majority of interventions targeting appearance stressed noncompliance-derived negative effects in attractiveness. For instance, exposure to young adults of UV photographs and photoaging information (for instance effect on wrinkles and age spots) appears promising as a brief and cheap strategy to increase young adults' sun protection intentions and attitudes, which may in turn reduce their chances of developing skin cancer [226]. A new paradigm encouraging the intake of fruits and vegetables as a means to improve appearance seems promising, since showing individuals images of how their faces improve because of the dietary change is motivating. For this strategy to be viable, more studies providing empirical data about how the diet impacts the appearance are neede. For instance investigations comparing the current appearance of the individual and that achievable through better or worse dietary patterns. As already discussed, some studies indicate that individuals from different ethnicities are expected to prefer facial skin colouration attributable to increased fruit and vegetable intakes [214,221].

10. Conclusions

The protection of skin is important in the context of health as a means of preventing disorders that eventually lead to harmful conditions. Additionally, the appearance of the skin, notably the face, is an attribute of great relevance in signalling, since it conveys information with impacts at the socioeconomic level.

The role of nutrition in skin health and appearance is undeniable and long-known, hence the efforts of the industry to innovate in cosmetics, cosmeceuticals, nutricosmetics, functional foods, or nutraceuticals.

Carotenoids are natural dietary products that have been shown to intervene in health-promoting actions and whose value in the context of nutricosmetics continues to grow. From the literature it can be inferred that a diet rich in carotenoid-containing products and the avoidance of stress factors have a beneficial impact on skin health and appearance, with other likely beneficial systemic effects.

Excellent recent original studies and reviews point to the fact that the positive perceivable effects that dietary carotenoids cause in the skin may be harnessed in the context of public health. For instance, they can be used to promote healthy dietary patterns rich in carotenoid containing products as a strategy to reduce the risk of developing serious diseases, including cancer, cardiovascular disease, eye disorders, osteoporosis, or metabolic diseases.

Among dietary carotenoids, the UVR-absorbing colourless carotenes phytoene and phytofluene have been largely overlooked, probably due to their lack of colour, which made their detection more challenging in the past compared to other carotenoids. Hence, the considerable lack of abundant data about their presence in foods and tissues, in contrast with the other major dietary carotenoids found in humans. However, it is well-established that they are major dietary carotenoids (found in products frequently consumed as tomatoes, carrots, citrus, and derivatives), present in plasma, human milk, skin, and other tissues, and involved in several health-promoting actions, as revealed by studies of different nature. Notably, evidence is accumulating that they could be involved in the health benefits traditionally associated to lycopene, since the latter seems to always occur along with

Nutrients **2019**, 11, 1093

the colourless carotenoids in foods. Being the unique major dietary carotenoids absorbing maximally in the UV region and possessing other distinctive characteristics within the carotenoid family, research and applications in the use of these carotenoids in the promotion of health and cosmetics is a timely and expanding area recently featuring in the carotenoid field.

Future studies should devote attention to generating more data about phytoene and phytofluene skin levels in order to provide accurate information about skin carotenoid status, since mainly coloured carotenoids are currently being considered. Additionally, mechanistic studies about the beneficial effects of colourless carotenoids in the skin (whitening, improvement of aging signs, etc.) are needed.

Acknowledgments: Research on colourless carotenoids by A.J.M.M.'s team has been funded by the Andalusian Council of Economy, Innovation, Science and Employment (ref. CAROTINCO-P12-AGR-1287), the Spanish State Secretariat of Research, Development and Innovation (Ministry of Economy and Competitiveness, project ref. AGL2012-37610, co-funded by FEDER, Carotenoid Network: from microbia and plants to food and health, CaRed, BIO2015-71703-REDT), CaRed: Spanish carotenoid network (BIO2017-90877-REDT) and IBR—Israeli Biotechnology Research, Ltd. The publication costs were covered by IBR—Israeli Biotechnology Research, Ltd.

Conflicts of Interest: A.J.M.-M. is a member of the advisory board of IBR—Israeli Biotechnology Research, Ltd.

References

1. Lai-Cheong, J.E.; McGrath, J.A. Structure and function of skin, hair and nails. *Medicine (Baltimore).* **2017**, *45*, 347–351. [CrossRef]
2. Rittie, L.; Fisher, G.J.; Rittié, L.; Fisher, G.J. Natural and sun-induced aging of human skin. *Cold Spring Harb. Perspect. Med.* **2015**, *5*, a015370. [CrossRef] [PubMed]
3. Presland, R.B.; Dale, B.A. Epithelial Structural Proteins of the Skin and Oral Cavity: Function in Health and Disease. *Crit. Rev. Oral Biol. Med.* **2000**, *11*, 383–408. [CrossRef] [PubMed]
4. Dąbrowska, A.K.; Spano, F.; Derler, S.; Adlhart, C.; Spencer, N.D.; Rossi, R.M. The relationship between skin function, barrier properties, and body-dependent factors. *Ski. Res. Technol.* **2018**, *24*, 165–174. [CrossRef]
5. Pérez-Sánchez, A.; Barrajón-Catalán, E.; Herranz-López, M.; Micol, V. Nutraceuticals for skin care: A comprehensive review of human clinical studies. *Nutrients* **2018**, *10*, 403. [CrossRef]
6. Fink, B.; Grammer, K.; Matts, P.J. Visible skin color distribution plays a role in the perception of age, attractiveness, and health in female faces. *Evol. Hum. Behav.* **2006**, *27*, 433–442. [CrossRef]
7. Matts, P.J.; Fink, B.; Grammer, K.; Burquest, M. Color homogeneity and visual perception of age, health, and attractiveness of female facial skin. *J. Am. Acad. Dermatol.* **2007**, *57*, 977–984. [CrossRef]
8. Little, A.C.; Jones, B.C.; Debruine, L.M. Facial attractiveness: Evolutionary based research. *Philos. Trans. R. Soc. B Biol. Sci.* **2011**, *366*, 1638–1659. [CrossRef] [PubMed]
9. Samson, N.; Fink, B.; Matts, P.J. Visible skin condition and perception of human facial appearance. *Int. J. Cosmet. Sci.* **2010**, *32*, 167–184. [CrossRef]
10. Lee, C.-M. Fifty years of research and development of cosmeceuticals: A contemporary review. *J. Cosmet. Dermatol.* **2016**, *15*, 527–539. [CrossRef] [PubMed]
11. Goralczyk, R.; Wertz, K. Skin Photoprotection by Carotenoids. In *Carotenoids. Volume 5. Nutrition and Health*; Britton, G., Liaaen-Jensen, S., Pfander, H., Eds.; Birkhäuser: Basel, Switzerland, 2009; pp. 335–362, ISBN 978-3-7643-7501-0.
12. Draelos, Z.D. Nutrition and enhancing youthful-appearing skin. *Clin. Dermatol.* **2010**, *28*, 400–408. [CrossRef]
13. Anunciato, T.P.; da Rocha Filho, P.A. Carotenoids and polyphenols in nutricosmetics, nutraceuticals, and cosmeceuticals. *J. Cosmet. Dermatol.* **2012**, *11*, 51–54. [CrossRef]
14. Madhere, S.; Simpson, P. A market overview of nutricosmetics. *Cosmet. Dermatology* **2010**, *23*, 268–274.
15. Britton, G.; Khachik, F. Carotenoids in Food. In *Carotenoids. Volume 5: Nutrition and Health*; Britton, G., Liaaen-Jensen, S., Pfander, H., Eds.; Birkhäuser: Basel, Switzerland, 2009; pp. 45–66, ISBN 978-3-7643-7501-0.
16. Mortensen, A. Supplements. In *Carotenoids: Volume 5: Nutrition and Health*; Britton, G., Pfander, H., Liaaen-Jensen, S., Eds.; Birkhäuser: Basel, Switzerland, 2009; pp. 67–82, ISBN 978-3-7643-7501-0.
17. Meléndez-Martínez, A.J.; Pérez-Gálvez, A.; Roca, M.; Estévez Santiago, R.; Olmedilla Alonso, B.; Mercadante, A.Z.; Jesús Ornelas-Paz, J.D. Biodisponibilidad de carotenoides, factores que la determinan y métodos de estimación. In *Carotenoides en Agroalimentación y Salud*; Meléndez-Martínez, A.J., Ed.; Editorial Terracota: Ciudad de México, México, 2017; pp. 574–608, ISBN 978-84-15413-35-6.

18. Britton, G.; Liaaen-Jensen, S.; Pfander, H. *Carotenoids. Volume 5: Nutrition and Health*; Birkhäuser: Basel, Switzerland, 2009; ISBN 978-3-7643-7501-0.

19. Krinsky, N.I.; Johnson, E.J. Carotenoid actions and their relation to health and disease. *Mol. Aspects Med.* **2005**, *26*, 459–516. [CrossRef] [PubMed]

20. Krinsky, N.; Mayne, S.T.; Sies, H. *Carotenoids in Health and Disease*; Marcel Dekker: New York, NY, USA, 2004.

21. Rodriguez-Concepcion, M.; Avalos, J.; Bonet, M.L.; Boronat, A.; Gomez-Gomez, L.; Hornero-Mendez, D.; Limon, M.C.; Meléndez-Martínez, A.J.; Olmedilla-Alonso, B.; Palou, A.; et al. A global perspective on carotenoids: Metabolism, biotechnology, and benefits for nutrition and health. *Prog. Lipid Res.* **2018**, *70*, 62–93. [CrossRef]

22. Engelmann, N.J.; Clinton, S.K.; Erdman, J.W. Nutritional aspects of phytoene and phytofluene, carotenoid precursors to lycopene. *Adv. Nutr.* **2011**, *2*, 51–61. [CrossRef]

23. Meléndez-Martínez, A.J.; Mapelli-Brahm, P.; Benítez-González, A.; Stinco, C.M. A comprehensive review on the colorless carotenoids phytoene and phytofluene. *Arch. Biochem. Biophys.* **2015**, *572*, 188–200. [CrossRef] [PubMed]

24. Meléndez-Martínez, A.J.; Mapelli-Brahm, P.; Stinco, C.M. The colourless carotenoids phytoene and phytofluene: From dietary sources to their usefulness for the functional foods and nutricosmetics industries. *J. Food Compos. Anal.* **2018**, *67*, 91–103. [CrossRef]

25. De Spirt, S.; Lutter, K.; Stahl, W. Carotenoids in Photooxidative Stress. *Curr. Nutr. Food Sci.* **2010**, *6*, 36–43. [CrossRef]

26. Brenner, M.; Hearing, V.J.V.J. The protective role of melanin against UV damage in human skin. *Photochem. Photobiol.* **2008**, *84*, 539–549. [CrossRef]

27. Darvin, M.E.; Fluhr, J.W.; Schanzer, S.; Richter, H.; Patzelt, A.; Meinke, M.C.; Zastrow, L.; Golz, K.; Doucet, O. Dermal carotenoid level and kinetics after topical and systemic administration of antioxidants: Enrichment strategies in a controlled in vivo study. *J. Dermatol. Sci.* **2011**, *64*, 53–58. [CrossRef]

28. Beckenbach, L.; Baron, J.M.; Merk, H.F.; Amann, P.M. Retinoid treatment of skin diseases. *Eur J Dermatol* **2015**, *25*, 384–391. [CrossRef]

29. Havas, F.; Krispin, S.; Meléndez-Martínez, A.J.; von Oppen-Bezalel, L. Preliminary data on the safety of phytoene and phytofluene-rich products for human use including topical application. *J. Toxicol.* **2018**, *2018*, 5475784. [CrossRef]

30. Sorg, O.; Antille, C.; Kaya, G.; Saurat, J.-H. Retinoids in cosmeceuticals. *Dermatol. Ther.* **2006**, *19*, 289–296. [CrossRef]

31. Bear, M.; Connors, B.W.; Paradiso, M.A. The somatic sensory system. In *Neuroscience. Exploring the Brain*; Lippincott Williams & Wilkins: Philadelphia, PA, USA, 2007; pp. 387–423.

32. Chu, D.H.; Haake, A.R.; Holbrook, K.; Loomis, C.A. Nerves and receptors of the skin in Chapter 6: The structure and development of skin. In *Fitzpatrick's Dermatology General Medicine*; Freedberg, I.M., Eisen, A.Z., Wolff, K., Austen, K.F., Goldsmith, L.A., Katz, S.I., Eds.; McGraw-Hill: New York, NY, USA, 2003; pp. 89–115.

33. Kandel, E.R.; Schwartz, J.H.; Jessell, T.M.; Siegelbaum, S.A.; Hudspeth, A.J. Touch. In *Principles of Neural Science*; McGraw-Hill: New York, NY, USA, 2013.

34. Bikle, D.D. Vitamin D and the skin: Physiology and pathophysiology. *Rev. Endocr. Metab. Disord.* **2012**, *13*, 3–19. [CrossRef]

35. Fitzpatrick, T.B. The validity and practicality of sun-reactive skin types I through VI. *Arch. Dermatol.* **1988**, *124*, 869. [CrossRef] [PubMed]

36. Halliwell, B. Reactive species and antioxidants. Redox biology is a fundamental theme of aerobic life. *Plant Physiol.* **2006**, *141*, 312–322. [CrossRef]

37. Sies, H.; Stahl, W. Nutritional protection against skin damage from sunlight. *Annu. Rev. Nutr.* **2004**, *24*, 173–200. [CrossRef]

38. Stahl, W.; Sies, H. Photoprotection by dietary carotenoids: Concept, mechanisms, evidence and future development. *Mol. Nutr. Food Res.* **2012**, *56*, 287–295. [CrossRef]

39. Mathews-Roth, M.M. Carotenoids in Erythropoietic Protoporphyria and Other Photosensitivity Diseases. *Ann. N. Y. Acad. Sci.* **1993**, *691*, 127–138. [CrossRef]

40. Von Laar, J.; Stahl, W.; Bolsen, K.; Goerz, G.; Sies, H. β-Carotene serum levels in patients with erythropoietic protoporphyria on treatment with the synthetic all-trans isomer or a natural isomeric mixture of β-carotene. *J. Photochem. Photobiol. B Biol.* **1996**, *33*, 157–162. [CrossRef]

41. Kou, J.; Dou, D.; Yang, L. Porphyrin photosensitizers in photodynamic therapy and its applications. *Oncotarget* **2017**, *8*, 81591–81603. [CrossRef] [PubMed]
42. Fitzpatrick, T.B. Soleil et peau. *J Med Esthet.* **1975**, *2*, 33–34.
43. Fitzpatrick, T.B. Ultraviolet-induced pigmentary changes: Benefits and hazards. *Curr. Probl. Dermatol.* **1986**, *15*, 25–38.
44. Pathak, M.A. Sunlight and melanin pigmentation. In *Photochemical and Photobiological Reviews*; Smith, K.C., Ed.; Plenum Press: New York, NY, USA, 1976; pp. 211–239.
45. Schagen, S.K.; Zampeli, V.A.; Makrantonaki, E.; Zouboulis, C.C. Discovering the link between nutrition and skin aging. *Dermatoendocrinol.* **2012**, *4*, 298–307. [CrossRef]
46. Terao, J.; Minami, Y.; Bando, N. Singlet molecular oxygen quenching activity of carotenoids: Relevance to protection of the skin from photoaging. *J. Clin. Biochem. Nutr.* **2011**, *48*, 57–62. [CrossRef]
47. Kochevar, I.E.; Pathak, M.A.; Parrish, J.A. Photophysics, photochemistry, and phobiology. In *Fitzpatrick's Dermatology General Medicine*; Freedberg, I., Eisen, A.Z., Wolff, K., Al, E., Eds.; McGraw-Hill: New York, NY, USA, 1999; pp. 220–229.
48. Akhalaya, M.Y.; Maksimov, G.V.; Rubin, A.B.; Lademann, J.; Darvin, M.E. Molecular action mechanisms of solar infrared radiation and heat on human skin. *Ageing Res. Rev.* **2014**, *16*, 1–11. [CrossRef]
49. Freitas, J.V.; Junqueira, H.C.; Martins, W.K.; Baptista, M.S.; Gaspar, L. Antioxidant role on the protection of melanocytes against visible light-induced photodamage. *Free Radic. Biol. Med.* **2019**, *131*, 399–407. [CrossRef]
50. Ascenso, A.; Ribeiro, H.; Marques, H.C.; Oliveira, H.; Santos, C.; Simões, S. Chemoprevention of photocarcinogenesis by lycopene. *Exp. Dermatol.* **2014**, *23*, 874–878. [CrossRef] [PubMed]
51. Pullar, J.; Carr, A.; Vissers, M. The Roles of Vitamin C in Skin Health. *Nutrients* **2017**, *9*, 866. [CrossRef]
52. Reich, P.; Shwachman, H.; Craig, J.M. Lycopenemia. *N. Engl. J. Med.* **1960**, *262*, 263–269. [CrossRef]
53. Maharshak, N.; Shapiro, J.; Trau, H. Carotenoderma - a review of the current literature. *Int. J. Dermatol.* **2003**, *42*, 178–181. [CrossRef]
54. Stahl, W.; Sies, H. β-Carotene and other carotenoids in protection from sunlight. *Am. J. Clin. Nutr.* **2012**, *96*, 1179–1184. [CrossRef]
55. Pezdirc, K.; Hutchesson, M.J.; Williams, R.L.; Rollo, M.E.; Burrows, T.L.; Wood, L.G.; Oldmeadow, C.; Collins, C.E. Consuming High-Carotenoid Fruit and Vegetables Influences Skin Yellowness and Plasma Carotenoids in Young Women: A Single-Blind Randomized Crossover Trial. *J. Acad. Nutr. Diet.* **2016**, *116*, 1257–1265. [CrossRef] [PubMed]
56. Darvin, M.E.; Sterry, W.; Lademann, J.; Vergou, T. The Role of Carotenoids in Human Skin. *Molecules* **2011**, *16*, 10491–10506. [CrossRef]
57. Darvin, M.E.; Fluhr, J.W.; Caspers, P.; van der Pool, A.; Richter, H.; Patzelt, A.; Sterry, W.; Lademann, J. In vivo distribution of carotenoids in different anatomical locations of human skin: Comparative assessment with two different Raman spectroscopy methods. *Exp. Dermatol.* **2009**, *18*, 1060–1063. [CrossRef] [PubMed]
58. Niu, C.; Aisa, H.A. Upregulation of Melanogenesis and Tyrosinase Activity: Potential Agents for Vitiligo. *Molecules* **2017**, *22*, 1303. [CrossRef] [PubMed]
59. Foo, Y.Z.; Simmons, L.W.; Rhodes, G. Predictors of facial attractiveness and health in humans. *Sci. Rep.* **2017**, *7*, 1–12. [CrossRef]
60. Hunter, M.L. If You're Light You're Alright: Light Skin Color as Social Capital for Women of Color. *Gend. Soc.* **2002**, *16*, 175–193. [CrossRef]
61. Hill, M.E. Skin color and the perception of attractiveness among African Americans: Does gender make a difference? *Soc. Psychol. Q.* **2002**, *65*, 77–91. [CrossRef]
62. Awad, G.H.; Norwood, C.; Taylor, D.S.; Martinez, M.; McClain, S.; Jones, B.; Holman, A.; Chapman-Hilliard, C. Beauty and Body Image Concerns Among African American College Women. *J Black Psychol.* **2015** **2015**, *41*, 1–20. [CrossRef]
63. Coley, M.K.; Alexis, A.F. Cosmetic concerns in skin of color, part 1. *Cosmet. Dermatology* **2009**, *22*, 360–370.
64. Tranggono, R.I. Adityarini Cosmeceuticals for Asians who are living in the tropics. *J. Appl. Cosmetol.* **2010**, *28*, 71–86.
65. Howlett, J. *Functional foods from science To health and claims*; ILSI Press: Brussels, Belgium, 2008; ISBN 0849313724.
66. DeFelice, S.L. The nutraceutical revolution: Its impact on food industry. *Trends Food Sci. Technol.* **1995**, *6*, 59–61. [CrossRef]

67. Santini, A.; Cammarata, S.M.; Capone, G.; Ianaro, A.; Tenore, G.C.; Pani, L.; Novellino, E. Nutraceuticals: Opening the debate for a regulatory framework. *Br. J. Clin. Pharmacol.* **2018**, *84*, 659–672. [CrossRef]

68. The European Parliament and the Council of the European Union (institution) Regulation (EC) no 1223/2009 of The European Parliament and of the Council of 30 November 2009 on cosmetic products. *Off. J. Eur. Union* **2009**, *342*, 59–209.

69. FDA. *Federal Food, Drug, and Cosmetic Act*; 2019. Available online: https://www.fda.gov/cosmetics/guidanceregulation/lawsregulations/ucm074201.htm (accessed on 14 May 2019).

70. Saint-Leger, D. 'Cosmeceuticals'. Of men, science and laws … . *Int. J. Cosmet. Sci.* **2012**, *34*, 396–401. [CrossRef]

71. Meléndez-Martínez, A.J. Carotenoides: Estructura, propiedades y funciones. In *Carotenoides en agroalimentación y salud*; Meléndez-Martínez, A., Ed.; Editorial Terracota: Ciudad de México, México, 2017; ISBN 978-84-15413-35-6.

72. Dias, M.G.; Olmedilla-Alonso, B.; Hornero-Méndez, D.; Mercadante, A.Z.; Osorio, C.; Vargas-Murga, L.; Meléndez-Martínez, A.J. Comprehensive Database of Carotenoid Contents in Ibero-American Foods. A Valuable Tool in the Context of Functional Foods and the Establishment of Recommended Intakes of Bioactives. *J. Agric. Food Chem.* **2018**, *66*, 5055–5107. [CrossRef]

73. Singh, J.; Fan, D.; Banskota, A.H.; Stefanova, R.; Khan, W.; Hafting, J.; Craigie, J.; Critchley, A.T.; Prithiviraj, B. Bioactive components of the edible strain of red alga, Chondrus crispus, enhance oxidative stress tolerance in Caenorhabditis elegans. *J. Funct. Foods* **2013**, *5*, 1180–1190. [CrossRef]

74. Zarekarizi, A.; Hoffmann, L.; Burritt, D. Approaches for the sustainable production of fucoxanthin, a xanthophyll with potential health benefits. *J. Appl. Phycol.* **2018**, 1–19. [CrossRef]

75. Rasmussen, H.M.; Muzhingi, T.; Eggert, E.M.R.; Johnson, E. Lutein, zeaxanthin, meso-zeaxanthin content in egg yolk and their absence in fish and seafood. *J. Food Compos. Anal.* **2012**, *27*, 139–144. [CrossRef]

76. Schlatterer, J.; Breithaupt, D.E. Xanthophylls in commercial egg yolks: Quantification and identification by HPLC and LC-(APCI)MS using a C30 phase. *J. Agric. Food Chem.* **2006**, *54*, 2267–2273. [CrossRef]

77. Álvarez, R.; Meléndez-Martínez, A.J.; Vicario, I.M.; Alcalde, M.J. Carotenoid and Vitamin A Contents in Biological Fluids and Tissues of Animals as an Effect of the Diet: A Review. *Food Rev. Int.* **2015**, *31*, 319–340. [CrossRef]

78. Maoka, T. Carotenoids in Marine Animals. *Mar. Drugs* **2011**, *9*, 278–293. [CrossRef] [PubMed]

79. Shahidi, F.; Metusalach; Brown, J.A.; Taylor, P. Carotenoid pigments in seafoods and aquaculture. *Crit. Rev. Food Sci. Nutr.* **1998**, *38*, 1–67. [CrossRef]

80. Lehto, S.; Buchweitz, M.; Klimm, A.; Straßburger, R.; Bechtold, C.; Ulberth, F. Comparison of food colour regulations in the EU and the US: A review of current provisions. *Food Addit. Contam. Part A Chem. Anal. Control. Expo. Risk Assess.* **2017**, *34*, 335–355. [CrossRef]

81. Martins, N.; Roriz, C.L.; Morales, P.; Barros, L.; Ferreira, I.C.F.R. Food colorants: Challenges, opportunities and current desires of agro-industries to ensure consumer expectations and regulatory practices. *Trends Food Sci. Technol.* **2016**, *52*, 1–15. [CrossRef]

82. Phelan, D.; Prado-Cabrero, A.; Nolan, J.M. Stability of Commercially Available Macular Carotenoid Supplements in Oil and Powder Formulations. *Nutrients* **2017**, *9*, 1. [CrossRef]

83. Khachik, F. Analysis of carotenoids in nutritional studies. In *Carotenoids. Volume 5: Nutrition and Health*; Britton, G., Liaaen-Jensen, S., Pfander, H., Eds.; Birkhäuser: Basel, Switzerland, 2009; pp. 7–44, ISBN 978-3-7643-7501-0.

84. Canene-Adams, K.; Erdman, J.W., Jr. Absorption, Transport, Distribution in Tissues and Bioavailability. In *Carotenoids. Volume 5. Nutrition and Health*; Britton, G., Liaaen-Jensen, S., Pfander, H., Eds.; Birkhäuser: Basel, Switzerland, 2009; pp. 115–148, ISBN 978-3-7643-7501-0.

85. Meléndez-Martínez, A.; Stinco, C.M.; Liu, C.; Wang, X.-D. A simple HPLC method for the comprehensive analysis of cis/trans (Z/E) geometrical isomers of carotenoids for nutritional studies. *Food Chem.* **2013**, *138*, 1341–1350. [CrossRef]

86. Burrows, T.; Rollo, M.; Williams, R.; Wood, L.; Garg, M.; Jensen, M.; Collins, C. A Systematic Review of Technology-Based Dietary Intake Assessment Validation Studies That Include Carotenoid Biomarkers. *Nutrients* **2017**, *9*, 140. [CrossRef]

87. Britton, G. Vitamin A and Vitamin A Deficiency. In *Carotenoids. Volume 5. Nutrition and Health*; Britton, G., Liaaen-Jensen, S., Pfander, H., Eds.; Birkhäuser: Basel, Switzerland, 2009; pp. 173–190, ISBN 978-3-7643-7501-0.

88. Giuliano, G. Provitamin A biofortification of crop plants: A gold rush with many miners. *Curr. Opin. Biotechnol.* **2017**, *44*, 169–180. [CrossRef]

89. Van Hoang, D.; Pham, N.; Lee, A.; Tran, D.; Binns, C. Dietary Carotenoid Intakes and Prostate Cancer Risk: A Case-Control Study from Vietnam. *Nutrients* **2018**, *10*, 70. [CrossRef]

90. Yamaguchi, M. β-Cryptoxanthin and bone metabolism: The preventive role in osteoporosis. *J. Heal. Sci.* **2008**, *54*, 356–369. [CrossRef]

91. Böhm, V. Lycopene and heart health. *Mol. Nutr. Food Res.* **2012**, *56*, 296–303. [CrossRef]

92. Nishino, H.; Murakoshi, M.; Tokuda, H.; Satomi, Y. Cancer prevention by carotenoids. *Arch. Biochem. Biophys.* **2009**, *483*, 165–168. [CrossRef]

93. Mares, J. Lutein and Zeaxanthin Isomers in Eye Health and Disease. *Annu. Rev. Nutr.* **2016**, *36*, 571–602. [CrossRef]

94. Johnson, E.J. Role of lutein and zeaxanthin in visual and cognitive function throughout the lifespan. *Nutr. Rev.* **2014**, *72*, 605–612. [CrossRef]

95. Bonet, M.; Canas, J.A.; Ribot, J.; Palou, A. Carotenoids and their conversion products in the control of adipocyte function, adiposity and obesity. *Arch. Biochem. Biophys.* **2015**, *572*, 112–125. [CrossRef]

96. Coyne, T.; Ibiebele, T.I.; Baade, P.D.; McClintock, C.S.; Shaw, J.E. Metabolic syndrome and serum carotenoids: Findings of a cross-sectional study in Queensland, Australia. *Br. J. Nutr.* **2009**, *102*, 1668–1677. [CrossRef] [PubMed]

97. Than, A.; Bramley, P.M.; Davies, B.H.; Rees, A.F. Stereochemistry of phytoene. *Phytochemistry* **1972**, *11*, 3187–3192. [CrossRef]

98. Britton, G. Structure and properties of carotenoids in relation to function. *FASEB J.* **1995**, *9*, 1551–1558. [CrossRef] [PubMed]

99. Meléndez-Martínez, A.J.; Britton, G.; Vicario, I.M.; Heredia, F.J. Relationship between the colour and the chemical structure of carotenoid pigments. *Food Chem.* **2007**, *101*, 1145–1150. [CrossRef]

100. Rodriguez-Amaya, D. *A Guide to Carotenoid Analysis in Foods*; ILSI Press: Washington, DC, USA, 2001; ISBN 1578810728.

101. Davies, B.H. Carotenoids. In *Chemistry and Biochemistry of Plant Pigments*; Goodwin, T.W., Ed.; Academic Press: London, UK, 1976; pp. 38–165.

102. Britton, G.; Liaaen-Jensen, S.; Pfander, H. *Carotenoids. Handbook*; Birkhäuser: Basel, Switzerland, 2004.

103. Martínez, A.; Stinco, C.M.; Meléndez-Martínez, A.J. Free radical scavenging properties of phytofluene and phytoene isomers as compared to lycopene: A combined experimental and theoretical study. *J. Phys. Chem. B* **2014**, *118*, 9819–9825. [CrossRef] [PubMed]

104. Cooperstone, J.L.; Francis, D.M.; Schwartz, S.J. Thermal processing differentially affects lycopene and other carotenoids in cis-lycopene containing, tangerine tomatoes. *Food Chem.* **2016**, *210*, 466–472. [CrossRef]

105. Meléndez-Martínez, A.J.; Paulino, M.; Stinco, C.M.; Mapelli-Brahm, P.; Wang, X.-D. Study of the Time-Course of cis/trans (Z/E) Isomerization of Lycopene, Phytoene, and Phytofluene from Tomato. *J. Agric. Food Chem.* **2014**, *62*, 12399–12406. [CrossRef] [PubMed]

106. Desmarchelier, C.; Borel, P. Overview of carotenoid bioavailability determinants: From dietary factors to host genetic variations. *Trends Food Sci. Technol.* **2017**, *69*, 270–280. [CrossRef]

107. Mapelli-Brahm, P.; Corte-Real, J.; Meléndez-Martínez, A.J.A.J.; Bohn, T. Bioaccessibility of phytoene and phytofluene is superior to other carotenoids from selected fruit and vegetable juices. *Food Chem.* **2017**, *229*, 304–311. [CrossRef]

108. Mapelli-Brahm, P.; Stinco, C.M.; Rodrigo, M.J.; Zacarías, L.; Meléndez-Martínez, A.J. Impact of thermal treatments on the bioaccessibility of phytoene and phytofluene in relation to changes in the microstructure and size of orange juice particles. *J. Funct. Foods* **2018**, *46*, 38–47. [CrossRef]

109. Mapelli-Brahm, P.; Desmarchelier, C.; Margier, M.; Reboul, E.; Meléndez Martínez, A.J.; Borel, P. Phytoene and Phytofluene Isolated from a Tomato Extract are Readily Incorporated in Mixed Micelles and Absorbed by Caco-2 Cells, as Compared to Lycopene, and SR-BI is Involved in their Cellular Uptake. *Mol. Nutr. Food Res.* **2018**, *1800703*. [CrossRef]

110. Biehler, E.; Alkerwi, A.; Hoffmann, L.; Krause, E.; Guillaume, M.; Lair, M.L.; Bohn, T. Contribution of violaxanthin, neoxanthin, phytoene and phytofluene to total carotenoid intake: Assessment in Luxembourg. *J. Food Compos. Anal.* **2012**, *25*, 56–65. [CrossRef]

111. Paetau, I.; Khachik, F.; Brown, E.D.; Beecher, G.R.; Kramer, T.R.; Chittams, J.; Clevidence, B.A. Chronic ingestion of lycopene-rich tomato juice or lycopene supplements significantly increases plasma concentrations of lycopene and related tomato carotenoids in humans. *Int. J. Cancer* **1998**, *68*, 1187–1195. [CrossRef]

112. Müller, H.; Bub, A.; Watzl, B.; Rechkemmer, G.; Contribution, O. Plasma concentrations of carotenoids in healthy volunteers after intervention with carotenoid-rich foods. *Eur. J. Nutr.* **1999**, *38*, 35–44. [CrossRef]

113. Aust, O.; Stahl, W.; Sies, H.; Tronnier, H.; Heinrich, U. Supplementation with tomato- based products increase lycopene, phytofluene, and phytoene levels in human serum and protects against UV-light-induced erythema. *Int. J. Vitam. Nutr. Res.* **2005**, *75*, 54–60. [CrossRef] [PubMed]

114. Khachik, F.; Spangler, C.J.; Smith, J.C., Jr.; Canfield, L.M.; Steck, A.; Pfander, H. Identification, quantification, and relative Concentrations of carotenoids and their metabolites in Human Milk and Serum. *Anal. Chem.* **1997**, *69*, 1873–1881. [CrossRef]

115. Khachik, F.; Carvalho, L.; Bernstein, P.S.; Muir, G.J.; Zhao, D.-Y.; Katz, N.B. Chemistry, distribution, and metabolism of tomato carotenoids and their impact on human health. *Exp. Biol. Med. (Maywood).* **2002**, *227*, 845–851. [CrossRef]

116. Basu, A.; Imrhan, V. Tomatoes versus lycopene in oxidative stress and carcinogenesis: Conclusions from clinical trials. *Eur. J. Clin. Nutr.* **2007**, *61*, 295–303. [CrossRef] [PubMed]

117. Jacques, P.F.; Lyass, A.; Massaro, J.M.; Vasan, R.S.; D'Agostino Sr, R.B. Relationship of lycopene intake and consumption of tomato products to incident CVD. *Br. J. Nutr.* **2013**, *110*, 545–551. [CrossRef]

118. Ben-dor, A.; Steiner, M.; Gheber, L.; Danilenko, M.; Dubi, N.; Linnewiel, K.; Zick, A.; Sharoni, Y.; Levy, J. Carotenoids activate the antioxidant response element transcription system Carotenoids activate the antioxidant response element transcription system. *Mol. Cancer Ther.* **2005**, *4*, 177–186.

119. Shaish, A.; Harari, A.; Kamari, Y.; Soudant, E.; Harats, D.; Ben-Amotz, A. A carotenoid algal preparation containing phytoene and phytofluene inhibited LDL oxidation in vitro. *Plant Foods Hum. Nutr.* **2008**, *63*, 83–86. [CrossRef] [PubMed]

120. Porrini, M.; Riso, P.; Brusamolino, A.; Berti, C.; Guarnieri, S.; Visioli, F. Daily intake of a formulated tomato drink affects carotenoid plasma and lymphocyte concentrations and improves cellular antioxidant protection. *Br. J. Nutr.* **2005**, *93*, 93. [CrossRef]

121. Simmons, D.L.; Botting, R.M.; Hla, T. Cyclooxygenase isozymes: The biology of prostaglandin synthesis and inhibition. *Pharmacol. Rev.* **2004**, *56*, 387–437. [CrossRef] [PubMed]

122. Fuller, B.; Smith, D.; Howerton, A.; Kern, D. Anti-inflammatory effects of CoQ10 and colorless carotenoids. *J. Cosmet. Dermatol.* **2006**, *5*, 30–38. [CrossRef]

123. Meléndez-Martínez, A.J.; Nascimento, A.F.; Wang, Y.; Liu, C.; Mao, Y.; Wang, X.-D. Effect of tomato extract supplementation against high-fat diet-induced hepatic lesions. *Hepatobiliary Surg. Nutr.* **2013**, *2*, 198–208. [CrossRef]

124. Kotake-Nara, E.; Kushiro, M.; Zhang, H.; Sugawara, T.; Miyashita, K.; Nagao, A. Carotenoids affect proliferation of human prostate cancer cells. *J. Nutr.* **2001**, *131*, 3303–3306. [CrossRef]

125. Nara, E.; Hayashi, H.; Kotake, M.; Miyashita, K.; Nagao, A. Acyclic Carotenoids and Their Oxidation Mixtures Inhibit the Growth of HL-60 Human Promyelocytic Leukemia Cells Acyclic Carotenoids and Their Oxidation Mixtures Inhibit the Growth of HL-60 Human Promyelocytic Leukemia Cells. *Nutr. Cancer* **2009**, *39*, 37–41. [CrossRef]

126. Hirsch, K.; Atzmon, A.; Danilenko, M.; Levy, J.; Sharoni, Y. Lycopene and other carotenoids inhibit estrogenic activity of 17-β-estradiol and genistein in cancer cells. *Breast Cancer Res. Treat.* **2007**, *104*, 221–230. [CrossRef]

127. Boileau, T.W.-M.; Liao, Z.; Kim, S.; Lemeshow, S.; Erdman, J.; Clinton, S.K. Prostate Carcinogenesis in N-methyl-N-nitrosourea (NMU)-Testosterone-Treated Rats Fed Tomato Powder, Lycopene, or Energy-Restricted Diets. *J. Natl. Cancer Inst.* **2003**, *95*, 1578–1586. [CrossRef] [PubMed]

128. Campbell, J.K.; Stroud, C.K.; Nakamura, M.T.; Lila, M.A.; Erdman, J.W. Serum testosterone is reduced following short-term phytofluene, lycopene, or tomato powder consumption in F344 rats. *J. Nutr.* **2006**, *136*, 2813–2819. [CrossRef] [PubMed]

129. FDA. *GRAS Notice 163*; 2005. Available online: https://www.accessdata.fda.gov/scripts/fdcc/index.cfm?set=GRASNotices&id=163 (accessed on 12 August 2018).

130. FDA. *GRAS Notice 185*; 2006. Available online: https://www.accessdata.fda.gov/scripts/fdcc/index.cfm?set=GRASNotices&id=185 (accessed on 12 August 2018).

131. EFSA Safety of Lycopene oleoresin from tomatoes. *EFSA J.* **2008**, *675*, 1–22. [CrossRef]

132. Darvin, M.E.; Patzelt, A.; Knorr, F.; Blume-Peytavi, U.; Sterry, W.; Lademann, J. One-year study on the variation of carotenoid antioxidant substances in living human skin: Influence of dietary supplementation and stress factors. *J. Biomed. Opt.* **2008**, *13*, 044028-1–044028-9. [CrossRef]

133. Mayne, S.T.; Cartmel, B.; Scarmo, S.; Lin, H.; Leffell, D.J.; Welch, E.; Ermakov, I.; Bhosale, P.; Bernstein, P.S.; Gellermann, W. Noninvasive assessment of dermal carotenoids as a biomarker of fruit and vegetable intake. *Am. J. Clin. Nutr.* **2010**, *92*, 794–800. [CrossRef] [PubMed]

134. Ermakov, I.V.; Gellermann, W. Validation model for Raman based skin carotenoid detection. *Arch. Biochem. Biophys.* **2010**, *504*, 40–49. [CrossRef]

135. Ermakov, I.V.; Ermakova, M.; Sharifzadeh, M.; Gorusupudi, A.; Farnsworth, K.; Bernstein, P.S.; Stookey, J.; Evans, J.; Arana, T.; Tao-Lew, L.; et al. Optical assessment of skin carotenoid status as a biomarker of vegetable and fruit intake. *Arch. Biochem. Biophys.* **2018**, *646*, 46–54. [CrossRef] [PubMed]

136. Stahl, W.; Heinrich, U.; Jungmann, H.; von Laar, J.; Schietzel, M.; Sies, H.; Tronnier, H. Increased Dermal Carotenoid Levels Assessed by Noninvasive Reflection Spectrophotometry Correlate with Serum Levels in Women Ingesting Betatene. *J. Nutr.* **1998**, *128*, 903–907. [CrossRef] [PubMed]

137. Stahl, W.; Heinrich, U.; Jungmann, H.; Tronnier, H.; Sies, H. Carotenoids in human skin: Noninvasive measurement and identification of dermal carotenoids and carotenol esters. *Methods Enzymol.* **2000**, *319*, 494–502. [CrossRef] [PubMed]

138. Lademann, J.; Meinke, M.C.; Sterry, W.; Darvin, M.E. Carotenoids in human skin. *Exp. Dermatol.* **2011**, *20*, 377–382. [CrossRef]

139. Ermakov, I.V.; Gellermann, W. Optical detection methods for carotenoids in human skin. *Arch. Biochem. Biophys.* **2015**, *572*, 101–111. [CrossRef]

140. Meinke, M.C.; Lohan, S.B.; Köcher, W.; Magnussen, B.; Darvin, M.E.; Lademann, J. Multiple spatially resolved reflection spectroscopy to monitor cutaneous carotenoids during supplementation of fruit and vegetable extracts in vivo. *Ski. Res. Technol.* **2017**, *23*, 459–462. [CrossRef]

141. Jilcott Pitts, S.B.; Jahns, L.; Wu, Q.; Moran, N.E.; Bell, R.A.; Truesdale, K.P.; Laska, M.N. A non-invasive assessment of skin carotenoid status through reflection spectroscopy is a feasible, reliable and potentially valid measure of fruit and vegetable consumption in a diverse community sample. *Public Health Nutr.* **2018**, *21*, 1664–1670. [CrossRef]

142. Ashton, L.M.; Pezdirc, K.B.; Hutchesson, M.J.; Rollo, M.E.; Collins, C.E. Is skin coloration measured by reflectance spectroscopy related to intake of nutrient-dense foods? A cross-sectional evaluation in Australian young adults. *Nutrients* **2018**, *10*, 11. [CrossRef] [PubMed]

143. Coyle, D.H.; Pezdirc, K.; Hutchesson, M.J.; Collins, C.E. Intake of specific types of fruit and vegetables is associated with higher levels of skin yellowness in young women: A cross-sectional study. *Nutr. Res.* **2018**, *56*, 23–31. [CrossRef]

144. Stinco, C.M.; Rodríguez-Pulido, F.J.; Escudero-Gilete, M.L.; Gordillo, B.; Vicario, I.M.; Meléndez-Martínez, A.J. Lycopene isomers in fresh and processed tomato products: Correlations with instrumental color measurements by digital image analysis and spectroradiometry. *Food Res. Int.* **2013**, *50*, 111–120. [CrossRef]

145. Meléndez-Martínez, A.J.; Gómez-Robledo, L.; Melgosa, M.; Vicario, I.M.; Heredia, F.J. Color of orange juices in relation to their carotenoid contents as assessed from different spectroscopic data. *J. Food Compos. Anal.* **2011**, *24*, 837–844. [CrossRef]

146. Meléndez-Martínez, A.J.; Vicario, I.M.; Heredia, F.J. Influence of white reference measurement and background on the colour specification of orange juices by means of diffuse reflectance spectrophotometry. *J. AOAC Int.* **2006**, *89*, 452–457.

147. Meléndez-Martínez, A.J.; Vicario, I.M.; Heredia, F.J. Instrumental measurement of orange juice colour: A review. *J. Sci. Food Agric.* **2005**, *85*, 894–901. [CrossRef]

148. Meléndez-Martínez, A.J.; Vicario, I.M.; Heredia, F.J. Rapid assessment of vitamin A activity through objective color measurements for the quality control of orange juices with diverse carotenoid profiles. *J. Agric. Food Chem.* **2007**, *55*, 2808–2815. [CrossRef] [PubMed]

149. Scarmo, S.; Cartmel, B.; Lin, H.; Leffell, D.J.; Ermakov, I.V.; Gellermann, W.; Bernstein, P.S.; Mayne, S.T. Single *v.* multiple measures of skin carotenoids by resonance Raman spectroscopy as a biomarker of usual carotenoid status. *Br. J. Nutr.* **2013**, *110*, 911–917. [CrossRef]

150. Aguilar, S.S.; Wengreen, H.J.; Lefevre, M.; Madden, G.J.; Gast, J. Skin Carotenoids: A Biomarker of Fruit and Vegetable Intake in Children. *J. Acad. Nutr. Diet.* **2014**, *114*, 1174–1180. [CrossRef] [PubMed]

151. Nguyen, L.M.; Scherr, R.E.; Linnell, J.D.; Ermakov, I.V.; Gellermann, W.; Jahns, L.; Keen, C.L.; Miyamoto, S.; Steinberg, F.M.; Young, H.M.; et al. Evaluating the relationship between plasma and skin carotenoids and reported dietary intake in elementary school children to assess fruit and vegetable intake. *Arch. Biochem. Biophys.* **2015**, *572*, 73–80. [CrossRef]

152. Holt, E.W.; Wei, E.K.; Bennett, N.; Zhang, L.M. Low skin carotenoid concentration measured by resonance Raman spectroscopy is associated with metabolic syndrome in adults. *Nutr. Res.* **2014**, *34*, 821–826. [CrossRef]

153. Stahl, W.; Heinrich, U.; Wiseman, S.; Eichler, O.; Sies, H.; Tronnier, H. Dietary Tomato Paste Protects against Ultraviolet Light–Induced Erythema in Humans. *J. Nutr.* **2001**, *131*, 1449–1451. [CrossRef]

154. Hata, T.R.; Scholz, T.A.; Ermakov, I.V.; McClane, R.W.; Khachik, F.; Gellermann, W.; Pershing, L.K. Non-invasive Raman spectroscopic detection of carotenoids in human skin. *J. Invest. Dermatol.* **2000**, *115*, 441–448. [CrossRef]

155. Wingerath, T.; Sies, H.; Stahl, W. Xanthophyll Esters in Human Skin. *Arch. Biochem. Biophys.* **1998**, *355*, 271–274. [CrossRef]

156. Wingerath, T.; Stahl, W.; Sies, H. β-Cryptoxanthin selectively increases in human chylomicrons upon ingestion of tangerine concentrate rich in beta-cryptoxanthin esters. *Arch. Biochem. Biophys.* **1995**, *324*, 385–390. [CrossRef]

157. Mercadante, A.Z.; Rodrigues, D.B.; Petry, F.C.; Mariutti, L.R.B. Carotenoid esters in foods—A review and practical directions on analysis and occurrence. *Food Res. Int.* **2017**, *99*, 830–850. [CrossRef]

158. Bohn, T.; McDougall, G.J.; Alegría, A.; Alminger, M.; Arrigoni, E.; Aura, A.-M.; Brito, C.; Cilla, A.; El, S.N.; Karakaya, S.; et al. Mind the gap-deficits in our knowledge of aspects impacting the bioavailability of phytochemicals and their metabolites-a position paper focusing on carotenoids and polyphenols. *Mol. Nutr. Food Res.* **2015**. [CrossRef] [PubMed]

159. Bohn, T.; Desmarchelier, C.; Dragsted, L.O.; Nielsen, C.S.; Stahl, W.; Rühl, R.; Keijer, J.; Borel, P. Host-related factors explaining interindividual variability of carotenoid bioavailability and tissue concentrations in humans. *Mol. Nutr. Food Res.* **2017**, *61*. [CrossRef]

160. Massenti, R.; Perrone, A.; Livrea, M.A.; Lo Bianco, R. Regular consumption of fresh orange juice increases human skin carotenoid content. *Int. J. Food Sci. Nutr.* **2015**, *66*, 718–721. [CrossRef]

161. Jahns, L.; Johnson, L.K.; Mayne, S.T.; Cartmel, B.; Sr, M.J.P.; Ermakov, I.V.; Gellermann, W.; Whigham, L.D.; Picklo, M.J.; Ermakov, I.V.; et al. Skin and plasma carotenoid response to a provided intervention diet high in vegetables and fruit: Uptake and depletion kinetics. *Am. J. Clin. Nutr.* **2014**, *100*, 930–937. [CrossRef] [PubMed]

162. Aguilar, S.S.; Wengreen, H.J.; Dew, J. Skin Carotenoid Response to a High-Carotenoid Juice in Children: A Randomized Clinical Trial. *J. Acad. Nutr. Diet.* **2015**, *115*, 1771–1778. [CrossRef]

163. Whitehead, R.D.; Re, D.; Xiao, D.; Ozakinci, G.; Perrett, D.I. You Are What You Eat: Within-Subject Increases in Fruit and Vegetable Consumption Confer Beneficial Skin-Color Changes. *PLoS ONE* **2012**, *7*, e32988. [CrossRef]

164. Sansone, R.A.; Sansone, L.A. Carrot Man: A Case of Excessive Beta-Carotene Ingestion. *Int. J. Eat. Disord.* **2012**, *45*, 816–818. [CrossRef]

165. Meinke, M.C.; Lauer, A.; Taskoparan, B.; Gersonde, I.; Lademann, J.; Darvin, M.E. Influence on the carotenoid levels of skin arising from age, gender, body mass index in smoking/non-smoking individuals. *Free Radicals Antioxidants* **2011**, *1*, 15–20. [CrossRef]

166. Biesalski, H.K.; Hemmes, C.; Hopfenmuller, W.; Schmid, C.; Gollnick, H.P. Effects of controlled exposure of sunlight on plasma and skin levels of beta-carotene. *Free Radic. Res.* **1996**, *24*, 215–224. [CrossRef]

167. Vandersee, S.; Beyer, M.; Lademann, J.; Darvin, M.E. Blue-violet light irradiation dose dependently decreases carotenoids in human skin, which indicates the generation of free radicals. *Oxid. Med. Cell. Longev.* **2015**, *2015*. [CrossRef]

168. Darvin, M.E.; Patzelt, A.; Meinke, M.; Sterry, W.; Lademann, J. Influence of two different IR radiators on the antioxidative potential of the human skin. *Laser Phys. Lett.* **2009**, *6*, 229–234. [CrossRef]

169. Darvin, M.E.; Fluhr, J.W.; Meinke, M.C.; Zastrow, L.; Sterry, W.; Lademann, J. Topical β-carotene protects against infra-red-light-induced free radicals. *Exp. Dermatol.* **2011**, *20*, 125–129. [CrossRef]

170. Lima, X.T.T.; Kimball, A.B.B. Skin carotenoid levels in adult patients with psoriasis. *J. Eur. Acad. Dermatol. Venereol.* **2011**, *25*, 945–949. [CrossRef] [PubMed]

171. Li, D.G.; LeCompte, G.; Golod, L.; Cecchi, G.; Irwin, D.; Harken, A.; Matecki, A. Dermal carotenoid measurement is inversely related to anxiety in patients with breast cancer. *J. Investig. Med.* **2018**, *66*, 329–333. [CrossRef] [PubMed]

172. Meinke, M.C.; Darvin, M.E.; Vollert, H.; Lademann, J. Bioavailability of natural carotenoids in human skin compared to blood. *Eur. J. Pharm. Biopharm.* **2010**, *76*, 269–274. [CrossRef]

173. Takaichi, S.; Mochimaru, M.; Sciences, M.L. Carotenoids and carotenogenesis in cyanobacteria: Unique ketocarotenoids and carotenoid glycosides. *Cell. Mol. Life Sci.* **2007**, *64*, 2607–2619. [CrossRef] [PubMed]

174. Britton, G. Functions of intact carotenoids. In *Carotenoids. Volume 4: Natural functions*; Britton, G., Liaaen-Jensen, S., Pfander, H., Eds.; Birkhäuser: Basel, Switzerland, 2008; pp. 189–212, ISBN 978-3-7643-7499-0.

175. Schalch, W.; Landrum, J.T.; Bone, R.A. The eye. In *Carotenoids. Volume 5: Nutrition and Health*; Britton, G., Liaaen-Jensen, S., Pfander, H., Eds.; Birkhäuser: Basel, Switzerland, 2009; pp. 301–334, ISBN 978-3-7643-7501-0.

176. Johnson, E.J.; Neuringer, M.; Russell, R.M.; Schalch, W.; Snodderly, D.M. Nutritional manipulation of primate retinas, III: Effects of lutein or zeaxanthin supplementation on adipose tissue and retina of xanthophyll-free monkeys. *Investig. Ophthalmol. Vis. Sci.* **2005**, *46*. [CrossRef]

177. Bhosale, P.; Larson, A.J.; Frederick, J.M.; Southwick, K.; Thulin, C.D.; Bernstein, P.S. Identification and characterization of a Pi isoform of glutathione S-transferase (GSTP1) as a zeaxanthin-binding protein in the macula of the human eye. *J. Biol. Chem.* **2004**, *279*, 49447–49454. [CrossRef]

178. Li, B.; Vachali, P.; Frederick, J.M.; Bernstein, P.S. Identification of StARD3 as a lutein-binding protein in the macula of the primate retina. *Biochemistry* **2011**, *50*, 2541–2549. [CrossRef]

179. Krinsky, N.I.; Landrum, J.T.; Bone, R. a Biologic mechanisms of the protective role of lutein and zeaxanthin in the eye. *Annu. Rev. Nutr.* **2003**, *23*, 171–201. [CrossRef]

180. Kopec, R.E.; Schick, J.; Tober, K.L.; Riedl, K.M.; Francis, D.M.; Young, G.S.; Schwartz, S.J.; Oberyszyn, T.M. Sex differences in skin carotenoid deposition and acute UVB-induced skin damage in SKH-1 hairless mice after consumption of tangerine tomatoes. *Mol. Nutr. Food Res.* **2015**, *59*, 2491–2501. [CrossRef]

181. Álvarez, R.; Vaz, B.; Gronemeyer, H.; ALera, Á.R.; Rosana, A.; Gronemeyer, H.; de Lera, A.R. Functions, therapeutic applications, and synthesis of retinoids and carotenoids. *Chem. Rev.* **2014**, *114*, 1–125. [CrossRef]

182. Niki, E. Lipid peroxidation: Physiological levels and dual biological effects. *Free Radic. Biol. Med.* **2009**, *47*, 469–484. [CrossRef] [PubMed]

183. Böhm, F.; Edge, R.; Truscott, G. Interactions of dietary carotenoids with activated (singlet) oxygen and free radicals: Potential effects for human health. *Mol. Nutr. Food Res.* **2012**, *56*, 205–216. [CrossRef] [PubMed]

184. Skibsted, L.H. Carotenoids in Antioxidant Networks. Colorants or Radical Scavengers. *J. Agric. Food Chem.* **2012**, *60*, 2409–2417. [CrossRef]

185. Meinke, M.C.; Friedrich, A.; Tscherch, K.; Haag, S.F.; Darvin, M.E.; Vollert, H.; Groth, N.; Lademann, J.; Rohn, S. Influence of dietary carotenoids on radical scavenging capacity of the skin and skin lipids. *Eur. J. Pharm. Biopharm.* **2013**, *84*, 365–373. [CrossRef]

186. Yeum, K.-J.; Aldini, G.; Russell, R.M.; Krinsky, N.I. Antioxidant/Pro-oxidant Actions of Carotenoids. In *Carotenoids. Volume 5. Nutrition and Health2*; Britton, G., Liaaen-Jensen, S., Pfander, H., Eds.; Birkhäuser: Basel, Switzerland, 2009; pp. 235–268.

187. Palozza, P. Prooxidant Actions of Carotenoids in Biologic Systems. *Nutr. Rev.* **2009**, *56*, 257–265. [CrossRef]

188. Ribeiro, D.; Freitas, M.; Silva, A.M.S.; Carvalho, F.; Fernandes, E. Antioxidant and pro-oxidant activities of carotenoids and their oxidation products. *Food Chem. Toxicol.* **2018**, *120*, 681–699. [CrossRef]

189. Grochot-Przeczek, A.; Dulak, J.; Jozkowicz, A. Haem oxygenase-1: Non-canonical roles in physiology and pathology. *Clin. Sci. (Lond).* **2012**, *122*, 93–103. [CrossRef]

190. Malemud, C. Matrix metalloproteinases (MMPs) in health and disease: An overview. *Front. Biosci.* **2006**, *11*, 1696–1701. [CrossRef] [PubMed]

191. Wertz, K.; Hunziker, P.B.; Seifert, N.; Riss, G.; Neeb, M.; Steiner, G.; Hunziker, W.; Goralczyk, R. beta-Carotene interferes with ultraviolet light A-induced gene expression by multiple pathways. *J. Invest. Dermatol.* **2005**, *124*, 428–434. [CrossRef]

192. Marini, A.; Jaenicke, T.; Stahl, W.; Krutmann, J.; Grether-Beck, S.; Marini, A.; Jaenicke, T.; Stahl, W.; Krutmann, J. Molecular evidence that oral supplementation with lycopene or lutein protects human skin against ultraviolet radiation: Results from a double-blinded, placebo-controlled, crossover study. *Br. J. Dermatol.* **2017**, *176*, 1231–1240. [CrossRef]

193. Yuan, J.P.; Peng, J.; Yin, K.; Wang, J.H. Potential health-promoting effects of astaxanthin: A high-value carotenoid mostly from microalgae. *Mol. Nutr. Food Res.* **2011**, *55*, 150–165. [CrossRef] [PubMed]

194. Davinelli, S.; Nielsen, M.; Scapagnini, G. Astaxanthin in Skin Health, Repair, and Disease: A Comprehensive Review. *Nutrients* **2018**, *10*, 522. [CrossRef]

195. EFSA Scientific Opinion on the re-evaluation of canthaxanthin (E 161 g) as a food additive. *EFSA J.* **2010**, *8*, 1852–1893. [CrossRef]

196. Heinrich, U.; Gärtner, C.; Wiebusch, M.; Eichler, O.; Sies, H.; Tronnier, H.; Stahl, W. Supplementation with beta-carotene or a similar amount of mixed carotenoids protects humans from UV-induced erythema. *J. Nutr.* **2003**, *133*, 98–101. [CrossRef] [PubMed]

197. Stahl, W.; Sies, H. Carotenoids and Protection against Solar UV Radiation. *Skin Pharmacol. Physiol.* **2002**, *15*, 291–296. [CrossRef]

198. Köpcke, W.; Krutmann, J. Protection from Sunburn with β-Carotene—A Meta-analysis. *Photochem. Photobiol.* **2008**, *84*, 284–288. [CrossRef]

199. Stahl, W.; Heinrich, U.; Jungmann, H.; Sies, H.; Tronnier, H. Carotenoids and carotenoids plus vitamin E protect against ultraviolet light-induced erythema in humans. *Am. Soc. Clin. Nutr.* **2000**, *71*, 795–798. [CrossRef]

200. Greenberger, S.; Harats, D.; Salameh, F.; Lubish, T.; Harari, A.; Trau, H.; Shaish, A. 9-cis-rich β-carotene powder of the alga Dunaliella reduces the severity of chronic plaque psoriasis: A randomized, double-blind, placebo-controlled clinical trial. *J. Am. Coll. Nutr.* **2012**, *31*, 320–326. [CrossRef]

201. Murillo, E.; Meléndez-Martínez, A.J.; Portugal, F. Screening of vegetables and fruits from Panama for rich sources of lutein and zeaxanthin. *Food Chem.* **2010**, *122*, 167–172. [CrossRef]

202. Rojas-Garbanzo, C.; Gleichenhagen, M.; Heller, A.; Esquivel, P.; Schulze-Kaysers, N.; Schieber, A. Carotenoid Profile, Antioxidant Capacity, and Chromoplasts of Pink Guava (Psidium guajava L. Cv. 'Criolla') during Fruit Ripening. *J. Agric. Food Chem.* **2017**, *65*. [CrossRef]

203. Beutner, S.; Bloedorn, B.; Frixel, S.; Blanco, I.H.; Hoffmann, T.; Martin, H.D.; Mayer, B.; Noack, P.; Ruck, C.; Schmidt, M.; et al. Quantitative assessment of antioxidant properties of natural colorants and phytochemicals: Carotenoids, flavonoids, phenols and indigoids. The role of β-carotene in antioxidant functions. *J. Sci. Food Agric.* **2001**, *81*, 559–568. [CrossRef]

204. Stahl, W.; Heinrich, U.; Aust, O.; Tronnier, H.; Sies, H. Lycopene-rich products and dietary photoprotection. *Photochem. Photobiol. Sci.* **2006**, *5*, 238–242. [CrossRef]

205. Cooperstone, J.L.; Tober, K.L.; Riedl, K.M.; Teegarden, M.D.; Cichon, M.J.; Francis, D.M.; Schwartz, S.J.; Oberyszyn, T.M. Tomatoes protect against development of UV-induced keratinocyte carcinoma via metabolomic alterations. *Sci. Rep.* **2017**, *7*, 1–9. [CrossRef]

206. Nishino, A.; Sugimoto, K.; Sambe, H.; Ichihara, T.; Takaha, T.; Kuriki, T. Effects of Dietary Paprika Xanthophylls on Ultraviolet Light-Induced Skin Damage: A Double-Blind Placebo-Controlled Study. *J. Oleo Sci.* **2018**, *67*, 863–869. [CrossRef] [PubMed]

207. Mathews Roth, M.M.; Pathak, M.A.; Mathews-Roth, M.M.; Pathak, M.A. Phytoene as a protective agent against sunburn (>280nm) radiation in guinea pigs. *Photochem. Photobiol.* **1975**, *21*, 261–263. [CrossRef] [PubMed]

208. Mathews-Roth, M.M. Antitumor activity of β-carotene, canthaxanthin and phytoene. *Oncology* **1982**, *39*, 33–37. [CrossRef]

209. Von Oppen-Bezalel, L.; Fishbein, D.; Havas, F.; Ben-Chitrit, O.; Khaiat, A. The photoprotective effects of a food supplement tomato powder rich in phytoene and phytofluene, the colorless carotenoids, a preliminary study. *Glob. Dermatol.* **2015**, *2*, 178–182.

210. Von Oppen-Bezalel, L.; Shaish, A. Application of the colorless carotenoids, phytoene, and phytofluene in cosmetics, wellness, nutrition, and therapeutics. In *The alga Dunaliella: Biodiversity, Physiology, Genomics & Biotechnology*; Ben-Amotz, A., Polle, J., Rao, S., Eds.; Science Publishers: Enfield, NH, USA, 2009; pp. 423–444, ISBN 978-1-5780-8545-3.

211. Blount, J.D.; McGraw, K.J. Signal functions of carotenoid colouration. In *Carotenoids. Volume 4: Natural functions*; Gritton, G., Liaaen-Jensen, S., Pfander, H., Eds.; Birkhäuser: Basel, Switzerland, 2008; pp. 213–236, ISBN 978-3-7643-7499-0.

212. Blount, J.D. Carotenoids and life-history evolution in animals. *Arch. Biochem. Biophys.* **2004**, *430*, 10–15. [CrossRef]

213. Britton, G.; Liaaen-Jensen, S.; Pfander, H. *Carotenoids: A Colourful History*; CaroteNature GmbH: Bern, Switzerland, 2017.
214. Stephen, I.D.; Coetzee, V.; Perrett, D.I. Carotenoid and melanin pigment coloration affect perceived. *Evol. Hum. Behav.* **2011**, *32*, 216–227. [CrossRef]
215. Lefevre, C.E.; Perrett, D.I.; Lefevre, C.E.; Perrett, D.I. Fruit over sunbed: Carotenoid skin colouration is found more attractive than melanin colouration. *Q. J. Exp. Psychol.* **2015**, *0218*, 1–10. [CrossRef]
216. Pezdirc, K.; Rollo, M.E.M.E.; Whitehead, R.; Hutchesson, M.J.M.J.; Ozakinci, G.; Perrett, D.; Collins, C.E.C.E. Perceptions of carotenoid and melanin colouration in faces among young Australian adults. *Aust. J. Psychol.* **2017**, *70*, 85–90. [CrossRef]
217. Desmedt, B.; Courselle, P.; De Beer, J.O.; Rogiers, V.; Grosber, M.; Deconinck, E.; De Paepe, K. Overview of skin whitening agents with an insight into the illegal cosmetic market in Europe. *J. Eur. Acad. Dermatol. Venereol.* **2016**, *30*, 943–950. [CrossRef]
218. Von Oppen-Bezalel, L. Colorless Carotenoids, phytoene and phytofluene for the skin: For prevention of aging/photo-aging from inside and out. *SOFW J.* **2007**, *133*, 1–3.
219. Von Oppen-Bezalel, L.; Havas, F.; Ramot, O.; Kalo, E.; Fishbein, D.; Ben-Chitrit, O. Phytoene and Phytofluene for (Photo) Protection, Anti Aging, Lightening and Evening of Skin Tone. *SOFW J.* **2014**, *140*, 8–12.
220. *CIE Recommendations on Uniform Color Spaces, Color-Difference Equations, Psychometric Color Terms*; CIE Publication No. 15 (E-1.3.1) 1971, Supplement 2; Bureau Central de la CIE: Vienna, Austria, 1978.
221. Stephen, I.D.; Law Smith, M.J.; Stirrat, M.R.; Perrett, D.I. Facial skin coloration affects perceived health of human faces. *Int. J. Primatol.* **2009**, *30*, 845–857. [CrossRef] [PubMed]
222. Chung, S.-J.; Hoerr, S.; Levine, R.; Coleman, G. Processes underlying young women's decisions to eat fruits and vegetables. *J. Hum. Nutr. Diet.* **2006**, *19*, 287–298. [CrossRef] [PubMed]
223. Whitehead, R.D.; Ozakinci, G.; Stephen, I.D.; Perrett, D.I. Appealing to Vanity: Could Potential Appearance Improvement Motivate Fruit and Vegetable Consumption? *Am. J. Public Health* **2012**, *102*, 207–211. [CrossRef] [PubMed]
224. Whitehead, R.D.; Ozakinci, G.; Perrett, D.I. Attractive skin coloration: Harnessing sexual selection to improve diet and health. *Evol. Psychol.* **2012**, *10*, 842–854. [CrossRef] [PubMed]
225. WHO. *Diet, Nutrition, and the Prevention of Chronic Diseases*; Report of the Joint WHO/FAO Expert Consultation WHO Technical Report Series, No. 916 (TRS 916); World Health Organization: Geneva, Switzerland, 2003.
226. Mahler, H.I.M.; Kulik, J.A.; Gerrard, M.; Gibbons, F.X. Long-term effects of appearance-based interventions on sun protection behaviors. *Heal. Psychol.* **2007**, *26*, 350–360. [CrossRef]

nutrients

MDPI

Review

Can Lycopene Impact the Androgen Axis in Prostate Cancer?: A Systematic Review of Cell Culture and Animal Studies

Catherine C. Applegate [1], Joe L. Rowles III [1] and John W. Erdman, Jr. [1,2,*]

[1] Division of Nutritional Sciences, University of Illinois at Urbana-Champaign, Urbana, IL 61801, USA; cca2@illinois.edu (C.C.A.); jrowles2@illinois.edu (J.L.R.)
[2] Department of Food Science and Human Nutrition, University of Illinois at Urbana-Champaign, Urbana, IL 61801, USA
* Correspondence: jwerdman@illinois.edu; Tel.: +1-217-333-2527; Fax: +1-217-333-9368

Received: 6 February 2019; Accepted: 11 March 2019; Published: 15 March 2019

Abstract: First-line therapy for advanced or metastatic prostate cancer (PCa) involves the removal of tumor-promoting androgens by androgen deprivation therapy (ADT), resulting in transient tumor regression. Recurrent disease is attributed to tumor adaptation to survive, despite lower circulating androgen concentrations, making the blockage of downstream androgen signaling a chemotherapeutic goal for PCa. Dietary intake of tomato and its predominant carotenoid, lycopene, reduce the risk for PCa, and preclinical studies have shown promising results that tomato and lycopene can inhibit androgen signaling in normal prostate tissue. The goal of this systematic review was to evaluate whether mechanistic evidence exists to support the hypothesis that tomato or lycopene interact with the androgen axis in PCa. Eighteen studies ($n = 5$ in vivo; $n = 13$ in vitro) were included in the final review. A formal meta-analysis was not feasible due to variability of the data; however, the overall estimated directions of effect for the compared studies were visually represented by albatross plots. All studies demonstrated either null or, more commonly, inhibitory effects of tomato or lycopene treatment on androgen-related outcomes. Strong mechanistic evidence was unable to be ascertained, but tomato and lycopene treatment appears to down-regulate androgen metabolism and signaling in PCa.

Keywords: prostate cancer; tomato; lycopene; androgen; cell culture; animal

---

1. Introduction

Despite average annual declines in incidence, prostate cancer (PCa) remains the most commonly diagnosed male cancer in the United States, with an estimated three million men currently living with PCa [1]. It is well-understood that primary PCa growth is strongly dependent upon the activity of androgens within the prostate gland, as evidenced by the observed rise of androgen-regulated prostate-specific antigen (PSA) in the serum of men diagnosed with PCa [2]. First-line therapy for advanced or metastatic disease involves androgen suppression through androgen deprivation therapy (ADT) [3]. ADT results in castrate levels of androgens in the bloodstream and subsequent success of initial tumor regression; however, the return of castration-resistant disease inevitably occurs within a few years after ADT and is thought to be the result of adaptive or persistent intratumoral androgen production, metabolism, and signaling [4]. While the mechanisms of androgen metabolism and signaling leading to prostate carcinogenesis and continued tumor growth are still under investigation, the blockade of androgen signaling, in addition to ADT, has been identified as a target goal for chemotherapy.

Abundant epidemiological evidence indicates that tomato consumption and blood levels of the predominant carotenoid found in tomatoes, lycopene, are inversely associated with PCa risk [5–9].

Additional evidence suggests that tomato and lycopene interact with the androgen axis to reduce blood levels of PSA [10,11], as well as reduce the risk of advanced stage, lethal PCa [8,12,13]. Animal and cell culture studies reveal an interaction between lycopene and androgen status and signaling, further indicating a potential protective role of tomato and lycopene intake for PCa patients.

Androgens, such as testosterone and dihydrotestosterone (DHT), are male sex hormones required for prostate differentiation and the maintenance of prostate structure and function throughout the lifespan [14]. Once delivered to the prostate from the testes via the bloodstream, androgens can either be converted to more active forms or metabolized to less active forms by a variety of hydroxysteroid dehydrogenase (HSD) enzymes. For example, testosterone is a potent ligand for the androgen receptor (AR), but is converted to DHT by two isoforms of 5-α-reductase (SRD5A1 and SRD5A2). DHT has a higher affinity for binding to the AR, which leads to AR nuclear translocation, DNA binding, and the transcription of genes related to growth and survival pathways [15].

Our laboratory has previously shown that castrated male F344 rats accumulated two times more lycopene in the liver than intact rats or castrated rats treated with testosterone repletion [16,17]. We have also shown that short-term tomato or tomato carotenoid feeding led to significant decreases of serum testosterone in rats, with carotenoid intake interacting with castration to further decrease serum testosterone [18]. In addition, both castration and carotenoid intake resulted in the regulation of prostatic androgen-related enzyme gene expression. Expression of HSD17β4 was significantly higher after castration, as well as between castrated rats fed tomato or lycopene diets when compared to castrated or intact, control-fed rats. HSD17β4 activity results in the metabolism of more potent androgens to less potent forms, and HSD17β4 silencing has been shown to increase AR nuclear localization and PSA expression [19]. This upregulation of HSD17β4 may indicate a switch from androgen signaling propagation to androgen deactivation. Modulation of androgen-related enzyme gene expression by tomato and lycopene is supported by the observed upregulation of HSD17β4 and downregulation of SRD5A2 in the prostate of Copenhagen rats supplemented with lycopene [20]. Supplementation with lycopene also decreased the prostatic expression of steroid target genes prostatic steroid binding chains C1 and C3, cystatin-related protein 2, and seminal vesicle secretion protein IV. In addition, lycopene supplementation of human primary prostatic epithelial cells (PrE) reduced the expression of AR chaperone heat shock protein 90 (HSP90) and protein DJ1, a positive regulator of AR-dependent transcription [21].

These results support the hypothesis that lycopene metabolism is affected by androgen status and that tomato and lycopene interact with the androgen axis in the normal prostate by modulating the expression of genes involved in androgen metabolism and signaling. As such, we hypothesized that tomato and lycopene could similarly interact with the androgen axis during PCa. Because interference with androgen signaling is a critical chemotherapeutic goal for PCa treatment, the primary objective of this review was to systematically evaluate whether mechanistic evidence exists to support a role for tomato or lycopene interaction with the androgen axis during PCa. To accomplish this objective, we included animal and cell culture studies exploring this relationship and evaluated the overall strength and comparability of the evidence. This study is novel in that there is a general dearth of systematic reviews of animal and cell culture studies, and, to our knowledge, no studies currently exist to mechanistically evaluate the relationship between tomato or lycopene and the androgen axis during PCa. While strong mechanistic evidence was unable to be ascertained, the results showed that tomato and lycopene appeared to down-regulate androgen metabolism and signaling in PCa tissues.

2. Materials and Methods

2.1. Study Selection Criteria

Currently, no validated guidelines or tools exist for conducting systematic reviews or for evaluating the validity and quality of mechanistic studies. As part of an effort to utilize a cohesive and standardized set of guidelines for systematically reviewing evidence from cell culture and animal studies, this systematic review was conducted in accordance with the framework outlined by the

World Cancer Research Fund (WCRF) International/University of Bristol (UoB) [22]. In addition, care was taken to follow PRISMA reporting guidelines as closely as possible [23].

Cell culture and animal studies that met the following criteria were included in this systematic review: (a) evaluated the relationship between tomatoes and/or their primary bioactive, lycopene, and androgen metabolism, androgen signaling, or androgen-mediated outcomes in PCa through cell culture studies evaluating direct androgen endpoints in PCa cell lines, cell culture studies comparing androgen-sensitive versus androgen-insensitive PCa cell lines, or animal studies evaluating direct androgen endpoints in experimental animal models of PCa (carcinogen-induced, xenograft, transplantable, or transgenic); (b) methodology was documented in replicable detail; (c) used whole tomato, tomato extract, or lycopene as a single supplement; (d) were written in English; and (e) were peer-reviewed original research articles or theses.

2.2. Literature Search

We conducted a comprehensive literature search of PubMed, Web of Science, and the Cochrane Library using a combination of the following keywords and their variants: tomato, lycopene, testosterone, androgen, dihydrotestosterone, DHT, prostate specific antigen, PSA, prostate cancer, and prostate neoplasm (up to 23 January 2019). Titles and abstracts of articles that were identified by the search results were screened against the study selection criteria. Full texts of potentially relevant articles identified by abstract screening were further reviewed for study inclusion or exclusion. To minimize the risk of excluding potentially relevant studies, we also conducted a reference list search (i.e., backward search) and cited reference search (i.e., forward search) from studies meeting the study selection criteria. Studies identified through this process were further screened and evaluated using the afore-mentioned criteria. We repeated reference searches on all newly identified studies until no further relevant studies were found. Two authors (CCA and JLR3) individually determined the inclusion/exclusion of all studies retrieved in full text, and discrepancies were resolved through discussion.

2.3. Data Extraction and Quality Assessment

Data extraction was performed according to study type (animal or cell culture) using the recommendations set forth by the WCRF/UoB framework as a guide [22]. The following information was extracted from each animal study: animal model, housing conditions and dietary information for experimental and control groups, experimental design (investigator blinding, randomization or grouping of animals, etc.), duration of follow-up, androgen-related outcomes analyzed, results of androgen-related outcomes, sample size, and *p*-values. The following information was extracted from each cell culture study: names of cell lines, whether cell lines were established patient-derived tumor cell lines or freshly isolated primary cells, whether cell lines were authenticated, culture conditions, treatment regime (dose and length of treatment), details of laboratory procedures, outcomes analyzed, results, sample size, and *p*-values.

There is a lack of validated quality assessment (QA) tools to evaluate the risk of bias associated with animal and cell culture studies. QA of animal studies was performed using the SYstematic Review Centre for Laboratory animal Experimentation (SYRCLE) risk of bias tool [24] adapted from the established and validated Cochrane tool [25] for human study risk of bias assessment. Risk of bias was determined to be "high," "low," or "unclear." Total scores were not evaluated using the SYRCLE tool to avoid inappropriate weighting of each category. QA of cell culture studies was performed using the criteria recommended by the WCRF/UoB framework (score range 0–6; a score of 0 was assigned for each parameter not fulfilled or not reported) [22]. Based on the score, studies were rated as low (0–2), moderate (3–4), or high (5–6) quality. QA scores were utilized to provide a measure of the strength of the evidence and to determine if a risk of bias was present for each study, but were not used to determine the inclusion of studies. QA scores were considered when making conclusions about whether the included studies supported the biological plausibility of the causal pathway being investigated.

2.4. Albatross Plot Generation

The extreme degree of variation between the methodologies and outcome measures of animal and cell culture studies prevented statistical analysis via a meta-analysis. Due to this variation, outcome measures were grouped according to indirect or direct androgen-related outcomes. In lieu of a meta-analysis, albatross plots for each outcome measure were generated to both graphically represent the data and create an effect estimate for each outcome category measured. An albatross plot, as described by Harrison et al. [26], scatters the *p*-values of each study according to their sample size and according to the observed direction of the effect (positive or negative). In the absence of exact *p*-values provided, the most conservative *p*-value was assigned to that outcome (e.g., if given $p < 0.05$, set $p = 0.05$; if no *p*-value given for a non-significant association [e.g., $p > 0.05$], set $p = 1$). If studies included multiple *p*-values for different outcomes, all *p*-values were included as separate data points. The contour lines extending over the plots represent estimated effect sizes (represented as standardized mean differences [SMD]) to allow for the estimation of the magnitude of treatment effects for individual studies and for their association as a whole. Visual inspection of the albatross plots was used to determine an overall estimated effect for each outcome. Because study outcomes were grouped according to indirect or direct androgen-related outcomes and albatross plots only provide an estimate of the effect of tomato or lycopene for these categories, the results provided by the plots are representative of this estimate, rather than a true statistical analysis. Albatross plots were generated using STATA/IC V14.2 (StataCorp LP, College Station, TX, USA).

3. Results

3.1. Literature Search

In total, 326 studies were identified from the library search engines. After duplicate removal and adding studies identified from reference lists, 192 studies remained for abstract screening. Subsequently, 28 studies were found to contain potentially relevant information to be further evaluated by full text review. Inclusion was determined according to the afore-mentioned inclusion criteria, with a resulting 18 studies included in the final review. Ten studies were excluded because they did not discuss PCa ($n = 2$), did not evaluate direct androgen outcomes or compare androgen-sensitive vs. androgen-insensitive PCa cell lines ($n = 7$), or did not evaluate lycopene as a single supplement ($n = 1$). Of the 18 included studies, five were animal studies [27–31] and 13 were cell culture studies [32–44] (Figure 1).

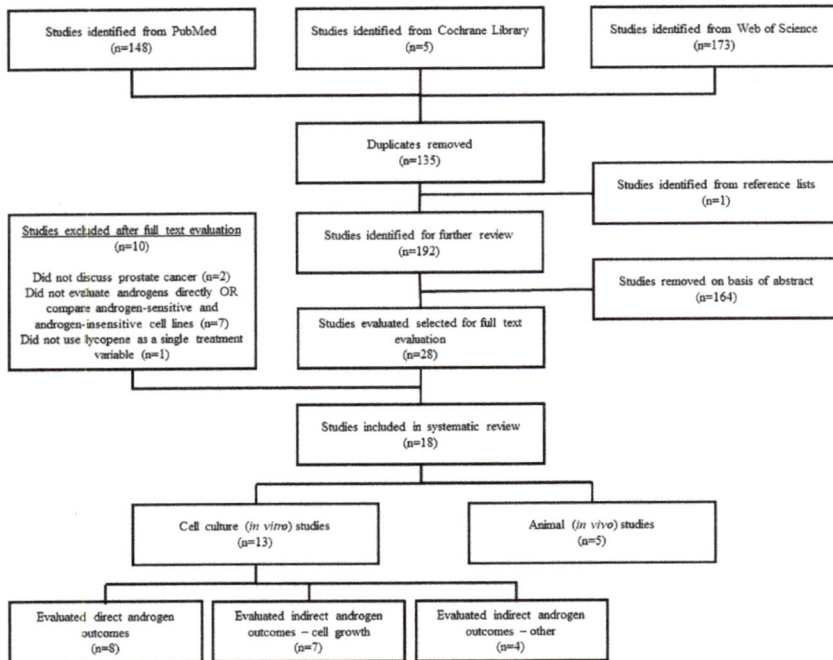

Figure 1. Literature search and study selection flow chart.

3.2. Study Characteristics

Of the five animal studies, three studies utilized rats (transplantable tumor models) [27,29,30] and two studies used mice (n = 1 xenograft model [28] and n = 1 transgenic model [31]). Animal study characteristics and results are summarized in Table 1. Of the 13 cell culture studies, 11 [32–34,36–40,42–44] used patient-derived PCa tumor cell lines, one [35] used primary PCa tumor cell lines, and one [41] used rat-derived PCa tumor cell lines. The cell culture studies were further stratified according to lycopene interaction with the androgen axis, as follows: (a) nine studies [32–34,36,40–44] evaluated androgen-related outcomes pertaining to indirect lycopene interaction with the androgen axis by comparing differential effects of lycopene between androgen-sensitive (AS) and androgen-insensitive (AI) PCa cell lines (indirect androgen outcomes); and (b) eight studies [32–39] evaluated androgen-related outcomes pertaining to direct lycopene interaction with androgen signaling, androgen metabolism, or androgen-regulated gene expression (direct androgen outcomes). Due to a range of indirect androgen outcomes reported, this group was further subdivided into studies that measured (i) growth [32,33,36,40–43] or (ii) other [32,34,40,43,44]. Cell culture study characteristics and results are summarized in Table 2.

QA of animal studies was carried out using the SYRCLE tool [24] and for cell culture studies using the criteria recommended by the WCRF/UoB framework [22]. The SYRCLE tool resulted in a largely unclear risk of bias for all animal studies (Appendix A, Table A1). The recommended criteria used to evaluate the quality of cell culture studies were vague, resulting in 10 cell culture studies considered to be high quality (5–6) [33–37,40–44] and three studies considered to be moderate quality (3–4) [32,38,39] (Appendix A, Table A2).

Table 1. Characteristics of included animal studies.

Author, Year	Animal Model	Baseline Diet(s)	Dietary Tomato and/or Lycopene Content *	Length of Intervention	Primary Findings
Canene-Adams, 2009 [27]	Dunning R3327-H transplantable tumors (Copenhagen rats)	AIN-93G-based diets fed; *ad libitum*	10% TP (providing 13 nmol lycopene/g diet and resulting in 511 nmol/g serum lycopene), 23 nmol/g diet supplemental lycopene beadlet (252 nmol/g serum lycopene), or 224 nmol/g diet supplemental lycopene beadlet (884 nmol/g serum lycopene)	18 weeks after tumor transplantation	No differences in serum testosterone or DHT between rats fed tomato, lycopene, or control diets
Limpens, 2006 [28]	Xenograft using PC-346C cells (athymic mice)	821077 CRM(P) low (but adequate) vitamin E rodent diet; *ad libitum*	5 or 50 mg/kg BW lycopene (from LycoVit) oral gavage = 0.5–1.5 mg lyco/day	42 days after tumor inoculation	No differences in plasma PSA between mice given control or lycopene gavages (PSA levels were proportional to tumor size regardless of intervention)
Lindshield, 2010 [29]	Dunning R3327-H transplantable tumors (Copenhagen rats)	AIN-93G-based diets; *ad libitum*	250 mg/kg diet supplemental lycopene beadlet = 5 mg lyco/day	18 weeks after tumor transplantation	No differences in serum testosterone or DHT between rats fed lycopene or control diets
Siler, 2004 [30]	MayLyLu Dunning transplantable tumors (Copenhagen rats)	Kliba #2019 with added coconut fat (6%), <5 ppm vitamin E, reduced (but adequate) vitamin A, and devoid of phytosterols; did not indicate if *ad libitum*	200 ppm lycopene (1.02 µM plasma lycopene)	4 weeks on diet prior to tumor transplantation, then 18 additional days	Lycopene supplementation reduced tumor expression levels of SRD5A1 and androgen-target genes (*cystatin related proteins 1 and 2; prostatic spermine binding protein; prostatic steroid-binding protein C1, C2, and C3; probasin*) (only fold reductions reported—no other statistical values reported)
Wan, 2014 [31]	Transgenic (TRAMP mice)	AIN-93G-based diets; did not indicate if *ad libitum*	10% TP (providing 384 mg lycopene/kg diet) or 462 mg/kg diet supplemental lycopene beadlet (0.36 µM plasma lycopene) = 3–4 mg lyco/day	4 weeks diet prior to surgery, then 12 additional days	Reduced prostatic expression of genes related to androgen metabolism by tomato feeding (SRD5A2 ($p = 0.04$), Pxn ($p = 0.04$), and Srebf1 ($p = 0.05$)) and lycopene supplementation (SRD5A1, $p = 0.03$)

* All treatments can lead to blood levels of lycopene within a physiological range (~1 µM). Abbreviations: TRAMP (transgenic adenocarcinoma of the mouse prostate); TP (tomato powder); BW (body weight); DHT (dihydrotestosterone); PSA (prostate-specific antigen); SRD5A1 and 2 (5 α-reductase type 1 and 2); Pxn (paxillin); Srebf1 (sterol regulatory element binding transcription factor 1).

Nutrients **2019**, *11*, 633

Table 2. Characteristics of included cell culture studies.

Author, Year	Cell Line(s)	Culture Conditions *†	Lycopene Dose(s) δ	Direct Androgen Outcomes: Primary Findings	Indirect Androgen Outcomes: Primary Findings	
					Growth	Other
Fu, 2014 [44]	LNCaP, PC-3	RPMI1640, 10% FBS	0, 5, 10, 20, 40 µM			10 µM lycopene inhibited GSTP1 methylation ($p < 0.05$), increased GSTP1 gene expression ($p < 0.05$), and reduced DNMT3A gene expression ($p < 0.01$) in PC-3 but not LNCaP cells
Gong, 2016 [40]	LNCaP, C4-2, PC-3, DU145	RPMI1640, 10% FBS	1 µM		1 µM lycopene inhibited growth of LNCaP ($p < 0.05$) but not in C4-2, PC-3, or DU145 cells	1 µM lycopene induced BCO2 gene expression in LNCaP ($p < 0.05$) but not DU145 cells
Gunasekera, 2007 [41]	AT3, DTE	RD (50% RPMI1640 + 50% DMEM), 2% FBS	0.02, 0.2, 5, 10, 20 µM		0.2 µM lycopene inhibited growth of AT3 ($p < 0.0001$) but not DTE cells	
Ivanov, 2007 [32]	LNCaP, PC-3	RPMI1640 or DMEM, 10% FBS	0.01–10 µM (cell proliferation) 0.2, 0.4 µM (protein expression) 0–100 µM (androgen responsiveness)	Lycopene did not inhibit reporter activity of ARE-Luc transfected LNCaP cells at any concentration (no statistical values reported)	0.2–0.8 µM lycopene inhibited growth of LNCaP and PC-3 cells ($p < 0.05$)	0.2–0.8 µM lycopene inhibited Akt phosphorylation, cyclins D1 and E, and CDK2 in LNCaP and PC-3 cells (no statistical values reported)
Linnewiel-Hermoni, 2015 [33]	LNCaP, PC-3, DU145	RPMI1640 or DMEM, 10% FCS, 10^{-9} DHT (for growth, stripped of steroid hormones prior to treatment)	1–5 µM (cell proliferation) 8 µM (ARE-Luc) 2.5 µM (PSA)	8 µM lycopene inhibited DHT-induced reporter activity of ARE-Luc transfected LNCaP cells ($p < 0.01$); non-significant decrease of DHT-induced PSA secretion by LNCaP cells treated with 2.5 µM lycopene	1–5 µM lycopene inhibited DHT-induced growth of LNCaP cells ($p < 0.01$)	
Liu, 2006 [34]	LNCaP, PC-3, DU145	RPMI1640 or Ham's F12K or EMEM, 10% FBS	1–1.48 µM	1.48 µM lycopene did not directly bind to the AR (no statistical values reported)		Uptake is highest in LNCaP ($p < 0.001$) with 1.48 µM lycopene compared to PC-3 or DU145 cells
Liu, 2008 [35]	6S, 6S + NPE	DMEM, 5% FBS	0.3, 1 µM	0.3, 1 µM lycopene increased CM-mediated cell death and reduced IGF-1 gene expression of 6S + NPE cells in the presence of DHT ($p < 0.01$); lycopene reduced DHT-induced total ($p < 0.05$) and nuclear ($p < 0.01$) AR protein expression in 6S cells		

Table 2. *Cont.*

Author, Year	Cell Line(s)	Culture Conditions *†	Lycopene Dose(s) δ	Direct Androgen Outcomes: Primary Findings	Indirect Androgen Outcomes: Primary Findings	
					Growth	Other
Peternac, 2008 [36]	LNCaP, C4-2	T medium (80% DMEM + 20% Ham's F12K) + 10% FCS	0.04, 0.4, 4 μg/mL (equivalent to 0.075, 0.75, and 7.5 μM)	Lycopene did not reduce PSA gene or protein expression in LNCaP or C4-2 cells at any concentration	Lycopene inhibited growth of LNCaP (0.04, 0.4, 4 μg/mL) and C4-2 (0.4, 4 μg/mL) cells ($p < 0.05$)	
Rafi, 2013 [37]	PC-3	RPMI1640, 10% FBS	25 μM	25 μM lycopene led to fold reductions of kallikrein peptidase family proteins gene expression in PC-3 cells (only fold reductions reported—no other statistical values reported)		
Richards, 2003 [38]	LNCaP	Did not report cell culture conditions	1, 10 μM	1, 10 μM lycopene appeared to reduce PSA in LNCaP cells (no statistical values reported)		
Tang, 2005 [42]	LNCaP, PC-3, DU145	DMEM + Ham's F12K, 10% FBS	Up to 50 μM		10–50 μM lycopene more potently inhibited growth of PC-3 and DU145 cells ($p < 0.01$) compared to LNCaP cells	
Tang, 2011 [43]	LNCaP, LAPC-4, PC-3, 22Rv1, DU145	RPMI1640, 10% FBS	1 μM		1 μM lycopene appeared to reduce growth of all cell lines (no statistical values reported)	1 μM lycopene more potently reduced Akt phosphorylation in DU145 (by 60%) than LNCaP (by 20%) cells (no statistical values reported)
Zhang, 2010 [39]	LNCaP	RPMI1640, no other conditions reported	0.5, 5, 10, 15 μM	0.5–15 μM lycopene appeared to reduce reporter activity and ARE protein expression in ARE-Luc transfected LNCaP cells (no statistical values reported)		

* All studies reported standard incubator conditions (5% CO_2, 37 °C) unless otherwise stated. † Media does not contain added androgens unless otherwise stated; FBS and FCS supply castrate levels of androgens. δ Compare to reference of ~1 μM in human plasma. Abbreviations: GSTP1 (glutathione S-transferase Pi 1); DNMT3A (DNA methyltransferase 3A); BCO2 (β-carotene 9′,10′-oxygenase 2); ARE (androgen receptor element); Luc (luciferase); CDK2 (cyclin-dependent kinase 2); DHT (dihydrotestosterone); PSA (prostate-specific antigen); AR (androgen receptor); CM (camptothecin); IGF-I (insulin-like growth factor-I).

3.3. Animal Studies

Animal studies did not compare differences between AS or AI PCa. As such, only animal studies measuring direct androgen outcomes were included in this review. Of the five animal studies included, one study [28] measured the effect of lycopene supplementation on plasma PSA levels, and two studies [27,29] measured whether lycopene or tomato feeding impacted serum testosterone or DHT. Limpens et al. [28] tested the effects of low- and high-dose lycopene supplementation on PSA levels in a xenograft PCa model. After 42 days of daily oral gavage, plasma PSA values did not differ between dietary treatments, suggesting that PSA levels were proportional to tumor size, regardless of dietary intervention. Canene-Adams et al. [27] analyzed the effect of a 10% tomato powder (TP) diet and two different supplemental doses of lycopene (similar dose to the lycopene content of the TP diet and a dose 10-fold higher) on serum testosterone and DHT levels in the Dunning R3327-H transplantable prostate adenocarcinoma model after 18 weeks of tumor growth. None of the interventions had any effect on serum testosterone or DHT levels. Using the same model, Lindshield et al. [29] similarly saw no effect of lycopene supplementation on serum testosterone or DHT levels.

Siler et al. [30] observed reductions in the expression of genes involved in androgen metabolism (SRD5A1) and signaling (cystatin related proteins 1 and 2; prostatic spermine binding protein; prostatic steroid-binding protein C1, C2, and C3; and probasin) in MayLyLu Dunning transplantable tumors with dietary lycopene intake. Only fold-changes in gene expression with no statistical measurements were reported; however, consistent with these results, Wan et al. [31] confirmed that tomato feeding and lycopene supplementation similarly impacted androgen-related gene expression in the prostate of the transgenic mouse model (TRAMP) at early stages of prostate carcinogenesis. Tomato and lycopene diets both decreased the expression of genes related to androgen metabolism (SRD5A2 by tomato and SRD5A1 by lycopene), while the tomato diet reduced the expression of androgen co-regulators paxillin (pxn) and sterol regulatory element binding transcription factor 1 (srebf1).

An albatross plot (Figure 2) was generated for these animal studies to integrate the data, and visual inspection of the plot provided an estimated standardized effect for tomato or lycopene intake on androgen-related outcomes. The effects given were not intended to be precise, as they only provide estimates of the treatment magnitude of effect. The overall SMD of −0.4 (range: 0 to −1.25) represented a reduction in androgen-related outcome measures by exposure to tomato or lycopene. Three studies [27–29] showed no effects (assigned $p = 1$) of tomato or lycopene on androgen-related outcomes and as such, cluster at the center (null) line of the plot. The remaining two studies [30,31] showed inverse associations, with SMDs of between −0.5 and −1 and between −1 and −1.5, indicating a reduction in androgen-related outcome measures by exposure to tomato or lycopene. It is important to note that no studies reported an increase in androgen levels or androgen-regulated gene expression with tomato or lycopene exposure, suggesting that neither tomato nor lycopene propagate androgen production or signaling.

3.4. Cell Culture Studies

We present the results of cell culture studies here as reported by the original studies with the caveat that care must be taken when interpreting results involving cell culture treatment with lycopene, as antioxidants such as lycopene are extremely labile and are readily oxidized in cell culture [45]. For this reason, lycopene source, purity, storage, delivery vehicle, air and light exposure, and length of time in culture can substantially affect the initial lycopene integrity. These factors vary by study, creating the immediate limitation that the observed results could be due in part to the oxidation products of lycopene, rather than solely to the parent compound.

In general, cell culture studies supported the wide breadth of existing in vivo evidence that tomato and lycopene inhibit PCa tumor growth. Cell culture studies also provide some support for the limited in vivo evidence that tomato and lycopene down-regulate the expression of genes related to androgen signaling and metabolism. The included studies provided mixed evidence to indicate that lycopene interacts with AS and AI PCa cell lines in a differential manner (indirect androgen outcomes) and that lycopene directly interacts with the androgen axis (direct androgen outcomes), as detailed below. Study results provided limited insight into the specific mechanisms by which lycopene might interact with the androgen axis.

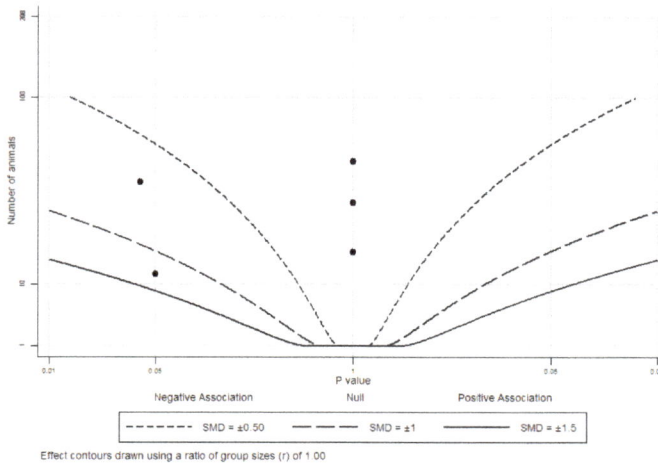

Figure 2. Albatross plot for animal studies. Each point represents a single study, with the effect estimate (represented as a *p*-value), plotted against the total given sample size (*n*) included within each study. Contour lines are standardized mean differences (SMD). *p*-values reported as <0.05 were plotted as 0.05 as a conservative estimate, while non-significant (null) *p*-values were plotted as 1.

3.4.1. Influence of Lycopene on Indirect Androgen Outcomes: Comparison of Androgen-Sensitive vs. Androgen-Insensitive Cell Lines

The limited pool of mechanistic studies makes it difficult to conclude if or how lycopene may directly interact with the androgen axis in PCa to exert growth inhibitory effects. In an effort to glean some insight into this question, it is important to consider how lycopene may impact human PCa cells of differing androgen-responsiveness. To this end, studies that evaluated the differences in outcomes between AS and AI cell lines were included as an indicator of how androgen signaling may influence lycopene activity. Appendix A Table A3 outlines the differences between each cell line in the included cell culture studies, which may impact the observed results.

(i) Cell Growth

Seven studies [32,33,36,40–43] evaluated the relationship between PCa cell growth and lycopene treatment. Despite large variation in study methodologies and lycopene doses, all seven studies demonstrated lycopene inhibition of PCa cell growth. All six studies [32,33,36,40,42,43] utilizing human PCa cells lines reported that lycopene treatment inhibited the growth of AS cell lines (LNCaP, LAPC-4), with three studies [33,36,40] reporting greater growth inhibition of an AS cell line (LNCaP) when compared to AI cell lines (C4-2, PC-3, DU145), one study [32] reporting growth inhibition regardless of androgen sensitivity, and two studies [42,43] reporting greater growth inhibition in AI cell lines (PC-3, DU145) than what was observed in an AS cell line (LNCaP). The single study [41] comparing an AS rat-derived PCa cell line (DTE) with its AI daughter cell line (AT-3) reported that lycopene inhibited AI cell growth, but not AS cell growth.

Gong et al. [40] showed that cell growth was reduced by lycopene and its metabolite, apo-10'-lycopenal, in AS LNCaP cells, but not in AI DU145, PC-3, or C4-2 cells. Peternac et al. [36] also found that lycopene inhibited cell proliferation in LNCaP cells and, to a slightly lesser extent, C4-2 cells. Comparably, Linnewiel-Hermoni et al. [33] showed that DHT-induced growth of LNCaP cells and serum-induced (castrate levels of androgens) growth of DU145 and PC-3 cells were inhibited by lycopene treatment, with LNCaP cells exhibiting a more profound response.

Ivanov et al. [32] observed that physiological doses of lycopene (0.2–0.8 μM) resulted in the dose-responsive inhibition of cell proliferation in both LNCaP and PC-3 cells. However, lycopene was observed to exert these effects in each cell line at different mitotic phases, with LNCaP cells mainly undergoing G_0/G_1 cell cycle arrest and subsequent apoptosis and PC-3 cells mainly undergoing S and G_2/M cell cycle arrest without an observed increase in the apoptotic index.

Tang et al. [42] (2005) observed that LNCaP cells resisted apoptosis by high-dose (up to 50 μM) lycopene, while PC-3 and DU145 cells were very responsive to apoptosis by high-dose lycopene. Tang et al. [43] (2011) tested the effects of lycopene on the growth inhibition of PCa cell types with varying androgen sensitivity (LAPC-4, LNCaP, 22Rv1, PC-3, DU145) and found DU145 cells to be the most inhibited by a physiological dose of lycopene (1 μM). Interestingly, PC-3 cells were completely unaffected by lycopene treatment, while LNCaP cells were only marginally affected. AI DU145 cells were not compared with AS LNCaP cells, but both Tang et al. [42] (2005) and Tang et al. [43] (2011) observed that DU145 cells mainly underwent cell cycle arrest at G_0/G_1, which contrasts with the S and G_2/M cell cycle arrest in PC-3 cells observed by Ivanov et al. [32]. The degree of androgen-insensitivity of DU145 cells is greater than that of PC-3 cells, making it possible that the level of androgen insensitivity affects at which stage of the cell cycle lycopene may interfere. More likely, these differing results best serve to highlight the inherent variability of cell culture studies, with different treatment variables potentially impacting outcomes.

Finally, Gunasekera et al. [41] showed significant concentration-dependent decreases of the cell proliferation of malignant, rat-derived, AI AT-3 cells, with lycopene concentrations as low as 0.2 μM and up to 10 μM when compared to the control treatment, with no effect on cell proliferation of AS DTE parent tumor cells by any concentration of lycopene.

An albatross plot summarizing the effects of lycopene on cell growth is presented in Figure 3A. The overall SMD of −2 (range: −0.6–<−2) indicates a reduction in cell growth by lycopene exposure. Two studies [33,36] showed reduced growth with smaller effect estimates (SMDs between −0.5 and −1), two studies [40,43] showed SMDs between −1 and −2, and three studies [32,41,42] showed an SMD < −2.

(ii) Other Outcomes

As introduced in the previous section, Ivanov et al. [32] observed that while physiological doses of lycopene inhibited cell proliferation in both AS LNCaP and AI PC-3 cells at different phases of the cell cycle, lycopene treatment led to similar dose-responsive changes in protein expression, regardless of androgen sensitivity. These effects were not differentially mediated by proteins involved in cell growth pathways in AS and AI cells; protein expression of cyclins D1 and E, cyclin-dependent kinase 2 (CDK2), and Akt phosphorylation were similarly inhibited by lycopene treatment in both cell types.

Tang et al. [43] (2011) attributed the growth inhibitory effects of lycopene in AI DU145 cells when compared to other AS and AI cell types to a correlation with higher levels of insulin-like growth factor-I receptor (IGF-IR) present in cells. The reported order of lycopene leading to the most growth inhibition to the least growth inhibition in cell types was: DU145 > LAPC-4 > 22Rv1 > LNCaP > PC-3; the order of highest to lowest IGF-IR expression in each cell type was: DU145 > PC-3 > LNCaP > 22Rv1 > LAPC-4. However, to determine this correlation, the group used calculated IC_{50} values (reported to be ordered as: LAPC-4 > LNCaP > 22Rv1 > PC-3 > DU145), but did not directly measure lycopene uptake into the cells. Liu et al. [34] (2006) showed lycopene uptake to be much higher in LNCaP cells when compared to PC-3 or DU145 cells ($2.5\times$ and $4.5\times$ higher, respectively) at a physiological concentration (1.48 μM). Because lycopene uptake differs between cell lines, using the IC_{50} values calculated from lycopene treatment effects on growth inhibition alone may not be the most appropriate approach. Lycopene was also found to inhibit Akt phosphorylation to a greater extent in DU145 cells (60%) than in LNCaP cells (20%). Because LNCaP cells have a phosphatase and tensin homolog (PTEN) mutation that may lead to enhanced Akt phosphorylation by phosphatidylinositol 3-kinase (PI3K) rather than IGF-IR, the greater inhibition of Akt phosphorylation by lycopene in the DU145 cells may be attributed to lycopene inhibition of the IGF-IR pathway.

Gong et al. [40] showed that lycopene uptake, lycopene cleaving enzyme β-carotene 9′,10′-oxygenase (BCO2) gene expression, and lycopene-induced BCO2 expression were greater in LNCaP cells than in DU145 cells; transfection with either a wild-type (active) or mutant (inactive) BCO2 expression vector led to reduced cell growth in each cell line with or without lycopene treatment. Increased BCO2 expression (wild-type or mutant) also inhibited nuclear factor-κB (NF-κB) luciferase reporter activity by hindering NF-κB p65 subunit nuclear translocation and DNA binding in response to lycopene treatment. Constitutive NF-κB signaling is observed in AI cells lines and is associated with enhanced cell proliferation. These results show that both wild-type (active) BCO2 and mutant (inactive) BCO2 inhibit proliferation, indicating that the anti-proliferative effects of BCO2 are independent of its enzymatic (lycopene cleavage) functions and instead rely on some structural element of BCO2. However, the cellular uptake of lycopene and lycopene-induced expression of BCO2 are dependent on androgen sensitivity, suggesting that lycopene may be less effective at reducing AI cell growth.

Fu et al. [44] compared the effects of lycopene on the methylation and expression of an enzyme involved in detoxification reactions and tumor suppression, glutathione *S*-transferase Pi (GSTP1), in PC-3 and LNCaP cell lines. Treatment of PC-3 cells with 10 μM lycopene significantly reduced levels of GSTP1 promoter methylation, significantly increased the mRNA and protein expression of GSTP1, and significantly decreased the protein expression of DNA methyltransferase 3A (DNMT3A) when compared to control cells. Increasing lycopene treatment to 40 μM showed no additional inhibition of DNMT3A protein expression than the inhibition observed at a dose of 10 μM. Alternatively, LNCaP cells treated with lycopene showed no changes in GSTP1 methylation or expression.

The albatross plot presented in Figure 3B shows a reduction in other androgen-mediated outcomes by lycopene exposure with an SMD of <−3, and one study [40] with an SMD ≈ −2.

3.4.2. Influence of Lycopene on Direct Androgen Outcomes

Liu et al. [34] (2006) reported that lycopene uptake was much higher in AS LNCaP cells when compared to AI PC-3 or DU145 cells. To evaluate whether this higher uptake resulted in direct lycopene binding to the AR, LNCaP cells were transfected with a plasmid containing the ligand-binding domain of L701H, the T877A double mutant, cortisone/cortisol-responsive AR with a broader ligand specificity (AR^{ccr}). No direct lycopene binding to the AR occurred, but subcellular fractionation revealed the majority of lycopene to localize within the nuclear membranes and nuclear matrix. Therefore, these data suggest that because lycopene uptake followed the order of AR expression in AS and AI cell lines, lycopene uptake and storage may be mediated by androgen signaling by some mechanism not involving direct binding to the AR.

As discussed in the previous section, Ivanov et al. [32] showed that physiological doses of lycopene decreased cell proliferation in both LNCaP and PC-3 cells at different phases of the cell cycle. These effects were not shown to rely on androgen signaling directly, as the transfection of LNCaP cells with a luciferase-containing androgen response element (ARE) reporter (ARR3-Luc) exhibited no effect of lycopene treatment on androgen-stimulated expression of the gene construct.

Linnewiel-Hermoni et al. [33] showed that DHT-induced growth of LNCaP cells and serum-induced growth of DU145 and PC-3 cells were inhibited by lycopene treatment. To determine whether these effects may be mediated by direct androgen responsiveness, LNCaP cells were transfected with a PSA enhancer luciferase reporter gene construct containing six AREs. Physiological levels of lycopene were not tested, but a high lycopene concentration (8 μM) was found to significantly lower the reporter activity after DHT treatment. Similarly, DHT-induced PSA secretion by LNCaP cells was reported to be ~40% reduced by a more physiological, albeit still high, dose of lycopene (2.5 μM), but statistical analysis did not indicate a significant reduction.

Zhang et al. [39] measured the effect of 0.5–15 μM lycopene on LNCaP cells transfected with a luciferase-containing ARE reporter. The authors reported that lycopene inhibited ARE reporter activity and ARE protein expression in a dose-dependent manner, but failed to include a statistical analysis. Visually, it appears as though 5 μM lycopene acted similarly to 15 μM lycopene, but effects were seen with doses as low as 0.5 μM lycopene. The three studies (Ivanov et al. [32], Linnewiel-Hermoni et al. [33], and Zhang et al. [39]) using luciferase-containing ARE reporters in LNCaP cells show some conflicting evidence. However, it could be insinuated that lycopene effects were directly related to the ARE rather than the AR, mainly at supraphysiological doses of lycopene.

Liu et al. [35] co-cultured primary human prostate cancer stromal (6S) cells with primary normal prostatic epithelial (NPE) cells to determine the effects of DHT on camptothecin (CM)-induced cell death by DNA fragmentation. DHT treatment increased the mRNA expression of IGF-I in 6S cells, which then led to the rescue of CM-induced NPE cell death. Treatment of this co-culture with physiological doses of lycopene (0.3 and 1 μM) inhibited the pro-survival effects of DHT in a dose-responsive manner, potentially due to the administration of lycopene decreasing DHT-induced IGF-I gene expression in 6S cells. Furthermore, lycopene treatment inhibited DHT-induced AR expression in both whole cell lysates and nuclear extracts of 6S cells.

Peternac et al. [36] found that lycopene inhibited cell proliferation in both AS LNCaP and AI C4-2 cell lines. However, these growth effects were not directly related to AR activation, as lycopene had no effect on PSA mRNA or protein levels in either cell line.

Rafi et al. [37] showed that while strictly androgen-regulated genes were largely unaffected in PC-3 cells treated with supraphysiological doses of lycopene (25 μM), some genes within the kallikrein-related peptidase family did show a fold-reduction in expression (klk1, 5, 9, 10, 14). These genes are regulated by members of the steroid hormone family and their expression is typically associated with carcinogenesis, so a slight reduction in gene expression by lycopene treatment may indicate some interference with steroid hormone-regulated gene activation.

Richards et al. [38] measured the effects of low (physiological)- (1 μM) and high-dose (10 μM) lycopene on the PSA secretion of LNCaP cells. While the results showed that PSA protein levels were decreased by about 50% in both treatment groups compared to the control group at all time points, the authors did not report any statistical values or make any comment about the treatments. They also failed to report cell incubation conditions and the mode of delivery of lycopene. Therefore, while these results suggesting that lycopene treatment had direct effects on PSA secretion of LNCaP cells appear promising at first glance, closer inspection revealing the lack of methodological and statistical reporting creates uncertainty when considering the accuracy of the results (refer to Appendix A Table A2 for study quality assessment).

The albatross plot presented in Figure 3C shows the overall effect estimate as an SMD of −2 (range: 0–<−3), indicating a reduction in direct androgen-mediated outcomes by lycopene exposure. Three studies [32,34,36] showed no effects (assigned $p = 1$) of lycopene on androgen-related outcomes and cluster at the center (null) line of the plot. Three studies [33,35,38] showed SMDs between −0.75 and −1.5, two studies [35,37] showed SMDs between −1.5 and −3, and one study [39] showed an SMD < −3. It is important to note that similar to the effect estimates seen in the animal studies, no cell culture studies reported an increase in androgen-regulated gene activity or expression with lycopene exposure, despite a wide range of effect sizes, suggesting that lycopene has either a neutral or muting impact on androgen signaling.

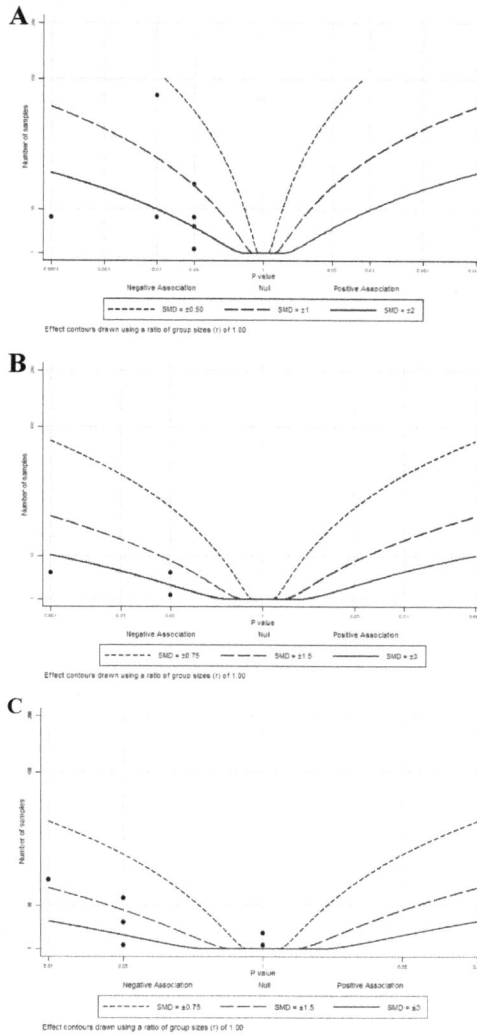

Figure 3. Albatross plots for each outcome of cell culture studies: (**A**) indirect effects (growth); (**B**) indirect effects (other); and (**C**) direct effects. Each point represents a single study, with the effect estimate (represented as a *p*-value), plotted against the total given sample size (*n*) included within each study. Contour lines are standardized mean differences (SMD). *p*-values reported as <0.05 were plotted as 0.05 as a conservative estimate, while non-significant (null) *p*-values were plotted as 1.

4. Discussion

This systematic review sought to determine whether there was any mechanistic evidence to demonstrate that lycopene directly interacts with the androgen axis during PCa. To our knowledge, this is the first systematic review synthesizing evidence to support the biological plausibility of a role for lycopene interaction with the androgen axis in PCa, a disease state in critical need of the identification of potential therapeutic interventions. The choice to conduct a systematic review was made in accordance with a growing need for the systematic evaluation of preclinical research to determine the strength of the evidence associated with a given topic. In addition, this systematic examination of all available evidence was intended to eliminate the risk of bias incurred by preparing a narrative review. In the absence of validated methods for performing systematic reviews of cell culture and animal studies, we used the guidelines recently set forth by the WCRF/UoB [22]. These guidelines serve as a good starting point for conducting mechanistic systematic reviews and meta-analyses; however, the suggested QA tool provides limited criteria for the critical evaluation of cell culture studies, leading to an inappropriate distribution of high range scores, highlighting a need for validated tools to assess the quality of these studies.

Unfortunately, performing a meta-analysis to statistically analyze the results was not feasible due to the inherent variability in the design and outcomes measured for the included studies. As such, data were grouped according to categories of androgen-related outcomes and graphically presented using albatross plots, a novel method described by Harrison et al. [26] by which to provide estimated SMDs of effects of the available data in the absence of sufficient homogenous data to perform a true meta-analysis. Because insufficient data were available to perform this statistical analysis and because reported outcomes were grouped according to their relationship with the androgen axis, it is important to note that these plots are not intended to provide an exact statistical evaluation of lycopene treatment on the androgen axis. Considering these limitations, the effect estimates shown are largely shifted to the left of the plots, suggesting that tomato or lycopene treatment decreased the effects of androgen signaling or metabolism for almost all outcomes measured. Therefore, while we were not able to determine mechanistically how lycopene interacts with the androgen axis during PCa, we have presented some proposed pathways by which lycopene exerts its anti-androgenic effects (Figure 4). These effects are complex and differ in lycopene uptake and growth pathway activation, depending on model used (animal or cell type), cell metabolism, and androgen signaling and metabolism.

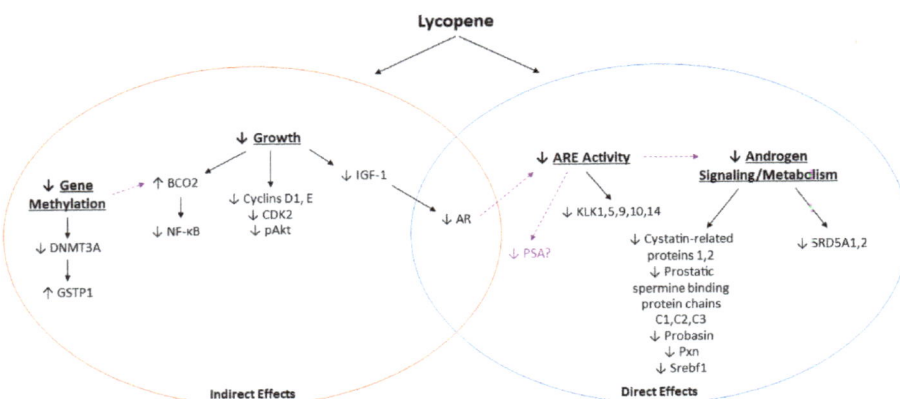

Figure 4. Summary of potential mechanisms by which lycopene may interact with the androgen axis in PCa. Solid lines represent outcomes reported by the reviewed studies, and dashed lines represent potential connections.

Based on the results of the limited animal studies available for inclusion in this systematic review, there was no evidence that tomato or lycopene intake impact circulating PSA, testosterone, or DHT during PCa. Previous research by our laboratory has shown reductions in serum testosterone by tomato and lycopene intake in a non-PCa rat model [18]. However, while previously thought to be a biomarker for advancing or aggressive PCa, discordance between circulating and intraprostatic levels of androgens, as well as observed adaptive changes in intratumoral steroidogenesis and metabolism, have made serum androgens unreliable markers of disease status [46–48]. Instead, changes in intratumoral androgen signaling may be a better indicator of androgen activity than serum testosterone or DHT levels. In accordance with this, there was evidence that tomato and lycopene intake decreased the expression of genes involved in androgen signaling (cystatin-related proteins 1 and 2; prostatic spermine binding protein; prostatic steroid-binding protein C1, C2, and C3; probasin; pxn; and srebf1) and metabolism (SRD5A1 and SRD5A2).

Results from cell culture studies were varied and provided weak mechanistic evidence to demonstrate that lycopene interacts with the androgen axis during PCa. To allow for all possible evidence to be considered, studies comparing differences in cell growth, gene, or protein expression in AS versus AI PCa cell lines were included. Overall, the results showed that cell growth was inhibited by lycopene, regardless of androgen sensitivity. However, the extent to which physiological levels of lycopene inhibit cell growth may be greater in AS cell lines than AI cell lines. In general, AS cell lines more readily accumulate lycopene than AI cell lines [34,40], as shown in Appendix A Table A3. This difference in lycopene accumulation may contribute to the differences seen in gene expression between the two cell types. AI cells express lower levels of the lycopene metabolizing enzyme, BCO2, which assists in inhibiting tumor promoting NF-κB signaling, independent of its lycopene metabolizing function. However, BCO2 is also inducible by lycopene [40], so the lower lycopene uptake by AI cell types may result in attenuated growth inhibitory effects by BCO2. IGF-1 expression was also shown to be higher in AI cells when compared to AS cells, with lycopene treatment resulting in the inhibition of IGF-1 expression and a resultant decrease in total and nuclear AR expression. Proteomic comparisons between AS and AI cells showed differences in the expression of proteins involved in metabolism, with AI cells exhibiting enhanced glycolysis [49]. Comparisons also revealed AI cells to have decreased poly[ADP-ribose] polymerase 1 (PARP-1) expression when compared to their AS counterparts; PARP-1 plays a role in promoting AR transcriptional activity, with a reduction of PARP-1 correlating with a reduced dependence on androgen signaling in AI cells. Combined, these results suggest that androgen signaling and cellular metabolism interact in PCa cells, with changes in lycopene uptake and metabolism, thereby having the potential to influence androgen signaling.

In addition to its growth inhibitory effects, DNA methylation was inhibited by lycopene in AI cells but not AS cells. DNA hypermethylation is an epigenetic modification that occurs more frequently as PCa progresses [50]. One such example of DNA hypermethylation occurs with GSTP1, a detoxifying enzyme with tumor suppressive activity. GSTP1 promoter hypermethylation and accompanying gene silencing have been consistently detected in more than 90% of PCa cases, with more advanced cases exhibiting higher levels of GSTP1 promoter methylation [51]. Lycopene treatment resulted in reduced GSTP1 methylation and associated suppression of DNA methylating enzyme DNMT3A expression in DU145 cells, but not LNCaP cells. Interestingly, Gong et al. [40] showed that the inhibition of methyltransferase activity resulted in a robust increase of BCO2 expression in all cell lines. This suggests that lycopene may have multiple beneficial effects in both cell types, regardless of androgen sensitivity.

When considering all available studies evaluating lycopene effects on PSA secretion, there is weak evidence, at best, to indicate that lycopene has a direct effect on PSA secretion. A meta-analysis to determine if clinical evidence shows an association between tomato or lycopene intake and PSA levels in humans is currently in progress by our laboratory. Despite a lack of evidence showing a direct impact of lycopene treatment on PSA levels, some evidence exists to suggest that while lycopene does not directly associate with the AR, it may reduce ARE activity. However, this may be at levels higher than the normal physiologic intake of lycopene. Reduction of ARE activity may be associated with lycopene accumulation in the nucleus, which could point to lycopene interference with AR co-regulators and subsequent DNA binding and expression. In addition, both in vivo and in vitro evidence has shown that lycopene influences the expression of genes associated with androgen signaling and metabolism. Combined, these data suggest that because lycopene uptake follows the order of AR expression in AS and AI cell lines, lycopene uptake may be mediated by androgen signaling through mechanisms independent of direct AR binding.

This review is novel in topic and sought to incorporate the use of standardized guidelines recommended for conducting a preclinical systematic review with the addition of quantitative estimates of effects using albatross plots. However, this review presents some limitations. First, few studies mechanistically evaluating the effects of lycopene or tomato intervention on endpoints directly related to androgen status, signaling, or metabolism exist. While cell culture studies measuring endpoints comparing AS and AI cell lines were included for a consideration of the potential differences of lycopene treatment with varying androgen sensitivity, these studies do not directly link lycopene to androgen signaling and must be considered separately from the studies evaluating the effects of lycopene treatment on direct androgen endpoints.

Second, the variability of study design and outcome measurement creates difficultly in comparing animal and cell culture studies. Models used must be carefully evaluated and follow the general progression of human disease. While extremely useful for studying mechanistic outcomes, cell culture models are generally not considered to mimic human disease progression. Cancer, particularly PCa, is considered to be a heterogenous disease. Multiple cell types expressing different mutations are available and considered within this review (e.g., LNCaP, PC-3, DU145), but these single cell lines only represent a small subset of individual tumor phenotypes and do not account for cell interaction with the surrounding tumor microenvironment. In addition, differences between cell culture medium, lycopene dose and delivery method, length of treatment, and laboratory tests conducted may significantly contribute to differences in the overall outcomes. Animal models are a step-up in study design, but only models mimicking the human progression of disease (i.e., tumors originate from the animal) are considered to be high quality models. To this end, only one of the five included studies used a transgenic animal model of PCa (TRAMP), while all others used some form of transplantable tumor model. Furthermore, differences in laboratory tests conducted and outcomes reported for both cell culture and animal studies necessitated the grouping of outcomes by indirect or direct interaction with the androgen axis. As a result, these groupings only served to create estimations of the effects of lycopene treatment on each outcome category, rather than a true statistical evaluation of the effects.

Finally, as previously discussed, lycopene is a potent antioxidant and, as such, is unstable when isolated and exposed to light and air, making it a difficult compound to work with in vitro [45]. Lycopene source, purity, storage, delivery vehicle, air and light exposure, and length of time in culture are all factors that vary by study. These factors can result in oxidation of the parent compound, making study outcomes the result of these oxidized lycopenoids. Studies generally addressed this issue by confirming cellular lycopene uptake, lycopene stability after media culture, or simply by refreshing cell media and treatment daily. However, the labile nature of lycopene and the inherent variability in the study methodology pose major limitations when considering results from cell culture studies.

To address these limitations, future studies should be designed specifically to probe the hypothesis that physiologically relevant doses of lycopene can impact the androgen axis by measuring changes related to androgen activity, signaling, or metabolism. Studies evaluating the effect of lycopene on androgen concentrations, androgen metabolizing enzyme activity, and androgen-regulated gene activity (such as PSA) in animal models, as well as cell lines representative of varying stages of PCa, would result in valuable additions to strengthen the current literature. In the interest of enabling systematic reviews of preclinical research to identify potential mechanisms whereby lycopene can modulate androgen status, future studies should also take care to report a detailed and comprehensive methodology and experimental results.

The reviewed evidence shows that lycopene potentially reduces androgen metabolism and signaling in PCa, thereby reducing the effects of one of the main factors driving PCa growth and progression. While the current pool of research is promising, there is a general lack of preclinical and clinical research relating to the effects or mechanisms of lycopene or other compounds present in tomatoes on androgens, their metabolites, and their downstream effectors at different stages of PCa. Androgen signaling is an important chemotherapeutic target for advanced PCa because intratumoral signaling persists, despite the removal of androgens through ADT. Regular and feasible dietary intake of tomatoes has been shown to reduce the risk for PCa. The mechanisms behind this risk reduction are unclear; however, a reduction of androgen signaling would suggest an important role for tomato and tomato carotenoids at all stages of PCa growth and progression. Therefore, it is essential to identify how simple and widely accepted dietary interventions such as increased tomato intake may act as adjuvant therapies to attenuate the adverse effects of persistent androgen signaling on tumor growth and, as a result, on patient outcome. Future research is needed to fill the large gap that still exists in the literature pertaining to the mechanisms by which tomatoes or lycopene may work to modulate androgen status and androgen signaling during various stages of PCa development and progression.

Author Contributions: Conceptualization, methodology design, data curation, and validation, C.C.A., J.L.R., and J.W.E.; investigation and formal analysis, C.C.A.; writing—original draft preparation, C.C.A.; writing—review and editing, C.C.A, J.L.R., and J.W.E. All authors approved the manuscript.

Funding: Research reported in this publication was supported by the National Institute of Biomedical Imaging and Bioengineering of the National Institutes of Health under Award Number T32EB019944. The content is solely the responsibility of the authors and does not necessarily represent the official views of the National Institutes of Health.

Acknowledgments: We would like to thank Sean Harrison for his extensive knowledge, patience, and kind guidance in the use and creation of the albatross plots.

Conflicts of Interest: The authors declare no conflict of interest. The funders had no role in the design of the study; in the collection, analyses, or interpretation of data; in the writing of the manuscript, or in the decision to publish the results.

Appendix A

Table A1. Quality assessment of included animal studies (SYRCLE tool) [24].

Author, Year	Selection Bias			Performance Bias		Detection Bias		Attrition Bias [8]	Reporting Bias [9]	Other Bias [10]
	Sequence Generation [1]	Baseline [2]	Allocation Concealment [3]	Random Housing [4]	Blinding [5]	Random Outcome Assessment [6]	Blinding [7]			
Canene-Adams, 2009 [27]	Unclear	Low	Unclear	Low	Unclear	Unclear	Unclear	Low	Low	Low
Limpens, 2006 [28]	Low	Low	Unclear	Low	Unclear	Unclear	Low	Low	Low	Low
Lindshield, 2010 [29]	Unclear	Low	Unclear	Low	Unclear	Unclear	Unclear	Low	Low	Low
Siler, 2004 [30]	Low	Low	Unclear	Low	Unclear	Unclear	Unclear	Low	Low	Low
Wan, 2014 [31]	Low	Low	Unclear	Low	Unclear	Unclear	Unclear	Low	Low	Low

Studies are given a risk of bias of either "high" (disagreement with parameters), "low" (agreement with parameters), or "unclear" (unclear is parameters were met/unmet) based on the following parameters: [1] random allocation of animals; [2] similarity of baseline characteristics; [3] allocation blinding; [4] random housing distribution within the room; [5] investigator blinding; [6] random animal selection for outcome assessment; [7] outcome assessor blinding; [8] incomplete outcome data addressed; [9] free from selective outcome reporting; [10] free from any other potential sources of bias (e.g., contamination, funding sources, unit of analysis errors). No summary score is given to avoid assigning weights to each category.

Table A2. Quality assessment of included cell culture studies (adapted from WCRF/UoB recommendations) [22].

Author, Year	Source [1]	Experimental Conditions				Selective Reporting [6]	Total
		Culture Conditions [2]	Replicates [3]	Controls [4]	Multiple Cell Lines [5]		
Fu, 2014 [44]	1	1	0	1	1	1	5
Gong, 2016 [40]	1	1	1	1	1	1	6
Gunasekera, 2007 [41]	1	1	1	1	1	1	6
Ivanov, 2007 [32]	1	1	0	1	1	0	4
Linnewiel-Hermoni, 2015 [33]	1	1	1	1	1	1	6
Liu, 2006 [34]	1	1	1	0	1	1	5
Liu, 2008 [35]	1	1	1	1	1	1	6
Peternac, 2008 [36]	1	1	1	1	1	1	6
Rafi, 2013 [37]	1	1	1	1	0	1	5
Richards, 2003 [38]	1	0	1	1	0	0	3
Tang, 2005 [42]	1	1	1	1	1	1	6
Tang, 2011 [43]	1	1	1	1	1	0	5
Zhang, 2010 [39]	1	1	0	1	0	0	3

Studies are given a score of 0 or 1 for each of the following parameters (a score of 0 was given for a lack of criteria fulfillment or failure to report): [1] cell lines are independently validated; [2] comparable culture conditions to other studies; [3] experiment performed in replicate(s); [4] appropriate controls included; [5] more than one cell line used; [6] all experimental results are reported (a score of 0 was given for missing statistical data).

Table A3. Characteristics of prostate cancer cell lines.

Cell Line	Source	Androgen Receptor Expression	Androgen Sensitivity (AS or AI)	Other Characteristics
LNCaP [52]	Human PCa left supraclavicular lymph node metastasis	AR+	AS	Broad AR ligand-specificity; PTEN mutation [52]
C4-2 [53]	Human PCa (subline of LNCaP cells derived from prostate epithelial cells of mouse xenograft cultured from osteosarcoma of mouse xenograft)	AR+ †	AI	High EGFR expression (5–10-fold higher) when compared to parental LNCaP cell line
LAPC-4 [54,55]	Human PCa (subline of LNCaP cells derived from mouse xenograft)	AR+	AS	
22Rv1 [56]	Human PCa (derived from prostate epithelial cells of muse xenograft after castration-induced regression and relapse of the parental, androgen-dependent CWR22 xenograft)	AR+	AS	Express AR splice variants, making this line potentially AI
PC-3 [57]	Human PCa bone metastasis	AR− †	AI	Low SRD5A activities; PTEN deletion [52]
DU145 [57]	Human PCa brain metastasis	AR-	AI	High IGF-I expression [58]
6S [35]	Primary human PCa stromal cells	AR+	AS	
DTE [59]	Rat PCa (Dunning-R3327 tumors)	AR+	AS	
AT-3 [59]	Rat PCa (derived from castration selection of parental DTE Dunning-R3327 tumors)	AR-	AI	Low SRD5A activities

Order of AR expression *   LNCaP > LAPC-4 > ? 22Rv1 > ? C4-2 > PC-3 > DU145
Order of lycopene uptake [34,40] *   LNCaP > PC-3 > DU145 > C4-2

* Comparison between all cell lines not reported. Question marks indicate suggested order of AR expression; variable results reported. ? Direct comparisons of AR expression levels between LAPC-4, C4-2, and 22Rv1 cell not reported. Abbreviations: AS (androgen-sensitive); AI (androgen-insensitive); AR (androgen receptor); PTEN (phosphatase and tensin homolog); EGFR (epidermal growth factor receptor); SRD5A (5α-reductase); IGF-I (insulin-like growth factor-I).

References

1. Negoita, S.; Feuer, E.J.; Mariotto, A.; Cronin, K.A.; Petkov, V.I.; Hussey, S.K.; Benard, V.; Henley, S.J.; Anderson, R.N.; Fedewa, S.; et al. Annual Report to the Nation on the Status of Cancer, part II: Recent changes in prostate cancer trends and disease characteristics. *Cancer* **2018**, *124*, 2801–2814. [CrossRef] [PubMed]

2. Pezaro, C.; Woo, H.H.; Davis, I.D. Prostate cancer: Measuring PSA. *Intern. Med. J.* **2014**, *44*, 433–440. [CrossRef] [PubMed]

3. Huggins, C.; Stevens, R.E., Jr.; Hodges, C.V. Studies on prostatic cancer: II. the effects of castration on advanced carcinoma of the prostate gland. *Arch. Surg.* **1941**, *43*, 209–223. [CrossRef]

4. Tilki, D.; Evans, C.P. The changing landscape of advanced and castration resistant prostate cancer: Latest science and revised definitions. *Can. J. Urol.* **2014**, *21*, 7–13. [PubMed]

5. Rowles, J.L., 3rd; Ranard, K.M.; Applegate, C.C.; Jeon, S.; An, R.; Erdman, J.W., Jr. Processed and raw tomato consumption and risk of prostate cancer: A systematic review and dose-response meta-analysis. *Prostate Cancer Prostatic. Dis.* **2018**, *21*, 319–336. [CrossRef] [PubMed]

6. Rowles, J.L., 3rd; Ranard, K.M.; Smith, J.W.; An, R.; Erdman, J.W., Jr. Increased dietary and circulating lycopene are associated with reduced prostate cancer risk: A systematic review and meta-analysis. *Prostate Cancer Prostatic. Dis.* **2017**, *20*, 361–377. [CrossRef] [PubMed]

7. Xu, X.; Li, J.; Wang, X.; Wang, S.; Meng, S.; Zhu, Y.; Liang, Z.; Zheng, X.; Xie, L. Tomato consumption and prostate cancer risk: A systematic review and meta-analysis. *Sci. Rep.* **2016**, *6*, 37091. [CrossRef] [PubMed]

8. Zu, K.; Mucci, L.; Rosner, B.A.; Clinton, S.K.; Loda, M.; Stampfer, M.J.; Giovannucci, E. Dietary lycopene, angiogenesis, and prostate cancer: A prospective study in the prostate-specific antigen era. *J. Natl. Cancer Inst.* **2014**, *106*, djt430. [CrossRef] [PubMed]

9. Wang, Y.; Cui, R.; Xiao, Y.; Fang, J.; Xu, Q. Effect of carotene and lycopene on the risk of prostate cancer: A systematic review and dose-response meta-analysis of observational studies. *PLoS ONE* **2015**, *10*, e0137427. [CrossRef]

10. Paur, I.; Lilleby, W.; Bohn, S.K.; Hulander, E.; Klein, W.; Vlatkovic, L.; Axcrona, K.; Bolstad, N.; Bjoro, T.; Laake, P.; et al. Tomato-based randomized controlled trial in prostate cancer patients: Effect on PSA. *Clin. Nutr.* **2017**, *36*, 672–679. [CrossRef] [PubMed]

11. Zhang, X.; Yang, Y.; Wang, Q. Lycopene can reduce prostate-specific antigen velocity in a phase II clinical study in Chinese population. *Chin. Med. J.* **2014**, *127*, 2143–2146. [PubMed]

12. Wang, Y.; Jacobs, E.J.; Newton, C.C.; McCullough, M.L. Lycopene, tomato products and prostate cancer-specific mortality among men diagnosed with nonmetastatic prostate cancer in the Cancer Prevention Study II Nutrition Cohort. *Int. J. Cancer* **2016**, *138*, 2846–2855. [CrossRef] [PubMed]

13. Key, T.J.; Appleby, P.N.; Travis, R.C.; Albanes, D.; Alberg, A.J.; Barricarte, A.; Black, A.; Boeing, H.; Bueno-de-Mesquita, H.B.; Chan, J.M.; et al. Carotenoids, retinol, tocopherols, and prostate cancer risk: A pooled analysis of 15 studies. *Am. J. Clin. Nutr.* **2015**, *102*, 1142–1157. [CrossRef] [PubMed]

14. Schrecengost, R.; Knudsen, K.E. Molecular pathogenesis and progression of prostate cancer. *Semin. Oncol.* **2013**, *40*, 244–258. [CrossRef] [PubMed]

15. Cai, C.; Balk, S.P. Intratumoral androgen biosynthesis in prostate cancer pathogenesis and response to therapy. *Endocr. Relat. Cancer* **2011**, *18*, R175–R182. [CrossRef] [PubMed]

16. Boileau, T.W.M.; Clinton, S.K.; Erdman, J.W., Jr. Tissue lycopene concentrations and isomer patterns are affected by androgen status and dietary lycopene concentration in male F344 rats. *J. Nutr.* **2000**, *130*, 1613–1618. [CrossRef] [PubMed]

17. Boileau, T.W.M.; Clinton, S.K.; Zaripheh, S.; Monaco, M.H.; Donovan, S.M.; Erdman, J.W., Jr. Testosterone and food restriction modulate hepatic lycopene isomer concentrations in male F344 rats. *J. Nutr.* **2001**, *131*, 1746–1752. [CrossRef] [PubMed]

18. Campbell, J.K.; Stroud, C.K.; Nakamura, M.T.; Lila, M.A.; Erdman, J.W., Jr. Serum testosterone is reduced following short-term phytofluene, lycopene, or tomato powder consumption in F344 rats. *J. Nutr.* **2006**, *136*, 2813–2819. [CrossRef] [PubMed]

19. Ko, H.-K.; Berk, M.; Chung, Y.-M.; Willard, B.; Bareja, R.; Rubin, M.; Sboner, A.; Sharifi, N. Loss of an androgen-inactivating and isoform-specific HSD17B4 splice form enables emergence of castration-resistant prostate cancer. *Cell Rep.* **2018**, *22*, 809–819. [CrossRef] [PubMed]

20. Herzog, A.; Siler, U.; Spitzer, V.; Seifert, N.; Denelavas, A.; Hunziker, P.B.; Hunziker, W.; Goralczyk, R.; Wertz, K. Lycopene reduced gene expression of steroid targets and inflammatory markers in normal rat prostate. *FASEB J.* **2005**, *19*, 272–274. [CrossRef] [PubMed]

21. Qiu, X.; Yuan, Y.; Vaishnav, A.; Tessel, M.A.; Nonn, L.; van Breemen, R.B. Effects of lycopene on protein expression in human primary prostatic epithelial cells. *Cancer Prev. Res.* **2013**, *6*, 419–427. [CrossRef] [PubMed]

22. Lewis, S.J.; Gardner, M.; Higgins, J.; Holly, J.M.P.; Gaunt, T.R.; Perks, C.M.; Turner, S.D.; Rinaldi, S.; Thomas, S.; Harrison, S.; et al. Developing the WCRF International/University of Bristol methodology for identifying and carrying out systematic reviews of mechanisms of exposure-cancer associations. *Cancer Epidemiol. Biomark. Prev.* **2017**, *26*, 1667–1675. [CrossRef] [PubMed]

23. Moher, D.; Liberati, A.; Tetzlaff, J.; Altman, D.G.; Group, P. Preferred reporting items for systematic reviews and meta-analyses: The PRISMA statement. *J. Clin. Epidemiol.* **2009**, *62*, 1006–1012. [CrossRef] [PubMed]

24. Hooijmans, C.R.; Rovers, M.M.; de Vries, R.B.M.; Leenaars, M.; Ritskes-Hoitinga, M.; Langendam, M.W. SYRCLE's risk of bias tool for animal studies. *BMC Med. Res. Methodol.* **2014**, *14*, 43. [CrossRef] [PubMed]

25. Higgins, J.P.T.; Altman, D.G.; Gøtzsche, P.C.; Jüni, P.; Moher, D.; Oxman, A.D.; Savović, J.; Schulz, K.F.; Weeks, L.; Sterne, J.A.C. The Cochrane Collaboration's tool for assessing risk of bias in randomised trials. *BMJ* **2011**, *343*, d5928. [CrossRef] [PubMed]

26. Harrison, S.; Jones, H.E.; Martin, R.M.; Lewis, S.J.; Higgins, J.P.T. The albatross plot: A novel graphical tool for presenting results of diversely reported studies in a systematic review. *Res. Synth. Methods* **2017**, *8*, 281–289. [CrossRef] [PubMed]

27. Canene-Adams, K.; Lindshield, B.L.; Wang, S.; Jeffery, E.H.; Clinton, S.K.; Erdman, J.W., Jr. Combinations of tomato and broccoli enhance antitumor activity in dunning r3327-h prostate adenocarcinomas. *Cancer Res.* **2007**, *67*, 836–843. [CrossRef] [PubMed]

28. Limpens, J.; Schroder, F.H.; de Ridder, C.M.A.; Bolder, C.A.; Wildhagen, M.F.; Obermuller-Jevic, U.C.; Kramer, K.; van Weerden, W.M. Combined lycopene and vitamin E treatment suppresses the growth of PC-346C human prostate cancer cells in nude mice. *J. Nutr.* **2006**, *136*, 1287–1293. [CrossRef] [PubMed]

29. Lindshield, B.L.; Ford, N.A.; Canene-Adams, K.; Diamond, A.M.; Wallig, M.A.; Erdman, J.W., Jr. Selenium, but not lycopene or vitamin E, decreases growth of transplantable dunning R3327-H rat prostate tumors. *PLoS ONE* **2010**, *5*, e10423. [CrossRef] [PubMed]

30. Siler, U.; Barella, L.; Spitzer, V.; Schnorr, J.; Lein, M.; Goralczyk, R.; Wertz, K. Lycopene and Vitamin E interfere with autocrine/paracrine loops in the Dunning prostate cancer model. *FASEB J.* **2004**, *18*, 1019–1021. [CrossRef] [PubMed]

31. Wan, L.; Tan, H.L.; Thomas-Ahner, J.M.; Pearl, D.K.; Erdman, J.W., Jr.; Moran, N.E.; Clinton, S.K. Dietary tomato and lycopene impact androgen signaling and carcinogenesis-related gene expression during early TRAMP prostate carcinogenesis. *Cancer Prev. Res.* **2014**, *7*, 1228–1239. [CrossRef] [PubMed]

32. Ivanov, N.I.; Cowell, S.P.; Brown, P.; Rennie, P.S.; Guns, E.S.; Cox, M.E. Lycopene differentially induces quiescence and apoptosis in androgen-responsive and -independent prostate cancer cell lines. *Clin. Nutr.* **2007**, *26*, 252–263. [CrossRef] [PubMed]

33. Linnewiel-Hermoni, K.; Khanin, M.; Danilenko, M.; Zango, G.; Amosi, Y.; Levy, J.; Sharoni, Y. The anti-cancer effects of carotenoids and other phytonutrients resides in their combined activity. *Arch. Biochem. Biophys.* **2015**, *572*, 28–35. [CrossRef] [PubMed]

34. Liu, A.; Pajkovic, N.; Pang, Y.; Zhu, D.; Calamini, B.; Mesecar, A.L.; van Breemen, R.B. Absorption and subcellular localization of lycopene in human prostate cancer cells. *Mol. Cancer Ther.* **2006**, *5*, 2879–2885. [CrossRef] [PubMed]

35. Liu, X.X.; Allen, J.D.; Arnold, J.T.; Blackman, M.R. Lycopene inhibits IGF-I signal transduction and growth in normal prostate epithelial cells by decreasing DHT-modulated IGF-I production in co-cultured reactive stromal cells. *Carcinogenesis* **2008**, *29*, 816–823. [CrossRef] [PubMed]

36. Peternac, D.; Klima, I.; Cecchini, M.G.; Schwaninger, R.; Studer, U.E.; Thalmann, G.N. Agents used for chemoprevention of prostate cancer may influence PSA secretion independently of cell growth in the LNCaP model of human prostate cancer progression. *Prostate* **2008**, *68*, 1307–1318. [CrossRef] [PubMed]

37. Rafi, M.M.; Kanakasabai, S.; Reyes, M.D.; Bright, J.J. Lycopene modulates growth and survival associated genes in prostate cancer. *J. Nutr. Biochem.* **2013**, *24*, 1724–1734. [CrossRef] [PubMed]

38. Richards, L.T.R.; Benghuzzi, H.; Tucci, M.; Hughes, J. The synergistic effect of conventional and sustained delivery of antioxidants on LNCaP prostate cancer cell line. *Biomed. Sci. Instrum.* **2003**, *39*, 402–407. [PubMed]

39. Zhang, X.; Wang, Q.; Neil, B.; Chen, X.A. Effect of lycopene on androgen receptor and prostate-specific antigen velocity. *Chin. Med. J.* **2010**, *123*, 2231–2236. [CrossRef] [PubMed]

40. Gong, X.; Marisiddaiah, R.; Zaripheh, S.; Wiener, D.; Rubin, L.P. Mitochondrial beta-carotene 9′,10′ oxygenase modulates prostate cancer growth via NF-kappaB inhibition: A lycopene-independent function. *Mol. Cancer Res.* **2016**, *14*, 966–975. [CrossRef] [PubMed]

41. Gunasekera, R.S.; Sewgobind, K.; Desai, S.; Dunn, L.; Black, H.S.; McKeehan, W.L.; Patil, B. Lycopene and lutein inhibit proliferation in rat prostate carcinoma cells. *Nutr. Cancer* **2007**, *58*, 171–177. [CrossRef] [PubMed]

42. Tang, L.L.; Jin, T.Y.; Zeng, X.B.; Wang, J.S. Lycopene inhibits the growth of human androgen-independent prostate cancer cells in vitro and in BALB/c nude mice. *J. Nutr.* **2005**, *135*, 287–290. [CrossRef] [PubMed]

43. Tang, Y.; Parmakhtiar, B.; Simoneau, A.R.; Xie, J.; Fruehauf, J.; Lilly, M.; Zi, X. Lycopene enhances docetaxel's effect in castration-resistant prostate cancer associated with insulin-like growth factor I receptor levels. *Neoplasia* **2011**, *13*, 108–119. [CrossRef] [PubMed]

44. Fu, L.J.; Ding, Y.B.; Wu, L.X.; Wen, C.J.; Qu, Q.; Zhang, X.; Zhou, H.H. The effects of lycopene on the methylation of the GSTP1 promoter and global methylation in prostatic cancer cell lines PC3 and LNCaP. *Int. J. Endocrinol.* **2014**, *2014*, 620125. [CrossRef] [PubMed]

45. Lin, C.Y.; Huang, C.S.; Hu, M.L. The use of fetal bovine serum as delivery vehicle to improve the uptake and stability of lycopene in cell culture studies. *Br. J. Nutr.* **2007**, *98*, 226–232. [CrossRef] [PubMed]

46. Cook, M.B.; Stanczyk, F.Z.; Wood, S.N.; Pfeiffer, R.M.; Hafi, M.; Veneroso, C.C.; Lynch, B.; Falk, R.T.; Zhou, C.K.; Niwa, S.; et al. Relationships between circulating and intraprostatic sex steroid hormone concentrations. *Cancer Epidemiol. Biomark. Prev.* **2017**, *26*, 1660–1666. [CrossRef] [PubMed]

47. Izumi, K.; Shigehara, K.; Nohara, T.; Narimoto, K.; Kadono, Y.; Mizokami, A. Both high and low serum total testosterone levels indicate poor prognosis in patients with prostate cancer. *Anticancer Res.* **2017**, *37*, 5559–5564. [CrossRef] [PubMed]

48. Wang, K.; Chen, X.; Bird, V.Y.; Gerke, T.A.; Manini, T.M.; Prosperi, M. Association between age-related reductions in testosterone and risk of prostate cancer-an analysis of patients' data with prostatic diseases. *Int. J. Cancer* **2017**, *141*, 1783–1793. [CrossRef] [PubMed]

49. Shu, Q.; Cai, T.; Chen, X.; Zhu, H.H.; Xue, P.; Zhu, N.; Xie, Z.; Wei, S.; Zhang, Q.; Niu, L.; et al. Proteomic comparison and MRM-based comparative analysis of metabolites reveal metabolic shift in human prostate cancer cell lines. *J. Proteome Res.* **2015**, *14*, 3390–3402. [CrossRef] [PubMed]

50. Jerónimo, C.; Bastian, P.J.; Bjartell, A.; Carbone, G.M.; Catto, J.W.F.; Clark, S.J.; Henrique, R.; Nelson, W.G.; Shariat, S.F. Epigenetics in prostate cancer: Biologic and clinical relevance. *Eur. Urol.* **2011**, *60*, 753–766. [CrossRef] [PubMed]

51. Lin, X.; Tascilar, M.; Lee, W.-H.; Vles, W.J.; Lee, B.H.; Veeraswamy, R.; Asgari, K.; Freije, D.; van Rees, B.; Gage, W.R.; et al. GSTP1 CpG island hypermethylation is responsible for the absence of GSTP1 expression in human prostate cancer cells. *Am. J. Pathol.* **2001**, *159*, 1815–1826. [CrossRef]

52. Russell, P.J.; Kingsley, E.A. Human prostate cancer cell lines. *Methods Mol. Med.* **2003**, *81*, 21–39. [CrossRef] [PubMed]

53. Wu, H.C.; Hsieh, J.T.; Gleave, M.E.; Brown, N.M.; Pathak, S.; Chung, L.W. Derivation of androgen-independent human LNCaP prostatic cancer sublines: Role of bone stromal cells. *Int. J. Cancer* **1994**, *57*, 406–412. [CrossRef] [PubMed]

54. Klein, K.A.; Reiter, R.E.; Redula, J.; Moradi, H.; Zhu, X.L.; Brothman, A.R.; Lamb, D.J.; Marcelli, M.; Belldegrun, A.; Witte, O.N.; et al. Progression of metastatic human prostate cancer to androgen independence in immunodeficient SCID mice. *Nat. Med.* **1997**, *3*, 402–408. [CrossRef] [PubMed]

55. Hoefer, J.; Akbor, M.; Handle, F.; Ofer, P.; Puhr, M.; Parson, W.; Culig, Z.; Klocker, H.; Heidegger, I. Critical role of androgen receptor level in prostate cancer cell resistance to new generation antiandrogen enzalutamide. *Oncotarget* **2016**, *7*, 59781–59794. [CrossRef] [PubMed]

56. Sramkoski, R.M.; Pretlow, T.G., II; Giaconia, J.M.; Pretlow, T.P.; Schwartz, S.; Sy, M.-S.; Marengo, S.R.; Rhim, J.S.; Zhang, D.; Jacobberger, J.W. A new human prostate carcinoma cell line, 22Rv1. *In Vitro Cell Dev. Biol. Anim.* **1999**, *35*, 403–409. [CrossRef] [PubMed]

57. Matsuoka, T.; Shigemura, K.; Yamamichi, F.; Fujisawa, M.; Kawabata, M.; Shirakawa, T. Detection of tumor markers in prostate cancer and comparison of sensitivity between real time and nested PCR. *Kobe J. Med. Sci.* **2012**, *58*, E51–E59. [PubMed]

58. Pietrzkowski, Z.; Mulholland, G.; Gomella, L.; Jameson, B.A.; Wernicke, D.; Baserga, R. Inhibition of growth of prostatic cancer cell lines by peptide analogues of insulin-like growth factor 1. *Cancer Res.* **1993**, *53*, 1102–1106. [PubMed]

59. Isaacs, J.T.; Isaacs, W.B.; Feitz, W.F.; Scheres, J. Establishment and characterization of seven dunning rat prostatic cancer cell lines and their use in developing methods for predicting metastatic abilities of prostatic cancers. *Prostate* **1986**, *9*, 261–281. [CrossRef] [PubMed]

nutrients

MDPI

Article

Dietary Carotenoids Intake and the Risk of Gastric Cancer: A Case—Control Study in Korea

Ji Hyun Kim [1], Jeonghee Lee [1], Il Ju Choi [2], Young-Il Kim [2], Oran Kwon [3], Hyesook Kim [3] and Jeongseon Kim [1,*]

[1] Graduate School of Cancer Science and Policy, National Cancer Center, 323 Ilsan-ro, Ilsandong-gu, Goyang-si, Gyeonggi-do 10408, Korea; 1601009@ncc.re.kr (J.H.K.); jeonghee@ncc.re.kr (J.L.)
[2] Center for Gastric Cancer, National Cancer Center Hospital, National Cancer Center, 323 Ilsan-ro, Ilsandong-gu, Goyang-si, Gyeonggi-do 10408, Korea; cij1224@ncc.re.kr (I.J.C.); 11996@ncc.re.kr (Y.-I.K.)
[3] Department of Nutritional Science and Food Management, Ewha Womans University, 52, Ewhayeodae-gil, Seodaemun-gu, Seoul 03760, Korea; orank@ewha.ac.kr (O.K.); khs7882@hanmail.net (H.K.)
* Correspondence: jskim@ncc.re.kr; Tel.: +82-31-920-2570; Fax: +82-31-920-2579

Received: 27 June 2018; Accepted: 4 August 2018; Published: 7 August 2018

Abstract: Although the incidence of gastric cancer (GC) has declined, it remains the second most common cancer in Korea. As a class of phytochemicals, carotenoids are fat-soluble pigments that are abundant in fruits and vegetables and have health-promoting properties, including cancer prevention effects. This case-control study investigated the effects of total dietary carotenoids, dietary carotenoid subclasses (α-carotene, β-carotene, β-cryptoxanthin, lutein/zeaxanthin, and lycopene), and foods contributing to the dietary intake of each carotenoid on the risk of GC. Four hundred and fifteen cases and 830 controls were recruited from the National Cancer Center Hospital in Korea between March 2011 and December 2014. A significant inverse association between total dietary carotenoids and GC risk was observed among women (odds ratio (OR) 0.56, 95% confidence interval (CI) 0.32–0.99). A higher intake of dietary lycopene was inversely associated with GC risk overall in the subjects (OR 0.60, 95% CI 0.42–0.85, *p* for trend = 0.012), men (OR 0.60, 95% CI 0.39–0.93), and women (OR 0.54, 95% CI 0.30–0.96, *p* for trend = 0.039). This significant association between dietary lycopene intake and GC risk was also observed in the subgroups of *Helicobacter pylori* (*H. pylori*)-positive subjects and those who had ever smoked. Among the major contributing foods of dietary lycopene, consumption of tomatoes and tomato ketchup was inversely associated with GC risk in the overall subjects, men, and women. Based on our findings, a higher intake of dietary lycopene and contributing foods of lycopene (tomatoes and tomato ketchup) may be inversely associated with the risk of GC.

Keywords: carotenoids; lycopene; gastric cancer; *H. pylori*; smoking; case-control study; Korea

1. Introduction

According to the GLOBOCAN estimates reported in 2012, although gastric cancer (GC) incidence is declining, it remains the fifth most common cancer worldwide [1]. GC is the second most common cause of cancer in Korea, as the estimated age-standardized incidence rate of GC was 35.8 per 100,000 persons in 2015 [2]. Therefore, the primary prevention of GC is a major priority in public health.

Several risk factors for GC have been identified, such as *Helicobacter pylori* (*H. pylori*) infection and tobacco smoking, which are classified as International Agency for Research on Cancer (IARC) group 1 carcinogens (carcinogenic to humans) [3,4]. Dietary factors are considered modifiable risk factors that account for approximately 35% of all causes of cancer, and therefore dietary factors related to cancer must be identified [5,6].

Differences in the incidence rates of GC subtypes by gender have also been observed [7]. The rate of cardia GC was approximately three times higher in males than in females, whereas the rate of non-cardia GC was two times higher in males than in females [7]. The explanations for the higher rates of GC in males are not known, but might be due to the male predominance of *H. pylori* infection, one of the major risk factors for GC [8]. This phenomenon is found in adults worldwide [8]. The higher incidence rate of GC in males might also be due to the higher consumption of tobacco [9]. However, the predominance of males in the male to female ratio were similar in both smokers and non-smokers; therefore, the higher rates of GC in males might not be completely explained by the smoking history [9].

According to the third expert report from the Continuous Update Project (CUP) published by the World Cancer Research Fund (WCRF) International in 2018, the intake of certain types of foods (e.g., fruits) might be closely related to the risk of GC [10]. Carotenoids are fat-soluble pigments that are highly abundant in fruits and vegetables, and belong to a class of phytochemicals that have health-promoting properties [11–13]. Among the more than 40 carotenoids that are derived from a variety of food sources, six types of primary dietary carotenoids are usually detected in human blood plasma (α-carotene, β-carotene, β-cryptoxanthin, lutein, zeaxanthin, and lycopene), suggesting the selective intestinal absorption of these carotenoids [11,12,14].

Several epidemiological studies have been conducted to determine the associations between dietary carotenoids and the risk of GC. However, until now, few studies have been conducted and the results of previous studies are conflicting. Thus, a clear association between dietary carotenoids intake and the risk of GC has not been identified to date. Dietary α-carotene intake was inversely associated with GC risk in some studies [15–18], whereas no significant associations were observed in other studies [19–21]. Dietary β-carotene intake was associated with a reduced risk of GC in some studies [15,16,18,22–24], while other studies did not report a significant association [17,19–21,25,26]. Studies that analyzed dietary β-cryptoxanthin intake did not identify a significant association with the risk of GC [16–18,21]. Dietary lutein/zeaxanthin intake was not significantly associated with the risk of GC [15–17,19,20]. Dietary lycopene intake was inversely associated with the risk of GC in one study [17], whereas other studies have found no association [15,16,18–21,26].

Accordingly, this case-control study aims to investigate the effects of total dietary carotenoids, dietary carotenoid subclasses (α-carotene, β-carotene, β-cryptoxanthin, lutein/zeaxanthin, and lycopene), and the contributing foods of each dietary carotenoid on the risk of GC.

2. Materials and Methods

2.1. Study Population

The subjects were recruited from the National Cancer Center Hospital in Korea between March 2011 and December 2014. Cases were subjects who had been histologically diagnosed with early GC within the preceding three months at the Center for GC. Early GC was defined as GC restricted to the mucosa or submucosa with or without lymph node metastasis, regardless of the tumor size [27]. Patients in the case group did not have advanced GC, diabetes mellitus, severe systemic/mental disease, or a history of cancer within the past five years, and women who were pregnant or currently breastfeeding were also excluded. Controls were subjects who underwent health-screening examinations at the Center for Cancer Prevention and Detection at the same hospital.

Among the 1727 subjects (500 cases and 1227 controls) who agreed to participate in the study, 26 cases and 30 controls were excluded due to an incomplete self-administered questionnaire or semi-quantitative food frequency questionnaire (SQFFQ). Of the 1671 subjects remaining, 5 cases and 10 controls were excluded due to the implausibility of a total energy intake of <500 kcal or ≥4000 kcal. Of the 1656 subjects remaining, cases and controls were matched at a ratio of 1:2 by the distribution of age within 5 years and gender. Ultimately, a total of 1245 subjects (415 cases and 830 matched controls; 810 men and 435 women) were selected for this study (Figure 1). Written informed consent was obtained from all participants, and the study protocol was approved by the Institutional Review Board of the National Cancer Center [IRB Number: NCCNCS-11-438].

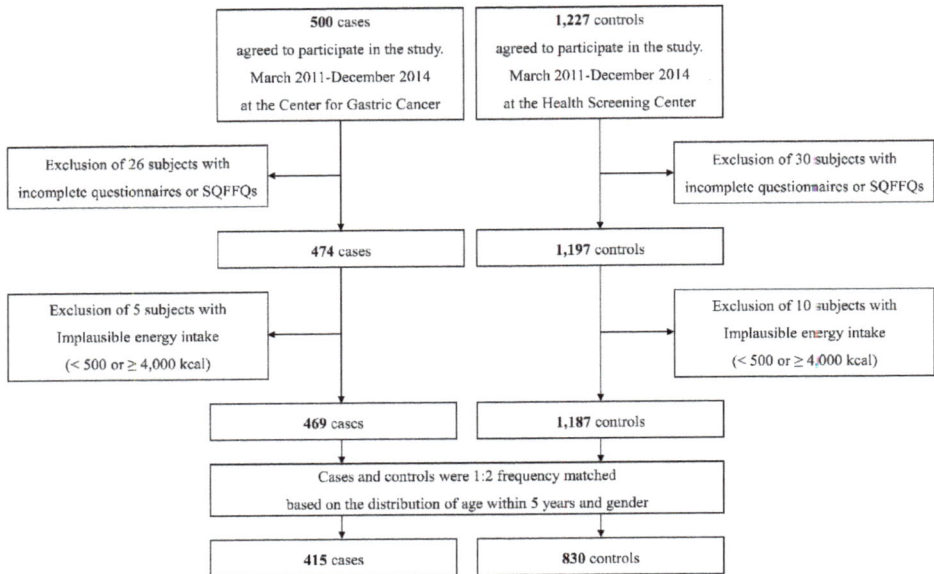

Figure 1. Flow chart of the study subjects. SQFFQ: semi-quantitative food frequency questionnaire.

2.2. Data Collection and Management

Participants were asked to complete a self-administered questionnaire that included demographic, lifestyle, and medical history information. Dietary intake was collected from the 106-item SQFFQ, which has been previously reported to be reliable and valid [28]. The study participants in the case group were surveyed with the self-administered questionnaire and SQFFQ by interviewers who were trained beforehand. Participants in the control group were initially asked to complete the survey by themselves, and interviewers asked any questions with missing answers from the self-completed survey during the second round of the survey.

After collecting the dietary information, the amount of each food item consumed was calculated using CAN-PRO 4.0 (Computer Aided Nutritional Analysis Program, The Korean Nutrition Society, Seoul, Korea). The 106 items of the SQFFQ were classified into 663 detailed food items. The overlapping food items were excluded, and 410 food items remained for the analysis. Then, the above food consumption information was merged with the database of carotenoid contents. The carotenoid database used in our study was composed of the United States Department of Agriculture (USDA) carotenoid database [29] and the Food Functional Composition Table provided by the Korea National Academy of Agricultural Science (NAAS) [30]. Additionally, by referring to the recipes from the NAAS Agricultural and Food Integrated Information System, information on seasoned vegetables and kimchi was also included. In total, this database contained 2903 food items. In terms of the carotenoid subclasses, it contained the five main carotenoids, namely, α-carotene, β-carotene, β-cryptoxanthin, lutein/zeaxanthin, and lycopene, and total carotenoids was defined as the sum of the five main carotenoid subclasses.

In terms of matching, except for meat, poultry, seafood, and dairy products, which rarely contain carotenoids (137 food items), the carotenoid database included 98.5% of all food items reported in the SQFFQ. Four food items were excluded due to a lack of information on the carotenoid content. The validity of the SQFFQ for dietary carotenoid intake has been tested using three-day dietary records from 207 people as a gold standard. The crude, energy-adjusted, and energy-adjusted and de-attenuated correlation coefficients for total carotenoids were 0.189, 0.244, and 0.307, respectively.

A rapid urease test (Pronto Dry; Medical Instruments Corporation, Solothurn, Switzerland) was conducted to examine the *H. pylori* infection status.

2.3. Statistical Analysis

To compare the general characteristics between cases and controls, Student's *t*-test was used for continuous variables and the chi-square test was used for categorical variables. A contribution analysis was conducted to select the foods contributing to total dietary carotenoids and the carotenoid subclasses. The food items contributing to total dietary carotenoids or carotenoid subclasses that represented up to 90% of the cumulative contribution were selected. All types of dietary carotenoids and their contributing foods were adjusted for total energy intake using the regression residual method [31]. Dietary carotenoids and their contributing foods were categorized by tertiles for the analysis based on the distribution of controls. The lowest tertile of each carotenoid and the foods contributing to dietary intake of each carotenoid were used as references. Odds ratios (ORs) and 95% confidence intervals (CIs) were calculated across the tertiles of dietary carotenoids and their contributing foods using the logistic regression model, after controlling for potential confounding factors. To test for trends, the median values of each tertile category of dietary carotenoids and their contributing foods were used as continuous variables. Model 1 was adjusted for age (as a continuous variable). Model 2 was adjusted for age, total caloric intake (as a continuous variable), a first-degree family history of GC (yes or no), smoking status (current, ex-, or non-smoker), regular exercise status (yes or no), education level (middle school or less, high school, or college or more), occupation (professional and administrative, office and sales/service, labor and agricultural, others, and unemployed), and monthly household income in units of 10,000 won/month (<200, 200–400, or ≥400). Model 3 was adjusted for the *H. pylori* infection status (yes or no) and the variables included in model 2. In the overall subjects, models 1, 2, and 3 were additionally adjusted for gender.

In the stratified analysis according to *H. pylori* infection status, model 1 was adjusted for age and gender. Model 2 was adjusted for age, gender, total caloric intake (as a continuous variable), a first-degree family history of GC (yes or no), smoking status (current, ex-, or non-smoker), regular exercise status (yes or no), education level (middle school or less, high school, or college or more), occupation (professional and administrative, office and sales/service, labor and agricultural, others, and unemployed), and monthly household income in units of 10,000 won/month (<200, 200–400, or ≥400). In the stratified analysis according to smoking status (ever-smoker and non-smoker), model 1 was adjusted for age and gender. Model 2 was adjusted for age, gender, total caloric intake (as a continuous variable), a first-degree family history of GC (yes or no), *H. pylori* infection status (yes or no), regular exercise status (yes or no), education level (middle school or less, high school, or college or more), occupation (professional and administrative, office and sales/service, labor and agricultural, others, and unemployed), and monthly household income in units of 10,000 won/month (<200, 200–400, or ≥400).

All statistical analyses were performed using SAS software (version 9.4, SAS Institute, Cary, NC, USA), and a two-sided *p*-value less than 0.05 was considered statistically significant.

3. Results

Table 1 describes the general characteristics of the 415 patients with early GC and 830 controls. Participants in the case group tended to have a higher proportion of *H. pylori* infection ($p < 0.001$) and first-degree family history of GC ($p = 0.001$), a lower percentage of those who had never smoked ($p < 0.001$) and those who exercise regularly ($p < 0.001$), a lower level of education ($p < 0.001$), a different occupational distribution ($p = 0.001$), and a lower level of monthly household income ($p < 0.001$) than those in the control group. Both men and women in the case group had a higher proportion of *H. pylori* infection, a lower proportion of those who never smoked and those who exercise regularly, a lower education level, a different occupational distribution, and a lower level of monthly household income than those in the control group. Additionally, men in the case group had a higher percentage of first-degree family history of GC than the men in the control group.

Table 2 describes the comparison of the consumption of total energy, total dietary carotenoids, and carotenoid subclasses. Subjects in the case group consumed more energy ($p < 0.001$), less total carotenoids ($p = 0.003$), less β-carotene ($p = 0.018$), less β-cryptoxanthin ($p = 0.007$), and less lycopene ($p < 0.001$) than subjects in the control group. Both men and women in the case group consumed less lycopene than the controls. Additionally, men in the case group consumed more energy, and women in the case group consumed less total carotenoids and β-cryptoxanthin compared to controls.

Table 3 shows the ORs and corresponding 95% CIs according to tertiles of total dietary carotenoids and carotenoid subclasses. Among women, higher total carotenoid intake was inversely associated with GC risk (model 2: OR 0.56, 95% CI 0.32–0.99), but the significant association disappeared in model 3, which was additionally adjusted for *H. pylori* infection. A higher lycopene intake was inversely associated with the risk of GC in the overall subjects (model 3: OR 0.60, 95% CI 0.42–0.85, p for trend = 0.012), men (model 3: OR 0.60, 95% CI 0.39–0.93), and women (model 2: OR 0.54, 95% CI 0.30–0.96, p for trend = 0.039).

Because dietary lycopene intake was significantly associated with GC risk in the overall subjects, a further analysis was performed on lycopene intake. Table 4 shows the ORs and 95% CIs of GC according to tertiles of dietary lycopene intake stratified by *H. pylori* infection status. Among the *H. pylori*-positive subjects, higher lycopene intake was inversely associated with the risk of GC (model 2: OR 0.61, 95% CI 0.42–0.90, p for trend = 0.037). When stratified by gender, higher intake of dietary lycopene was associated with a decreased risk of GC among *H. pylori*-positive males (model 2: OR 0.57, 95% CI 0.36–0.91, p for trend = 0.043) (Table S1). Table 5 shows the ORs and 95% CIs of GC according to tertiles of dietary lycopene intake stratified by smoking status (ever-smoker and non-smoker). Subjects who currently smoke or previously smoked were combined as ever-smokers. Among ever-smokers, a significantly reduced risk of GC was observed for the subjects who consumed greater amounts of dietary lycopene (model 2: OR 0.39, 95% CI 0.23–0.65, p for trend = 0.001). When stratified by gender, higher lycopene intake was inversely associated with the risk of GC among males who has ever smoked (model 2: OR 0.45, 95% CI 0.27–0.73, p for trend = 0.005) (Table S2).

Table 6 shows the comparison of the consumption of foods contributing to the dietary intake of lycopene. Compared to controls, cases consumed less tomato in the overall subjects ($p < 0.001$), men ($p = 0.001$), and women ($p < 0.001$); consumed less tomato ketchup in the overall subjects ($p < 0.001$), men ($p = 0.001$), and women ($p < 0.001$); and consumed less watermelon in the overall subjects ($p = 0.004$) and women ($p = 0.028$). Table 7 shows the associations between foods contributing to dietary lycopene intake and the risk of GC. A higher tomato intake was associated with a decreased risk of GC in the overall subjects (model 3: OR 0.59, 95% CI 0.41–0.85, p for trend = 0.016), men (model 3: OR 0.58, 95% CI 0.37–0.92, p for trend = 0.043), and women (model 2: OR 0.47, 95% CI 0.25–0.87, p for trend = 0.010). Higher tomato ketchup intake was associated with a decreased risk of GC in the overall subjects (model 3: OR 0.55, 95% CI 0.38–0.80, p for trend = 0.005), men (model 3: OR 0.62, 95% CI 0.39–0.97), and women (model 2: OR 0.47, 95% CI 0.25–0.88, p for trend = 0.011).

Table 1. General characteristics of the study subjects [a].

	Total (n = 1245)			Men (n = 810)			Women (n = 435)		
	Controls (n = 830)	Cases (n = 415)	p [b]	Controls (n = 540)	Cases (n = 270)	p [b]	Controls (n = 290)	Cases (n = 145)	p [b]
Age (years)	53.7 ± 9.0	53.8 ± 9.3	0.892	54.8 ± 8.4	54.9 ± 8.7	0.905	51.6 ± 9.8	51.7 ± 10.0	0.942
<50	285 (34.3)	139 (33.5)	0.816	153 (28.3)	77 (28.5)	>0.999	132 (45.5)	62 (42.8)	0.658
≥50	545 (65.7)	276 (66.5)		387 (71.7)	193 (71.5)		158 (54.5)	83 (57.2)	
Male, n (%)	540 (65.1)	270 (65.1)	>0.999						
BMI (kg/m²)	23.9 ± 2.9	23.9 ± 3.0	0.627	24.4 ± 2.7	24.2 ± 3.0	0.390	23.1 ± 3.1	23.2 ± 3.0	0.533
<23	314 (37.8)	159 (38.3)	0.975	161 (29.8)	91 (33.7)	0.509	153 (52.8)	68 (46.9)	
23–25	249 (30.0)	122 (29.4)		170 (31.5)	78 (28.9)		79 (27.2)	44 (30.3)	
≥25	266 (32.1)	133 (32.1)		209 (38.7)	101 (37.4)		57 (19.7)	32 (22.1)	
H. pylori infection			<0.001			<0.001			<0.001
Positive	486 (58.6)	382 (92.1)		333 (61.7)	252 (93.3)		153 (52.8)	130 (89.7)	
Negative	320 (38.6)	33 (8.0)		187 (34.6)	18 (6.7)		133 (45.9)	15 (10.3)	
First-degree family history of GC			0.001			0.003			0.155
Yes	103 (12.4)	82 (19.8)		74 (13.7)	60 (22.2)		29 (10.0)	22 (15.2)	
No	725 (87.4)	332 (80.0)		464 (85.9)	209 (77.4)		261 (90.0)	123 (84.8)	
Smoking status, n (%)			<0.001			<0.001			0.021
Current-smoker	162 (19.5)	128 (30.8)		157 (29.1)	121 (44.8)		5 (1.7)	7 (4.8)	
Ex-smoker	284 (34.2)	119 (28.7)		277 (51.3)	110 (40.7)		7 (2.4)	9 (6.2)	
Non-smoker	384 (46.3)	167 (40.2)		106 (19.6)	39 (14.4)		278 (95.9)	128 (88.3)	
Alcohol intake			0.243			0.282			0.819
Current-drinker	534 (64.3)	254 (61.2)		404 (74.8)	193 (71.5)		130 (44.8)	61 (42.1)	
Ex-drinker	60 (7.2)	41 (9.9)		47 (8.7)	33 (12.2)		13 (4.5)	8 (5.5)	
Non-drinker	236 (28.4)	119 (28.7)		89 (16.5)	44 (16.3)		147 (50.7)	75 (51.7)	
Regular exercise			<0.001			<0.001			<0.001
Yes	466 (56.1)	147 (35.4)		303 (56.1)	109 (40.4)		163 (56.2)	38 (26.2)	
No	361 (43.5)	268 (64.6)		234 (43.3)	161 (59.6)		127 (43.8)	107 (73.8)	
Education, n (%)			<0.001			<0.001			<0.001
Middle school or less	119 (14.3)	142 (34.2)		71 (13.2)	91 (33.7)		48 (16.6)	51 (35.2)	
High school	253 (30.5)	174 (41.9)		140 (25.9)	112 (41.5)		113 (39.0)	62 (42.8)	
College or more	426 (51.3)	97 (23.4)		301 (55.7)	66 (24.4)		125 (43.1)	31 (21.4)	
Marital status, n (%)			0.674			0.553			>0.999
Married	716 (86.3)	361 (87.0)		478 (88.5)	243 (90.0)		238 (82.1)	118 (81.4)	
Others	113 (13.6)	52 (12.5)		61 (11.3)	26 (9.6)		52 (17.9)	26 (17.9)	

Table 1. *Cont.*

	Total (n = 1245)			Men (n = 810)			Women (n = 435)		
	Controls (n = 830)	Cases (n = 415)	p [b]	Controls (n = 540)	Cases (n = 270)	p [b]	Controls (n = 290)	Cases (n = 145)	p [b]
Occupation, n (%)			0.001			0.010			0.002
Professional, administrative	156 (18.8)	70 (16.9)		117 (21.7)	59 (21.9)		39 (13.5)	11 (7.6)	
Office, sales and service	266 (32.1)	122 (29.4)		203 (37.6)	81 (30.0)		63 (21.7)	41 (28.3)	
Laborer, agricultural	128 (15.4)	104 (25.1)		111 (20.6)	83 (30.7)		17 (5.9)	21 (14.5)	
Others, unemployed	277 (33.4)	117 (28.2)		106 (19.6)	46 (17.0)		171 (59.0)	71 (49.0)	
Monthly household income, 10,000 won/month, n (%)			<0.001			<0.001			0.016
<200	149 (18.0)	133 (32.1)		85 (15.7)	85 (31.5)		64 (22.1)	48 (33.1)	
200–400	341 (41.1)	148 (35.7)		232 (43.0)	106 (39.3)		109 (37.6)	42 (29.0)	
>400	273 (32.9)	96 (23.1)		168 (31.1)	55 (20.4)		105 (36.2)	41 (28.3)	
Histological subtype of GC (Lauren's classification)			-			-			-
Intestinal	-	158 (38.1)		-	132 (48.9)		-	26 (17.9)	
Diffuse	-	164 (39.5)		-	77 (28.5)		-	87 (60.0)	
Mixed	-	59 (14.2)		-	40 (14.8)		-	19 (13.1)	
Indeterminate	-	4 (1.0)		-	3 (1.1)		-	1 (0.7)	

Missing data are included in the total %. [a] Values are presented as the means ± standard deviations (SD) or n (%); [b] p-values for continuous variables and categorical variables were calculated using Student's t-test and the chi-square test, respectively. BMI: body mass index; GC: gastric cancer.

Table 2. Comparison of the consumption of total energy, total dietary carotenoids, and carotenoid subclasses [a].

	Total (n = 1245)			Men (n = 810)			Women (n = 435)		
	Controls (n = 830)	Cases (n = 415)	p [b]	Controls (n = 540)	Cases (n = 270)	p [b]	Controls (n = 290)	Cases (n = 145)	p [b]
Total energy intake (kcal)	1713.6 ± 545.5	1924.1 ± 612.9	<0.001	1760.6 ± 541.5	2038.5 ± 634.8	<0.001	1626.0 ± 543.1	1711.1 ± 507.0	0.116
Total carotenoid intake (µg/day)									
Total carotenoids	12,121.5 ± 7762.4	10,799.2 ± 6987.1	<0.001	10,785.7 ± 6632.5	9990.7 ± 6294.5	0.102	14,608.7 ± 9014.1	12,304.7 ± 7927.0	0.009
α-Carotene	947.5 ± 913.8	925.3 ± 966.0	0.003	839.8 ± 805.4	833.1 ± 861.7	0.913	1148.0 ± 1059.9	1096.9 ± 1118.1	0.642
β-Carotene	5075.9 ± 3276.2	4632.6 ± 3009.4	0.692	4529.7 ± 2682.8	4226.4 ± 2568.0	0.124	6093.1 ± 3971.2	5388.9 ± 3582.4	0.073
β-Cryptoxanthin	393.7 ± 438.8	330.5 ± 364.6	0.018	310.0 ± 269.7	314.3 ± 390.1	0.872	549.6 ± 615.7	360.6 ± 310.5	<0.001
Lutein/Zeaxanthin	3531.7 ± 2584.6	3455.2 ± 2764.4	0.007	3256.6 ± 2200.2	3268.1 ± 2726.8	0.952	4044.0 ± 3119.2	3803.7 ± 2809.2	0.435
Lycopene	2218.0 ± 3847.2	1439.0 ± 2135.7	<0.001	1869.8 ± 3433.0	1312.1 ± 2110.4	0.005	2866.4 ± 4452.9	1675.3 ± 2169.5	<0.001

[a] Adjusted for total energy intake using the residuals method; [b] p-values were calculated using Student's t-test.

Table 3. Odds ratios (ORs) and 95% confidence intervals (CIs) of gastric cancer (GC) according to the tertiles of total dietary carotenoids and carotenoid subclasses [a].

	Median Intake (µg/day)	No. of Controls/Cases	Model 1		Model 2		Model 3	
			OR	(95% CI)	OR	(95% CI)	OR	(95% CI)
Total Carotenoids								
Total (n = 1245)								
T1	6064.75	276/162	1.00		1.00		1.00	
T2	10,171.44	277/146	0.88	(0.67–1.17)	0.98	(0.71–1.34)	1.01	(0.72–1.42)
T3	17,946.94	277/107	0.64	(0.47–0.86)	0.75	(0.53–1.06)	0.79	(0.55–1.15)
p for trend [b]			0.003		0.082		0.185	
Men (n = 810)								
T1	5603.87	180/106	1.00		1.00		1.00	
T2	8921.20	180/88	0.83	(0.58–1.17)	0.95	(0.63–1.43)	1.09	(0.70–1.68)
T3	15,850.62	180/76	0.71	(0.49–1.02)	0.78	(0.51–1.20)	0.84	(0.54–1.33)
p for trend [b]			0.073		0.236		0.386	
Women (n = 435)								
T1	7151.14	96/74	1.00		1.00		1.00	
T2	12,377.35	97/38	0.50	(0.31–0.81)	0.51	(0.30–0.88)	0.51	(0.29–0.91)
T3	21,678.14	97/33	0.43	(0.26–0.71)	0.56	(0.32–0.99)	0.65	(0.35–1.18)
p for trend [b]			0.001		0.051		0.169	
α-Carotene								
Total (n = 1245)								
T1	321.92	276/151	1.00		1.00		1.00	
T2	665.17	277/136	0.90	(0.67–1.19)	0.99	(0.71–1.36)	0.94	(0.67–1.33)
T3	1527.25	277/128	0.84	(0.62–1.12)	0.99	(0.71–1.38)	1.00	(0.70–1.41)
p for trend [b]			0.269		0.969		0.957	
Men (n = 810)								
T1	291.35	180/100	1.00		1.00		1.00	
T2	587.52	180/78	0.78	(0.54–1.12)	0.95	(0.63–1.44)	0.95	(0.61–1.48)
T3	1404.67	180/92	0.92	(0.65–1.31)	0.98	(0.65–1.48)	0.95	(0.61–1.47)
p for trend [b]			0.879		0.972		0.839	
Women (n = 435)								
T1	427.91	96/52	1.00		1.00		1.00	
T2	853.57	97/42	0.80	(0.49–1.31)	0.84	(0.48–1.48)	0.83	(0.46–1.50)
T3	1704.44	97/51	0.97	(0.60–1.57)	1.48	(0.85–2.58)	1.49	(0.83–2.70)
p for trend [b]			0.961		0.113		0.125	
β-Carotene								
Total (n = 1245)								
T1	2574.39	276/162	1.00		1.00		1.00	
T2	4339.20	277/134	0.82	(0.61–1.08)	0.81	(0.58–1.11)	0.84	(0.60–1.18)
T3	7353.32	277/119	0.71	(0.53–0.96)	0.77	(0.55–1.08)	0.85	(0.59–1.22)
p for trend [b]			0.030		0.158		0.426	

Table 3. *Cont.*

	Median Intake (µg/day)	No. of Controls/Cases	Model 1		Model 2		Model 3	
			OR	(95% CI)	OR	(95% CI)	OR	(95% CI)
Men (n = 810)								
T1	2447.55	180/100	1.00		1.00		1.00	
T2	3791.98	180/93	0.93	(0.65–1.32)	1.00	(0.67–1.50)	1.10	(0.72–1.70)
T3	6504.49	180/77	0.76	(0.53–1.10)	0.76	(0.49–1.17)	0.89	(0.56–1.41)
p for trend [b]			0.136		0.179		0.535	
Women (n = 435)								
T1	3178.03	96/68	1.00		1.00		1.00	
T2	5104.57	97/35	0.50	(0.31–0.83)	0.52	(0.30–0.92)	0.53	(0.29–0.97)
T3	8683.53	97/42	0.60	(0.37–0.97)	0.73	(0.42–1.25)	0.79	(0.44–1.41)
p for trend [b]			0.062		0.351		0.551	
β-Cryptoxanthin								
Total (n = 1245)								
T1	128.90	276/170	1.00		1.00		1.00	
T2	270.81	277/136	0.79	(0.60–1.05)	0.90	(0.66–1.24)	0.94	(0.67–1.32)
T3	594.01	277/109	0.62	(0.46–0.84)	0.79	(0.56–1.10)	0.77	(0.54–1.10)
p for trend [b]			0.003		0.172		0.142	
Men (n = 810)								
T1	116.88	180/104	1.00		1.00		1.00	
T2	229.41	180/86	0.83	(0.58–1.18)	0.86	(0.57–1.30)	0.96	(0.62–1.49)
T3	483.12	180/80	0.77	(0.54–1.10)	0.85	(0.56–1.29)	0.92	(0.59–1.43)
p for trend [b]			0.180		0.502		0.717	
Women (n = 435)								
T1	180.27	96/68	1.00		1.00		1.00	
T2	369.16	97/46	0.67	(0.42–1.07)	0.92	(0.54–1.57)	0.84	(0.47–1.48)
T3	851.83	97/31	0.45	(0.27–0.75)	0.60	(0.34–1.07)	0.55	(0.30–1.02)
p for trend [b]			0.003		0.076		0.054	
Lutein/Zeaxanthin								
Total (n = 1245)								
T1	1716.04	276/152	1.00		1.00		1.00	
T2	2832.71	277/135	0.88	(0.66–1.17)	0.93	(0.67–1.28)	1.00	(0.71–1.41)
T3	5188.69	277/128	0.83	(0.62–1.12)	0.87	(0.63–1.22)	0.91	(0.64–1.30)
p for trend [b]			0.245		0.440		0.575	
Men (n = 810)								
T1	1676.82	180/91	1.00		1.00		1.00	
T2	2691.75	180/89	0.98	(0.68–1.40)	1.18	(0.78–1.79)	1.41	(0.91–2.20)
T3	4670.81	190/100	0.99	(0.69–1.41)	1.03	(0.68–1.58)	1.13	(0.72–1.76)
p for trend [b]			0.954		0.996		0.788	

Table 3. *Cont.*

	Median Intake (µg/day)	No. of Controls/Cases	Model 1		Model 2		Model 3	
			OR	(95% CI)	OR	(95% CI)	OR	(95% CI)
Women (n = 435)								
T1	1894.17	96/61	1.00		1.00		1.00	
T2	3268.31	97/39	0.63	(0.38–1.03)	0.60	(0.35–1.05)	0.64	(0.36–1.15)
T3	5678.59	97/45	0.71	(0.44–1.17)	0.83	(0.48–1.45)	0.82	(0.45–1.47)
p for trend [b]			0.240		0.636		0.602	
Lycopene								
Total (n = 1245)								
T1	327.68	276/209	1.00		1.00		1.00	
T2	1105.49	277/112	0.53	(0.40–0.70)	0.67	(0.49–0.92)	0.67	(0.48–0.94)
T3	3666.52	277/94	0.44	(0.33–0.59)	0.57	(0.41–0.80)	0.60	(0.42–0.85)
p for trend [b]			<0.001		0.003		0.012	
Men (n = 810)								
T1	268.48	180/138	1.00		1.00		1.00	
T2	934.05	180/65	0.47	(0.33–0.68)	0.51	(0.34–0.78)	0.55	(0.35–0.86)
T3	2963.59	180/67	0.49	(0.34–0.69)	0.60	(0.40–0.91)	0.60	(0.39–0.93)
p for trend [b]			0.001		0.056		0.062	
Women (n = 435)								
T1	442.87	96/69	1.00		1.00		1.00	
T2	1528.34	97/46	0.66	(0.41–1.05)	0.77	(0.45–1.31)	0.81	(0.46–1.42)
T3	4843.59	97/30	0.43	(0.26–0.72)	0.54	(0.30–0.96)	0.60	(0.32–1.11)
p for trend [b]			0.002		0.039		0.113	

[a] Total dietary carotenoids and carotenoid subclasses were categorized into tertiles according to the distribution of the control groups: Total carotenoids for overall subjects (T1: 7932.26, T2: 7932.26–13,062.36, and T3: ≥13,062.36), men (T1: <7393.71, T2: 7393.71–11,605.58, and T3: ≥11,605.58), and women (T1: <10,177.56, T2: 10,177.56–16,039.47, and T3: ≥16,039.47); α-carotene for overall subjects (T1: <480.37, T2: 480.37–956.11, and T3: ≥956.11), men (T1: <435.49, T2: 435.49–799.48, and T3: ≥799.48), and women (T1: <587.17, T2: 587.17–1160.14, and T3: ≥1160.14); β-carotene for overall subjects (T1: <3370.16, T2: 3370.16–5327.21, and T3: ≥5327.21), men (T1: <3165.51, T2: 3165.51–4843.61, and T3: ≥4843.61), and women (T1: <4238.05, T2: 4238.05–6436.30, and T3: ≥6436.30); β–cryptoxanthin for overall subjects (T1: <190.36, T2: 190.36–381.30, and T3: ≥381.30), men (T1: <166.94, T2: 166.94–322.32, and T3: ≥322.32), and women (T1: <254.69, T2: 254.69–510.98, and T3: ≥510.98); lutein/zeaxanthin for overall subjects (T1: <2242.79, T2: 2242.79–3,707.97, and T3: ≥3,707.97), men (T1: <2122.45, T2: 2122.45–3379.04, and T3: ≥3379.04), and women (T1: <2507.60, T2: 2507.60–4218.61, and T3: ≥4218.61); and lycopene for overall subjects (T1: <683.61, T2: 683.61–1881.13, and T3: ≥1881.13), men (T1: <574.69, T2: 574.69–1590.15, and T3: ≥1590.15), and women (T1: <898.55, T2: 898.55–2572.63, and T3: ≥2572.63). [b] To test for a trend across tertiles, the median intake for each tertile category was used as a continuous variable. Model 1: Adjusted for age; Model 2: Adjusted for age, total caloric intake, family history of GC, smoking status, regular exercise, education level, occupation, and monthly household income; Model 3: Additionally adjusted for *H. pylori* infection. In the overall subjects, models 1, 2, and 3 were additionally adjusted for gender.

Table 4. ORs and 95% CIs of GC according to tertiles of dietary lycopene intake stratified by *H. pylori* infection status [a].

| | | H. pylori-Positive (n = 868) | | | | | H. pylori-Negative (n = 353) | | | |
| | Median Intake (μg/day) | No. of Controls/Cases | Model 1 | | Model 2 | | No. of Controls/Cases | Model 1 | | Model 2 | |
			OR	(95% CI)	OR	(95% CI)		OR	(95% CI)	OR	(95% CI)
Lycopene Total (n = 1221)											
T1	327.68	169/194	1.00		1.00		100/15	1.00		1.00	
T2	1105.49	166/100	0.52	(0.37–0.72)	0.60	(0.42–0.86)	102/12	0.75	(0.33–1.69)	1.23	(0.48–3.18)
T3	3666.52	151/88	0.49	(0.35–0.69)	0.61	(0.42–0.90)	118/6	0.30	(0.11–0.81)	0.50	(0.16–1.54)
p for trend [b]			<0.001		0.037			0.017		0.168	

[a] Dietary lycopene intake was categorized into tertiles according to the distribution of the control group among overall subjects (T1: <683.61, T2: 683.61–1881.13, and T3: ≥1881.13). [b] To test for trend across tertiles, the median intake for each tertile category was used as a continuous variable. Model 1: Adjusted for age and gender; Model 2: Adjusted for age, gender, total caloric intake, family history of GC, smoking status, regular exercise, education level, occupation, and monthly household income.

Table 5. ORs and 95% CIs of GC according to tertiles of dietary lycopene intake stratified by smoking status [a].

| | | Ever-Smoker (n = 693) [b] | | | | | Non-Smoker (n = 551) | | | |
| | Median Intake (μg/day) | No. of Controls/Cases | Model 1 | | Model 2 | | No. of Controls/Cases | Model 1 | | Model 2 | |
			OR	(95% CI)	OR	(95% CI)		OR	(95% CI)	OR	(95% CI)
Lycopene Total (n = 1244)											
T1	327.68	166/138	1.00		1.00		110/70	1.00		1.00	
T2	1105.49	152/63	0.49	(0.34–0.71)	0.49	(0.31–0.77)	125/49	0.61	(0.39–0.96)	0.88	(0.52–1.48)
T3	3666.52	128/46	0.40	(0.26–0.60)	0.39	(0.23–0.65)	149/48	0.49	(0.32–0.77)	0.75	(0.45–1.27)
p for trend [c]			<0.001		0.001			0.006		0.307	

[a] Dietary lycopene intake was categorized into tertiles according to the distribution of the control group among overall subjects (T1: <683.61, T2: 683.61–1881.13, and T3: ≥1881.13). [b] Subjects who currently smoke or previously smoked were combined as ever-smokers. [c] To test for trend across tertiles, the median intake for each tertile category was used as a continuous variable. Model 1: Adjusted for age and gender; Model 2: Adjusted for age, gender, total caloric intake, family history of GC, *H. pylori* infection, regular exercise, education level, occupation, and monthly household income.

Table 6. Comparison of the consumption of lycopene contributing foods [a].

Food Consumption (g/day)	Cumulative (%) [b]	Total (n = 1245)			Men (n = 810)			Women (n = 435)		
		Controls (n = 830)	Cases (n = 415)	p [c]	Controls (n = 540)	Cases (n = 270)	p [c]	Controls (n = 290)	Cases (n = 145)	p [c]
Watermelon	34.56	21.02 ± 69.39	13.45 ± 21.73	0.004	15.25 ± 36.47	11.27 ± 21.19	0.050	31.75 ± 105.62	17.50 ± 22.22	0.028
Tomato	66.68	32.61 ± 90.50	14.71 ± 31.94	<0.001	27.01 ± 91.31	12.68 ± 32.60	0.001	43.05 ± 88.19	18.50 ± 30.43	<0.001
Tomato ketchup	90.26	5.64 ± 15.31	2.62 ± 5.67	<0.001	4.71 ± 15.51	2.28 ± 5.79	0.001	7.38 ± 14.81	3.25 ± 5.40	<0.001

[a] Adjusted for total energy intake using the residuals method. [b] Food items contributing to dietary lycopene intake that represented up to 90% of the cumulative contribution were selected. [c] p-values were calculated using Student's t-test.

Table 7. ORs and 95% CIs of GC according to tertiles of the consumption of lycopene contributing foods.

	Range (g/day)	Median Intake (g/day)	No. of Controls/Cases	Model 1		Model 2		Model 3	
				OR	(95% CI)	OR	(95% CI)	OR	(95% CI)
Watermelon									
Total (n = 1245)									
T1	<3.91	1.47	276/188	1.00		1.00		1.00	
T2	3.91–13.59	7.23	277/116	0.61	(0.46–0.81)	0.75	(0.54–1.04)	0.73	(0.52–1.03)
T3	≥13.59	29.32	277/111	0.58	(0.43–0.77)	0.77	(0.55–1.07)	0.71	(0.50–1.02)
p for trend [a]				0.002		0.233		0.130	
Men (n = 810)									
T1	<3.16	1.30	180/121	1.00		1.00		1.00	
T2	3.16–10.63	5.34	180/75	0.62	(0.44–0.88)	0.75	(0.50–1.13)	0.78	(0.51–1.20)
T3	≥10.63	24.63	180/74	0.61	(0.43–0.87)	0.84	(0.56–1.27)	0.78	(0.50–1.21)
p for trend [a]				0.034		0.657		0.391	
Women (n = 435)									
T1	<6.14	2.50	96/63	1.00		1.00		1.00	
T2	6.14–22.08	11.35	97/42	0.66	(0.41–1.07)	0.80	(0.46–1.40)	0.83	(0.46–1.51)
T3	≥22.08	43.54	97/40	0.63	(0.39–1.02)	0.76	(0.44–1.34)	0.77	(0.42–1.39)
p for trend [a]				0.117		0.421		0.440	
Tomato									
Total (n = 1245)									
T1	<4.70	1.70	276/217	1.00		1.00		1.00	
T2	4.70–20.12	10.15	277/120	0.55	(0.41–0.72)	0.66	(0.48–0.90)	0.66	(0.47–0.91)
T3	≥20.12	47.87	277/78	0.34	(0.25–0.47)	0.54	(0.38–0.77)	0.59	(0.41–0.85)
p for trend [a]				<0.001		0.002		0.016	

Table 7. Cont.

	Range (g/day)	Median Intake (g/day)	No. of Controls/Cases	Model 1		Model 2		Model 3	
				OR	(95% CI)	OR	(95% CI)	OR	(95% CI)
Men (n = 810)									
T1	<4.03	1.33	180/144	1.00		1.00		1.00	
T2	4.03–16.68	8.75	180/72	0.50	(0.35–0.71)	0.67	(0.45–1.00)	0.64	(0.42–0.99)
T3	≥16.68	35.92	180/54	0.37	(0.26–0.54)	0.57	(0.37–0.88)	0.58	(0.37–0.92)
p for trend [a]				<0.001		0.022		0.043	
Women (n = 435)									
T1	<6.13	2.74	96/67	1.00		1.00		1.00	
T2	6.13–30.90	14.03	97/54	0.79	(0.50–1.25)	0.98	(0.58–1.65)	0.97	(0.56–1.68)
T3	≥30.90	63.63	97/24	0.34	(0.20–0.60)	0.47	(0.25–0.87)	0.57	(0.30–1.12)
p for trend [a]				<0.001		0.010		0.084	
Tomato ketchup									
Total (n = 1245)									
T1	<0.84	0.30	276/217	1.00		1.00		1.00	
T2	0.84–3.58	1.81	277/123	0.56	(0.42–0.74)	0.66	(0.49–0.90)	0.66	(0.48–0.92)
T3	≥3.58	8.34	277/75	0.33	(0.24–0.45)	0.51	(0.36–0.72)	0.55	(0.38–0.80)
p for trend [a]				<0.001		0.001		0.005	
Men (n = 810)									
T1	<0.71	0.24	180/140	1.00		1.00		1.00	
T2	0.71–2.98	1.52	180/75	0.54	(0.38–0.76)	0.72	(0.49–1.08)	0.69	(0.45–1.06)
T3	≥2.98	6.45	180/55	0.39	(0.27–0.57)	0.61	(0.40–0.94)	0.62	(0.39–0.97)
p for trend [a]				<0.001		0.039		0.067	
Women (n = 435)									
T1	<1.08	0.50	96/67	1.00		1.00		1.00	
T2	1.08–5.19	2.43	97/54	0.79	(0.50–1.25)	0.97	(0.58–1.64)	0.96	(0.56–1.67)
T3	≥5.19	10.94	97/24	0.34	(0.20–0.60)	0.47	(0.25–0.88)	0.58	(0.30–1.12)
p for trend [a]				<0.001		0.011		0.089	

[a] Adjusted for total energy intake using the residuals method. Model 1: Adjusted for age; Model 2: Adjusted for age, total caloric intake, family history of GC, smoking status, regular exercise, education level, occupation, and monthly household income; Model 3: Additionally adjusted for *H. pylori* infection. In the overall subjects, models 1, 2, and 3 were additionally adjusted for gender.

4. Discussion

In our study, an inverse association was observed between higher dietary lycopene intake and the risk of GC in the overall subjects. This significant association remained in the subgroups of gender, *H. pylori*-positive subjects, and those who had ever smoked. The protective effect of dietary lycopene intake on the risk of GC among *H. pylori*–positive subjects and those who had ever smoked was particularly evident in males. Among the contributing foods of dietary lycopene, tomatoes and tomato ketchup exerted protective effects on the risk of GC.

However, previous epidemiological studies of the association between dietary lycopene intake and the risk of GC revealed that lycopene intake is less likely to be associated with GC risk. Among the five case-control studies regarding dietary lycopene intake, a significantly reduced GC risk was observed in only one study conducted in Uruguay (OR 0.37, 95% CI 0.19–0.73) [17]. In other studies, conducted in the US [20], Spain [19], Poland [15], and Italy [16], no significant association was observed. No significant associations between dietary lycopene intake and the risk of GC were observed in three cohort studies conducted in the Netherlands [21], Sweden [18], and in male participants who smoked in Finland [26]. Furthermore, in a meta-analysis that included only studies with validated food frequency questionnaires (FFQs), no statistically significant association was observed between higher dietary lycopene intake with the risk of GC [32]. Another meta-analysis of five case-control studies did not identify any association between higher dietary lycopene intake and the risk of GC [33].

The possible explanation for this trend is that dietary sources and trends of consumption differ among countries and, thus, do not perfectly overlap between studies. Another potential contributing factor is that the studies utilized diverse carotenoid content databases to estimate the consumption of dietary carotenoids, which were linked to various FFQ models. The carotenoid content of each food item was estimated using a different food composition database, such as the USDA database [20,24], the Nutrition Coding Center Nutrient Data System from the University of Minnesota [23], both American and Polish databases [15], Swedish [18,25], Italian [16], North American [17], Spanish [19], Finnish [26], and Dutch databases [21].

In our study, nine lycopene-containing food items were included, in order from the largest contribution to dietary lycopene intake: Watermelons, tomatoes, tomato ketchup, hamburgers, pizza, persimmons, red cabbage, carrots, and pepper powder (data not shown). Three of the nine food items were selected as the primary foods contributing to dietary lycopene intake. Unlike other dietary carotenoid subclasses and total carotenoids, which consist of contributing foods that were different from other studies, foods contributing to dietary lycopene intake were consistent with those from other studies because they included tomatoes and tomato-based products. Tomatoes and tomato-based products accounted for 81.2% of all dietary sources of lycopene in the US, and tomato-based soups and stews accounted for 3.8% [34]. Similarly, in European countries, tomatoes and tomato-based products constituted the major foods contributing to dietary lycopene intake: Tomatoes (25%), canned tomatoes (16%), and pizza (16%) in France; canned tomatoes (23%), tomato soup (17%), and pizza (16%) in Ireland; tomatoes (21%), canned tomatoes (20%), and pizza (15%) in the UK; tomato soup (29%), tomatoes (16%), and pizza (16%) in the Netherlands; and tomatoes (55%) and tomato puree (42%) in Spain [35]. On the other hand, in a previous study conducted in Korea, watermelons, tomatoes, and tomato ketchup contributed to 53.6%, 36.9%, and 5.7% of lycopene intake, respectively [36]. In our study, the major contributors of dietary lycopene intake were the same, but the only difference was the contribution percentage of each food item: 34.56% for watermelons, 32.12% for tomatoes, and 23.58% for tomato ketchup. Although the proportion of each food item that contributed to dietary intake of lycopene differed, the foods contributing to lycopene overlapped, indicating that some country-specific trends in the food items that contribute to lycopene intake might exist.

Among the foods that contribute to dietary lycopene intake, tomatoes and tomato ketchup were significantly associated with the risk of GC. The lycopene content (mg lycopene/100 g) of tomato ketchup is greater than fresh tomatoes [37–39]. Lycopene from fresh tomatoes is not readily bioavailable, and thus, by processing tomatoes, the bioavailability of lycopene is increased due to a breakdown of the tissue matrix [40]. This characteristic is due to the structure of the lycopene molecule: Lycopene in raw tomatoes is mainly in the *trans*-isomer form, but, during heat processing, the structure undergoes isomerization to a *cis* form and is thereby more efficiently absorbed [40]. This result is consistent with our finding that the intake of both tomatoes and tomato ketchup were inversely associated with the risk of GC.

Previously, several epidemiological studies on tomato consumption and the risk of GC have been conducted. In a case-control study conducted in the US, an inverse association was observed among only African Americans (OR 0.56, 95% CI 0.34–0.90), but not in Caucasians [41]. In a case-control study conducted in Sweden, a significant inverse association was observed between higher tomato intake (consumed more than 2.9 times per month) during adolescence (OR 0.36, 95% CI 0.23–0.58, *p* for trend < 0.0001) [42]. However, other case-control studies conducted in Spain [43], Sweden [22], Japan [44], and the US [45] did not identify any significant associations. In a cohort study conducted in the Netherlands, a borderline positive association was found between tomato consumption and non-cardia GC risk (relative risk (RR) per 25 g/day increase in the amount of tomatoes consumed: 1.13, 95% CI 1.00–1.28) [46]. However, when a meta-analysis was conducted with those seven studies listed above that have validated FFQs, tomato consumption was significantly associated with a decreased risk of GC (OR 0.73, 95% CI 0.60–0.90), with moderate heterogeneity (I^2 = 47.92%) [32]. This finding is consistent with our results because we also observed the protective effects of tomatoes and tomato ketchup on the risk of GC.

In our study, lycopene was inversely associated with the risk of GC, and the association remained significant in *H. pylori*-positive subjects and those who had ever smoked. There is a possible explanation of those risk factors and the effect of lycopene on GC prevention. Smoking, inflammation, and *H. pylori* infection may increase oxidative stress in the gastrointestinal tract, which leads to DNA damage, extracellular signal-regulated kinase (ERK) activation and p53 induction, decreased activities of antioxidant enzymes (glutathione, GSH; glutathione-S-transferase, GST; and glutathione peroxidase, GPx), and impaired immune function [46]. Lycopene might scavenge reactive oxygen species (ROS) and stimulate antioxidant enzyme activities, which protect gastric mucosa from oxidative stress-induced ERK activation, p53 induction, cell cycle disturbances, and impaired immune function, thereby, preventing gastric carcinogenesis [47]. Lycopene might scavenge ROS and stimulate antioxidant enzyme activities, which protect the gastric mucosa from oxidative stress-induced ERK activation, p53 induction, cell cycle disturbances, and impaired immune function, thereby, preventing gastric carcinogenesis [47].

However, regarding contributing foods, it would be presumptuous to assume that certain foods are representative of a specific nutrient, and to conclude that a food item protects against GC because of the specific nutrient. Foods contributing to dietary carotenoid intake are mostly fruits and vegetables, but they are also good sources of bioactive phytochemicals [48]. In a Spanish case-control study, a higher intake of dietary flavonoids, particularly kaempferol, exerted a protective effect on the risk of GC, while no significant association was observed for carotenoids [19]. Tomatoes are rich in carotenoids, but they also contain lower concentrations of polyphenols, such as hydroxycinnamic acids, flavanones, flavonols, anthocyanins, and flavonol glycosides [49]. In addition, tomatoes are a relatively rich source of vitamin C [32]. In vivo and in vitro studies have suggested that bioactive compounds in tomatoes may work additively or synergistically to reduce the growth of cancer cells [12]. However, since both higher dietary lycopene consumption and higher tomato intake were inversely associated with the risk of GC in our study, we may still suggest a possible effect of dietary lycopene intake by consuming tomatoes and tomato ketchup, which contributed to 55.7% of the total lycopene intake.

Our study has certain strengths: (1) A comprehensive and validated 106-item SQFFQ was used; (2) the study participants in the case group were surveyed by trained interviewers. Subjects in the control group were initially asked to complete the survey by themselves, and trained interviewers asked any questions with missing answers in a second session. Therefore, the quality of data was improved; (3) the carotenoid database included the contents of kimchi and seasoned vegetables, and covered 98.5% of all items reported in the SQFFQ; (4) information on the prevalence of *H. pylori* infection and smoking status, which are known risk factors for GC according to the IARC, were available [3,4]; and (5) to our knowledge, this is the first study conducted in Korea to investigate the association between dietary carotenoids and GC risk.

Several limitations should also be mentioned: (1) In this hospital-based, case-control study, selection bias might have occurred because the controls were those who had participated in the health screening. Subjects who chose to undergo screening may have had healthier lifestyles and dietary habits (e.g., greater consumption of fruits and vegetables) than individuals who did not choose to undergo screening; thus, the controls might be less representative of the general population; (2) our findings might be prone to recall bias because subjects were required to report their dietary intake for the past 12 months, which is a relatively long period. Additionally, when the SQFFQ was assessed, cases had already received a confirmed diagnosis; thus, the recall ability may have differed between cases and controls. However, in our study, cases were only patients who were diagnosed with early GC. Thus, compared to advanced GC cases, the influence of dietary changes on GC symptoms will be negligible. Moreover, cases with other health factors that might have affected their diet (subjects with diabetes mellitus, a severe systematic/mental disease, a history of cancer within five years, and women who were pregnant or currently breastfeeding) were excluded. Therefore, the recall bias in this study might be minimal; and (3) the sample size was relatively small in our study; in particular, few *H. pylori*-negative subjects and subjects with cardia GC were included. However, the higher percentage of cases with non-cardia GC among the overall GC incidence is a unique trend in Asian countries, including Korea [7]; nevertheless, a larger sample size is needed to increase the statistical power.

5. Conclusions

Based on our findings, higher dietary lycopene intake might be inversely associated with the risk of GC in the overall subjects. The association remained significant in the subgroups of gender, *H. pylori*-positive subjects, and those who had ever smoked. Foods contributing to dietary lycopene that exerted protective effects on the risk of GC were tomatoes and tomato ketchup. Further studies with larger sample sizes, including sufficient numbers of *H. pylori*-negative subjects and patients with cardia GC, are needed.

Supplementary Materials: The following are available online at http://www.mdpi.com/2072-6643/10/8/1031/s1, Table S1: Gender-specific ORs and 95% CIs of GC according to tertiles of dietary lycopene intake stratified by *H. pylori* infection status, Table S2: Gender-specific ORs and 95% CIs of GC according to tertiles of dietary lycopene intake stratified by smoking status.

Author Contributions: Formal analysis, J.L. and J.H.K.; Writing original draft, J.H.K.; Data curation, I.J.C., O.K., J.K., H.K. and J.L.; Investigation, I.J.C. and Y.-I.K.; Methodology, I.J.C., Y.-I.K. and J.K.; Funding acquisition, J.K. and O.K.; Project administration, J.K.; Supervision, J.K.

Funding: This research was supported by Grants-in-Aid for Cancer Research and Control from the National Cancer Center, Korea (no. 1410260, 1810090, and 1810980) and the Ministry of Science, ICT, and Future Planning, Korea (Bio-synergy Research Project no. NRF2012M3A9C4048761).

Conflicts of Interest: The authors declare no conflict of interest.

References

1. Ferlay, J.; Soerjomataram, I.; Dikshit, R.; Eser, S.; Mathers, C.; Rebelo, M.; Parkin, D.M.; Forman, D.; Bray, F. Cancer incidence and mortality worldwide: Sources, methods and major patterns in GLOBOCAN 2012. *Int. J. Cancer* **2015**, *136*, E359–E386. [CrossRef] [PubMed]
2. Jung, K.W.; Won, Y.J.; Kong, H.J.; Lee, E.S.; Kim, C.H.; Yoo, C.I.; Park, J.H.; Nam, H.S.; Huh, J.S.; Youm, J.H.; et al. Cancer statistics in Korea: Incidence, mortality, survival, and prevalence in 2015. *Cancer Res. Treat.* **2018**, *50*, 303–316. [CrossRef] [PubMed]

3. International Agency for Research on Cancer, Schistosomes, Liver Flukes and *Helicobacter pylori*: IARC Monographs on the Evaluation of Carcinogenic Risks to Humans. Volume 61. Available online: https://monographs.iarc.fr/ENG/Monographs/vol61/mono61.pdf (accessed on 31 March 2018).
4. International Agency for Research on Cancer, Tobacco Smoke and Involuntary Smoking: IARC Monographs on the Evaluation of Carcinogenic Risks to Humans. Volume 83. Available online: https://monographs.iarc.fr/ENG/Monographs/vol83/mono83.pdf (accessed on 31 March 2018).
5. Doll, R.; Peto, R. The causes of cancer: Quantitative estimates of avoidable risks of cancer in the United States today. *J. Natl. Cancer Inst.* **1981**, *66*, 1192–1308. [CrossRef]
6. Willett, W.C. Diet, nutrition, and avoidable cancer. *Environ. Health Perspect.* **1995**, *103*, 165–170. [CrossRef] [PubMed]
7. Colquhoun, A.; Arnold, M.; Ferlay, J.; Goodman, K.; Forman, D.; Soerjomataram, I. Global patterns of cardia and non-cardia gastric cancer incidence in 2012. *Gut* **2015**, *64*, 1881–1888. [CrossRef] [PubMed]
8. De Martel, C.; Parsonnet, J. *Helicobacter pylori* infection and gender: A meta-analysis of population-based prevalence surveys. *Dig. Dis. Sci.* **2006**, *51*, 2292–2301. [CrossRef] [PubMed]
9. Freedman, N.; Derakhshan, M.; Abnet, C.; Schatzkin, A.; Hollenbeck, A.; McColl, K. Male predominance of upper gastrointestinal adenocarcinoma cannot be explained by differences in tobacco smoking in men versus women. *Eur. J. Cancer* **2010**, *46*, 2473–2478. [CrossRef] [PubMed]
10. World Cancer Research Fund International. *Diet, Nutrition, Physical Activity and Stomach Cancer*; World Cancer Research Fund International: London, UK, 2018.
11. Stange, C. *Carotenoids in Nature: Biosynthesis, Regulation and Function*; Springer Nature: Basel, Switzerland, 2016.
12. Tanumihardjo, S.A. *Carotenoids and Human Health*; Humana Press: New York, NY, USA, 2013.
13. Wardlaw, G.M.; Smith, A.M.; Lindeman, A.K. *Contemporary Nutrition: A Functional Approach*; McGraw-Hill: New York, NY, USA, 2010; Volume 2.
14. Tanaka, T.; Shnimizu, M.; Moriwaki, H. Cancer chemoprevention by carotenoids. *Molecules* **2012**, *17*, 3202–3242. [CrossRef] [PubMed]
15. Lissowska, J.; Gail, M.H.; Pee, D.; Groves, F.D.; Sobin, L.H.; Nasierowska Guttmejer, A.; Sygnowska, E.; Zatonski, W.; Blot, W.J.; Chow, W.H. Diet and stomach cancer risk in Warsaw, Poland. *Nutr. Cancer* **2004**, *48*, 149–159. [CrossRef] [PubMed]
16. Pelucchi, C.; Tramacere, I.; Bertuccio, P.; Tavani, A.; Negri, E.; La Vecchia, C. Dietary intake of selected micronutrients and gastric cancer risk: An Italian case-control study. *Ann. Oncol.* **2008**, *20*, 160–165. [CrossRef] [PubMed]
17. De S., E.; Boffetta, P.; Brennan, P.; Deneo Pellegrini, H.; Carzoglio, J.; Ronco, A.; Mendilaharsu, M. Dietary carotenoids and risk of gastric cancer: A case-control study in Uruguay. *Eur. J. Cancer Prev.* **2000**, *9*, 329–334.
18. Larsson, S.C.; Bergkvist, L.; Näslund, I.; Rutegård, J.; Wolk, A. Vitamin A, retinol, and carotenoids and the risk of gastric cancer: A prospective cohort study. *Am. J. Clin. Nutr.* **2007**, *85*, 497–503. [CrossRef] [PubMed]
19. Garcia Closas, R.; Gonzalez, C.A.; Agudo, A.; Riboli, E. Intake of specific carotenoids and flavonoids and the risk of gastric cancer in Spain. *Cancer Cause Control* **1999**, *10*, 71–75. [CrossRef]
20. Harrison, L.E.; Zhang, Z.F.; Karpeh, M.S.; Sun, M.; Kurtz, R.C. The role of dietary factors in the intestinal and diffuse histologic subtypes of gastric adenocarcinoma: A case-control study in the US. *Cancer* **1997**, *80*, 1021–1028. [CrossRef]
21. Botterweck, A.A.; Van den Brandt, P.A.; Goldbohm, R.A. Vitamins, carotenoids, dietary fiber, and the risk of gastric carcinoma: Results from a prospective study after 6.3 years of follow-up. *Cancer* **2000**, *88*, 737–748. [CrossRef]
22. Ekström, A.M.; Serafini, M.; Nyrén, O.; Hansson, L.E.; Ye, W.; Wolk, A. Dietary antioxidant intake and the risk of cardia cancer and noncardia cancer of the intestinal and diffuse types: A population-based case-control study in Sweden. *Int. J. Cancer* **2000**, *87*, 133–140. [CrossRef]
23. Mayne, S.T.; Risch, H.A.; Dubrow, R.; Chow, W.H.; Gammon, M.D.; Vaughan, T.L.; Farrow, D.C.; Schoenberg, J.B.; Stanford, J.L.; Ahsan, H. Nutrient intake and risk of subtypes of esophageal and gastric cancer. *Cancer Epidemiol. Biomark.* **2001**, *10*, 1055–1062.
24. Nomura, A.M.; Hankin, J.H.; Kolonel, L.N.; Wilkens, L.R.; Goodman, M.T.; Stemmermann, G.N. Case-control study of diet and other risk factors for gastric cancer in Hawaii (United States). *Cancer Cause Control* **2003**, *14*, 547–558. [CrossRef]

25. Terry, P.; Lagergren, J.; Ye, W.; Nyrén, O.; Wolk, A. Antioxidants and cancers of the esophagus and gastric cardia. *Int. J. Cancer* **2000**, *87*, 750–754. [CrossRef]

26. Nouraie, M.; Pietinen, P.; Kamangar, F.; Dawsey, S.M.; Abnet, C.C.; Albanes, D.; Virtamo, J.; Taylor, P.R. Fruits, vegetables, and antioxidants and risk of gastric cancer among male smokers. *Cancer Epidemiol. Biomark.* **2005**, *14*, 2087–2092. [CrossRef] [PubMed]

27. Hamilton, S.R.; Aaltonen, L.A. *World Health Organization Classification of Tumors: Pathology and Genetics of Tumours of the Digestive System*; IARC Press: Lyon, France, 2000.

28. Ahn, Y.; Kwon, E.; Shim, J.; Park, M.; Joo, Y.; Kim, K.; Park, C.; Kim, D. Validation and reproducibility of food frequency questionnaire for Korean genome epidemiologic study. *Eur. J. Clin. Nutr.* **2007**, *61*, 1435–1441. [CrossRef] [PubMed]

29. United States Department of Agriculture, USDA Food Composition Databases. Available online: https://ndb.nal.usda.gov/ndb/nutrients/index (accessed on 31 March 2018).

30. *Tables of Food Functional Composition*, 1st ed.; Korea National Academy of Agricultural Science: Suwon, Korea, 2009.

31. Willett, W.C. *Nutritional Epidemiology*; Oxford University Press: New York, NY, USA, 2013; Volume 3.

32. Yang, T.; Yang, X.; Wang, X.; Wang, Y.; Song, Z. The role of tomato products and lycopene in the prevention of gastric cancer: A meta-analysis of epidemiologic studies. *Med. Hypotheses* **2013**, *80*, 383–388. [CrossRef] [PubMed]

33. Zhou, Y.; Wang, T.; Meng, Q.; Zhai, S. Association of carotenoids with risk of gastric cancer: A meta-analysis. *Clin. Nutr.* **2016**, *35*, 109–116. [CrossRef] [PubMed]

34. Murphy, M.M.; Barraj, L.M.; Herman, D.; Bi, X.; Cheatham, R.; Randolph, R.K. Phytonutrient intake by adults in the United States in relation to fruit and vegetable consumption. *J. Am. Diet Assoc.* **2012**, *112*, 222–229. [CrossRef]

35. O'neill, M.; Carroll, Y.; Corridan, B.; Olmedilla, B.; Granado, F.; Blanco, I.; Van den Berg, H.; Hininger, I.; Rousell, A.; Chopra, M. A European carotenoid database to assess carotenoid intakes and its use in a five-country comparative study. *Br. J. Nutr.* **2001**, *85*, 499–507. [CrossRef] [PubMed]

36. Lee, H.S.; Cho, Y.H.; Park, J.; Shin, H.R.; Sung, M.K. Dietary intake of phytonutrients in relation to fruit and vegetable consumption in Korea. *J. Acad. Nutr. Diet* **2013**, *113*, 1194–1199. [CrossRef] [PubMed]

37. Alda, L.M.; Gogoasa, I.; Bordean, D.M.; Gergen, I.; Alda, S.; Moldovan, C.; Nita, L. Lycopene content of tomatoes and tomato products. *J. Agroaliment. Process. Technol.* **2009**, *15*, 540–542.

38. Tonucci, L.H.; Holden, J.M.; Beecher, G.R.; Khachik, F.; Davis, C.S.; Mulokozi, G. Carotenoid content of thermally processed tomato-based food products. *J. Agric. Food Chem.* **1995**, *43*, 579–586. [CrossRef]

39. Story, E.N.; Kopec, R.E.; Schwartz, S.J.; Harris, G.K. An update on the health effects of tomato lycopene. *Annu. Rev. Food Sci. Technol.* **2010**, *1*, 189–210. [CrossRef] [PubMed]

40. Rao, A.; Waseem, Z.; Agarwal, S. Lycopene content of tomatoes and tomato products and their contribution to dietary lycopene. *Food Res. Int.* **1998**, *31*, 737–741. [CrossRef]

41. Correa, P.; Fontham, E.; Pickle, L.W.; Chen, V.; Lin, Y.; Haenszel, W. Dietary determinants of gastric cancer in south Louisiana inhabitants. *J. Natl. Cancer Inst.* **1985**, *75*, 645–654. [PubMed]

42. Hansson, L.E.; Nyrén, O.; Bergström, R.; Wolk, A.; Lindgren, A.; Baron, J.; Adami, H.O. Diet and risk of gastric cancer. A population-based case-control study in Sweden. *Int. J. Cancer* **1993**, *55*, 181–189. [CrossRef] [PubMed]

43. González, C.A.; Sanz, J.M.; Marcos, G.; Pita, S.; Brullet, E.; Saigi, E.; Badia, A.; Riboli, E. Dietary factors and stomach cancer in Spain: A multi-centre case-control study. *Int. J. Cancer* **1991**, *49*, 513–519. [CrossRef] [PubMed]

44. Takezaki, T.; Gao, C.M.; Wu, J.Z.; Ding, J.H.; Liu, Y.T.; Zhang, Y.; Li, S.P.; Su, P.; Liu, T.K.; Tajima, K. Dietary protective and risk factors for esophageal and stomach cancers in a low-epidemic area for stomach cancer in Jiangsu Province, China: Comparison with those in a high-epidemic area. *Jpn. J. Cancer Res.* **2001**, *92*, 1157–1165. [CrossRef] [PubMed]

45. Navarro Silvera, S.A.; Mayne, S.T.; Risch, H.; Gammon, M.D.; Vaughan, T.L.; Chow, W.H.; Dubrow, R.; Schoenberg, J.B.; Stanford, J.L.; West, A.B. Food group intake and risk of subtypes of esophageal and gastric cancer. *Int. J. Cancer* **2008**, *123*, 852–860. [CrossRef] [PubMed]

46. Steevens, J.; Schouten, L.J.; Goldbohm, R.A.; Van den Brandt, P.A. Vegetables and fruits consumption and risk of esophageal and gastric cancer subtypes in the Netherlands cohort study. *Int. J. Cancer* **2011**, *129*, 2681–2693. [CrossRef] [PubMed]

47. Kim, M.J.; Kim, H. Anticancer effect of lycopene in gastric carcinogenesis. *J. Cancer Prev.* **2015**, *20*, 92–96. [CrossRef] [PubMed]

48. Rao, A.V.; Rao, L.G. Carotenoids and human health. *Pharmacol. Res.* **2007**, *55*, 207–216. [CrossRef] [PubMed]

49. Martí, R.; Roselló, S.; Cebolla Cornejo, J. Tomato as a source of carotenoids and polyphenols targeted to cancer prevention. *Cancers* **2016**, *8*, 58. [CrossRef] [PubMed]

nutrients

MDPI

Article

Effect of Tomato Nutrient Complex on Blood Pressure: A Double Blind, Randomized Dose–Response Study

Talia Wolak [1,†], Yoav Sharoni [2,*], Joseph Levy [2], Karin Linnewiel-Hermoni [3], David Stepensky [2] and Esther Paran [1]

[1]   Hypertension Unit, Soroka University Medical center and Ben-Gurion University of the Negev, Beer Sheva 84101, Israel; twolak@bgu.ac.il (T.W.); paran@bgu.ac.il (E.P.)
[2]   Department of Clinical Biochemistry and Pharmacology, Ben-Gurion University of the Negev, Beer Sheva 84105, Israel; lyossi@bgu.ac.il (J.L.); davidst@bgu.ac.il (D.S.)
[3]   Lycored Ltd., Secaucus, NJ 07094, USA; Karin.Hermoni@lycored.com
*   Correspondence: yoav@bgu.ac.il; Tel.: +972-52-483-0883
†   Present address: Internal medicine D, Shaare Zedek Medical Center, Jerusalem 9103102, Israel.

Received: 27 February 2019; Accepted: 24 April 2019; Published: 26 April 2019

Abstract: Oxidative stress is implicated in the pathogenesis of essential hypertension, a risk factor for cardiovascular morbidity and mortality. Tomato carotenoids such as lycopene and the colorless carotenoids phytoene and phytofluene induce the antioxidant defense mechanism. This double-blind, randomized, placebo-controlled study aimed to find effective doses of Tomato Nutrient Complex (TNC) to maintain normal blood pressure in untreated hypertensive individuals. The effect of TNC treatment (5, 15 and 30 mg lycopene) was compared with 15 mg of synthetic lycopene and a placebo over eight weeks. Results indicate that only TNC treatment standardized for 15 or 30 mg of lycopene was associated with significant reductions in mean systolic blood pressure (SBP). Treatment with the lower dose standardized for 5 mg of lycopene or treatment with 15 mg of synthetic lycopene as a standalone had no significant effect. To test carotenoid bioavailability, volunteers were treated for four weeks with TNC providing 2, 5 or 15 mg lycopene. The increase in blood levels of lycopene, phytoene, and phytofluene was dose dependent. Results suggest that only carotenoid levels achieved by the TNC dose of 15 mg lycopene or higher correlate to a beneficial effect on SBP in hypertensive subjects while lower doses and lycopene alone do not.

Keywords: hypertension; carotenoids; tomato extract; lycopene; phytoene; phytofluene; bioavailability

1. Introduction

Hypertension (HT) is a major risk factor for cardiovascular morbidity and mortality. Prior studies provide strong evidence that oxidative stress and inflammatory processes correlate with the pathogenesis of HT. Although many drugs related to this chronic pathology are available, the high profile of side effects of anti-HT drugs warrant alternative and complementary treatment for blood pressure (BP) control [1,2]. This includes lifestyle modifications, especially dietary interventions such as consumption of more fruits and vegetables [3,4]. In line with these, the antioxidant lycopene has gained considerable attention in improving vascular function and regulating BP [5]. However, intervention trials investigating the role of lycopene supplementation or lycopene-containing foods in BP regulation have produced conflicting results. Several studies demonstrate that oral supplementation with tomato extract or tomato juice significantly decreases BP [5–9], while other studies show no relation [10] or no obvious association [11,12]. Interestingly one study even shows that lycopene can elevate BP [13].

In a meta-analysis, Ried et al. [14] reviewed four studies investigating the effect of lycopene-containing foods on BP and concluded that lycopene treatment could effectively decrease systolic BP (SBP), but had no statistically significant effect on diastolic BP (DBP). In an updated meta-analysis of intervention trials, Li et al. [15] similarly concluded a significant reduction in SBP by lycopene, but no significant effect on DBP. An important finding in this report is that lycopene treatment is more efficient in decreasing higher SBP (above 120 mmHg), but has no effect on normal BP.

Tomato products or extracts are used as the lycopene source in most of these studies. The extracts contain lycopene, β-carotene and the colorless carotenoids phytoene and phytofluene, in addition to a myriad of other active nutrients such as tocopherols and polyphenols. Many of these are strong antioxidants and known inducers of the antioxidant defense pathway, which is one of the mechanisms for the cardiovascular protective effect. The fact that these preparations are not pure lycopene has been frequently ignored, for example, in meta-analysis publications [14,15]. Thus, one of the major aims of the current work was to compare the effect on BP of a pure synthetic lycopene to a proprietary tomato extract. To complement our previous results showing the improvement of BP by natural antioxidants from tomato extract in pre-HT patients [6,9], we tested its BP-lowering effects in Stage 1 and 2 HT patients, attempting to establish a dose dependency and to show bioavailability of the major carotenoids present in tomato extract.

2. Materials and Methods

2.1. Study Population

The original sample size was 26 subjects in each study arm (see Section 2.5). However, following completion of treatment of the first 46 subjects (9–10 subjects in each treatment arm), an interim analysis was performed. Based on this, the study recruitment was stopped, and all study data were analyzed.

Sixty-one hypertensive individuals (according to the 2017 ACC guidelines) [16] without anti-HT treatment, aged 35–60 years, were recruited from primary-care clinics and through advertisements posted in local newspapers. The recruited patients had BP values in the range of 130 < SBP < 145 mmHg or 80 < DBP < 95 mmHg, measured as detailed in Section 2.3. Subjects treated for HT or dyslipidemia; who had any suspected allergy to tomato, carotenoids or α-tocopherol; or were taking vitamins or other food additives were excluded from the study. Smokers, persons with diabetes or cardiovascular, gastrointestinal, hepatic or malignant diseases were also excluded. All included subjects signed an informed consent form presented to them by the researchers after receiving an explanation regarding the course of the trial. The local Helsinki ethics committee approved the study protocol (approval No. 4594, 2008). The study was registered at ClinicalTrials.gov with the Identifier: NCT00637858.

2.2. Materials

Tomato Nutrient Complex, a proprietary tomato extract, was supplied by Lycored Ltd., Beer Sheva, Israel at doses corresponding to 5, 15 and 30 mg lycopene. Lycored also supplied identical-looking capsules with 15 mg synthetic lycopene (18–20% cis isomers) and placebo capsules containing soybean oil. The TNC 15-mg capsule contained the indicated amounts of the following tomato phytonutrients: lycopene (6%; 8–10% cis isomers), phytoene (1%), phytofluene (1%); β-carotene (0.15–0.2%), vitamin E (2%), and other fat-soluble phytonutrients naturally present in the tomato, suspended in tomato oleoresin oil.

2.3. Procedure

At the screening visit, held at the Hypertension Outpatient Clinic of the Soroka University Medical Center, a thorough physical examination was performed, and a comprehensive medical and dietary history was collected for each participant. Blood pressure, pulse rate, height and weight were measured, and body mass index (BMI) was calculated. BP was measured after 10 min of rest in a sitting position using a new and calibrated Omron HEM–705CP electronic semiautomatic sphygmomanometer (Tokyo, Japan), which was used only for this study. Recorded BP was calculated as the average of three serial measurements if the difference between them was <8 mmHg for SBP and <5 mmHg for DBP. A trained research nurse blinded to the study periods and treatment took all measurements. BP measurements were taken at the same hour of the morning after abstinence from food and caffeine for a minimum of 30 min.

Allocated participants started treatment with a four-week single-blind placebo run-in period, in which they were asked to consume one placebo capsule a day. This phase was used to confirm that participants indeed had HT. Only subjects with BP values of 130 < SBP < 145 mmHg or 80 < DBP < 95 mmHg were randomized for the double-blind placebo-controlled treatment phase. The study was performed under complete blind conditions (single-blind or double-blind, in accordance with the study phase). Knowledge of the randomization list was limited to the independent study statistician responsible for the creation of the list (Dr. Michael Friger, Department of Epidemiology, Ben-Gurion University) and the LycoRed designee responsible for preparation of the study medication. The subjects were allocated sequential subject numbers, and the investigator entered the subject randomization number into each subject's file. There were five treatment arms: TNC 5 mg, TNC 15 mg, TNC 30 mg, synthetic lycopene (15 mg) and a placebo administered once daily for eight weeks. Participants were blinded to the different study periods and were instructed to take the capsules with the main meal of the day to improve absorption of its ingredients. No other dietary supplements were allowed throughout the study, and participants were instructed to keep their usual dietary habits. Follow-up visits were held every two weeks and included a short clinical evaluation, and BP and pulse rate measurements. Study medications were dispensed, and compliance was verified by counting the remaining capsules and giving reinforcement at each visit.

Blood was drawn after an overnight fast at the end of the run-in phase and the end of the eight-week treatment phase. Blood analysis for safety included: complete blood count; plasma levels of glucose, urea, uric acid, sodium, potassium, creatinine, aspartate amino transferase (AST), alanine amino transferase (ALT), alkaline phosphatase (ALP), lactate dehydrogenase (LDH) and bilirubin; and blood lipid profile: cholesterol, high-density lipoprotein (HDL), low-density lipoprotein (LDL) and triglycerides. Additional plasma samples were stored at −70 °C. Stored samples were extracted with ethanol and hexane/dichloromethane, and then analyzed for lycopene using a high-performance liquid chromatography (HPLC)-based analytical method.

2.4. Steady-State Bioavailability of Tomato Carotenoids TNC Dose–Response

The study was performed with a different group of subjects from the ones participating in the BP study. It was conducted at the Endocrine Laboratory, Department of Clinical Biochemistry and Pharmacology of the Ben-Gurion University and the Soroka University Medical Center and was approved by the Helsinki Committee of the medical center (approval No. 0012-15-SOR). Twenty-five healthy volunteers, aged 20–40, were recruited from students of Ben-Gurion University through advertisements posted on the university electronic bulletin board. Written informed consent was obtained from all volunteers prior to entry into the double-blind, randomized, cross-over study. In each experiment, baseline blood samples were collected after overnight fasting, and treatment with TNC at 2, 5 or 15 mg of lycopene was begun for four weeks, followed by a four-week washout period between experiments without consumption of any supplement. The capsules with the different doses were identical in appearance, and the different doses were randomly distributed to about one-third of the volunteers in each experiment. Both the volunteers and the researchers were blinded to the received doses. Participants were instructed to consume one capsule per day with their main meal and to keep their usual dietary habits throughout the experimental and washout periods. Blood samples were taken after two, three and four weeks of treatment, after overnight fasting and 24 h after taking the last supplement. Compliance was verified by counting the remaining capsules and it was 90% or higher.

Carotenoid analysis was performed after ethanol and hexane/dichloromethane (4/1) extraction by dedicated HPLC method using a C30 reverse phase column, gradient elution and detection at multiple wavelengths using a photodiode array detector.

2.5. Statistical Methods

Sample size calculations were performed by using appropriate formulas based on 80% power and a two-sided $\alpha = 0.05$ with assumption of a standard deviation of DBP equal to 3.8 mmHg. A clinically significant difference in DBP was determined at 3.0 mmHg. For this determination, the sample size was 26 patients in each treatment arm—in all, 130 participants.

The paired t-test or non-parametric signed-rank test was applied for testing the differences of the continuous assessments between all visits to the baseline. An ANOVA model using the Duncan method was applied for testing the differences in blood pressure changes between all study groups. All tests applied were two-tailed, and a p-value of 5% or less was considered statistically significant. The data were analyzed using the SAS® version 9.1 for Windows (SAS Institute, Cary, North Carolina). The missing data for early withdrawals who were not replaced and who attended at least four weeks of the double-blind placebo-controlled phase of the study were handled as LOCFs (last observation carried forward).

For the bioavailability study, summary statistics were calculated using a GraphPad Prism 5.0 program. Paired t-tests were used to compare the carotenoid concentrations at different time points. A p-value of <0.05 was deemed statistically significant.

3. Results

3.1. Demographics and Baseline Characteristics

Sixty-one patients with BP values in the range of 130 < SBP < 145 mmHg or 80 < DBP < 95 mmHg were enrolled in the study and began treatment with a four-week single-blind placebo run-in phase. These patients were randomized for the double-blind placebo-controlled treatment phase. At enrollment, there were 12 subjects in each of the following arms: TNC 5 mg, TNC 15 mg, synthetic lycopene 15 mg and placebo, in addition to 13 subjects in the TNC 30-mg arm. Forty-six subjects completed the eight-week treatment period and 15 (3 in each arm) dropped out of the study prematurely. There were neither adverse effects reported during the entire study period nor any significant changes in glucose, urea, creatinine, uric acid, sodium, potassium, chloride, cholesterol, triglycerides, AST, ALT, ALP, LDH, HDL, LDL levels, or in the blood count parameters.

The treatment arms were comparable with respect to all demographic and baseline characteristics (Table 1). Baseline SBP and DBP measurements were not statistically significantly different Table 1). The mean age was 52.4 ± 8.2 years, and 73.8% were male. The mean SBP and DBP were 135.2 ± 7.4 mmHg and 82.0 ± 11.7 mmHg, respectively. There were no statistically significant differences between the five arms in the baseline plasma lycopene concentrations; however, the mean concentration of the TNC 15-mg arm was somewhat higher than the other arms.

Table 1. Baseline characteristics of the study population.

Parameter	TNC 5 mg	TNC 15 mg	TNC 30 mg	Synthetic Lycopene 15 mg	Placebo	All
Age (years) [1]	53.5 ± 8.0 (12)	54.0 ± 8.6 (12)	51.0 ± 10.1 (13)	52.1 ± 8.4 (12)	51.8 ± 6.3 (12)	52.4 ± 8.2 (61)
Males (%)	66.7% (8)	75.0% (9)	69.2% (9)	66.7% (8)	83.3% (10)	72.1% (44)
Females (%)	33.3% (4)	25.0% (3)	30.8 (4)	33.3% (4)	16.7% (2)	27.9% (17)
SBP (mmHg) [1]	133.6 ± 7.8 (12)	137.4 ± 5.6 (12)	136.4 ± 7.8 (13)	132.8 ± 9.3 (12)	135.7 ± 6.0 (12)	135.2 ± 7.4 (61)
DBP (mmHg) [1]	82.9 ± 9.3 (12)	83.8 ± 6.3 (12)	77.5 ± 21 (13)	82.3 ± 6.1 (12)	83.7 ± 8.7 (12)	82.0 ± 11.7 (61)
Plasma lycopene (μM)	0.97 ± 0.49 (7)	1.55 ± 0.86 (7)	0.80 ± 0.74 (10)	0.91 ± 0.95 (8)	1.01 ± 1.24 (10)	0.93 ± 0.68 (42)
TG (mg/dL)	152.0 ± 70.4 (11)	164.3 ± 65.3 (10)	124.5 ± 75.3 (12)	115.1 ± 50.5 (10)	166.0 ± 103.1 (12)	
HDL chol (mg/dL)	52.0 ± 12.2 (10)	49.5 ± 10.1 (10)	54.4 ± 10.0 (12)	59.1 ± 18.2 (10)	52.1 ± 11.7 (11)	
LDL chol (mg/dL)	142.1 ± 34.6 (10)	107.4 ± 26.0 (9)	104.8 ± 31.1 (11)	134.2 ± 29.5 (10)	112.9 ± 30.1 (9)	

[1] Age, BP, lycopene, TG, HDL, and LDL values are mean ± SD (N).

3.2. Changes in BP Values during the 8-Week Double-Blind Placebo-Controlled Treatment Phase

The average SBP in the five treatment arms during eight weeks of treatment is presented in Table 2. The change in SBP in the treatment arms during this same period are presented in Figure 1. Treatment with TNC containing 15 or 30 mg lycopene was associated with statistically significant reductions in mean SBP at almost all time points from two to eight weeks. Similar effects were not observed following treatment with 5 mg of TNC, 15 mg of synthetic lycopene or the placebo. A comparison between treatment arms revealed that the reduction in SBP in the TNC 15-mg arm following eight weeks of treatment was statistically different from that of the TNC 5-mg arm, the placebo arm and the synthetic lycopene arm ($p = 0.0495$). Remarkably, the rate of reduction in SBP with time of treatment for the 15-mg and 30-mg arms is almost parallel (Figure 1), suggesting that TNC containing 15 mg lycopene is both necessary and sufficient to normalize SBP.

Table 2. SBP during eight weeks of treatment.

SBP [1] (mmHg)	TNC 5 mg	TNC 15 mg	TNC 30 mg	Synthetic Lycopene 15 mg	Placebo
Baseline	133.6 ± 7.8 (12)	137.4 ± 5.6 (12)	136.4 ± 7.8 (13)	132.8 ± 9.3 (12)	135.7 ± 6.0 (12)
Week 2	132.7 ± 8.9 (10)	133.7 ± 5.3 (12)	131.5 ± 8.0 (13)	133.2 ± 8.9 (11)	137.1 ± 10.3 (11)
Week 4	133.1 ± 9.2 (9)	130.7 ± 4.7 (10)	131.7 ± 10.5 (11)	133.4 ± 7.2 (9)	134.5 ± 8.7 (10)
Week 6	132.8 ± 9.2 (9)	130.2 ± 10.6 (9)	129.9 ± 13.9 (10)	130.1 ± 9.8 (9)	133.7 ± 9.9 (9)
Week 8	131.4 ± 7.7 (9)	127.2 ± 6.3 (9)	130.0 ± 13.0 (10)	132.6 ± 5.1 (9)	133.0 ± 6.7 (9)

[1] SBP values are mean ± SD (N).

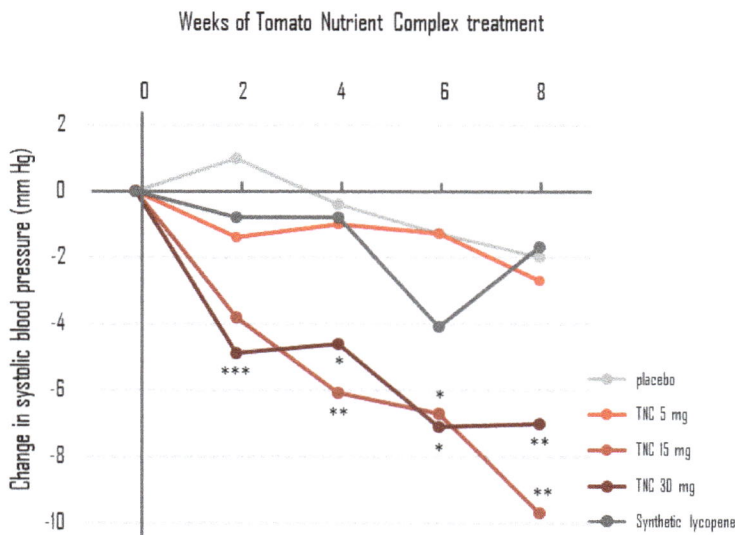

Figure 1. Tomato Nutrient Complex containing 15 mg and 30 mg lycopene reduces systolic blood pressure. Results are the mean of the individual changes from baseline to each time point from two to eight weeks. SDs are not shown for graph simplicity. N is the same as in Table 2. * $p < 0.05$; ** $p < 0.01$; *** $p < 0.001$ for changes to baseline within treatment arm.

The average DBP in the five treatment arms during the eight-week treatment is presented in Table 3. The changes in DBP in the five arms during this period are presented in Table 4. A significant reduction in DBP from baseline to eight weeks was observed only in the TNC 15-mg arm ($-4.1 ± 5.0$, $p = 0.038$, Table 4). There were no significant differences in DBP when treatment arms were compared.

Table 3. DBP during eight weeks of treatment.

DBP [1] (mmHg)	TNC 5 mg	TNC 15 mg	TNC 30 mg	Synthetic Lycopene 15 mg	Placebo
Baseline	82.9 ± 9.3 (12)	83.8 ± 6.3 (12)	77.5 ± 21 (13)	82.3 ± 6.1 (12)	83.7 ± 8.7 (12)
Week 2	81.4 ± 8.2 (10)	80.7 ± 7.1 (12)	80.5 ± 7.9 (13)	84.5 ± 6.1 (11)	84.2 ± 8.2 (11)
Week 4	82.8 ± 9.9 (9)	81.3 ± 3.9 (10)	82.6 ± 9.1 (11)	84.9 ± 6.6 (9)	83.0 ± 8.8 (10)
Week 6	82.6 ± 6.4 (9)	78.8 ± 7.5 (9)	72.3 ± 23 (10)	81.6 ± 6.0 (9)	83.2 ± 7.9 (9)
Week 8	81.7 ± 7.9 (9)	78.6 ± 7.9 (9)	74.4 ± 23 (10)	84.6 ± 5.0 (9)	85.3 ± 7.0 (9)

[1] DBP values are mean ± SD (*N*).

Table 4. Change in DBP from baseline during eight weeks of treatment.

DBP (mmHg)	TNC 5 mg	TNC 15 mg	TNC 30 mg	Synthetic Lycopene 15 mg	Placebo
	Mean ± SD (*p* [1])	Mean ± SD (*p*)	Mean ± SD (*p*)	Mean ± SD (*p*)	Mean ± SD (*p*)
Week 2	−2.9 ± 7.7 (0.265)	−3.2 ± 4.5 (0.032)	3.0 ± 4.9 (0.672)	0.9 ± 2.9 (0.331)	0.0 ± 5.0 (1.000)
Week 4	−0.9 ± 5.1 (0.613)	−1.9 ± 5.1 (0.268)	6.8 ± 7.0 (0.422)	1.2 ± 4.6 (0.449)	−0.4 ± 8.6 (0.887)
Week 6	−1.1 ± 5.8 (0.582)	−3.9 ± 6.5 (0.112)	−3.5 ± 5.9 (0.092)	−2.1 ± 4.5 (0.198)	−1.4 ± 9.8 (0.671)
Week 8	−2.0 ± 6.3 (0.371)	−4.1 ± 5.0 (0.038)	−1.4 ± 5.7 (0.454)	0.9 ± 5.6 (0.645)	0.7 ± 5.8 (0.740)

[1] *p*-value for changes to baseline within treatment arms. *N* is the same as in Table 3.

3.3. Changes in BP Values during the Single-Blind 16-Week Study Extension

Of the 46 subjects who completed the first phase of eight weeks of treatment, 31 agreed to continue for a single-blind four-month extension study with TNC containing 15 mg of lycopene. Four subjects ended this phase prematurely. The average SBP and DBP of the 27 subjects who completed the extension phase were compared to the average baseline values of the same subjects, measured at the randomization stage before the eight weeks of treatment. The results of SBP and DBP are presented in Table 5 and Figure 2. A statistically significant reduction in mean SBP was evident at all time points from one to four months. This result suggests that the SBP-reducing effect of TNC is long lasting. The changes in DBP from baseline to Month 4 were not significant.

Table 5. Mean SBP and DBP and changes from baseline in BP values during four-month study extension.

	SBP (mmHg) [1]	Change of SBP [1] from Baseline (mmHg)	*p*-Value for Change	DBP (mmHg) [1]	Change of DBP [1] from Baseline (mmHg)	*p*-Value for Change
Baseline	135.1 ± 8.76			85.1 ± 6.6		
Month 1	130.9 ± 8.6	−4.2 ± 6.5	0.0024	84.3 ± 6.2	−0.9 ± 5.7	0.422
Month 2	131.7 ± 10	−3.5 ± 7.4	0.0211	81.6 ± 15.5	−3.6 ± 15.8	0.252
Month 3	130.6 ± 8.8	−4.5 ± 7.0	0.0024	82.7 ± 7.7	−2.4 ± 6.3	0.056
Month 4	130.4 ± 10.9	−4.7 ±6.7	0.0012	83.9 ± 7.3	−1.4 ± 4.3	0.362

[1] BP values and changes in BP values are mean ± SD for the 27 subjects who completed the four-month extension phase.

3.4. Changes from Baseline of Plasma Levels of Lycopene

There was large variability in plasma lycopene baseline values (Table 6), and in the changes from baseline to eight weeks of treatment (Figure 3). Due to this variability and the small number of tested samples, there was no significant change in lycopene concentration in any of the arms, but the highest change was observed in the synthetic lycopene arm (Figure 3).

Months of Tomato Nutrient Complex extension treatment

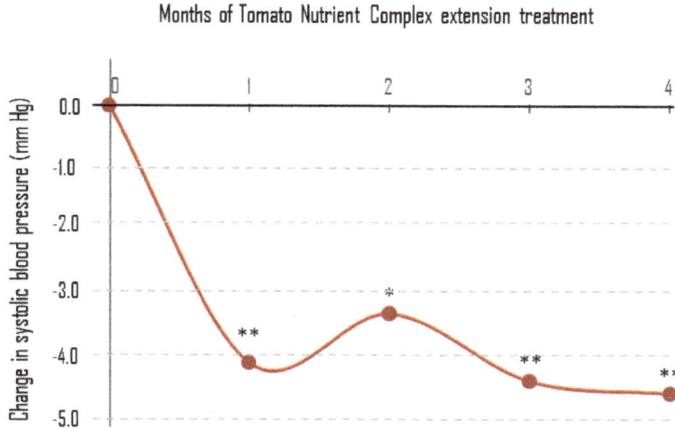

Figure 2. Changes in SBP values during the single-blind 16-week study extension. SBP values are the mean for the 27 subjects who completed this phase. * $p < 0.05$; ** $p < 0.01$ for changes to baseline within treatment arm.

Table 6. Plasma lycopene value at baseline and at eight weeks of treatment.

SBP [1] (mmHg)	TNC 5 mg	TNC 15 mg	TNC 30 mg	Synthetic Lycopene 15 mg	Placebo
Baseline	0.97 ± 0.49 (7)	1.55 ± 0.86 (7)	0.80 ± 0.74 (9)	0.91 ± 0.95 (7)	1.01 ± 1.24 (9)
Week 8	1.13 ± 0.97 (7)	1.89 ± 1.22 (7)	1.23 ±1.07 (9)	1.84 ± 1.14 (7)	0.72 ± 0.72 (9)

[1] Plasma lycopene values are mean ± SD (*N*).

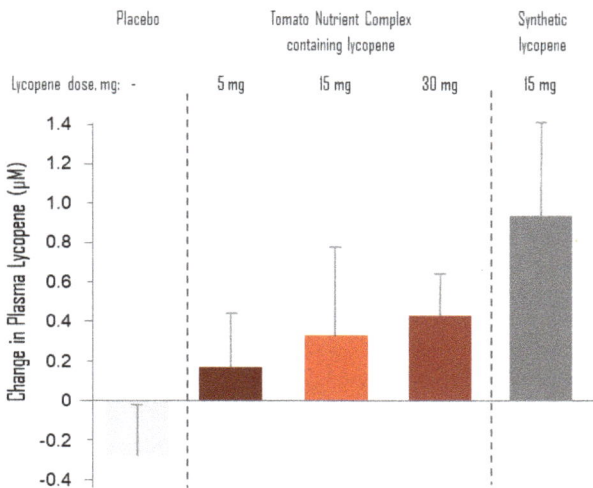

Figure 3. Changes of plasma levels of lycopene from baseline to eight weeks. Results are the mean ± SD of the changes from baseline to Week 8 within treatment arms. *N* was 9, 7, 7, 9, and 7 for the placebo, TNC 5 mg, TNC 15 mg, TNC 30 mg and synthetic lycopene arms, respectively.

3.5. Bioavailability of Tomato Carotenoids during 4 Weeks of Daily TNC Supplementation—A Dose–Response Study

Because of the large variability in the results of carotenoid plasma concentrations in the BP study patients (Section 3.4) and the low number of available samples, we conducted a separate bioavailability study of TNC with a different group of volunteers to evaluate the dose–response of steady-state carotenoid plasma concentrations during daily TNC treatment. A statistically significant increase in plasma lycopene was evident from Week 2 for the three supplemented doses (Table 7). For phytofluene, such an increase was not detected at all time points for the lowest dose (TNC 2 mg), but was evident at the highest dose (TNC 15 mg), and at Weeks 3 and 4 with TNC 5 mg. For phytoene, a significant increase was already detected for the two higher doses at Week 2 and at the lowest dose only at Week 4. No significant change was detected in the plasma concentrations of β-carotene, a control carotenoid that is found in very low amounts in the TNC preparations (e.g., 0.8 mg in the TNC 15 mg).

Table 7. Carotenoid plasma concentrations during 4-week treatment with TNC containing 2, 5 and 15 mg lycopene.

Carotenoid Plasma Concentration (µM)	Baseline Mean ± SD	Week 2 Mean ± SD (p)[1]	Week 3 Mean ± SD (p)[1]	Week 4 Mean ± SD (p)[1]
Lycopene				
TNC 2 mg[2]	1.021 ± 0.437	1.483 ± 0.481 (0.001)	1.354 ± 0.408 (0.009)	1.656 ± 0.537 (<0.000x)
TNC 5 mg[2]	1.149 ± 0.322	1.629 ± 0.444 (<0.000x)	1.677 ± 0.470 (<0.000x)	1.767 ± 0.473 (<0.000x)
TNC 15 mg[2]	1.036 ± 0.398	1.794 ± 0.453 (<0.000x)	1.754 ± 0.449 (<0.000x)	2.008 ± 0.627 (<0.000x)
Phytofluene				
TNC 2 mg[2]	0.310 ± 0.147	0.382 ± 0.211 (0.171)	0.300 ± 0.126 (0.805)	0.398 ± 0.242 (0.128)
TNC 5 mg[2]	0.354 ± 0.130	0.429 ± 0.190 (0.120)	0.456 ± 0.201 (0.046)	0.530 ± 0.263 (0.005)
TNC 15 mg[2]	0.270 ± 0.111	0.503 ± 0.235 (<0.000x)	0.536 ± 0.228 (<0.000x)	0.650 ± 0.275 (<0.000x)
Phytoene				
TNC 2 mg[2]	0.057 ± 0.040	0.073 ± 0.060 (0.300)	0.059 ± 0.037 (0.868)	0.091 ± 0.061 (0.028)
TNC 5 mg[2]	0.046 ± 0.032	0.096 ± 0.050 (<0.000x)	0.092 ± 0.057 (0.001)	0.124 ± 0.058 (<0.000x)
TNC 15 mg[2]	0.048 ± 0.031	0.133 ± 0.062 (<0.000x)	0.127 ± 0.068 (<0.000x)	0.158 ± 0.091 (<0.000x)
β-carotene				
TNC 2 mg[2]	0.842 ± 0.391	0.975 ± 0.484 (0.289)	0.872 ± 0.486 (0.811)	0.998 ± 0.502 (0.224)
TNC 5 mg[2]	0.947 ± 0.598	0.954 ± 0.580 (0.965)	0.919 ± 0.473 (0.866)	1.037 ± 0.533 (0.590)
TNC 15 mg[2]	0.885 ± 0.534	0.968 ± 0.609 (0.618)	0.970 ± 0.553 (0.587)	1.057 ± 0.631 (0.315)

[1] p-value for difference from baseline within the same dose arm. [2] N = 25 for TNC 2 mg and 15 mg. N = 24 for TNC 5 mg.

To calculate the average net increase in carotenoid concentrations, individual baseline concentrations were subtracted from values at other time points (Figure 4). The increase in plasma lycopene and phytofluene concentrations were significantly lower with TNC 5 mg and TNC 2 mg than with TNC 15 mg. This result suggests that the lower increase in these carotenoid concentrations for TNC 5 mg as compared to TNC 15 mg was not enough to drive the reduction in SBP in the HT subjects (Table 3). For phytoene, there was a significant difference between TNC 15 and TNC 2 mg, but not TNC 5 mg. When comparing the changes in carotenoid concentrations between TNC 5 mg and TNC 2 mg, a significantly higher increase was evident at the three time points only for phytoene concentrations, but not for lycopene and phytofluene.

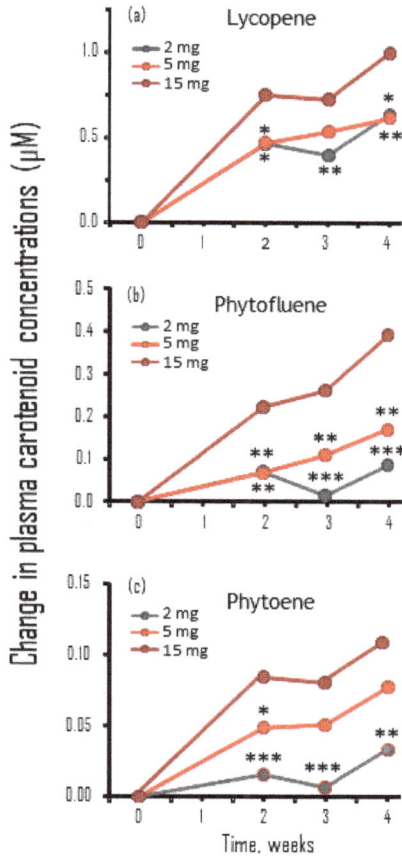

Figure 4. The increase in lycopene, phytofluene and phytoene concentrations after 2–4 weeks of treatment with TNC were significantly lower with TNC 5 mg and TNC 2 mg than with TNC 15 mg: (**a**) lycopene; (**b**) phytofluene; and (**c**) phytoene. Please note that the scale in the Y-axis is different for the three carotenoids. Results are the mean of individual changes in carotenoid concentrations of 25 subjects (TNC 2 mg and 15 mg) or 24 subjects (TNC 5 mg). SDs are not shown for graph simplicity. * $p < 0.05$; ** $p < 0.01$; *** $p < 0.001$ for the difference at each time point between TNC 15 mg to TNC 2 mg or TNC 5 mg.

A small continuous increase in the three carotenoid concentrations was observed from Week 2, to Weeks 3 and 4. However, there were no statistically significant differences between the concentrations at these time points, which suggests that steady-state concentrations were already attained after two weeks of TNC supplementation.

4. Discussion

The aims of the current study were to perform a dose–response analysis and uncover the optimal effective dose of a proprietary Tomato Nutrient Complex supplement in maintaining blood pressure within a normal range among untreated HT individuals and to compare the effect of TNC to that of synthetic lycopene. The bioavailability of the major tomato carotenoids was studied in a separate group of volunteers to gain insight into their relative contribution to the BP lowering effect of TNC.

Our study has several limitations. The minimum time between food and caffeine intake was determined to be 30 min. However, this time may be too short to prevent any effect of food intake on the BP measurement. In addition, the exact time and the type of food/beverages taken were not recorded; thus, it was not possible to estimate the effect of the food on BP analysis. The study lacks quantitative information on consumption of fruit and vegetables at baseline and at follow-up times. However, the lycopene plasma concentrations at baseline, which partially reflect this consumption, were not statistically different between study arms (Table 1), suggesting that this did not have a major effect on the study results. Although the participants were encouraged to maintain their dietary habits, there is no quantitative information on their food consumption during the study period to confirm that they did. In addition, there is no information on BMI or physical activity before and during the study, factors that may affect BP. These limitations should be considered in future studies, but it should be emphasized that, despite these limitations, the study resulted in showing significant effects of TNC on SBP.

Results of this double-blind, randomized, placebo-controlled study indicate that treatment with TNC standardized to contain 15 or 30 mg of lycopene was associated with statistically significant reductions in mean SBP. Treatment with the lower dose of TNC standardized for 5 mg of lycopene or treatment with 15 mg of synthetic lycopene did not produce a significant effect. DBP was not significantly different from the baseline with any treatment or control arm. Subjects from all treatment arms participated in an additional single-blind extension period for four months with TNC containing 15 mg lycopene, which resulted in a significant reduction of SBP that was persistent throughout the extension period. Thus, the reduction of SBP appears to be of a long-term nature. During the study and extension period, all treatments were safe and well tolerated, and no treatment-related adverse effects were reported.

The current results are in line with the recent meta-analysis of the effects of tomato products on BP [14,15] demonstrating a significant reduction of SBP, but no statistically significant effect on DBP. Subgroup analyses within the six studies [1,6–9,13], which met the inclusion criteria in the more recent study [15], demonstrated that higher dosage of tomato-derived supplements (containing more than 12 mg lycopene per day) could significantly lower SBP, whereas lower doses did not show a significant effect. This is similar to the results of the current study, which did not indicate a significant SBP reduction with TNC corresponding to 5 mg lycopene per day, but did show an effect at 15 and 30 mg. The low dose is probably the reason for the negative results in Paterson et al.'s study [13], which used only 4.5 mg/day lycopene, an amount close to the non-effective lycopene dose in the current study. Another important factor obtained in the meta-analysis was related to the baseline SBP. A significant reduction of SBP was observed if participants had higher baseline SBP (SBP $\geq$ 120 mmHg). Although this issue was not examined in the current study, it is important to note that the baseline SBP of all participants was above the threshold level (120 mmHg) suggested by the meta-analysis.

Remarkably, treatment with 15 mg of synthetic lycopene as a standalone did not cause a significant reduction in SBP. Apparently, this was not due to lower bioavailability of lycopene in the synthetic preparation. Actually, the mean increase in blood lycopene between the baseline and eight weeks was higher with the synthetic lycopene as compared to treatment with TNC containing 15 and 30 mg lycopene. Synthetic lycopene is different from tomato-derived lycopene in the ratio of cis:trans isomers—about 20% cis-isomers are in the synthetic lycopene and 10% in TNC. However, this difference cannot explain the better efficiency of TNC in lowering SBP, as we showed previously that the cis:trans ratio in supplemented lycopene does not affect the ratio in plasma and tissues [17]. The effects of purified lycopene was tested in another study on biomarkers of oxidative stress [18]. In that study, healthy volunteers were treated with 0, 6.5, 15 or 30 mg lycopene/day for eight weeks, which significantly increased plasma lycopene levels in agreement with our results. Interestingly, it was found that pure lycopene supplementation at all doses did not affect biomarkers of lipid peroxidation, whereas biomarkers of DNA damage were affected only at the highest dose of 30 mg/day of pure lycopene. This is similar to the lack of effect of 15 mg pure lycopene on BP in the current study.

One explanation for the lower efficacy of the synthetic lycopene is the presence of other active nutrients in the tomato extract preparation. Indeed, in a previous study [19], we found that the anti-cancer effects of carotenoids and other phytonutrients present in tomato extract (e.g., lycopene, phytoene and phytofluene) resides in their combined activity, which is synergistically higher than the activity of each compound alone. To assess the potential contribution of other tomato constituents to the BP-lowering effect, we analyzed the bioavailability of these carotenoids following treatment with different doses of TNC in a group of 25 healthy volunteers. Although the characteristics of these volunteers are different from those of the participants of the BP study, the information about the differences in the dose–response of the three major tomato carotenoids can shed light on the results of the BP study, as discussed below. The increase in blood levels of lycopene, phytoene and phytofluene was dose dependent, and was significantly higher with TNC 15 mg as compared to TNC 5 mg. A large increase in plasma concentrations of lycopene, phytofluene, and phytoene was already evident after two weeks of treatment. This correlates well with the reduction of SBP with TNC 15 mg, which was statistically significant after two weeks of treatment, suggesting that the increase in these tomato constituents was the reason and the driving force for the SBP reduction. It should be noted that the changes in phytofluene and phytoene plasma concentrations between the non-effective dose of TNC 5 mg and the effective TNC 15 mg was more pronounced than that of lycopene, which suggests that these carotenoids (and possibly other components of TNC) have an important role in reducing SBP. Such a role for phytofluene and phytoene can also explain the lower effect of pure synthetic lycopene, which does not contain these carotenoids or any other tomato component. This option is supported by a study showing that a reduction in UV-induced erythema by tomato extract enriched with phytofluene and phytoene was larger than that achieved with non-enriched tomato extract, whereas the lowest erythema reduction was found with synthetic lycopene [20].

An important question not addressed in the current study is what the mechanism might be by which the supplemented tomato carotenoids reduce BP. A possible answer is suggested by a study in endothelial cells [21], in which Di Tumo et al. found that lycopene and β-carotene reduce TNFα-induced inflammation in endothelial cells in culture. This effect was associated with reduced reactive oxygen species and nitrotyrosine [an index of interaction of NO with superoxide anion (O_2^-)] and an increase in NO and cGMP, which are known to cause vascular relaxation. The antioxidant effect of the carotenoids, which led to the observed effects in the endothelial cells, and possibly to lower SBP in the current human study, probably resulted from activation of the antioxidant response element and the Nrf2 transcription system that is induced by carotenoids, as we reported previously [22,23].

5. Conclusions

TNC containing 15 mg and 30 mg lycopene was well tolerated and showed efficacy in reducing SBP in the HT population, while lower doses and standalone pure lycopene were not sufficient to induce similar effects. The content of lycopene in raw tomatoes is 2.5–4 mg per 100 g [24]. Thus, reasonable consumption of tomatoes, e.g., 100–200 g per day, will supply 4–8 mg lycopene, but not enough to drive SBP reduction; therefore, inclusion in the daily diet of other lycopene-rich foods such as tomato products and supplements is recommended. It is not clear whether a supplement such as TNC can be used as a standalone treatment or in conjunction with other treatments. However, our previous study that used tomato extract in treated but uncontrolled hypertensive patients [9] suggests that patients treated with various antihypertensive drugs can benefit from the addition of TNC.

Author Contributions: Conceptualization, Y.S., J.L. and E.P.; Formal analysis, D.S. and E.P.; Funding acquisition, Y.S., J.L. and E.P.; Investigation, T.W., Y.S., J.L. and E.P.; Methodology, T.W., Y.S., J.L. and E.P.; Project administration, Y.S., J.L. and E.P.; Resources, Y.S., K.L.-H. and E.P.; Supervision, Y.S., J.L. and E.P.; Visualization, Y.S., J.L., K.L.-H. and D.S.; Writing—original draft, Y.S. and J.L.; and Writing—review and editing, T.W., Y.S., J.L., K.L.-H., D.S. and E.P.

Funding: This research was funded by Lycored Ltd., Beer Sheva, Israel to E.P. (LCR-2008-01) and to Y.S., J.L. and D.S. (87476811).

Nutrients **2019**, *11*, 950

Acknowledgments: We thank Arnon Aharon and Iris Alroy for initial analysis of the BP data, Michael Friger for the randomization process and for sample size calculation, Gil Harari, Medistat, Ltd. for the statistical analysis of the BP data, and Robin Miller for English editing.

Conflicts of Interest: J.L. and Y.S. are consultants for Lycored Ltd., Beer Sheva, Israel. K.L.-H. is employed by Lycored, whereas E.P., J.L., Y.S. and D.S. received research funding from the company. All other authors declare no conflicts of interest. Lycored Ltd. is a supplier to the dietary supplement and functional food industries worldwide. The company had a role in the design of the study, but had no part in the collection, analysis or interpretation of the experimental data.

References

1. Thies, F.; Masson, L.F.; Rudd, A.; Vaughan, N.; Tsang, C.; Brittenden, J.; Simpson, W.G.; Duthie, S.; Horgan, G.W.; Duthie, G. Effect of a tomato-rich diet on markers of cardiovascular disease risk in moderately overweight, disease-free, middle-aged adults: A randomized controlled trial. *Am. J. Clin. Nutr.* **2012**, *95*, 1013–1022. [CrossRef] [PubMed]

2. Yeh, G.Y.; Davis, R.B.; Phillips, R.S. Use of complementary therapies in patients with cardiovascular disease. *Am. J. Cardiol.* **2006**, *98*, 673–680. [CrossRef] [PubMed]

3. Chobanian, A.V.; Bakris, G.L.; Black, H.R.; Cushman, W.C.; Green, L.A.; Izzo, J.L., Jr.; Jones, D.W.; Materson, B.J.; Oparil, S.; Wright, J.T., Jr.; et al. The seventh report of the joint national committee on prevention, detection, evaluation, and treatment of high blood pressure: The jnc 7 report. *JAMA* **2003**, *289*, 2560–2572. [CrossRef]

4. Svetkey, L.P.; Simons-Morton, D.; Vollmer, W.M.; Appel, L.J.; Conlin, P.R.; Ryan, D.H.; Ard, J.; Kennedy, B.M. Effects of dietary patterns on blood pressure: Subgroup analysis of the dietary approaches to stop hypertension (dash) randomized clinical trial. *Arch. Intern. Med.* **1999**, *159*, 285–293. [CrossRef]

5. John, J.H.; Ziebland, S.; Yudkin, P.; Roe, L.S.; Neil, H.A.; Oxford, F.; Vegetable Study, G. Effects of fruit and vegetable consumption on plasma antioxidant concentrations and blood pressure: A randomised controlled trial. *Lancet* **2002**, *359*, 1969–1974. [CrossRef]

6. Engelhard, Y.N.; Gazer, B.; Paran, E. Natural antioxidants from tomato extract reduce blood pressure in patients with grade-1 hypertension: A double-blind, placebo-controlled pilot study. *Am. Heart J.* **2006**, *151*, 100. [CrossRef]

7. Kim, J.Y.; Paik, J.K.; Kim, O.Y.; Park, H.W.; Lee, J.H.; Jang, Y.; Lee, J.H. Effects of lycopene supplementation on oxidative stress and markers of endothelial function in healthy men. *Atherosclerosis* **2011**, *215*, 189–195. [CrossRef]

8. Ried, K.; Frank, O.R.; Stocks, N.P. Dark chocolate or tomato extract for prehypertension: A randomised controlled trial. *BMC Complement. Altern. Med.* **2009**, *9*, 22. [CrossRef] [PubMed]

9. Paran, E.; Novack, V.; Engelhard, Y.N.; Hazan-Halevy, I. The effects of natural antioxidants from tomato extract in treated but uncontrolled hypertensive patients. *Cardiovasc. Drugs Ther.* **2009**, *23*, 145–151. [CrossRef]

10. Hozawa, A.; Jacobs, D.R., Jr.; Steffes, M.W.; Gross, M.D.; Steffen, L.M.; Lee, D.H. Circulating carotenoid concentrations and incident hypertension: The coronary artery risk development in young adults (cardia) study. *J. Hypertens.* **2009**, *27*, 237–242. [CrossRef]

11. Itsiopoulos, C.; Brazionis, L.; Kaimakamis, M.; Cameron, M.; Best, J.D.; O'Dea, K.; Rowley, K. Can the mediterranean diet lower hba1c in type 2 diabetes? Results from a randomized cross-over study. *Nutr. Metab. Cardiovasc. Dis.* **2011**, *21*, 740–747. [CrossRef]

12. Upritchard, J.E.; Sutherland, W.H.; Mann, J.I. Effect of supplementation with tomato juice, vitamin e, and vitamin c on ldl oxidation and products of inflammatory activity in type 2 diabetes. *Diabetes Care* **2000**, *23*, 733–738. [CrossRef]

13. Paterson, E.; Gordon, M.H.; Niwat, C.; George, T.W.; Parr, L.; Waroonphan, S.; Lovegrove, J.A. Supplementation with fruit and vegetable soups and beverages increases plasma carotenoid concentrations but does not alter markers of oxidative stress or cardiovascular risk factors. *J. Nutr.* **2006**, *136*, 2849–2855. [CrossRef]

14. Ried, K.; Fakler, P. Protective effect of lycopene on serum cholesterol and blood pressure: Meta-analyses of intervention trials. *Maturitas* **2011**, *68*, 299–310. [CrossRef] [PubMed]

15. Li, X.; Xu, J. Lycopene supplement and blood pressure: An updated meta-analysis of intervention trials. *Nutrients* **2013**, *5*, 3696–3712. [CrossRef]

16. Whelton, P.K.; Carey, R.M.; Aronow, W.S.; Casey, D.E., Jr.; Collins, K.J.; Dennison Himmelfarb, C.; DePalma, S.M.; Gidding, S.; Jamerson, K.A.; Jones, D.W.; et al. 2017 acc/aha/aapa/abc/acpm/ags/apha/ash/aspc/ nma/pcna guideline for the prevention, detection, evaluation, and management of high blood pressure in adults: A report of the american college of cardiology/american heart association task force on clinical practice guidelines. *J. Am. Coll. Cardiol.* **2018**, *71*, e127–e248.

17. Walfisch, Y.; Walfisch, S.; Agbaria, R.; Levy, J.; Sharoni, Y. Lycopene in serum, skin and adipose tissues after tomato-oleoresin supplementation in patients undergoing haemorrhoidectomy or peri-anal fistulotomy. *Br. J. Nutr.* **2003**, *90*, 759–766. [CrossRef]

18. Devaraj, S.; Mathur, S.; Basu, A.; Aung, H.H.; Vasu, V.T.; Meyers, S.; Jialal, I. A dose-response study on the effects of purified lycopene supplementation on biomarkers of oxidative stress. *J. Am. Coll. Nutr.* **2008**, *27*, 267–273. [CrossRef]

19. Linnewiel-Hermoni, K.; Khanin, M.; Danilenko, M.; Zango, G.; Amosi, Y.; Levy, J.; Sharoni, Y. The anti-cancer effects of carotenoids and other phytonutrients resides in their combined activity. *Arch. Biochem. Biophys.* **2015**, *572*, 28–35. [CrossRef]

20. Aust, O.; Stahl, W.; Sies, H.; Tronnier, H.; Heinrich, U. Supplementation with tomato-based products increases lycopene, phytofluene, and phytoene levels in human serum and protects against uv-light-induced erythema. *Int. J. Vitam Nutr. Res.* **2005**, *75*, 54–60. [CrossRef]

21. Di Tomo, P.; Canali, R.; Ciavardelli, D.; Di Silvestre, S.; De Marco, A.; Giardinelli, A.; Pipino, C.; Di Pietro, N.; Virgili, F.; Pandolfi, A. Beta-carotene and lycopene affect endothelial response to tnf-alpha reducing nitro-oxidative stress and interaction with monocytes. *Mol. Nutr. Food Res.* **2012**, *56*, 217–227. [CrossRef] [PubMed]

22. Ben-Dor, A.; Steiner, M.; Gheber, L.; Danilenko, M.; Dubi, N.; Linnewiel, K.; Zick, A.; Sharoni, Y.; Levy, J. Carotenoids activate the antioxidant response element transcription system. *Mol. Cancer Ther.* **2005**, *4*, 177–186. [PubMed]

23. Linnewiel, K.; Ernst, H.; Caris-Veyrat, C.; Ben-Dor, A.; Kampf, A.; Salman, H.; Danilenko, M.; Levy, J.; Sharoni, Y. Structure activity relationship of carotenoid derivatives in activation of the electrophile/antioxidant response element transcription system. *Free Radic. Biol. Med.* **2009**, *47*, 659–667. [CrossRef] [PubMed]

24. Khachik, F.; Goli, M.B.; Beecher, G.R.; Holden, J.; Lusby, W.R.; Tenorio, M.D.; Barrera, M.R. Effect of food preparation on qualitative and quantitative distribution of major carotenoid constituents of tomatoes and several green vegetables. *J. Agric. Food Chem.* **1992**, *40*, 390–398. [CrossRef]

nutrients

MDPI

Review

Dietary Antioxidants, Macular Pigment, and Glaucomatous Neurodegeneration: A Review of the Evidence

Thomas Lawler [1], Yao Liu [2], Krista Christensen [3], Thasarat S. Vajaranant [4] and Julie Mares [3,*]

[1] Department of Nutritional Sciences, University of Wisconsin, 1415 Linden Drive, Madison, WI 53706, USA; tlawler2@wisc.edu
[2] Department of Ophthalmology and Visual Sciences, University of Wisconsin, 2870 University Avenue, Madison, WI 53705, USA; yao.liu2@wisc.edu
[3] Department of Ophthalmology and Visual Sciences, University of Wisconsin, 610 N. Walnut Street, 1069 WARF Building, Madison, WI 53726, USA; krista.christensen@wisc.edu
[4] Department of Ophthalmology and Visual Sciences, University of Illinois – Chicago, 1855 W. Taylor Street, Chicago, IL 60612, USA; thasarat@uic.edu
* Correspondence: jmarespe@wisc.edu; Tel.: +1-608-262-8044

Received: 10 April 2019; Accepted: 29 April 2019; Published: 1 May 2019

Abstract: Primary open-angle glaucoma (POAG) is a leading cause of irreversible blindness worldwide, and the prevalence is projected to increase to 112 million worldwide by 2040. Intraocular pressure is currently the only proven modifiable risk factor to treat POAG, but recent evidence suggests a link between antioxidant levels and risk for prevalent glaucoma. Studies have found that antioxidant levels are lower in the serum and aqueous humor of glaucoma patients. In this review, we provide a brief overview of the evidence linking oxidative stress to glaucomatous pathology, followed by an in-depth discussion of epidemiological studies and clinical trials of antioxidant consumption and glaucomatous visual field loss. Lastly, we highlight a possible role for antioxidant carotenoids lutein and zeaxanthin, which accumulate in the retina to form macular pigment, as evidence has emerged supporting an association between macular pigment levels and age-related eye disease, including glaucoma. We conclude that the evidence base is inconsistent in showing causal links between dietary antioxidants and glaucoma risk, and that prospective studies are needed to further investigate the possible relationship between macular pigment levels and glaucoma risk specifically.

Keywords: glaucoma; antioxidants; oxidative stress; macular pigment; lutein

1. Introduction

Glaucoma is a progressive neurodegenerative disease in which the death of retinal ganglion cells (RGCs) in the inner retina results in characteristic visual field defects that can progress to blindness [1]. It is a leading cause of irreversible blindness worldwide and the incidence of glaucoma is expected to increase as the population ages, with a projected prevalence of 112 million worldwide by 2040 [2]. Healthcare spending related to glaucoma is estimated at roughly $5.8 billion per year in the U.S. alone [3]. Moreover, vision loss from glaucoma may make it difficult to perform activities of daily living, limit independence, and increase the risk of falls [4], depression [5], and cognitive decline [6]. Due to the lack of symptoms in the early stages, detection and treatment may be delayed until advanced stages of disease [7]. It is estimated that half of patients with glaucoma are unaware of their diagnosis [8].

Nonmodifiable glaucoma risk factors include older age, race/ethnicity (e.g., African, Asian, or Hispanic heritage), family history of glaucoma, and central corneal thickness [9]. Current treatments for glaucoma are limited to lowering of intraocular pressure (IOP) using medications, laser treatment, or surgery [7], but these interventions are not always sufficient. Some patients continue to experience disease progression and severe vision loss can occur despite achieving therapeutic IOP lowering. The identification of novel modifiable risk factors has significant implications for glaucoma prevention and may lead to new treatment modalities. In recent years, new evidence has emerged of an association between markers of oxidative stress and risk for primary open-angle glaucoma (POAG) [10,11], suggesting that consumption of dietary antioxidants may be a factor capable of modifying disease risk.

We review the growing body of evidence linking oxidative stress to the pathophysiology of POAG and review human studies linking dietary intake of antioxidants to risk for POAG. In later sections, we focus specifically on dietary carotenoids lutein and zeaxanthin (L/Z), given their well-established antioxidant activities in the retina and the ability to noninvasively estimate the density of these carotenoids within the macular region of the retina (termed macular pigment (MP)), where they accumulate at the highest density. More recent cross-sectional studies have linked MP to a reduced risk for prevalent POAG, suggesting that dietary modifications to increase MP may be one approach to lower glaucoma risk in individuals with elevated IOP [12,13].

2. Pathophysiology of Glaucoma

POAG is the most common form of glaucoma in the U.S. and worldwide [1]. In POAG, the iridocorneal angle where drainage of the aqueous humor occurs remains visibly open, yet IOP may be elevated due to either excess production of aqueous humor by the ciliary body or an impediment of aqueous humor outflow through the trabecular meshwork into the canal of Schlemm [1]. The optic nerve head, where RGC axons pass through the porous connective tissue plates of the lamina cribrosa and leave the eye to form the optic nerve in the brain, is likely the initial site of IOP-related RGC injury [14]. Elevated IOP may interfere with retrograde transport of neurotropic factors, such as brain derived neurotrophic factor [15,16]; alter cytoskeletal structure [16]; and/or induce mitochondrial dysfunction [17]. These consequences of elevated IOP can induce a bioenergetic crisis, disturb normal axonal transport, and activate RGC cell death pathways.

Although IOP reduction through the use of topical medications and/or surgery is effective, many individuals continue to develop glaucoma despite having normal eye pressures and incur progressive glaucomatous vision loss despite treatment, suggesting that there are other mechanisms contributing to RGC death. These other biological insults may include chronic inflammation, hypoxia, excitotoxicity, loss of neurotrophic factors, and oxidative stress [18] that may be independent of pressure related injury. Consequently, there may be opportunities to develop novel treatment approaches to use in conjunction with pharmaceutical or surgical approaches to reduce IOP. A large body of evidence implicates oxidative stress, as there is evidence for oxidative damage (and depletion of antioxidants) in the aqueous humor [19,20] and trabecular meshwork [21,22] of glaucoma cases compared with age and sex-matched controls.

3. Oxidative Stress in Primary Open-Angle Glaucoma Pathogenesis

Current evidence suggests that oxidative stress is relevant to the pathophysiology of POAG and may contribute to damage in the retina and the anterior chamber of the eye (the space between the iris and the corneal endothelium containing aqueous humor). An imbalance between reactive oxygen species (ROS) and antioxidants in these tissues can result in excessive generation of ROS, including hydrogen peroxide, superoxide, and peroxynitrate, tipping the balance in favor of oxidative stress. These insults may damage cellular macromolecules (nucleic acids, structural proteins, and lipids [23]) and organelles, including mitochondria, and may promote cell death via apoptosis [24]. Prior studies have observed diminished total antioxidant capacity in plasma or serum of glaucoma cases compared to matched controls [10,11,25,26], as well as increased levels of oxidation products in the serum (e.g., malonyldialdehyde, a lipid peroxidation product) [25–27]. The total antioxidant capacity in aqueous humor has been observed to be lower in POAG patients compared to age- and sex-matched controls [20,25,26], along with levels of some specific antioxidants including vitamins C and E [28]. There is evidence for greater activity of antioxidant enzymes (glutathione peroxidase and superoxide dismutase) in the aqueous humor of glaucoma cases compared to control individuals [19,20,28], which may signal the body's attempt to combat oxidative stress [20]. In a recent meta-analysis containing pooled data from 22 case–control studies, there was consistent evidence for a significant increase in oxidative stress markers (e.g., malonyldialdehyde) and a decrease in antioxidant level in both serum and aqueous humor in cases from several glaucoma subtypes (including but not limited to POAG) compared to age- and sex-matched controls [29].

Mechanisms by which oxidative stress may contribute to glaucomatous neurodegeneration are complex and multifactorial. One mechanism by which oxidative stress may contribute to glaucoma is by damaging cells in the trabecular meshwork [30], leading to interference with normal drainage of aqueous humor [31] and an increase in IOP [21]. Consistent with oxidative stress in the aqueous humor, there is evidence that cells within the trabecular meshwork of glaucoma cases have higher levels of 8-hydroxy-2′-deoxyguanosine (8-OHdG) [21,22], a marker of oxidative damage to the DNA. Oxidative damage to trabecular meshwork cells may cause remodeling of the cytoskeleton [30] and decrease outflow facility [31] and it has been observed that presence of 8-OHdG in trabecular meshwork cells correlates with several measures of glaucoma disease severity, including degree of visual field loss and IOP [21,22].

Oxidative stress may also contribute directly to the apoptotic death of RGCs of the inner retina [32] and activation of the immune system [33–35]. Animal models of glaucoma suggest that elevated IOP causes oxidative stress in RGCs [32,36,37]. In a rat model of elevated IOP, induced by injection of hyaluronic acid in the anterior chamber, retinal levels of antioxidant enzymes SOD and catalase were decreased over 10 weeks, as were levels of glutathione, while markers of lipid peroxidation were markedly increased [36]. Another group reported that elevated IOP (induced in rats by episcleral vein cauterization) can increase levels of malonyldialdehyde and the free radical superoxide [32], while others have reported that acute elevation of IOP or hydrostatic pressure can induce markers of oxidative stress in both mouse retina and RGC-5 cell culture in a matter of hours [38].

Unstable perfusion of the optic nerve head and retina, characterized by large fluctuations in the ocular perfusion pressure, has emerged as another source of oxidative stress that may cause glaucomatous damage, evidenced by new imaging technologies that allow measurement of retinal vessel caliper, ocular blood flow, and ocular perfusion pressure [39]. Vascular diseases, including stroke, cardiovascular disease, migraine, and hypertension, are associated with increased risk for glaucoma, strongly suggesting that vascular abnormalities and impaired blood flow to the retina play a role in development of the disease [39]. Decreased ocular blood flow is a risk factor for prevalent glaucoma in large epidemiological studies [40,41], as well as progression of glaucoma over time [42]. Similar associations are seen for vessel narrowing [43]. It has also been noted that glaucoma is associated with primary vascular dysregulation [44,45], leading to speculation that unstable perfusion of the retina may be an important mechanism in damage to the optic nerve head. Unstable retinal blood flow may lead to oxidative stress-related injury, as periods of relative ischemia may be succeeded by periods of reperfusion in which blood supply to the retina exceeds the need for oxygen, contributing to the formation of ROS [46].

Mitochondrial dysfunction in RGCs may also contribute to the generation of ROS [47], as mitochondria are abundant in RGC axons to provide ATP for sodium/potassium ATPase activity needed to maintain electrical conductivity, and to support axonal transport of organelles and other materials [48]. Mitochondria are the cell's primary consumers of oxygen, and hence contribute significantly to generation of reactive oxygen species. Glaucoma patients have significantly lower rates of complex-1 driven ATP synthesis, which is inversely associated with disease severity [49]. Further, glaucoma cases may have higher rates of mitochondrial DNA (mtDNA) mutations and lower mean mitochondrial respiratory activity than controls [50], consistent with mitochondrial dysfunction and oxidative stress.

Oxidative stress may promote RGC death directly by activating apoptotic cell signaling pathways [51], or indirectly by interacting with the retinal immune system [33–35,52]. Oxidative stress arising from glaucomatous stimuli can cause retinal glial cells (including Mueller cells and astrocytes) to adapt an 'immune-activated' phenotype, characterized by greater expression of major histocompatibility complex (MHC) II [52] and toll-like receptors [33]. In cell culture experiments, oxidative stress increases the expression of MHCII by glial cells, consequently improving their ability to stimulate T cell proliferation and secretion of cytokines (including TNF-α) that can induce ganglion cell apoptosis [52]. Activation of the innate and adaptive immune systems under conditions of cell stress may help to promote homeostasis and the removal of dead or dying cells. However, it has also been suggested that chronic immune activation under pathologic conditions may also exacerbate the disease state in glaucoma [34].

4. Dietary Antioxidants and Risk for POAG

Given the body of evidence linking oxidative stress to the etiology of POAG, several researchers have proposed the potential utility of antioxidant treatments [53,54]. However, there is limited evidence that long-term antioxidant intake lowers risk for glaucoma incidence or severity. Four large-scale epidemiological studies have examined this relationship using estimated dietary intake and a handful of intervention trials have been conducted (see Table 1).

Table 1. Summary of human studies of dietary antioxidants (and antioxidant food sources) and primary open-angle glaucoma.

Author, Year	Study Design	Participants (age)	Exposure/Treatment	Outcome	Follow-Up Period	Significant Findings ($p < 0.05$)
Kang, 2003 [55]	Cohort	116,505 health professionals free from glaucoma (≥40 years)	Dietary antioxidant vitamins and carotenoids: Vitamins A, C, E, α and β carotene, lutein/zeaxanthin, lycopene, and β-cryptoxanthin (Total and from foods only)	Incident self-reported primary open-angle glaucoma confirmed by medical record review (n = 474)	9.3 years (average)	Vitamin E (foods only) (p-trend = 0.02) OR: 0.67 (quintile 5 vs. 1)
Wang, 2013 [56]	Cross-sectional	2912 NHANES participants (≥40 years)	Vitamins A, C, E (from supplements) Vitamins A, C, E (in serum)	Self-reported glaucoma (n = 203)	N/A	Vitamin C from supplements (p = 0.13) OR: 0.34 (quartile 1 vs. no consumption)
Garcia-Medina, 2015 [57]	RCT	117 POAG cases	Placebo (n = 63) Supplement containing vitamins, A, C, E, and B-complex, lutein and zeaxanthin, zinc, copper, selenium, and omega 3 fatty acids (n = 54)	-Visual fields: Mean deviation, pattern standard deviation -Retinal nerve fiber layer thickness -Ganglion cell complex thickness	2 years	No effect of treatment was observed for any of the three outcomes.
Kang, 2016 [58]	Cohort	104,987 health professionals (≥40 years)	Total green leafy vegetables	Incident self-reported open-angle glaucoma confirmed by medical record review (n = 1483)	16.0 years (average)	Green leafy vegetables inversely associated with glaucoma risk (p-trend = 0.02) OR 0.82, (quintile 5 vs. 1)
Ramdas, 2012 [59]	Cohort	3502 adults without POAG (≥55 years)	Dietary antioxidant vitamins and carotenoids: α and β carotene, lutein/zeaxanthin, β-cryptoxanthin, retinol, vitamins B1, B6, B12, C, and E.	Incident POAG, detected by visual field testing (n = 91)	9.7 years (average)	Vitamin B1 (p = 0.03) HR: 0.31 (tertile 3 vs. 1) Retinol equivalents (p = 0.01) HR: 0.33 (tertile 3 vs. 1)

Table 1. *Cont.*

Author, Year	Study Design	Participants (age)	Exposure/Treatment	Outcome	Follow-Up Period	Significant Findings (*p* < 0.05)
Coleman, 2008 [60]	Cross-sectional	1155 women (67–97 years)	Dietary fruits and vegetables Dietary vitamins and carotenoids (vitamins A, C, E, B-complex, α and β carotene, lutein/zeaxanthin, cryptoxanthin, and lycopene	Prevalent POAG, detected by visual field testing and optic disc examination (*n* = 95)	N/A	Canned/dried peaches (*p* = 0.04) OR: 0.53 (≥1 serving/week, vs. <1/month) Fresh carrots (*p* = 0.009) OR: 0.36, (>2 servings/week vs. <1/month) -Collard greens/kale (*p* = 0.34) OR: 0.31 (≥1 serving/month vs. <1/month) -Vitamin B2 (*p* = 0.02) OR: 0.19 (≥2 mg/day vs. <1 mg/day)
Giaconi, 2012 [61]	Cross-sectional	584 African American women	Dietary fruits and vegetables Dietary vitamins and carotenoids (vitamins A, C, E, B-complex, α and β carotene, lutein/zeaxanthin, cryptoxanthin, and lycopene	Prevalent POAG, detected by visual field testing and optic disc examination (*n* = 77)	N/A	Retinol (*p*-trend 0.11) Vitamin C (*p*-trend = 0.018) α-carotene (*p*-trend = 0.21) All fruits and fruit juices (*p*-trend = 0.023) Oranges (*p*-trend = 0.002) Peaches (*p*-trend = 0.002) Collard greens/kale (*p*-trend = 0.014)

4.1. Epidemiological Studies

A cross-sectional analysis of data from the National Health and Nutrition Examination Survey showed a statistically significant decrease in the odds of self-reported glaucoma among those currently using vitamin C supplements, with no association seen for use of vitamin A or E supplements, and no association for levels of vitamins A, C, or E in serum; carotenoid use in supplements or serum levels of carotenoids were not assessed in this study [56]. These analyses were limited by reliance on self-reported glaucoma as the main outcome measure, which is known to be of limited accuracy and many individuals with glaucoma are unaware of the disease, limiting the ability to detect an association if it exists [8]. Also, given that glaucoma may take decades to develop, it is unlikely that the use of antioxidant containing supplements at any single point in time would be an etiologically relevant exposure.

In the Nurses' Health Study (NHS) and Health Professionals Follow-up Study (HPFS), which included 116,484 participants ≥40 years of age, there was no strong evidence that dietary intake of carotenoids (α-carotene, β-carotene, β-cryptoxanthin, lycopene, and L/Z) or antioxidant vitamins (A, C, and E) was associated with risk for incident glaucoma over an average follow-up period of 9.1 years, except for vitamin E from foods (odds ratio (OR) of 0.67 comparing highest to lowest quintile of intake) [55]. The study authors made a considerable effort to prevent misclassification, by excluding individuals who did not attend regular eye checkups, and by confirming cases of self-reported glaucoma by reviewing medical records (including visual fields). Although this should be considered a strength of this study, it is still possible that misclassification of participants (especially those unaware that they had glaucoma) may have biased the results towards the null. It is also not clear that results from a cohort of healthcare professionals generalize to other populations due to differences in diet and other health-related behaviors; perhaps most importantly, dietary intake of antioxidants may not reflect the quantity available to the eye due to differences in absorption and transport of antioxidants related to genetic heterogeneity [62,63], bioavailability [64,65], and reliance on circulating lipoproteins for transport of lipid soluble molecules (e.g., carotenoids) [66–68]. In this study, it was not determined whether serum antioxidant levels were predictive of incident glaucoma.

In the Rotterdam Study, glaucoma was assessed by visual field testing and 91 of 3502 eligible participants, age 55 years and older, developed glaucoma over an average of 9.7 years of follow-up [59]. There was no association between the intake of the majority of antioxidant carotenoids and vitamins evaluated (including carotenoids α-carotene, β-carotene, lutein, zeaxanthin, and β cryptoxanthin, and vitamins C and E) and risk for glaucoma, although protective associations were observed for thiamine and retinol equivalents. This study is limited by lack of serum antioxidant measures, and may have been underpowered given the relatively small number of glaucoma cases.

A cross-sectional analysis of African American women greater than 65 years old participating in the Study of Osteoporotic Fractures found that specific antioxidant-rich fruits and vegetables (including oranges, peaches, and collard greens/kale) were associated with reduced risk for prevalent glaucoma ($p < 0.05$), ascertained via visual field testing and the appearance of the optic nerve head [61]. Specific dietary antioxidant vitamins and carotenoids (vitamins A and C and α-carotene) were also associated with significantly reduced risk after adjustment for potential confounders. Trends were also observed for β-carotene and L/Z, although the trends did not reach significance ($p = 0.05$–0.08). In an earlier report from the Study of Osteoporotic Fractures, including 1155 participants of Caucasian or African American background, there was a significant protective association for collard greens/kale (OR: 0.31 for ≥1 serving per month) [60]. α-carotene and β-carotene were also inversely associated with glaucoma risk in this sample, although the association between L/Z consumption and risk for glaucoma did not approach significance in this sample.

In summary, there is limited evidence from epidemiological studies that consumption of dietary antioxidants is associated with reduced risk for incident or prevalent glaucoma. Inconsistencies may arise from differences in diagnostic criteria to ascertain glaucoma cases (i.e., visual field testing vs. self-report) and heterogeneity in the racial/genetic background of participants, which may influence genetic disease risks and dietary habits. Measuring diet during the lengthy, etiologically relevant time frame for glaucoma remains a persistent challenge in nutritional epidemiology, and this may be considered a limitation in cross-sectional studies of glaucoma prevalence [56,60,61].

4.2. Clinical Trials

There have been few randomized controlled trials (RCTs) to determine whether the administration of high dose antioxidant supplements can prevent new cases of glaucoma or slow the progression of visual field loss in glaucoma patients. In one trial, a supplement containing B-vitamins; vitamins A, C, and E; carotenoids; and antioxidant minerals with or without omega 3 fatty acids consumed over 2 years, did not prevent visual field loss or thinning of the retinal nerve fiber layer or ganglion cell complex (GCC) in a study of older adults with glaucoma [57]. There is more evidence from trials of Ginkgo Biloba extract (GBE) indicating that this substance may be a useful therapeutic tool for those with glaucoma, possibly slowing the rate of visual field loss and thinning of the neuronal retina [69–74]. Results from animal studies are also supportive, as the administration of GBE can prevent RGC death related to chronic elevation of IOP and destruction of the optic nerve [70,72]. GBE has approximately 60 bioactive substances, including antioxidants, and may possibly influence glaucoma progression through several unique mechanisms, including effects on ocular blood flow [73].

5. Antioxidant Effects of Lutein and Zeaxanthin in the Retina

Although research consistently demonstrates increased oxidative stress and depletion of antioxidants in the serum and aqueous humor of adults with glaucoma [29], there is not consistent evidence from epidemiological or clinical studies to implicate specific antioxidants. However, a growing body of evidence from animal models and cell culture indicates that antioxidant carotenoids L/Z, which uniquely accumulate in the retina in humans, may prevent RGC death related to glaucomatous stimuli [75–79].

L/Z are polar dietary carotenoids and potent antioxidants that may hold special relevance for retinal health, integrity, and long-term risk for age-related eye disease [80]. L/Z are unique among other major carotenoids found in serum in that they cross the blood–retinal barrier and bind to specific binding proteins in the retina [81,82]. Their highest concentration in human body is located in the macular region located centrally in the retina where they contribute to the pigmentation level (hence the term macular pigment), but macular concentrations of these carotenoids are also correlated with levels in peripheral retinal areas [83]. MP peaks in the center fovea and rapidly declines toward the periphery, and is generally undetectable at 7 degrees from the foveal center [84]. Although the majority of MP is concentrated in the outer retina, in the photoreceptor outer segments and outer plexiform layer, considerable amounts of MP are detectable in the inner plexiform layer [85], adjacent to the ganglion cell layer that is damaged in glaucoma. Due to a backbone of conjugated carbon–carbon double bonds, L/Z act as potent retinal antioxidants [86], able to scavenge ROS including superoxide, hydroxyl radical, and hydrogen peroxide without being 'used up' in the process [80]. L/Z also have the unique capacity to filter short-wavelength blue light before it reaches the outer retina [87], thought to protect the retinal pigment epithelium and photoreceptors from the effects of photooxidative stress [88,89]. MP is modifiable with diet and supplement interventions [90,91], and has become an attractive target to treat or prevent age-related eye diseases that have high levels of oxidative stress.

6. Lutein and Zeaxanthin and POAG

6.1. Protective Mechanisms of L/Z in the Retina

Evidence, which links these carotenoids to diseases of the inner retina, like glaucoma, is now emerging. A recent study of Chinese adults demonstrated that serum concentrations of L/Z, but not other carotenoids, were positively associated with measures of retinal blood flow [92]. This provides preliminary evidence that L/Z may help to stabilize retinal perfusion, which has been linked to glaucoma risk [45]. Additional evidence has emerged that L/Z may also have protective effects on the inner retina in response to glaucomatous stimuli in animal models [75,76,79], with additional evidence that the mechanism of protection involves suppression of apoptotic signaling pathways. Lutein reduced the number of TUNEL-positive RGCs (a marker of apoptotic cell death) in mouse retina after ischemia/reperfusion injury (2 h of retinal ischemia followed by 22 h of reperfusion) and increased the number of viable cells to approximately control levels [75]. These results corroborate those of an earlier study in rats, in which intravitreous or intraperitoneal lutein almost completely prevented RGC loss after 1 h of ischemia and 24 h of reperfusion [93]. Likewise, lutein significantly increased the number of viable RGCs after intravitreal injection of an NMDA receptor agonist, meant to stimulate excitotoxic cell death in the inner retina, although cell viability did not approach control levels [79]. Reduced expression of the proapoptotic Bax protein and increased expression of antiapoptotic Bcl2 in the inner retina mirrored this finding, as well as reduced release of proapoptotic cytochrome C from the mitochondria, a key player in the activation of caspases, the primary effectors of apoptotic cell death. Similar effects on Bax and Bcl2 expression were seen in a cell culture model of rat Mueller cells (a prominent glial cell type in the retina whose activity is linked to onset of glaucomatous pathology), where it was also observed that lutein could reduce levels of cleaved caspase 3 after simulated hypoxia [94]. Antiapoptotic effects of lutein were also apparent in a model of retinal detachment, as lutein treatment reduced levels of cleaved caspase 3 and 8 [95].

While the results from animal models indicate that administration of L/Z may protect the inner retina from glaucomatous insults, the relevance for humans is unclear. Unlike humans, rodents do not accumulate MP in appreciable quantities (although mouse models of MP have recently been developed) [96], and animal models of RGC death based on short-term elevations in IOP or ischemia do not perfectly mimic the complex mechanisms that underlie human glaucoma. Cell culture models of physiologically relevant L/Z concentrations show that L/Z can maintain murine RGC-5 cell viability after cell stress, including simulated hypoxia and serum deprivation [77,97], helping to corroborate results from animal studies. However, cell culture does not perfectly mimic disease pathophysiology, and RGC-5 cells lack several key characteristics of human RGCs, including excitability [98].

6.2. L/Z and POAG—Epidemiological Studies

Several groups have evaluated the association between L/Z consumption and risk for POAG in epidemiological studies, with mixed results. In data pooled from the NHS and HPFS, individuals in the highest quintile of dietary L/Z had lower risk than individuals in the lowest quintile OR = 0.68. CI: 0.49–0.93), after excluding the four most recent years of dietary data (to account for a lag in disease diagnosis) [55]. Although there was no significant linear trend, these results are consistent with a possible threshold effect. In contrast, there was no association between for L/Z and glaucoma in the Rotterdam Study [59]. More recently, greater consumption of green leafy vegetables, the predominant source of L/Z in the diet, was linearly associated with reduced risk for incident POAG (*p*-trend = 0.01) in the NHS and HPFS [58]. However, the authors attribute this beneficial association to the presence of nitrates in leafy green vegetables, hypothesized to influence ocular blood flow and regulation of IOP, rather than L/Z content. There was a significant negative association between dietary nitrate consumption (OR: 0.67 for Q5 vs. Q1) and risk of POAG that maintained statistical significance after adjustment for dietary L/Z and other antioxidants. However, it may be difficult to identify independent effects among highly correlated nutrient intakes, and differences in the degree of measurement error between nutrients exacerbate this difficulty. Other groups have also observed a negative association for leafy green vegetables, and it is not clear whether this is attributable to the L/Z content, nitrates, and/or another bioactive component [60,61].

Accumulation of L/Z in the retina as MP may be a more relevant exposure metric than L/Z in the diet, as the quantity of L/Z that reaches the retina is highly dependent on genetic heterogeneity [99,100] and other factors, including body composition and metabolic health [101,102]. Several low-cost, noninvasive approaches to reliably measure MP have been developed in recent years [103], making it feasible to investigate MP level as a potential modifiable risk factor for POAG. Currently, the literature contains only four studies directly addressing the association between MP and the presence of POAG and glaucomatous optic neuropathy (see Table 2). In the first, a study of 40 glaucoma cases and 54 normal controls, it was determined that MP optical density (MPOD), assessed using customized heterochromatic flicker photometry (cHFP), was approximately 36% lower in cases, although there was no association with disease severity [12]. However, in a follow-up study of 88 glaucoma cases, they discovered that MPOD at multiple eccentricities was significantly lower (*p* < 0.0001) in cases with foveal GCC loss (an indicator of greater disease severity) compared to cases without foveal involvement [13]. Although the associations with disease severity differ in these studies, the authors suggest that the differences in sample size and methodology may explain the different results, as the first study did not include optical coherence tomography scans to assess foveal GCC loss. In the second study, MPOD was also significantly and positively associated with glaucoma-related structural parameters including optic disc rim area, GCC thickness, and thickness of the inner retina, although the results were only marginally significant (*p* = 0.02–0.05). MPOD was negatively associated with cup to disc ratio, suggesting maintenance of RGC number. The results obtained by this group were later corroborated by another group utilizing the one-wavelength reflectometry approach for measuring MP, as both maximum and mean MPOD were reduced in 30 POAG patients relative to 52 controls, by 19% and 14%, respectively [104]. However, in the recent San Diego Macular Pigment Study, MP volume did not differ between 85 glaucoma patients and 22 controls when measured utilizing the two-wavelength autofluorescence technique [105]. There was also no association between MP volume and retinal nerve fiber layer thickness, as assessed by optical coherence tomography imaging of the retina.

Table 2. Summary of cross-sectional studies of macular pigment optical density (MPOD) and primary open-angle glaucoma.

Author, Year	Participants	MPOD Measurement Technique	Exposure Variable	Mean ± SD (Controls)	Mean ± SD (Cases)	*p*-Value
Igras, 2013 [12]	40 POAG cases 54 healthy controls	HFP	MPOD at 0.5°	0.36 (median)	0.23 (median)	0.03
Siah, 2015 [13]	52 foveal-involved POAG cases 33 controls without foveal involvement	HFP	MPOD at 0.5° [1]	0.24 ± 0.12	0.15 ± 0.10	<0.001
Ji, 2016 [104]	30 POAG cases 52 healthy controls	Reflectometry	MPOD mean over 7° [2]	0.14 ± 0.03	0.12 ± 0.03	<0.001
Daga, 2018 [105]	85 POAG cases 22 healthy controls	2-wavelength autofluorescence	MP volume over 7°	7661 ± 3750	8717 ± 3903	0.44

[1] Significant results also observed for MPOD at 0.25° and 1° targets ($p < 0.001$); [2] Significant results also obtained for maximum MPOD ($p < 0.001$).

It is not clear why there have been inconsistent results, but it is possible that reliance on different testing approaches for MPOD may have led to different conclusions [103]. Some approaches are biased by the health of the lens and retina [103]. While cHFP is reliable in older adults [106] and controls for differences in lens density and retinal health [103], it is time-consuming, possibly leading to participant fatigue, and otherwise relies on the skill of the examiner and patient to produce valid measurements. It is not known whether the peculiar changes to retinal health and vision function that occur in glaucoma could also bias results of MPOD testing, or whether bias would be limited to cases of late-stage, foveal-involved glaucoma, as MP accumulates almost exclusively in the fovea [107]. Given the inconsistent results of MP levels in relation to glaucoma and the small samples used to evaluate this association, this is an interesting area for further investigation.

As all of these are studies cross-sectional, it is not possible to determine whether high MPOD is associated with risk for incident glaucoma over time. To help clarify this issue, long-term prospective studies of MP and glaucomatous visual field loss and changes to the retina are required. In addition, reverse causality cannot be ruled out, as glaucomatous changes to the retina, including thinning of the GCC and impaired blood flow to the retina, may make it difficult to accrue MP. Although MP attenuates oxidative stress by filtering short wavelength light and by directly neutralizing ROS [80], the biological rationale for a protective effect is not entirely clear, as the primary site of RGC injury is believed to be located at the optic nerve head [14], rather, not within the retina. Further, although L/Z accumulate throughout the macula, concentrations are highest in the photoreceptors of the outer retina and in the fovea [107], which usually remains unaffected until later stages of glaucoma [108]. Despite this, multiple studies showing an inverse association between MPOD and risk for glaucoma provide a strong rationale for prospective studies, as interventions to increase MPOD would be a low cost and low risk approach to prevent disease.

7. Conclusions and Future Directions

Given the increasing prevalence of glaucoma among older adults and lack of known modifiable risk factors other than lowering of IOP, interventions, such as increasing consumption of antioxidants targeted to those who are at high risk for disease development or progression, may have a considerable population health impact. Despite the evidence supporting oxidative stress in glaucoma, there is still inconsistent evidence as to whether dietary antioxidant intake predicts the incidence or prevalence of POAG, and no prospective evidence that interventions to increase antioxidant consumption can prevent or slow glaucomatous visual field loss or thinning of the neural retina. This may reflect persistent challenges in accurately measuring dietary intake, as well as high levels of interindividual variability in the absorption of antioxidants. Future prospective studies of serum antioxidants and incident visual field loss, or thinning of the inner retina, may help to clarify the effect of antioxidants on the health of the neural retina. Older adults at increased risk for glaucoma are currently advised to consume an overall healthy diet rich in fruits and vegetables, as this diet pattern is linked to reduced risk [61,109], and because maintaining a healthy body weight and good metabolic health through diet and exercise may help to prevent disease.

Despite the limitations of the current evidence, there is a strong rationale for continuing to investigate MP as risk factor for POAG, given its unique antioxidant/protective activities in the retina [80], and ease of modification with diet [91,110] and supplements [90]. The majority of recent cross-sectional research, made possible by low-cost, noninvasive approaches to MPOD testing, suggests an association between higher MPOD and reduced risk for POAG [12,13,104,105]. A protective association is also supported by observations that L/Z may protect the neural retina from glaucomatous stressors, at least in animal models [75,79]. There is currently a need for prospective studies of MPOD and glaucomatous visual field loss to further elucidate whether the relationship is likely to be causal in humans. There is a strong precedent for this type of research, given longstanding interest in MPOD as a modifiable risk factor for AMD (a disease characterized by oxidative stress), and evidence from large-scale clinical trials that high doses of L/Z (12 mg/day) may reduce risk for progression of AMD in some individuals [111] without increased risk for side effects [112]. Prospective studies showing reduced risk for visual field loss with higher MPOD may support the development of low-cost interventions to increase MPOD in individuals at high risk for POAG, administered alongside treatment to lower IOP, and monitoring of MPOD in a clinical setting.

Author Contributions: T.L.; Writing—Original Draft Preparation, K.C., J.M., Y.L., and T.S.V.; Writing—Review & Editing, J.M. and Y.L.; Supervision and funding acquisition.

Funding: This paper was supported by NIH grants T32 DK007665 and R01 EY025292, and in part by an institutional grant from Research to Prevent Blindness, Inc. to the Department of Ophthalmology and Visual Sciences at the University of Wisconsin–Madison. The sponsors had no role in the design, execution, interpretation, or writing of the study.

Conflicts of Interest: The authors declare no conflicts of interest.

References

1. Quigley, H.A. Glaucoma. *Lancet* **2011**, *377*, 1367–1377. [CrossRef]
2. Tham, Y.C.; Li, X.; Wong, T.Y.; Quigley, H.A.; Aung, T.; Cheng, C.Y. Global prevalence of glaucoma and projections of glaucoma burden through 2040: A systematic review and meta-analysis. *Ophthalmology* **2014**, *121*, 2081–2090. [CrossRef] [PubMed]
3. John Wittenborn, D.R. *Cost of Vision Problems: The Economic Burden of Vision Loss and Eye Disorders in the United States*; University of Chicago: Chicago, IL, USA, 2013; p. 8.
4. Kannus, P.; Sievanen, H.; Palvanen, M.; Jarvinen, T.; Parkkari, J. Prevention of falls and consequent injuries in elderly people. *Lancet* **2005**, *366*, 1885–1893. [CrossRef]
5. Zhang, X.; Bullard, K.M.; Cotch, M.F.; Wilson, M.R.; Rovner, B.W.; McGwin, G., Jr.; Owsley, C.; Barker, L.; Crews, J.E.; Saaddine, J.B. Association between depression and functional vision loss in persons 20 years of age or older in the United States, NHANES 2005–2008. *JAMA Ophthalmol.* **2013**, *131*, 573–581. [CrossRef] [PubMed]
6. Rogers, M.A.; Langa, K.M. Untreated poor vision: A contributing factor to late-life dementia. *Am. J. Epidemiol.* **2010**, *171*, 728–735. [CrossRef]
7. Cohen, L.P.; Pasquale, L.R. Clinical characteristics and current treatment of glaucoma. *Cold Spring Harb. Perspect. Med.* **2014**, *4*. [CrossRef]
8. Vajaranant, T.; Maki, P.; Pasquale, L.; Khan, F.; Mares, J.; Meyer, K.; Haan, M. The accuracy of self-reported glaucoma in the Women's Health Initiative. In Proceedings of the American Glaucoma Society Annual Meeting, Fort Lauderdale, FL, USA, 3–6 March 2016.
9. Cook, C.; Foster, P. Epidemiology of glaucoma: What's new? *Can. J. Ophthalmol. J. Can. Ophtalmol.* **2012**, *47*, 223–226. [CrossRef]
10. Abu-Amero, K.K.; Kondkar, A.A.; Mousa, A.; Osman, E.A.; Al-Obeidan, S.A. Decreased total antioxidants in patients with primary open angle glaucoma. *Curr. Eye Res.* **2013**, *38*, 959–964. [CrossRef] [PubMed]

11. Mousa, A.; Kondkar, A.A.; Al-Obeidan, S.A.; Azad, T.A.; Sultan, T.; Osman, E.; Abu-Amero, K.K. Association of total antioxidants level with glaucoma type and severity. *Saudi Med. J.* **2015**, *36*, 671–677. [CrossRef] [PubMed]

12. Igras, E.; Loughman, J.; Ratzlaff, M.; O'Caoimh, R.; O'Brien, C. Evidence of lower macular pigment optical density in chronic open angle glaucoma. *Br. J. Ophthalmol.* **2013**, *97*, 994–998. [CrossRef] [PubMed]

13. Siah, W.F.; Loughman, J.; O'Brien, C. Lower Macular Pigment Optical Density in Foveal-Involved Glaucoma. *Ophthalmology* **2015**, *122*, 2029–2037. [CrossRef]

14. Nickells, R.W.; Howell, G.R.; Soto, I.; John, S.W. Under pressure: Cellular and molecular responses during glaucoma, a common neurodegeneration with axonopathy. *Annu. Rev. Neurosci.* **2012**, *35*, 153–179. [CrossRef]

15. Morrison, J.C.; Johnson, E.C.; Cepurna, W.; Jia, L. Understanding mechanisms of pressure-induced optic nerve damage. *Prog. Retin. Eye Res.* **2005**, *24*, 217–240. [CrossRef]

16. Chidlow, G.; Ebneter, A.; Wood, J.P.; Casson, R.J. The optic nerve head is the site of axonal transport disruption, axonal cytoskeleton damage and putative axonal regeneration failure in a rat model of glaucoma. *Acta Neuropathol.* **2011**, *121*, 737–751. [CrossRef]

17. Wu, J.H.; Zhang, S.H.; Nickerson, J.M.; Gao, F.J.; Sun, Z.; Chen, X.Y.; Zhang, S.J.; Gao, F.; Chen, J.Y.; Luo, Y.; et al. Cumulative mtDNA damage and mutations contribute to the progressive loss of RGCs in a rat model of glaucoma. *Neurobiol. Dis.* **2015**, *74*, 167–179. [CrossRef]

18. Almasieh, M.; Wilson, A.M.; Morquette, B.; Cueva Vargas, J.L.; Di Polo, A. The molecular basis of retinal ganglion cell death in glaucoma. *Prog. Retin. Eye Res.* **2012**, *31*, 152–181. [CrossRef]

19. Ghanem, A.A.; Arafa, L.F.; El-Baz, A. Oxidative stress markers in patients with primary open-angle glaucoma. *Curr. Eye Res.* **2010**, *35*, 295–301. [CrossRef]

20. Ferreira, S.M.; Lerner, S.F.; Brunzini, R.; Evelson, P.A.; Llesuy, S.F. Oxidative stress markers in aqueous humor of glaucoma patients. *Am. J. Ophthalmol.* **2004**, *137*, 62–69. [CrossRef]

21. Sacca, S.C.; Pascotto, A.; Camicione, P.; Capris, P.; Izzotti, A. Oxidative DNA damage in the human trabecular meshwork: Clinical correlation in patients with primary open-angle glaucoma. *Arch. Ophthalmol.* **2005**, *123*, 458–463. [CrossRef]

22. Izzotti, A.; Sacca, S.C.; Cartiglia, C.; De Flora, S. Oxidative deoxyribonucleic acid damage in the eyes of glaucoma patients. *Am. J. Med.* **2003**, *114*, 638–646. [CrossRef]

23. Aslan, M.; Cort, A.; Yucel, I. Oxidative and nitrative stress markers in glaucoma. *Free Radic. Biol. Med.* **2008**, *45*, 367–376. [CrossRef]

24. Pinazo-Duran, M.D.; Zanon-Moreno, V.; Gallego-Pinazo, R.; Garcia-Medina, J.J. Oxidative stress and mitochondrial failure in the pathogenesis of glaucoma neurodegeneration. *Prog. Brain Res.* **2015**, *220*, 127–153. [CrossRef] [PubMed]

25. Nucci, C.; Di Pierro, D.; Varesi, C.; Ciuffoletti, E.; Russo, R.; Gentile, R.; Cedrone, C.; Pinazo Duran, M.D.; Coletta, M.; Mancino, R. Increased malondialdehyde concentration and reduced total antioxidant capacity in aqueous humor and blood samples from patients with glaucoma. *Mol. Vis.* **2013**, *19*, 1841–1846.

26. Sorkhabi, R.; Ghorbanihaghjo, A.; Javadzadeh, A.; Rashtchizadeh, N.; Moharrery, M. Oxidative DNA damage and total antioxidant status in glaucoma patients. *Mol. Vis.* **2011**, *17*, 41–46.

27. Yildirim, O.; Ates, N.A.; Ercan, B.; Muslu, N.; Unlu, A.; Tamer, L.; Atik, U.; Kanik, A. Role of oxidative stress enzymes in open-angle glaucoma. *Eye* **2005**, *19*, 580–583. [CrossRef]

28. Goyal, A.; Srivastava, A.; Sihota, R.; Kaur, J. Evaluation of oxidative stress markers in aqueous humor of primary open angle glaucoma and primary angle closure glaucoma patients. *Curr. Eye Res.* **2014**, *39*, 823–829. [CrossRef]

29. Benoist d'Azy, C.; Pereira, B.; Chiambaretta, F.; Dutheil, F. Oxidative and Anti-Oxidative Stress Markers in Chronic Glaucoma: A Systematic Review and Meta-Analysis. *PLoS ONE* **2016**, *11*, e0166915. [CrossRef]

30. Zhou, L.; Li, Y.; Yue, B.Y. Oxidative stress affects cytoskeletal structure and cell-matrix interactions in cells from an ocular tissue: The trabecular meshwork. *J. Cell. Physiol.* **1999**, *180*, 182–189. [CrossRef]

31. Kahn, M.G.; Giblin, F.J.; Epstein, D.L. Glutathione in calf trabecular meshwork and its relation to aqueous humor outflow facility. *Investig. Ophthalmol. Vis. Sci.* **1983**, *24*, 1283–1287.

32. Ko, M.L.; Peng, P.H.; Ma, M.C.; Ritch, R.; Chen, C.F. Dynamic changes in reactive oxygen species and antioxidant levels in retinas in experimental glaucoma. *Free Radic. Biol. Med.* **2005**, *39*, 365–373. [CrossRef]

33. Luo, C.; Yang, X.; Kain, A.D.; Powell, D.W.; Kuehn, M.H.; Tezel, G. Glaucomatous tissue stress and the regulation of immune response through glial Toll-like receptor signaling. *Investig. Ophthalmol. Vis. Sci.* **2010**, *51*, 5697–5707. [CrossRef]

34. Tezel, G. The immune response in glaucoma: A perspective on the roles of oxidative stress. *Exp. Eye Res.* **2011**, *93*, 178–186. [CrossRef]

35. Tezel, G.; Yang, X.; Luo, C.; Kain, A.D.; Powell, D.W.; Kuehn, M.H.; Kaplan, H.J. Oxidative stress and the regulation of complement activation in human glaucoma. *Investig. Ophthalmol. Vis. Sci.* **2010**, *51*, 5071–5082. [CrossRef]

36. Moreno, M.C.; Campanelli, J.; Sande, P.; Sanez, D.A.; Keller Sarmiento, M.I.; Rosenstein, R.E. Retinal oxidative stress induced by high intraocular pressure. *Free Radic. Biol. Med.* **2004**, *37*, 803–812. [CrossRef]

37. Muller, A.; Pietri, S.; Villain, M.; Frejaville, C.; Bonne, C.; Culcas, M. Free radicals in rabbit retina under ocular hyperpressure and functional consequences. *Exp. Eye Res.* **1997**, *64*, 637–643. [CrossRef]

38. Liu, Q.; Ju, W.K.; Crowston, J.G.; Xie, F.; Perry, G.; Smith, M.A.; Lindsey, J.D.; Weinreb, R.N. Oxidative stress is an early event in hydrostatic pressure induced retinal ganglion cell damage. *Investig. Ophthalmol. Vis. Sci.* **2007**, *48*, 4580–4589. [CrossRef]

39. Nakazawa, T. Ocular Blood Flow and Influencing Factors for Glaucoma. *Asia-Pac. J. Ophthalmol.* **2016**, *5*, 38–44. [CrossRef]

40. Tielsch, J.M.; Katz, J.; Sommer, A.; Quigley, H.A.; Javitt, J.C. Hypertension, perfusion pressure, and primary open-angle glaucoma. A population-based assessment. *Arch. Ophthalmol.* **1995**, *113*, 216–221. [CrossRef]

41. Leske, M.C.; Connell, A.M.; Wu, S.Y.; Hyman, L.G.; Schachat, A.P. Risk factors for open-angle glaucoma. The Barbados Eye Study. *Arch. Ophthalmol.* **1995**, *113*, 918–924. [CrossRef]

42. Zeitz, O.; Galambos, P.; Wagenfeld, L.; Wiermann, A.; Wlodarsch, P.; Praga, R.; Matthiessen, E.T.; Richard, G.; Klemm, M. Glaucoma progression is associated with decreased blood flow velocities in the short posterior ciliary artery. *Br. J. Ophthalmol.* **2006**, *90*, 1245–1248. [CrossRef]

43. Wang, S.; Xu, L.; Wang, Y.; Wang, Y.; Jonas, J.B. Retinal vessel diameter in normal and glaucomatous eyes: The Beijing eye study. *Clin. Exp. Ophthalmol.* **2007**, *35*, 800–807. [CrossRef]

44. Grieshaber, M.C.; Mozaffarieh, M.; Flammer, J. What is the link between vascular dysregulation and glaucoma? *Surv. Ophthalmol.* **2007**, *52* (Suppl. 2), S144–S154. [CrossRef]

45. Leske, M.C. Ocular perfusion pressure and glaucoma: Clinical trial and epidemiologic findings. *Curr. Opin. Ophthalmol.* **2009**, *20*, 73–78. [CrossRef]

46. Mozaffarieh, M.; Grieshaber, M.C.; Flammer, J. Oxygen and blood flow: Players in the pathogenesis of glaucoma. *Mol. Vis.* **2008**, *14*, 224–233.

47. Osborne, N.N.; Nunez-Alvarez, C.; Joglar, B.; Del Olmo-Aguado, S. Glaucoma: Focus on mitochondria in relation to pathogenesis and neuroprotection. *Eur. J. Pharmacol.* **2016**, *787*, 127–133. [CrossRef]

48. Osborne, N.N. Mitochondria: Their role in ganglion cell death and survival in primary open angle glaucoma. *Exp. Eye Res.* **2010**, *90*, 750–757. [CrossRef]

49. Lee, S.; Sheck, L.; Crowston, J.G.; Van Bergen, N.J.; O'Neill, E.C.; O'Hare, F.; Kong, Y.X.; Chrysostomou, V.; Vincent, A.L.; Trounce, I.A. Impaired complex-I-linked respiration and ATP synthesis in primary open-angle glaucoma patient lymphoblasts. *Investig. Ophthalmol. Vis. Sci.* **2012**, *53*, 2431–2437. [CrossRef]

50. Abu-Amero, K.K.; Morales, J.; Bosley, T.M. Mitochondrial abnormalities in patients with primary open-angle glaucoma. *Investig. Ophthalmol. Vis. Sci.* **2006**, *47*, 2533–2541. [CrossRef]

51. Martindale, J.L.; Holbrook, N.J. Cellular response to oxidative stress: Signaling for suicide and survival. *J. Cell. Physiol.* **2002**, *192*, 1–15. [CrossRef]

52. Tezel, G.; Yang, X.; Luo, C.; Peng, Y.; Sun, S.L.; Sun, D. Mechanisms of immune system activation in glaucoma: Oxidative stress-stimulated antigen presentation by the retina and optic nerve head glia. *Investig. Ophthalmol. Vis. Sci.* **2007**, *48*, 705–714. [CrossRef]

53. Grover, A.K.; Samson, S.E. Antioxidants and vision health: Facts and fiction. *Mol. Cell. Biochem.* **2014**, *388*, 173–183. [CrossRef]

54. Mozaffarieh, M.; Grieshaber, M.C.; Orgul, S.; Flammer, J. The potential value of natural antioxidative treatment in glaucoma. *Surv. Ophthalmol.* **2008**, *53*, 479–505. [CrossRef]

55. Kang, J.H.; Pasquale, L.R.; Willett, W.; Rosner, B.; Egan, K.M.; Faberowski, N.; Hankinson, S.E. Antioxidant intake and primary open-angle glaucoma: A prospective study. *Am. J. Epidemiol.* **2003**, *158*, 337–346. [CrossRef]

56. Wang, S.Y.; Singh, K.; Lin, S.C. Glaucoma and vitamins A, C, and E supplement intake and serum levels in a population-based sample of the United States. *Eye* **2013**, *27*, 487–494. [CrossRef]

57. Garcia-Medina, J.J.; Garcia-Medina, M.; Garrido-Fernandez, P.; Galvan-Espinosa, J.; Garcia-Maturana, C.; Zanon-Moreno, V.; Pinazo-Duran, M.D. A two-year follow-up of oral antioxidant supplementation in primary open-angle glaucoma: An open-label, randomized, controlled trial. *Acta Ophthalmol.* **2015**, *93*, 546–554. [CrossRef]

58. Kang, J.H.; Willett, W.C.; Rosner, B.A.; Buys, E.; Wiggs, J.L.; Pasquale, L.R. Association of Dietary Nitrate Intake with Primary Open-Angle Glaucoma: A Prospective Analysis from the Nurses' Health Study and Health Professionals Follow-up Study. *JAMA Ophthalmol.* **2016**, *134*, 294–303. [CrossRef]

59. Ramdas, W.D.; Wolfs, R.C.; Kiefte-de Jong, J.C.; Hofman, A.; de Jong, P.T.; Vingerling, J.R.; Jansonius, N.M. Nutrient intake and risk of open-angle glaucoma: The Rotterdam Study. *Eur. J. Epidemiol.* **2012**, *27*, 385–393. [CrossRef]

60. Coleman, A.L.; Stone, K.L.; Kodjebacheva, G.; Yu, F.; Pedula, K.L.; Ensrud, K.E.; Cauley, J.A.; Hochberg, M.C.; Topouzis, F.; Badala, F.; et al. Glaucoma risk and the consumption of fruits and vegetables among older women in the study of osteoporotic fractures. *Am. J. Ophthalmol.* **2008**, *145*, 1081–1089. [CrossRef]

61. Giaconi, J.A.; Yu, F.; Stone, K.L.; Pedula, K.L.; Ensrud, K.E.; Cauley, J.A.; Hochberg, M.C.; Coleman, A.L. The association of consumption of fruits/vegetables with decreased risk of glaucoma among older African-American women in the study of osteoporotic fractures. *Am. J. Ophthalmol.* **2012**, *154*, 635–644. [CrossRef]

62. Borel, P.; Desmarchelier, C.; Nowicki, M.; Bott, R.; Morange, S.; Lesavre, N. Interindividual variability of lutein bioavailability in healthy men: Characterization, genetic variants involved, and relation with fasting plasma lutein concentration. *Am. J. Clin. Nutr.* **2014**, *100*, 168–175. [CrossRef]

63. Borel, P. Genetic variations involved in interindividual variability in carotenoid status. *Mol. Nutr. Food Res.* **2012**, *56*, 228–240. [CrossRef]

64. Failla, M.L.; Chitchumronchokchai, C.; Ferruzzi, M.G.; Goltz, S.R.; Campbell, W.W. Unsaturated fatty acids promote bioaccessibility and basolateral secretion of carotenoids and alpha-tocopherol by Caco-2 cells. *Food Funct.* **2014**, *5*, 1101–1112. [CrossRef]

65. Chung, H.Y.; Rasmussen, H.M.; Johnson, E.J. Lutein bioavailability is higher from lutein-enriched eggs than from supplements and spinach in men. *J. Nutr.* **2004**, *134*, 1887–1893. [CrossRef]

66. Clevidence, B.A.; Bieri, J.G. Association of carotenoids with human plasma lipoproteins. *Methods Enzymol.* **1993**, *214*, 33–46.

67. Connor, W.E.; Duell, P.B.; Kean, R.; Wang, Y. The prime role of HDL to transport lutein into the retina: Evidence from HDL-deficient WHAM chicks having a mutant ABCA1 transporter. *Investig. Ophthalmol. Vis. Sci.* **2007**, *48*, 4226–4231. [CrossRef] [PubMed]

68. Loane, E.; Nolan, J.M.; Beatty, S. The respective relationships between lipoprotein profile, macular pigment optical density, and serum concentrations of lutein and zeaxanthin. *Investig. Ophthalmol. Vis. Sci.* **2010**, *51*, 5897–5905. [CrossRef] [PubMed]

69. Cybulska-Heinrich, A.K.; Mozaffarieh, M.; Flammer, J. Ginkgo biloba: An adjuvant therapy for progressive normal and high tension glaucoma. *Mol. Vis.* **2012**, *18*, 390–402.

70. Hirooka, K.; Tokuda, M.; Miyamoto, O.; Itano, T.; Baba, T.; Shiraga, F. The Ginkgo biloba extract (EGb 761) provides a neuroprotective effect on retinal ganglion cells in a rat model of chronic glaucoma. *Curr. Eye Res.* **2004**, *28*, 153–157. [CrossRef]

71. Lee, J.; Sohn, S.W.; Kee, C. Effect of Ginkgo biloba extract on visual field progression in normal tension glaucoma. *J. Glaucoma* **2013**, *22*, 780–784. [CrossRef] [PubMed]

72. Ma, K.; Xu, L.; Zhang, H.; Zhang, S.; Pu, M.; Jonas, J.B. The effect of ginkgo biloba on the rat retinal ganglion cell survival in the optic nerve crush model. *Acta Ophthalmol.* **2010**, *88*, 553–557. [CrossRef] [PubMed]

73. Ritch, R. Potential role for Ginkgo biloba extract in the treatment of glaucoma. *Med. Hypotheses* **2000**, *54*, 221–235. [CrossRef] [PubMed]

74. Shim, S.H.; Kim, J.M.; Choi, C.Y.; Kim, C.Y.; Park, K.H. Ginkgo biloba extract and bilberry anthocyanins improve visual function in patients with normal tension glaucoma. *J. Med. Food* **2012**, *15*, 818–823. [CrossRef] [PubMed]

75. Li, S.Y.; Fu, Z.J.; Ma, H.; Jang, W.C.; So, K.F.; Wong, D.; Lo, A.C. Effect of lutein on retinal neurons and oxidative stress in a model of acute retinal ischemia/reperfusion. *Investig. Ophthalmol. Vis. Sci.* **2009**, *50*, 836–843. [CrossRef] [PubMed]
76. Li, S.Y.; Fung, F.K.; Fu, Z.J.; Wong, D.; Chan, H.H.; Lo, A.C. Anti-inflammatory effects of lutein in retinal ischemic/hypoxic injury: In vivo and in vitro studies. *Investig. Ophthalmol. Vis. Sci.* **2012**, *53*, 5976–5984. [CrossRef] [PubMed]
77. Li, S.Y.; Lo, A.C. Lutein protects RGC-5 cells against hypoxia and oxidative stress. *Int. J. Mol. Sci.* **2010**, *11*, 2109–2117. [CrossRef] [PubMed]
78. Ozawa, Y.; Sasaki, M.; Takahashi, N.; Kamoshita, M.; Miyake, S.; Tsubota, K. Neuroprotective effects of lutein in the retina. *Curr. Pharm. Des.* **2012**, *18*, 51–56. [CrossRef] [PubMed]
79. Zhang, C.; Wang, Z.; Zhao, J.; Li, Q.; Huang, C.; Zhu, L.; Lu, D. Neuroprotective Effect of Lutein on NMDA-Induced Retinal Ganglion Cell Injury in Rat Retina. *Cell. Mol. Neurobiol.* **2016**, *36*, 531–540. [CrossRef] [PubMed]
80. Bernstein, P.S.; Li, B.; Vachali, P.P.; Gorusupudi, A.; Shyam, R.; Henriksen, B.S.; Nolan, J.M. Lutein, zeaxanthin, and meso-zeaxanthin: The basic and clinical science underlying carotenoid-based nutritional interventions against ocular disease. *Prog. Retin. Eye Res.* **2016**, *50*, 34–66. [CrossRef]
81. Li, B.; Vachali, P.; Frederick, J.M.; Bernstein, P.S. Identification of StARD3 as a lutein-binding protein in the macula of the primate retina. *Biochemistry* **2011**, *50*, 2541–2549. [CrossRef]
82. Bhosale, P.; Larson, A.J.; Frederick, J.M.; Southwick, K.; Thulin, C.D.; Bernstein, P.S. Identification and characterization of a Pi isoform of glutathione S-transferase (GSTP1) as a zeaxanthin-binding protein in the macula of the human eye. *J. Biol. Chem.* **2004**, *279*, 49447–49454. [CrossRef]
83. Bone, R.A.; Landrum, J.T.; Dixon, Z.; Chen, Y.; Llerena, C.M. Lutein and zeaxanthin in the eyes, serum and diet of human subjects. *Exp. Eye Res.* **2000**, *71*, 239–245. [CrossRef]
84. Chen, S.F.; Chang, Y.; Wu, J.C. The spatial distribution of macular pigment in humans. *Curr. Eye Res.* **2001**, *23*, 422–434. [CrossRef] [PubMed]
85. SanGiovanni, J.P.; Neuringer, M. The putative role of lutein and zeaxanthin as protective agents against age-related macular degeneration: Promise of molecular genetics for guiding mechanistic and translational research in the field. *Am. J. Clin. Nutr.* **2012**, *96*, 1223s–1233s. [CrossRef] [PubMed]
86. Khachik, F.; Bernstein, P.S.; Garland, D.L. Identification of lutein and zeaxanthin oxidation products in human and monkey retinas. *Investig. Ophthalmol. Vis. Sci.* **1997**, *38*, 1802–1811.
87. Ahmed, S.S.; Lott, M.N.; Marcus, D.M. The macular xanthophylls. *Surv. Ophthalmol.* **2005**, *50*, 183–193. [CrossRef]
88. Sparrow, J.R.; Nakanishi, K.; Parish, C.A. The lipofuscin fluorophore A2E mediates blue light-induced damage to retinal pigmented epithelial cells. *Investig. Ophthalmol. Vis. Sci.* **2000**, *41*, 1981–1989.
89. Leung, I.Y.; Sandstrom, M.M.; Zucker, C.L.; Neuringer, M.; Snodderly, D.M. Nutritional manipulation of primate retinas, II: Effects of age, n-3 fatty acids, lutein, and zeaxanthin on retinal pigment epithelium. *Investig. Ophthalmol. Vis. Sci.* **2004**, *45*, 3244–3256. [CrossRef] [PubMed]
90. Ma, L.; Liu, R.; Du, J.H.; Liu, T.; Wu, S.S.; Liu, X.H. Lutein, Zeaxanthin and Meso-zeaxanthin Supplementation Associated with Macular Pigment Optical Density. *Nutrients* **2016**, *8*, 426. [CrossRef] [PubMed]
91. Hammond, B.R., Jr.; Johnson, E.J.; Russell, R.M.; Krinsky, N.I.; Yeum, K.J.; Edwards, R.B.; Snodderly, D.M. Dietary modification of human macular pigment density. *Investig. Ophthalmol. Vis. Sci.* **1997**, *38*, 1795–1801.
92. Kumari, N.; Cher, J.; Chua, E.; Hamzah, H.; Wong, T.Y.; Cheung, C.Y. Association of serum lutein and zeaxanthin with quantitative measures of retinal vascular parameters. *PLoS ONE* **2018**, *13*, e0203868. [CrossRef]
93. Choi, J.S.; Kim, D.; Hong, Y.M.; Mizuno, S.; Joo, C.K. Inhibition of nNOS and COX-2 expression by lutein in acute retinal ischemia. *Nutrition* **2006**, *22*, 668–671. [CrossRef]
94. Fung, F.K.; Law, B.Y.; Lo, A.C. Lutein Attenuates Both Apoptosis and Autophagy upon Cobalt (II) Chloride-Induced Hypoxia in Rat Muller Cells. *PLoS ONE* **2016**, *11*, e0167828. [CrossRef]
95. Woo, T.T.; Li, S.Y.; Lai, W.W.; Wong, D.; Lo, A.C. Neuroprotective effects of lutein in a rat model of retinal detachment. *Graefe's Arch. Clin. Exp. Ophthalmol.* **2013**, *251*, 41–51. [CrossRef]
96. Li, B.; Rognon, G.T.; Mattinson, T.; Vachali, P.P.; Gorusupudi, A.; Chang, F.Y.; Ranganathan, A.; Nelson, K.; George, E.W.; Frederick, J.M.; et al. Supplementation with macular carotenoids improves visual performance of transgenic mice. *Arch. Biochem. Biophys.* **2018**. [CrossRef]

97. Nakajima, Y.; Shimazawa, M.; Otsubo, K.; Ishibashi, T.; Hara, H. Zeaxanthin, a retinal carotenoid, protects retinal cells against oxidative stress. *Curr. Eye Res.* **2009**, *34*, 311–318. [CrossRef] [PubMed]

98. Moorhouse, A.J.; Li, S.; Vickery, R.M.; Hill, M.A.; Morley, J.W. A patch-clamp investigation of membrane currents in a novel mammalian retinal ganglion cell line. *Brain Res.* **2004**, *1003*, 205–208. [CrossRef]

99. Meyers, K.J.; Johnson, E.J.; Bernstein, P.S.; Iyengar, S.K.; Engelman, C.D.; Karki, C.K.; Liu, Z.; Igo, R.P., Jr.; Truitt, B.; Klein, M.L.; et al. Genetic determinants of macular pigments in women of the Carotenoids in Age-Related Eye Disease Study. *Investig. Ophthalmol. Vis. Sci.* **2013**, *54*, 2333–2345. [CrossRef] [PubMed]

100. Liew, S.H.; Gilbert, C.E.; Spector, T.D.; Mellerio, J.; Marshall, J.; van Kuijk, F.J.; Beatty, S.; Fitzke, F.; Hammond, C.J. Heritability of macular pigment: A twin study. *Investig. Ophthalmol. Vis. Sci.* **2005**, *46*, 4430–4436. [CrossRef]

101. Mares, J.A.; LaRowe, T.L.; Snodderly, D.M.; Moeller, S.M.; Gruber, M.J.; Klein, M.L.; Wooten, B.R.; Johnson, E.J.; Chappell, R.J.; Group, C.M.P.S.; et al. Predictors of optical density of lutein and zeaxanthin in retinas of older women in the Carotenoids in Age-Related Eye Disease Study, an ancillary study of the Women's Health Initiative. *Am. J. Clin. Nutr.* **2006**, *84*, 1107–1122.

102. Bovier, E.R.; Lewis, R.D.; Hammond, B.R., Jr. The relationship between lutein and zeaxanthin status and body fat. *Nutrients* **2013**, *5*, 750–757. [CrossRef] [PubMed]

103. Hammond, B.R., Jr.; Wooten, B.R.; Smollon, B. Assessment of the validity of in vivo methods of measuring human macular pigment optical density. *Optom. Vis. Sci. Off. Publ. Am. Acad. Optom.* **2005**, *82*, 387–404. [CrossRef]

104. Ji, Y.; Zuo, C.; Lin, M.; Zhang, X.; Li, M.; Mi, L.; Liu, B.; Wen, F. Macular Pigment Optical Density in Chinese Primary Open Angle Glaucoma Using the One-Wavelength Reflectometry Method. *J. Ophthalmol.* **2016**, *2016*, 2792103. [CrossRef] [PubMed]

105. Daga, F.B.; Ogata, N.G.; Medeiros, F.A.; Moran, R.; Morris, J.; Zangwill, L.M.; Weinreb, R.N.; Nolan, J.M. Macular Pigment and Visual Function in Patients With Glaucoma: The San Diego Macular Pigment Study. *Investig. Ophthalmol. Vis. Sci.* **2018**, *59*, 4471–4476. [CrossRef]

106. Snodderly, D.M.; Mares, J.A.; Wooten, B.R.; Oxton, L.; Gruber, M.; Ficek, T. Macular pigment measurement by heterochromatic flicker photometry in older subjects: The carotenoids and age-related eye disease study. *Investig. Ophthalmol. Vis. Sci.* **2004**, *45*, 531–538. [CrossRef]

107. Bone, R.A.; Landrum, J.T.; Fernandez, L.; Tarsis, S.L. Analysis of the macular pigment by HPLC: Retinal distribution and age study. *Investig. Ophthalmol. Vis. Sci.* **1988**, *29*, 843–849.

108. Rapp, L.M.; Maple, S.S.; Choi, J.H. Lutein and zeaxanthin concentrations in rod outer segment membranes from perifoveal and peripheral human retina. *Investig. Ophthalmol. Vis. Sci.* **2000**, *41*, 1200–1209.

109. Moise, M.M.; Benjamin, L.M.; Doris, T.M.; Dalida, K.N.; Augustin, N.O. Role of Mediterranean diet, tropical vegetables rich in antioxidants, and sunlight exposure in blindness, cataract and glaucoma among African type 2 diabetics. *Int. J. Ophthalmol.* **2012**, *5*, 231–237. [CrossRef]

110. Wenzel, A.J.; Gerweck, C.; Barbato, D.; Nicolosi, R.J.; Handelman, G.J.; Curran-Celentano, J. A 12-wk egg intervention increases serum zeaxanthin and macular pigment optical density in women. *J. Nutr.* **2006**, *136*, 2568–2573. [CrossRef] [PubMed]

111. Chew, E.Y.; Clemons, T.E.; Sangiovanni, J.P.; Danis, R.P.; Ferris, F.L., 3rd; Elman, M.J.; Antoszyk, A.N.; Ruby, A.J.; Orth, D.; Bressler, S.B.; et al. Secondary analyses of the effects of lutein/zeaxanthin on age-related macular degeneration progression: AREDS2 report No. 3. *JAMA Ophthalmol.* **2014**, *132*, 142–149. [CrossRef] [PubMed]

112. Chew, E.Y.; Clemons, T.E.; SanGiovanni, J.P.; Danis, R.; Ferris, F.L.; Elman, M.; Antoszyk, A.; Ruby, A.; Orth, D.; Bressler, S.; et al. Lutein + zeaxanthin and omega-3 fatty acids for age-related macular degeneration: The Age-Related Eye Disease Study 2 (AREDS2) randomized clinical trial. *JAMA* **2013**, *309*, 2005–2015. [CrossRef]

nutrients

MDPI

Article

Dietary Carotenoids and Non-Alcoholic Fatty Liver Disease among US Adults, NHANES 2003–2014

Krista Christensen [1,*], Thomas Lawler [2] and Julie Mares [1]

[1]   Department of Ophthalmology and Visual Sciences, University of Wisconsin, 610 N. Walnut Street, 1069 WARF Building, Madison, WI 53726, USA; jmarespe@wisc.edu
[2]   Department of Nutritional Sciences, University of Wisconsin, 1415 Linden Drive, Madison, WI 53706, USA; tlawler2@wisc.edu
*   Correspondence: krista.christensen@wisc.edu; Tel.: +1-60-8265-3192

Received: 9 April 2019; Accepted: 10 May 2019; Published: 17 May 2019

Abstract: Non-alcoholic fatty liver disease (NAFLD) is highly prevalent worldwide. Oxidative stress is thought to be a major mechanism, and previous epidemiological studies found higher serum levels of antioxidant carotenoids were associated with reduced risk for development and progression of NAFLD. The objective of this analysis is to examine cross-sectional associations between dietary and serum levels of carotenoids in relation to NAFLD among a nationally representative sample of US adults. We used data from the 2003–2014 National Health and Nutrition Examination Survey (NHANES). Dietary carotenoid intake was estimated from a 24-hour recall, while serum carotenoids were measured from 2003 to 2006. The NAFLD status was determined based upon US Fatty Liver Index (FLI) value ≥30. Regression models were used to estimate associations between carotenoids and NAFLD by controlling for covariates and adjusting for survey design variables. Overall, 33% of participants were classified as having NAFLD. Intake of all carotenoids, with the exception of lycopene, was lower among those with NAFLD. This association was significant for the highest quartiles of intake of α-carotene, β-carotene, β-cryptoxanthin, and lutein/zeaxanthin. For serum measures, the highest level of all carotenoids was associated with significantly reduced odds of NAFLD. In conclusion, higher intake and serum levels of most carotenoids were associated with lower odds of having NAFLD. Identification of such modifiable lifestyle factors provide an opportunity to limit or prevent the disease and its progression.

Keywords: nutrition; chronic disease; lutein; zeaxanthin; lycopene; beta-carotene; alpha-carotene; beta-cryptoxanthin

1. Introduction

Non-alcoholic fatty liver disease (NAFLD) is highly prevalent worldwide, and is the most common liver disease in the United States with an estimated prevalence of 30% among adults [1], and 3% to 12% among children [2]. NAFLD is conceptualized as the hepatic manifestation of the metabolic syndrome, and comprises a range of conditions across the clinical spectrum. The American Association for the Study of Liver Diseases defines NAFLD as presence of both *"evidence of hepatic steatosis (HS), either by imaging or histology"* and *"lack of secondary causes of hepatic fat accumulation such as significant alcohol consumption, long-term use of a steatogenic medication, or monogenic hereditary disorders"* [3]. Hepatic steatosis increases risk for progression to non-alcoholic steatohepatitis (NASH), which is characterized by activation of hepatic macrophages, infiltration of inflammatory immune cells, hepatocyte ballooning, and cell death. NASH, in turn, drastically increases risk for fibrosis, cirrhosis leading to liver failure, and hepatocellular carcinoma [3,4]. Since there is currently no accepted pharmaceutical or surgical treatment for NAFLD [5], lifestyle modifications such as dietary changes are typically recommended to prevent the development of NAFLD, and to mitigate severity once disease has developed.

NAFLD is thought to develop through a series of 'hits' or stressors to the liver ('multiple hit' hypothesis). Although the pathophysiology of NAFLD is complex, abdominal adiposity, aberrant lipid metabolism, insulin resistance, oxidative stress, and inflammation likely contribute to NAFLD by overlapping and mutually reinforcing pathways. The proximal cause of NAFLD is accumulation of fat in liver cells (hepatocytes) combined with oxidative stress and other insults to the liver. Hence, antioxidant activity by carotenoids may reduce risk, severity, or progression of this disease. A growing body of epidemiological research indicates that higher levels of carotenoids in serum are associated with reduced risk for NAFLD or for disease progression [6–11]. Previous studies found increased risk for NAFLD and NASH with lower levels of serum β-carotene, α-carotene, β-cryptoxanthin, and lycopene [7,10,12], and lower total serum carotenoids [6]. Higher levels of both specific and total serum carotenoids were prospectively associated with improvements in NAFLD as assessed by abdominal ultrasonography in a cohort of Chinese adults [11]. However, there is little evidence for association with dietary intake of carotenoids, and no identified trials of carotenoid interventions to treat NAFLD or slow its progression.

Despite the incomplete body of evidence from epidemiological studies, there are animal studies to support the association of carotenoid exposure and risk of NAFLD. Multiple rodent studies demonstrate that both lycopene [13–16] and lutein [17] can slow the rate of hepatic lipid accumulation in mice fed a high-fat diet. Proposed mechanisms for this effect include activation of SIRT1, which is a master regulator of mitochondrial biogenesis and lipid oxidation, and improvement in insulin sensitivity. Lycopene feeding also reduced markers of liver damage including ALT and AST [14,16], and attenuated circulating levels of pro-inflammatory cytokine TNF-α [14], implicated in NAFLD progression. The antioxidant capacity of carotenoids may directly prevent liver damage and progression of hepatic steatosis to NASH, by mitigating the injurious effects of oxidative stress in hepatocytes. Likewise, carotenoids may attenuate pro-inflammatory signaling through transcription factor NFKβ [18], and inhibit activation of macrophages to an M1 phenotype (characteristic of NAFLD progression) [19]. A recent review concluded that carotenoids may confer protection via multiple pathways, including decreased hepatic levels of cholesterol, glucose, MDA (malondialdehyde, marker of lipid peroxidation), TNF-α, and NFKβ binding activity [20].

In this analysis, we examine cross-sectional associations between dietary and serum carotenoids in relation to NAFLD-related outcomes among a nationally representative sample of US adults. We hypothesize that greater exposure to carotenoids in diet and serum will be associated with lower odds of NAFLD, as assessed when utilizing the US Fatty Liver Index (FLI).

2. Materials and Methods

2.1. Design

The National Health and Nutrition Examination Survey (NHANES) is a cross-sectional survey designed to provide a representative sample of the US non-institutionalized civilian population [21]. This work is not considered human subjects research since it relies on free, publicly available datasets only, and is, thus, not subject to IRB review.

For these analyses, we used information on dietary (2003–2014) and serum (2003–2006) carotenoids, and factors needed to construct the US FLI (race/ethnicity, age, gammaglutamyl transferase level, waist circumference, insulin level, and glucose level). Participants were excluded for the following reasons: missing information needed to calculate US FLI (e.g., non-participation in the fasting sample) or on carotenoid dietary intake, age less than 20 years, presence of hepatitis B or C antibody, missing or elevated alcohol intake (≥10 g or 1 drink/day for women or 20 g or 2 drinks/day for men), and self-reported liver disease.

2.2. Carotenoids in Diet and Supplements and in Serum

NHANES participants provide detailed dietary intake information for up to two 24-h periods. These recalls are used to estimate intakes of energy, nutrients, and other food components. The first dietary recall is collected in-person during the NHANES visit, while the second recall is collected by telephone 3 to 10 days later. For these analyses, total estimated dietary carotenoid intake (micrograms, μg) was averaged over the two recall periods (if only the first day was available, that value was used). For the NHANES cycles used, information was available on α-carotene, β-carotene, β-cryptoxanthin, lycopene, and lutein/zeaxanthin (combined). While the main focus of this analysis is on dietary intake, participants are also queried about supplement use for the same two 24-h periods with respect to carotenoids and supplement intake of lycopene and lutein/zeaxanthin were available for later NHANES cycles (2007 onward). For participants reporting supplement use, total intake of lycopene and of lutein/zeaxanthin was calculated as the sum of dietary and supplement intake for sensitivity analyses.

For two of the NHANES cycles in this analysis (2003/2004, 2005/2006), serum measurements of carotenoids were also available. In these two cycles, participants aged 6 years and older provided serum samples for measurement of six carotenoids (α-carotene, trans-β-carotene, cis β-carotene, β-cryptoxanthin, combined lutein/zeaxanthin, trans-lycopene, and total lycopene) using high performance liquid chromatography (HPLC). We evaluated serum levels of carotenoids in relation to NAFLD outcomes as a sub-analysis.

2.3. NAFLD-Related Outcomes

Outcome status is based upon the US Fatty Liver Index (FLI), as described in Ruhl et al. [22]. The FLI was developed using NHANES III data and evaluated against hepatic steatosis, as diagnosed by an abdominal ultrasound. A cutoff of 30 was used to define NAFLD as suggested by the authors.

2.4. Statistical Analysis

All data analysis was performed using SAS/STAT software (Version 9.4, SAS Institute Inc., Cary, NC, USA). Regression models were used to identify associations between carotenoid intake and NAFLD-related outcomes, controlling for potential confounders. The primary models included adjustment for survey design variables, and for survey cycle, sex, and age in years. To evaluate whether the effects of carotenoid intake might be attributable more generally to a healthy diet, we also looked at models additionally adjusted for the healthy eating index (HEI) 2015 score. This score was only available for 2005 onwards. As a sensitivity analysis, we evaluated regression models including primary adjustment variables, as well as poverty income ratio (PIR) and educational attainment. These additional factors may be related to both carotenoid intake and risk for NAFLD, but it is likely that their association with NAFLD is mediated through diet. Serum cotinine as a proxy for smoke exposure was evaluated, but was not associated with carotenoid intake nor with an NAFLD status. To evaluate potential non-linear effects, carotenoid exposures were treated as both continuous and categorical (quartiles) variables. Statistically significant was defined as p-value < 0.05 for categorical variables and for a linear trend.

Both logistic (binary outcome of NAFLD, SURVEYLOGISTIC procedure) and linear (continuous outcome of US FLI, SURVEYREG procedure) regression models were used to evaluate associations with carotenoid intake. All statistical analyses were adjusted for survey design and weighting variables. For the full dataset analysis (dietary intake of carotenoids), we created 12-year weights as one-sixth of the value of the fasting subsample MEC weight (WTSAF2YR * 1/6) since this represented the smallest subsample of the study population. For sensitivity analyses of total (diet and supplement) lycopene and lutein/zeaxanthin intake, eight-year weights were calculated as one fourth of the value (WTSAF2YR * 1/4). When including HEI 2015 score as an adjustment variable, we used 10-year weights calculated as one-fifth the value (WTSAF2YR * 1/5). For the sub-analysis of serum carotenoids, we created 4-year weights as one-half of the value (WTSAF2YR * 1/2), and for analysis of serum carotenoids including HEI 2015 score (2005–2006 serum data only), the original 2-year weights were used.

3. Results

Table 1 displays characteristics of the study population stratified by NAFLD status. About one-third (33%) were classified as having NAFLD based on a US FLI score. Not surprisingly, those with NAFLD tended to be older and were more often male, Mexican-American, or non-Hispanic white, and were more likely to be obese or have diabetes or hypertensive factors.

Intake of all carotenoids with the exception of lycopene, was lower among those with NAFLD. The dietary intakes of the specific carotenoids did show correlations in some cases. When accounting for survey weighting, Pearson correlations were highest between intakes of α and β-carotene ($\rho = 0.75$), α-carotene, and lutein/zeaxanthin ($\rho = 0.48$), and β-carotene and lutein/zeaxanthin ($\rho = 0.69$). Correlations between other carotenoids were lower but significant. When looking at the NHANES cycles where serum carotenoids were measured, all were highly correlated with each other ($p < 0.001$). An especially high correlation was noted between serum α and β-carotene levels ($\rho = 0.66$), while other correlations were weaker ($\rho = 0.19–0.45$) but still highly significant. Dietary intake levels for each carotenoid were significantly associated with corresponding serum levels ($p < 0.001$ for all) although Pearson correlation coefficients were modest ($\rho = 0.25–0.37$), which may reflect an error in assessment of carotenoid intake, or inter-individual variation in carotenoid absorption.

Table 2 shows odd ratios (ORs, 95% confidence intervals [CIs]) for the presence of NAFLD as a binary outcome. Analogous results from modeling the US FLI score as a continuous outcome are shown in Supplementary Table S1. In both tables, results are presented treating carotenoid intake as a categorical variable (quartiles) since there was evidence of non-linearity in some cases. However, we do include a *p*-value for the trend calculated from models including the carotenoid as a continuous exposure, for reference. In adjusted logistic models, the highest quartiles of intake of α-carotene, β-carotene, β-cryptoxanthin, and lutein/zeaxanthin were all associated with lower odds of NAFLD compared to the lowest quartile of intake. Associations were attenuated when including the HEI 2015 score in the model, but the direction of association was unchanged and remained significant for β-carotene and lutein/zeaxanthin. In sensitivity analyses including education and income, results were largely similar but somewhat attenuated. Similarly, significant associations in linear models for US FLI as a continuous outcome were seen for the highest quartiles of intake of each carotenoid with the exception of lycopene. In logistic regression models, results were attenuated when including the HEI 2015 score (or education and income) but unchanged with respect to the direction of association.

Table 1. Characteristics of NHANES participants, 2003–2014, by the NAFLD status.

Group	No NAFLD		NAFLD		p-Value * for Difference between Those With and Without NAFLD
	Weighted n (SD)	Weighted Percent (SE)	Weighted n (SD)	Weighted Percent (SE)	
Total	61,456,846 (2,218,754)	100 (0)	30,295,373 (1,301,957)	100 (0)	
Survey Cycle					0.02
2003–2004	10,566,907 (1,252,551)	17.2 (1.8)	4,397,048 (603,680)	14.5 (1.8)	
2005–2006	9,983,650 (701,413)	16.2 (1.1)	4,560,556 (428,466)	15.1 (1.3)	
2007–2008	10,630,944 (843,068)	17.3 (1.3)	5,188,078 (602,156)	17.1 (1.8)	
2009–2010	9,483,325 (665,022)	15.4 (1.1)	5,996,470 (478,479)	19.8 (1.5)	
2011–2012	10,439,332 (1,024,219)	17.0 (1.5)	4,980,308 (513,756)	16.4 (1.6)	
2013–2014	10,352,685 (812,379)	16.8 (1.2)	5,172,910 (539,998)	17.1 (1.6)	
Age Group (years)					<0.0001
20–39	20,935,959 (928,501)	35.8 (1.2)	6,195,884 (420,329)	21.1 (1.2)	
40–59	20,750,665 (1,067,685)	35.5 (1.2)	11,253,296 (747,129)	38.4 (1.6)	
60–79	13,098,637 (712,268)	22.4 (0.8)	10,320,193 (568,536)	35.2 (1.5)	
≥80	3,682,338 (262,507)	6.3 (0.4)	1,547,004 (175,185)	5.3 (0.6)	
Sex					<0.0001
Male	26,036,711 (1,090,038)	42.4 (0.9)	16,577,926 (822,807)	54.7 (1.2)	
Female	35,420,134 (1,373,568)	57.6 (0.9)	13,717,446 (675,055)	45.3 (1.2)	
Race/Ethnicity					<0.0001
Mexican American	4,040,549 (353,152)	6.6 (0.7)	3,214,449 (336,258)	10.6 (1.3)	
Other Hispanic	2,502,869 (294,586)	4.1 (0.5)	1,169,495 (146,235)	3.9 (0.5)	
Non-Hispanic White	42,084,359 (2,357,817)	68.5 (1.6)	22,580,348 (1,364,103)	74.5 (1.8)	
Non-Hispanic Black	8,406,747 (589,390)	13.7 (1.1)	1,986,698 (189,877)	6.6 (0.7)	
Other/Multiracial	4,422,319 (386,390)	7.2 (0.6)	1,344,381 (187,613)	4.4 (0.6)	
BMI Category					<0.0001
Underweight (<18.5)	1,267,523 (190,732)	2.1 (0.3)	11,483 (11,483)	0 (0)	
Normal (18.5–24.9)	23,267,000 (1,199,452)	37.9 (1.1)	1,067,429 (166,431)	3.5 (0.5)	
Overweight (25–29.9)	24,039,650 (975,583)	39.1 (1)	6,865,896 (515,913)	22.7 (1.3)	
Obese (≥30)	12,882,671 (625,090)	21.0 (0.8)	22,350,563 (1,000,989)	73.8 (1.4)	

Table 1. *Cont.*

Group	No NAFLD		NAFLD		p-Value * for Difference between Those With and Without NAFLD
	Weighted *n* (SD)	Weighted Percent (SE)	Weighted *n* (SD)	Weighted Percent (SE)	
Smoking Status					<0.0001
Never	39,936,202 (1,411,228)	65 (1)	17,112,945 (831,979)	56.5 (1.5)	
Current	8,772,599 (631,562)	14.3 (0.8)	3,937,101 (329,377)	13 (1)	
Former	12,734,901 (761,119)	20.7 (0.9)	9,242,180 (599,355)	30.5 (1.2)	
Missing	13,143 (7800)	0 (0)	3146 (3146)	0 (0)	
Education Level					<0.0001
Less than HS	996,9649 (566,262)	16.2 (0.9)	6,847,959 (414,383)	22.6 (1.3)	
HS or some college	31,596,068 (1,513,799)	51.4 (1.3)	17,238,418 (1,037,220)	56.9 (1.7)	
College or more	19875667 (1,131,428)	32.3 (1.4)	6,208,995 (464,059)	20.5 (1.4)	
Missing	15,459 (9780)	0 (0)	0 (0)	0 (0)	
Diabetes (doctor's diagnosis)					<0.0001
No	57,768,923 (2,130,004)	94 (0.5)	24,365,926 (1,187,983)	80.4 (1.3)	
Yes	3687,922 (301,760)	6 (0.5)	5,929,446 (413,180)	19.6 (1.3)	
Diabetes (plasma fasting glucose ≥127 mg/dL)					<0.0001
No	58,992,373 (2,196,856)	96 (0.4)	23,910,800 (1,125,079)	78.9 (1.2)	
Yes	2,464,472 (221,631)	4 (0.4)	6,384,573 (417,802)	21.1 (1.2)	
Diabetes (either criteria)					<0.0001
No	56,999,582 (2,117,147)	92.7 (0.5)	22,113,112 (1,076,534)	73 (1.3)	
Yes	4,457,263 (331,259)	7.3 (0.5)	8,182,260 (487,535)	27 (1.3)	
High blood pressure at exam (systolic ≥130 and/or diastolic ≥85 mm Hg)					<0.0001
No	44,620,732 (1,833,006)	74.5 (0.9)	17,861,138 (883,010)	61.1 (1.4)	
Yes	15,263,826 (689,875)	25.5 (0.9)	11,373,199 (622,176)	38.9 (1.4)	

Table 1. *Cont.*

Group	No NAFLD		NAFLD		p-Value * for Difference between Those With and Without NAFLD
	Weighted *n* (SD)	Weighted Percent (SE)	Weighted *n* (SD)	Weighted Percent (SE)	
	Mean (SE)		Mean (SE)		
Age (years)	48.28 (0.4)		53.77 (0.47)		<0.0001
Family Income (poverty income ratio, PIR)	3.02 (0.04)		2.83 (0.06)		0.001
Body Mass Index (kg/m^2)	26.59 (0.1)		34.66 (0.23)		<0.0001
Waist Circumference (cm)	92.81 (0.25)		114.7 (0.47)		<0.0001
Dietary Measures					
Total Energy (kcal/day)	2018.42 (17.58)		2070.52 (23.51)		0.05
Dietary α-carotene (mcg)	489.85 (51.53)		387.73 (25.77)		0.16
Dietary β-carotene (mcg)	2547.38 (170.97)		1933.4 (70.65)		0.0003
Dietary β-cryptoxanthin (mcg)	110.22 (4.6)		93.48 (5.15)		0.05
Dietary lycopene (mcg)	5372.32 (190.38)		5624.4 (228.24)		0.37
Dietary lutein/zeaxanthin (mcg)	1748.17 (109.42)		1275.8 (49.83)		<0.0001
Laboratory Measures					
Insulin (pmol/L)	47.3 (0.62)		137.44 (3.38)		<0.0001
Glucose, plasma (mg/dL)	98.98 (0.43)		119.59 (1.17)		<0.0001
Cholesterol (mg/dL)	196.21 (1)		194.76 (1.33)		0.35
Triglycerides (mg/dL)	107.73 (2.15)		175.63 (4.18)		<0.0001
LDL–cholesterol (mg/dL)	116.69 (0.78)		115.08 (1.13)		0.24
Direct HDL–Cholesterol (mg/dL)	56.4 (0.38)		45.29 (0.34)		<0.0001
Gamma glutamyl transferase (U/L)	18.58 (0.26)		35.82 (0.85)		<0.0001
US Fatty Liver Index	12.39 (0.17)		52.8 (0.52)		<0.0001
Serum Measures (2003–2006 only)					
Serum α-carotene (µg/dL)	5.03 (0.28)		2.96 (0.14)		<0.0001
Serum β-carotene (µg/dL)	23.25 (1.14)		12.96 (0.6)		<0.0001
Serum β-cryptoxanthin (µg/dL)	10.32 (0.39)		7.6 (0.29)		<0.0001

Table 2. Odds ratios * (95% confidence interval) for association between the quartile of carotenoid intake, relative to Quartile 1, and presence of NAFLD. The median (range) for each quartile of carotenoid intake is also displayed (µg/day).

	Quartile 1	Quartile 2	Quartile 3	Quartile 4	p-Value for Trend
α-carotene					
Median (Range)	10.0 (0–27)	49.8 (27–95.5)	217.4 (96–477)	1007.4 (477.5–72,037)	
Model 1	Referent	0.99 (0.84, 1.16)	0.95 (0.79, 1.15)	**0.75 (0.63, 0.89)**	0.11
Model 2	Referent	**1.23 (1.04, 1.46)**	1.19 (0.98, 1.44)	0.99 (0.81, 1.21)	0.37
β-carotene					
Median (Range)	248.8 (0–435)	682.6 (435.5–1092.5)	1738.8 (1094–2726.5)	4691.2 (2728–246,122)	
Model 1	Referent	0.99 (0.82, 1.20)	**0.82 (0.70, 0.97)**	**0.63 (0.52, 0.76)**	<0.001
Model 2	Referent	1.04 (0.84, 1.28)	0.92 (0.76, 1.12)	**0.75 (0.60, 0.94)**	0.02
β-cryptoxanthin					
Median (Range)	5.7 (0–14.5)	26.3 (15–47)	76.6 (47.5–124)	229.5 (124.5–6088.5)	
Model 1	Referent	0.86 (0.72, 1.03)	0.82 (0.64, 1.04)	**0.64 (0.52, 0.80)**	0.04
Model 2	Referent	0.91 (0.74, 1.11)	0.97 (0.73, 1.28)	0.84 (0.65, 1.09)	0.50
Lycopene (Diet)					
Median (Range)	4.1 (0–564.5)	1273.5 (566.5–2233)	3792.1 (2234.5–6417.5)	12110 (6419–108,852)	
Model 1	Referent	0.94 (0.78, 1.14)	0.89 (0.74, 1.09)	1.12 (0.91, 1.37)	0.29
Model 2	Referent	0.99 (0.80, 1.23)	0.93 (0.75, 1.16)	1.07 (0.84, 1.35)	0.97
Lutein/Zeaxanthin (Diet)					
Median (Range)	277.0 (0–430.5)	600.2 (431–786.5)	1050.7 (787–1506)	2570.3 (1506.5–146,912)	
Model 1	Referent	0.94 (0.75, 1.17)	0.88 (0.73, 1.06)	**0.63 (0.51, 0.77)**	<0.001
Model 2	Referent	1.02 (0.81, 1.28)	1.02 (0.83, 1.26)	**0.78 (0.63, 0.96)**	<0.001
Total Lycopene (Diet and Supplement)					
Median (Range)	35.9 (0–590.5)	1313.8 (591–2276)	3838.2 (2280–6451.5)	12,203 (6455–108,852)	
Model 1	Referent	0.97 (0.81, 1.17)	0.92 (0.76, 1.1)	1.12 (0.92, 1.37)	0.42
Model 2	Referent	1.04 (0.86, 1.27)	0.96 (0.79, 1.18)	1.08 (0.86, 1.35)	0.72
Total Lutein/Zeaxanthin (Diet and Supplement)					
Median (Range)	281.6 (0–444.5)	627.6 (445–816.5)	1096.9 (817–1556)	2751.3 (1556.5–161,912)	
Model 1	Referent	0.92 (0.75, 1.14)	0.85 (0.71, 1.02)	**0.59 (0.48, 0.73)**	<0.001
Model 2	Referent	1.02 (0.82, 1.28)	1.01 (0.81, 1.25)	**0.73 (0.59, 0.91)**	0.002

* Model 1 is adjusted for age, sex, and the survey cycle year. Model 2 is adjusted for age, sex, and survey year, along with HEI 2015 score. All analyses are adjusted for survey design variables. Bold text indicates statistically significant associations.

Table 3 displays results from logistic regression modeling of serum carotenoid exposures and the NAFLD status as a binary outcome, while analogous results for US FLI as a continuous outcome are shown in Supplementary Table S2. As shown above for dietary intakes, serum carotenoids were treated as categorical exposures, with p-value for the trend calculated from models including serum carotenoids as a continuous variable. For serum carotenoid levels, a dose-response pattern was more apparent (compared with dietary intake exposures), with most of the *p*-values for the trend being statistically significant. In the fully adjusted model, the highest quartile was associated with significantly reduced risk for NAFLD among all carotenoids except lycopene. Statistically significant inverse associations were observed for Quartiles 2–4 of serum β-carotene, and Quartiles 3 and 4 of serum α-carotene, β-cryptoxanthin, and lutein/zeaxanthin. In general, estimates were largely unaffected by adjusting the HEI 2015 score, or for income and education. Similar results were seen in fully adjusted linear models for US FLI as a continuous outcome, but the highest quartile of serum lycopene was associated with lower US FLI in the fully adjusted model (see Supplementary Table S2).

Table 3. Logistic regression results (odds ratio (95% confidence interval)) for association between the quartile of serum carotenoids, relative to Quartile 1, and presence of NAFLD. The median (range) for each quartile of serum concentration is also displayed (μg/dL).

	Quartile 1	Quartile 2	Quartile 3	Quartile 4	p-Value for Trend
α-carotene					
Median (Range)	1.0 (0.2–1.5)	2.1 (1.6–3.0)	4.0 (3.0–5.4)	8.6 (5.4–96.5)	
Model 1	Referent	0.81 (0.62, 1.06)	**0.48 (0.33, 0.69)**	**0.33 (0.22, 0.5)**	<0.001
Model 2	Referent	0.7 (0.43, 1.13)	**0.35 (0.17, 0.71)**	**0.3 (0.15, 0.6)**	<0.001
β-carotene					
Median (Range)	6.0 (0.6–8.3)	10.5 (8.3–14.1)	18.2 (14.1–25.1)	36.1 (25.1–292.9)	
Model 1	Referent	**0.40 (0.28, 0.57)**	**0.31 (0.20, 0.46)**	**0.14 (0.09, 0.21)**	<0.001
Model 2	Referent	**0.28 (0.18, 0.45)**	**0.26 (0.14, 0.48)**	**0.13 (0.07, 0.23)**	<0.001
β-cryptoxanthin					
Median (Range)	3.5 (0.6–4.9)	6.4 (4.9–8.2)	10.3 (8.3–13.8)	19.4 (13.8–93.1)	
Model 1	Referent	0.73 (0.52, 1.03)	**0.49 (0.35, 0.68)**	**0.37 (0.26, 0.52)**	<0.001
Model 2	Referent	0.75 (0.44, 1.28)	**0.53 (0.31, 0.91)**	**0.45 (0.22, 0.92)**	0.02
Lycopene					
Median (Range)	19.2 (0.7–26.2)	32.6 (26.2–37.9)	43.5 (37.9–51.1)	64.0 (51.2–148.0)	
Model 1	Referent	0.91 (0.62, 1.32)	0.8 (0.53, 1.21)	**0.63 (0.41, 0.96)**	0.03
Model 2	Referent	0.99 (0.66, 1.49)	0.75 (0.4, 1.41)	0.68 (0.39, 1.21)	0.17
Lutein/Zeaxanthin					
Median (Range)	8.6 (2.4–10.9)	13.0 (10.6–15.2)	17.7 (15.2–27.0)	26.0 (20.7–113.1)	
Model 1	Referent	0.72 (0.51, 1.01)	**0.54 (0.36, 0.79)**	**0.33 (0.24, 0.47)**	<0.001
Model 2	Referent	0.78 (0.44, 1.38)	**0.53 (0.29, 0.97)**	**0.31 (0.2, 0.5)**	<0.01

Model 1 is adjusted for age, sex, and the survey cycle year. Model 2 is adjusted for age, sex, and survey year, along with the HEI 2015 score. All analyses are adjusted for survey design variables. Bold text indicates statistically significant associations.

4. Discussion

In this nationally representative sample of the US adult population, higher intake of certain carotenoids, and higher concentrations of carotenoids in the serum were associated with lower odds of having NAFLD, and with a better score of the US Fatty Liver Index. This is very promising given that there are no currently accepted therapeutic interventions for NAFLD. The dietary modification to increase carotenoid intake represents a possible route for prevention and amelioration of a common chronic health condition.

These findings build on a body of evidence suggesting relationships between carotenoid status to lipid metabolism and obesity, and are consistent with other epidemiological studies that demonstrate a prospective association between serum carotenoid levels and NAFLD [6,7,11], or between serum carotenoids and serum ALT [8,9]. These previous studies included a variety of designs. In a cross-sectional prevalence study (similar to the design of the NHANES analysis reported here), Cao et al. demonstrated that individuals with the highest serum levels of α and β-carotene, β-cryptoxanthin, lycopene, and lutein/zeaxanthin, had significantly reduced risk (ORs 0.32–0.62) for NAFLD, as diagnosed using abdominal ultrasonography [6]. A prospective study from the same group revealed that serum carotenoids were approximately 9% to 30% higher in individuals whose NAFLD improved compared to those who experienced disease progression [11]. Lastly, a case-control study demonstrated that serum lutein, zeaxanthin, lycopene, and α and β-carotene levels were 30% to 60% lower in NASH cases compared to controls [7]. Based on our knowledge, the present study is the first to report a statistically significant inverse association between dietary carotenoid intake and both risk of NAFLD and score on the US FLI, which is consistent with reduced risk for NAFLD. Our results are supported by animal studies in which the administration of carotenoids (including lycopene and lutein) attenuated the hepatic accumulation of lipids on a high-fat diet, as well as increased expression of SIRT1, which is a key regulator of fatty acid oxidation [15,17]. Higher carotenoid intake may reduce risk for NAFLD, and, in particular, the progression of simple hepatic steatosis to NASH, through several different pathways, including the attenuation of oxidative stress in the liver, with downstream effects on secretion of pro-inflammatory cytokines by hepatic macrophages, immune infiltration, and insulin sensitivity [20].

Many of the carotenoid measures were correlated with each other, which can make it difficult to distinguish individual effects (for example, correlations between intakes of α and β-carotene, and each of these with lutein/zeaxanthin, were between 0.78 to 0.94). For the logistic model (NAFLD as a binary outcome), when comparing the fully adjusted model with one carotenoid included at a time versus all included in the model, effect estimates were generally similar but attenuated (and confidence intervals increased in width). We investigated the possibility that beneficial associations of higher carotenoid intake are due to benefits of eating foods high in carotenoids (e.g., fruits and vegetables). Inclusion of the healthy eating index 2015 score did attenuate—but did not eliminate—associations with dietary carotenoids. Conversely, associations for serum carotenoids were largely unaffected by adjusting the healthy eating index score. It is likely that some of the protective association is due to healthier eating patterns. In addition, healthy diet patterns may enhance absorption of carotenoids from the diet. However, some effect may be due to carotenoid mechanisms specifically. This would be consistent with the observation that inverse associations were generally stronger and more consistent for serum carotenoids compared with dietary intake, which may reflect heterogeneity in the absorption of carotenoids into the bloodstream and greater measurement error or misclassification in the self-reported dietary data [23].

Limitations of this analysis included the use of self-reported dietary data, which may be due to a greater measurement error. The NAFLD status was inferred based upon a previously validated index, but is not a perfect proxy for NAFLD diagnosis based upon liver biopsy or other more invasive methods. However, evaluation of the FLI in the NHANES has shown good sensitivity, with the area under the receiver operator curve reported to be 78% [22], which suggests less potential for the outcome misclassification error. Furthermore, use of this index with the NHANES population made it impossible to adjust for potentially important covariates including body composition and markers of insulin sensitivity. Since the FLI is a surrogate marker for NAFLD based on markers of metabolic health, it is not clear whether carotenoid exposure is correlated with NAFLD independently of these markers. Another limitation is that the dietary analysis does not include an estimation of intake for certain carotenoids—such as astaxanthin and fucoxanthin—which may also have associations with NAFLD. Strengths include the use of multiple years of data from a large and nationally representative sample, consideration of multiple potential confounders, and examination of multiple carotenoids singly and in combination.

5. Conclusions

We found that higher intake and serum levels of certain carotenoids were cross-sectionally associated with lower odds of NAFLD. These findings are promising given the limited therapeutic options for treating NAFLD, in that identification of modifiable lifestyle factors provide an opportunity to limit or prevent the disease and its progression.

Supplementary Materials: The following are available online at http://www.mdpi.com/2072-6643/11/5/1101/s1. Table S1: Beta coefficients (standard error), *p*-values, for association between quartile of carotenoid intake, relative to quartile 1, and the US Fatty Liver Index. The median (range) for each quartile of carotenoid intake is also displayed (μg/day). Table S2: Beta coefficients (standard error), *p*-values, for association between quartile of serum carotenoids, relative to quartile 1, and the US Fatty Liver Index. The median (range) for each quartile of serum concentration is also displayed (μg/dL).

Author Contributions: Conceptualization, K.C. Formal analysis, K.C. and T.L. Funding acquisition, J.M. Investigation, K.C. Methodology, K.C. and J.M. Resources, J.M. Writing—original draft, K.C. and T.L. Writing—review & editing, J.M.

Funding: This work is supported by National Institutes of Health grant EY025292, the Department of Ophthalmology and Visual Sciences at the University of Wisconsin-Madison (in part, by an unrestricted grant to the department from the Research to Prevent Blindness), and the Department of Nutritional Sciences by a NIH institutional training grant (T32 DK007665). The funding sources had no role in the collection, analysis, or interpretation of data, in the writing of the report, or in the decision to submit the article for publication.

Conflicts of Interest: The authors declare no conflict of interest.

References

1. Le, M.H.; Devaki, P.; Ha, N.B.; Jun, D.W.; Te, H.S.; Cheung, R.C.; Nguyen, M.H. Prevalence of non-alcoholic fatty liver disease and risk factors for advanced fibrosis and mortality in the United States. *PLoS ONE* **2017**, *12*, e0173499. [CrossRef] [PubMed]

2. Bush, H.; Golabi, P.; Younossi, Z.M. Pediatric Non-Alcoholic Fatty Liver Disease. *Children* **2017**, *4*, 48. [CrossRef] [PubMed]

3. Chalasani, N.; Younossi, Z.; Lavine, J.E.; Charlton, M.; Cusi, K.; Rinella, M.; Harrison, S.A.; Brunt, E.M.; Sanyal, A.J. The diagnosis and management of nonalcoholic fatty liver disease: Practice guidance from the American Association for the Study of Liver Diseases. *Hepatology* **2018**, *67*, 328–357. [CrossRef]

4. Brunt, E.M.; Wong, V.W.; Nobili, V.; Day, C.P.; Sookoian, S.; Maher, J.J.; Bugianesi, E.; Sirlin, C.B.; Neuschwander-Tetri, B.A.; Rinella, M.E. Nonalcoholic fatty liver disease. *Nat. Rev. Dis. Primers* **2015**, *1*, 15080. [CrossRef] [PubMed]

5. Rinella, M.E.; Loomba, R.; Caldwell, S.H.; Kowdley, K.; Charlton, M.; Tetri, B.; Harrison, S.A. Controversies in the Diagnosis and Management of NAFLD and NASH. *Gastroenterol. Hepatol.* **2014**, *10*, 219–227.

6. Cao, Y.; Wang, C.; Liu, J.; Liu, Z.M.; Ling, W.H.; Chen, Y.M. Greater serum carotenoid levels associated with lower prevalence of nonalcoholic fatty liver disease in Chinese adults. *Sci. Rep.* **2015**, *5*, 12951. [CrossRef]

7. Erhardt, A.; Stahl, W.; Sies, H.; Lirussi, F.; Donner, A.; Haussinger, D. Plasma levels of vitamin E and carotenoids are decreased in patients with Nonalcoholic Steatohepatitis (NASH). *Eur. J. Med. Res.* **2011**, *16*, 76–78. [CrossRef]

8. Ruhl, C.E.; Everhart, J.E. Relation of elevated serum alanine aminotransferase activity with iron and antioxidant levels in the United States. *Gastroenterology* **2003**, *124*, 1821–1829. [CrossRef]

9. Sugiura, M.; Nakamura, M.; Ogawa, K.; Ikoma, Y.; Yano, M. High serum carotenoids are associated with lower risk for developing elevated serum alanine aminotransferase among Japanese subjects: The Mikkabi cohort study. *Br. J. Nutr.* **2016**, *115*, 1462–1469. [CrossRef]

10. Villaca Chaves, G.; Pereira, S.E.; Saboya, C.J.; Ramalho, A. Non-alcoholic fatty liver disease and its relationship with the nutritional status of vitamin A in individuals with class III obesity. *Obes. Surg.* **2008**, *18*, 378–385. [CrossRef]

11. Xiao, M.L.; Chen, G.D.; Zeng, F.F.; Qiu, R.; Shi, W.Q.; Lin, J.S.; Cao, Y.; Li, H.B.; Ling, W.H.; Chen, Y.M. Higher serum carotenoids associated with improvement of non-alcoholic fatty liver disease in adults: A prospective study. *Eur. J. Nutr.* **2018**. [CrossRef]

12. Park, S.K.; Lee, H.J.; Lee, D.H.; Lee, S.K.; Chun, B.Y.; Kim, S.A.; Lee, H.S.; Son, H.K.; Kim, S.H. Associations of non alcoholic fatty liver with the metabolic syndrome and serum carotenoids. *J. Prev. Med. Public Health* **2008**, *41*, 39–44. [CrossRef] [PubMed]

13. Ahn, J.; Lee, H.; Jung, C.H.; Ha, T. Lycopene inhibits hepatic steatosis via microRNA-21-induced downregulation of fatty acid-binding protein 7 in mice fed a high-fat diet. *Mol. Nutr. Food Res.* **2012**, *56*, 1665–1674. [CrossRef]

14. Bahcecioglu, I.H.; Kuzu, N.; Metin, K.; Ozercan, I.H.; Ustundag, B.; Sahin, K.; Kucuk, O. Lycopene prevents development of steatohepatitis in experimental nonalcoholic steatohepatitis model induced by high-fat diet. *Vet. Med. Int.* **2010**. [CrossRef]

15. Chung, J.; Koo, K.; Lian, F.; Hu, K.Q.; Ernst, H.; Wang, X.D. Apo-10′-lycopenoic acid, a lycopene metabolite, increases sirtuin 1 mRNA and protein levels and decreases hepatic fat accumulation in ob/ob mice. *J. Nutr.* **2012**, *142*, 405–410. [CrossRef]

16. Jiang, W.; Guo, M.H.; Hai, X. Hepatoprotective and antioxidant effects of lycopene on non-alcoholic fatty liver disease in rat. *World J. Gastroenterol.* **2016**, *22*, 10180–10188. [CrossRef]

17. Qiu, X.; Gao, D.H.; Xiang, X.; Xiong, Y.F.; Zhu, T.S.; Liu, L.G.; Sun, X.F.; Hao, L.P. Ameliorative effects of lutein on non-alcoholic fatty liver disease in rats. *World J. Gastroenterol.* **2015**, *21*, 8061–8072. [CrossRef]

18. Kim, J.H.; Na, H.J.; Kim, C.K.; Kim, J.Y.; Ha, K.S.; Lee, H.; Chung, H.T.; Kwon, H.J.; Kwon, Y.G.; Kim, Y.M. The non-provitamin A carotenoid, lutein, inhibits NF-kappaB-dependent gene expression through redox-based regulation of the phosphatidylinositol 3-kinase/PTEN/Akt and NF-kappaB-inducing kinase pathways: Role of H(2)O(2) in NF-kappaB activation. *Free Radic. Biol. Med.* **2008**, *45*, 885–896. [CrossRef] [PubMed]

19. Ni, Y.; Zhuge, F.; Nagashimada, M.; Ota, T. Novel Action of Carotenoids on Non-Alcoholic Fatty Liver Disease: Macrophage Polarization and Liver Homeostasis. *Nutrients* **2016**, *8*, 391. [CrossRef]

20. Murillo, A.G.; DiMarco, D.M.; Fernandez, M.L. The Potential of Non-Provitamin A Carotenoids for the Prevention and Treatment of Non-Alcoholic Fatty Liver Disease. *Biology* **2016**, *5*, 42. [CrossRef] [PubMed]

21. Centers for Disease Control and Prevention (CDC); National Center for Health Statistics (NCHS). National Health and Nutrition Examination Survey. NHANES Questionnaires, Datasets, and Related Documentation. Available online: https://wwwn.cdc.gov/nchs/nhanes/Default.aspx (accessed on 30 May 2018).

22. Ruhl, C.E.; Everhart, J.E. Fatty liver indices in the multiethnic United States National Health and Nutrition Examination Survey. *Aliment. Pharmacol. Ther.* **2015**, *41*, 65–76. [CrossRef] [PubMed]

23. Borel, P. Genetic variations involved in interindividual variability in carotenoid status. *Mol. Nutr. Food Res.* **2012**, *56*, 228–240. [CrossRef] [PubMed]

nutrients

MDPI

Article

Serum Lutein is related to Relational Memory Performance

Corinne N. Cannavale [1], Kelsey M. Hassevoort [2,3], Caitlyn G. Edwards [4], Sharon V. Thompson [4], Nicholas A. Burd [4,5], Hannah D. Holscher [4,5,6], John W. Erdman Jr. [4,6], Neal J. Cohen [1,2,3,7] and Naiman A. Khan [1,4,5,*]

[1] Neuroscience Program, University of Illinois at Urbana-Champaign, Champaign, IL 61801, USA; cannava2@illinois.edu (C.N.C.); njc@illinois.edu (N.J.C.)
[2] Beckman Institute for Advanced Science and Technology, University of Illinois at Urbana-Champaign, Champaign, IL 61801, USA; kelseyhassevoort@gmail.com
[3] Center for Brain Plasticity, University of Illinois at Urbana-Champaign, Champaign, IL 61801, USA
[4] Division of Nutritional Sciences, University of Illinois at Urbana-Champaign, Champaign, IL 61801, USA; cgedwar2@illinois.edu (C.G.E.); svthomp2@illinois.edu (S.V.T.); naburd@illinois.edu (N.A.B.); hholsche@illinois.edu (H.D.H.); jwerdman@illinois.edu (J.W.E.J.)
[5] Department of Kinesiology and Community Health, University of Illinois at Urbana-Champaign, Champaign, IL 61801, USA
[6] Department of Food Science and Human Nutrition, University of Illinois at Urbana-Champaign, Champaign, IL 61801, USA
[7] Department of Psychology, University of Illinois at Urbana-Champaign, Champaign, IL 61801, USA
* Correspondence: nakhan2@illinois.edu; Tel.: +1-217-300-1667

Received: 28 February 2019; Accepted: 29 March 2019; Published: 2 April 2019

Abstract: Dietary carotenoids, plant pigments with anti-oxidant properties, accumulate in neural tissue and are often found in lower concentrations among individuals with obesity. Given previous evidence of negative associations between excess adiposity and memory, it is possible that greater carotenoid status may confer neuroprotective effects among persons with overweight or obesity. This study aimed to elucidate relationships between carotenoids assessed in diet, serum, and the macula (macular pigment optical density (MPOD)) and relational memory among adults who are overweight or obese. Adults aged 25–45 years (N = 94) completed a spatial reconstruction task. Task performance was evaluated for accuracy of item placement during reconstruction relative to the location of the item during the study phase. Dietary carotenoids were assessed using 7-day diet records. Serum carotenoids were measured using high-performance liquid chromatography. Hierarchical linear regression analyses were used to determine the relationship between carotenoids and task performance. Although initial correlations indicated that dietary lutein, beta-carotene, and serum beta-carotene were positively associated with memory performance, these relationships were not sustained following adjustment for age, sex, and BMI. Serum lutein remained positively associated with accuracy in object binding and inversely related to misplacement error after controlling for covariates. Macular carotenoids were not related to memory performance. Findings from this study indicate that among the carotenoids evaluated, lutein may play an important role in hippocampal function among adults who are overweight or obese.

Keywords: obesity; hippocampus; nutrition; overweight; carotenoids

1. Introduction

Overweight and obesity are conditions that increase metabolic risk and are characterized by a body mass index (BMI) greater than 25 kg/m^2 [1]. Obesity can lead to a variety of metabolic and cardiovascular concerns such as type 2 diabetes, non-alcoholic fatty liver disease, and cardiovascular disease [1]. According to the National Health and Nutrition Examination Survey (NHANES) data from 2015–2016, approximately 40% of adults in the United States have obesity, with women having a slightly higher prevalence than men (41.1% vs. 37.9%) [1]. In addition to the cardiometabolic concerns that accompany obesity, excess fat mass or adiposity, as well as the associated metabolic complications, have been associated with poorer cognitive function and brain structure [2]. One specific brain region thought to be affected by obesity and other associated disorders is the hippocampus [2–4]. The hippocampus is a highly plastic region of the brain, which may explain why hippocampal-dependent memory function can be susceptible to behavior modulation through environmental factors [5,6]. Obesity, as well as a "western" diet characterized by high saturated fat and added sugar intake, has been related to poorer hippocampal function [3,4]. However, research examining the influence of specific dietary components on relational memory among individuals with overweight or obesity has been limited.

Relational memory is a hippocampal-dependent process that involves the flexible binding of arbitrary elements within an episode, and the subsequent reactivation of these relations [7]. In a practical sense, relational memory allows us to put a name to a face or re-tell the story of a trip in any order one chooses. One approach for assessing relational memory is to use a spatial reconstruction task. Spatial reconstruction tasks require participants to return objects to locations they have previously studied. This task design requires the participant to encode the arbitrary relationships between objects and locations (or other objects) within each trial and subsequently use these bindings to reconstruct the studied display successfully. Previous studies have shown that obesity and diet are associated with relational memory performance; however, this has not yet been thoroughly investigated [4,8].

Carotenoids have been recently found to have relevance for hippocampal-dependent memory performance [7]. Carotenoids are plant pigments that represent red, yellow, and orange color, and can be found in egg yolks and a variety of fruits and vegetables such as avocados, carrots, sweet potatoes, spinach, and other leafy green vegetables [9]. Previous studies have shown that, among the numerous carotenoids in nature, only a handful accumulate in human neural tissue. These include lutein, beta-carotene, beta-cryptoxanthin, and zeaxanthin, all of which accumulate in all cortices and the hippocampus in humans and non-human primates [10–12]. However, lutein accumulation in neural tissue is up to 5-fold greater than other carotenoids, inferring a potentially unique role for this carotenoid in cognitive function and brain health [10]. Further, lutein, its stereoisomer, zeaxanthin, and the lutein intermediate *meso*-zeaxanthin, collectively belonging to a group known as xanthophylls, selectively accumulate in the macula of the human eye. The structural properties of xanthophylls allow them to serve as blue light filters and antioxidants in the eye and protect retinal tissue from photo-oxidative damage [9,13,14]. Opportunely, the macular concentration of lutein, zeaxanthin, and *meso*-zeaxanthin can be non-invasively assessed as macular pigment optical density (MPOD) [15]. Macular xanthophylls have previously been related to brain carotenoid concentrations in non-human primates [16]. These macular carotenoids, assessed through MPOD, are also positively associated with relational memory performance in children and with intellectual ability and executive function in adults [7,17,18]. Additionally, higher MPOD scores have been associated with better memory performance assessed using a delayed recall task [11,19]. However, given that MPOD represents a composite metric of all three xanthophylls in the retina, it is incorrect to isolate carotenoid-related cognitive benefits to lutein alone. Individual carotenoid quantification in serum provides an opportunity to study the influence of lutein in a manner that would be separable from other carotenoids in vivo. While serum does not provide a direct assessment of neuronal lutein, it provides us with a more precise measure of individual carotenoids than what is determined using MPOD or dietary assessment. In previous studies quantifying serum lutein rather than the composite MPOD measure, it was shown that serum lutein mediated the relationship between the parahippocampal cortex and crystallized intelligence [20]. Serum lutein is also associated with multiple domains of cognition, including memory [11]. However, a potential limitation of relying on serum alone is that the content in serum is transient and may not reflect xanthophyll status in neural tissue. Therefore, additional

Nutrients **2019**, *11*, 768

studies that assess both MPOD and serum carotenoids are necessary to clarify the impact of lutein and other carotenoids on relational memory. To our knowledge, there have been no previous studies investigating associations between both MPOD and serum xanthophylls on relational memory abilities among adults with overweight or obesity. Accordingly, this study aimed to understand whether dietary, serum, or macular levels of carotenoids were associated with relational memory function in adults that are overweight and obese. We hypothesized that macular carotenoids, assessed as MPOD, would be positively associated with relational memory performance. Additionally, we anticipated that a greater serum concentration of lutein would be associated with greater relational memory performance.

2. Materials and Methods

2.1. Participants

Eligible participants were adults aged 25–45 years who were overweight or obese (BMI $\geq$ 25 kg/m^2) and no prior history of physician-diagnosed metabolic or gastrointestinal disease (e.g., Crohn's Disease, Diabetes, CVD, etc.), or neurological or cognitive disorders. Participants were also excluded if they were using any tobacco products.

2.2. Ethical Approval

At the first appointment, participants were informed of the overall procedure and written informed consent was obtained before collection of any data. The University of Illinois Institutional Review Board approved consent before recruitment, and the study was conducted following the Declaration of Helsinki.

2.3. Procedure

Data were collected over the course of two appointments. At the first appointment, participants were screened using medical history and demographic questionnaires. The second appointment followed a 10-h overnight fast. The participants subsequently underwent adiposity assessment using dual-energy X-ray absorptiometry (DXA) (Hologic, Bedford, MA, USA) and IQ was assessed using the Kaufman brief intelligence test-2 (KBIT-2). A spatial reconstruction task was administered to assess hippocampal-dependent relational memory ability. Fasted blood was collected at the conclusion of the appointment and serum lutein levels were determined using high-performance liquid chromatography. All participants were provided with a 7-day diet record to document their regular carotenoid intake, which was completed within one week of relational memory assessment.

2.4. Relational Memory Assessment

Hippocampal-dependent relational memory ability was evaluated through a computerized spatial reconstruction task (Figure 1). This task was completed using Presentation Software (Neurobehavioral Systems, Berkeley, CA, USA). Participants were first shown 6 abstract shapes in the center of the screen to introduce the shapes used for the reconstruction phase. After 6 s, the stimuli disappeared and reappeared in a randomized array on the screen. The identities of the objects were then masked by small squares and participants had 18 s to click the boxes and one-by-one learn the locations of each shape. The boxes then disappeared and, after a 2 s fixation, they reappeared at the top of the screen. Participants were instructed to reconstruct the array they previously studied. Reconstruction was self-paced and, once the participant was satisfied with their reconstruction, they moved on to the subsequent trial, for a total of 20 trials (4 blocks of 5). Each participant's performance was assessed using two error metrics: misplacement and object-location binding (Figure 2). Misplacement was calculated as the average measure of distance (in pixels) between the objects' studied and reconstructed locations, with a higher score indicating poorer performance. Object-location binding was defined as the number of times the participant correctly placed an item within a pre-defined radius around its studied location. For each trial, a participant received a score of 0–6 for this metric, with a higher score indicating better performance, and performance was averaged across trials. More detailed explanations of these error metrics can be found in Horecka et al. (2018) [21].

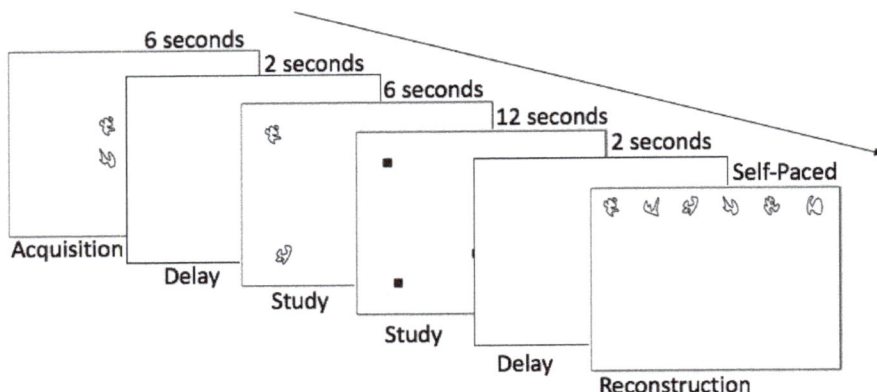

Figure 1. Spatial reconstruction task description.

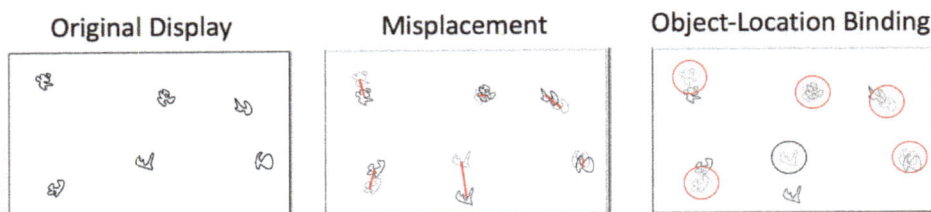

Figure 2. Spatial reconstruction task error metrics.

2.5. Intelligence Assessment

The KBIT-2 was administered by a trained staff member to estimate IQ. The assessment is comprised of 3 subtests: verbal knowledge, matrices, and riddles. Correct answers are given a score of 1 and each subtest score is then transformed to a standardized score normed for ages 4–90 years [22,23]. The verbal knowledge subtest includes 60 questions where the participant chooses which of six images is most associated with a word or question spoken by the researcher. The matrices subtest has 46 logic problems where the participant must choose an image that is most associated with a single stimulus picture or which picture best completes the pattern of a 2×2, 2×3 or 3×3 matrix. The riddle subtest consists of 48 riddles spoken by the researcher where the participant gives a single word response.

2.6. Dietary Assessment

All participants recorded regular dietary intake for 7 days in a provided food diary after instruction by a trained staff member. Diet record data were recorded and analyzed using the Nutrition Data System for Research (NDSR) 2015 (Minneapolis, MN, USA) by trained research staff. There was not enough dietary information in the available nutrient databases to extract levels of individual xanthophylls; therefore, dietary lutein and zeaxanthin were ascertained as an aggregate measure (i.e., lutein + zeaxanthin). Consumption amounts of lutein + zeaxanthin, beta-carotene, and beta-cryptoxanthin were extracted from NDSR.

2.7. Serum Carotenoid Assessment

Serum carotenoid levels were assessed using high-performance liquid chromatography (HPLC). Serum carotenoids were extracted using 3 consecutive hexane extraction processes using a previously published protocol [24]. Briefly, the hexane layers were combined, dried under nitrogen, taken up into 90% MTBE, 8% methanol, and 2% ammonium acetate in water solution (1.5% solution) and then analyzed for carotenoid concentrations using the Alliance HPLC system (e2695 Separation Module) equipped with 2998 photodiode array detector (Waters, Milford, MA, USA) and a reverse-phase C30 column (4.6 × 150 nm, 3 micron, YMC, Wilmington, NC, USA). Serum levels of carotenoids previously found in human neural tissue were assessed including lutein, zeaxanthin, beta-carotene, and cryptoxanthin.

Carotenoid standards were obtained from Carotenature, Ostermundigen, Switzerland. For quantification, standard curves were run for each carotenoid, and serum carotenoids were quantified by use of the following extinction coefficients (all 1% solution): lutein 2550 in ethanol; zeaxanthin 2540 in ethanol; beta carotene 2592 in hexane; cryptoxanthin 2565 in hexane. The Erdman laboratory routinely participates in the National Institutes for Standards and Testing micronutrient proficiency testing program, and our serum carotenoid values for blinded serum samples consistently were within 1–2 SD of the medians.

2.8. Retinal Carotenoid Assessment

MPOD was assessed using a macular densitometer (Macular Metrics Corporation, Rehoboth, MA, USA) via a customized hetero-flicker photometry (cHFP) technique. This two-step process first required participants to focus on a flickering stimulus in their central line of vision, where the macular pigment is at its highest concentration, and 7 degrees parafoveally, where the macular pigment is at its lowest concentration. The stimulus flickered between 460 nm and 570 nm wavelengths at a rate that has been optimized for the width of the null zone for the participant. Participants then adjusted the radiance to identify the point at which they could not detect a flicker (the null flicker zone). MPOD was calculated by subtracting the foveal from the parafoveal log sensitivity measurements after normalizing at 570 nm. More detailed information on the principles behind this technique were described in Wooten, et al. (1999) [25].

2.9. Weight Status and Adiposity

Height and weight were measured three times using a stadiometer (model 240; SECA, Hamburg, Germany) and a digital scale (WB-300 Plus; Tanita, Tokyo, Japan) while participants were wearing light clothing and no shoes. Mean height and weight values were used to calculate BMI. Whole body adiposity (%Fat) was measured using a Hologic Horizon W bone densitometer (software version 13.4.2, Bedford, MA, USA) DXA scanner.

2.10. Statistical Analysis

All statistical analyses were conducted using SPSS 2016 v24 (IBM Corp., Armonk, NY, USA). Normality was assessed using the Shapiro-Wilk test and variables that did not display normal distribution were log transformed. Pearson's correlations were run to determine initial relationships between hippocampal task performance, carotenoids, MPOD, Age, Sex, and %Fat. Hierarchical linear regression modeling was used to further investigate the relationships between carotenoids with statistically significant correlations to relational memory metrics. Step 1 of each regression model included covariates that were found to be significantly related to the dependent, relational memory, variable via bivariate correlations. Separate step 2's were conducted for each carotenoid found to be related to the respective dependent variable from bivariate correlations. A one-tailed approach was used due to the positive directionality of our hypothesis and outcomes seen in the previous literature [17,18,20,26].

3. Results

Descriptive statistics for participant demographics can be found in Table 1.

Table 1. Participant characteristics and memory performance.

Variable	Mean ± SD
Sex, F/M	45, 54
Age, years	34.9 ± 6.1
BMI, kg/m^2	33.3 ± 6.6
Fat, %	40.2 ± 8.42
Intelligence Quotient	107 ± 12.1
Macular Pigment Optical Density	0.438 ± 0.20
Dietary Lutein + Zeaxanthin, mcg/day	2283 ± 3382
Serum Lutein, μmol/L	0.129 ± 0.06
Misplacement, pixels	219.3 ± 77.0
Object-location binding	2.67 ± 0.83

Bivariate correlations revealed that misplacement was positively related to age ($r = 0.33$, $p = 0.001$) and %Fat ($r = 0.19$, $p = 0.03$), and inversely related to IQ ($r = -0.37$, $p < 0.001$). Further, misplacement was negatively associated with dietary lutein + zeaxanthin ($r = -0.22$, $p = 0.02$) and serum lutein ($r = -0.25$, $p = 0.005$) concentrations. Similarly, dietary and serum beta-carotene were negatively associated with misplacement ($r = -0.24$, $p = 0.01$; $r = -0.20$, $p = 0.03$). Object-location binding was negatively related to age ($r = -0.25$, $p = 0.007$) and positively related to IQ ($r = 0.32$, $p = 0.001$). Object-location binding was not associated with %Fat ($r = -0.11$, $p = 0.1$), likely due to the decreased range of performance. Additionally, object-location binding was positively associated with dietary lutein + zeaxanthin ($r = 0.21$, $p = 0.02$), serum lutein ($r = 0.22$, $p = 0.02$), and dietary beta-carotene ($r = 0.20$, $p = 0.03$). No other carotenoids were statistically significantly related to relational memory measures in either diet or serum (all p's > 0.09).

MPOD was not related to relational memory measures (all p's > 0.4), nor was it related to dietary lutein + zeaxanthin ($p > 0.1$). MPOD was, however, related to serum lutein ($r = 0.30$, $p = 0.002$), dietary ($r = 0.20$, $p = 0.03$) and serum ($r = 0.25$, $p = 0.007$) beta-carotene, serum cryptoxanthin ($r = 0.29$, $p = 0.007$), and serum zeaxanthin ($r = 0.22$, $p = 0.017$). Bivariate correlations are summarized in Table 2.

Table 2. Bivariate correlations (Pearson's *r*) between relational memory, participant characteristics, and carotenoids.

	Misplacement	Object-Location Binding
Age	0.33 **	−0.25 **
Sex	−0.17	0.10
%Fat	0.19 *	−0.11
IQ	−0.37 **	0.32 **
Dietary L + Z	−0.21 *	0.21 *
Dietary Beta-Carotene	−0.24 **	0.20 *
Dietary Beta-Cryptoxanthin	−0.06	0.05
MPOD	−0.110	0.083
Serum Lutein	−0.266 **	0.223 *
Serum Zeaxanthin	−0.027	0.002
Serum Beta-Carotene	−0.202 *	0.137
Serum Cryptoxanthin	−0.103	0.076

$* p < 0.05$, $** p < 0.01$.

Misplacement was modeled using hierarchical linear regression modeling with dietary lutein + zeaxanthin, serum lutein, and dietary and serum beta-carotene. Each carotenoid was modeled in a separate step 2. Step 1 of each misplacement model controlled for covariates that were statistically significantly related to misplacement in bivariate correlations (Age, %Fat, IQ).

Statistically significant regressions for misplacement can be found in Table 3. Serum lutein was the only carotenoid significantly related to misplacement after covariate adjustment ($\beta = -0.15$, $p = 0.05$). Dietary lutein + zeaxanthin ($\beta = -0.07$, $p = 0.2$), dietary beta-carotene ($\beta = -0.07$, $p = 0.2$), and serum beta-carotene ($\beta = -0.05$, $p = 0.3$) were no longer related after controlling for covariates in step 1.

Table 3. Hierarchical linear regression modeling of misplacement and carotenoids.

Step & Variable		β	ΔR^2
	Age	0.179 **	
Step 1	%Fat	0.019	0.142 **
	IQ	−0.363 **	
Step 2	Serum Lutein	−0.152 *	0.207 *

*$p < 0.05$, ** $p < 0.01$.

Object-location binding was modeled with serum lutein, dietary lutein + zeaxanthin, and dietary beta-carotene. Similarly, serum lutein was the only carotenoid that was still statistically significantly related to object-location binding after controlling for covariates in step 1 ($\beta = 0.16$, $p = 0.05$). Dietary lutein + zeaxanthin ($\beta = 0.10$, $p = 0.1$), and dietary beta-carotene ($\beta = 0.08$, $p = 0.2$) were no longer related to object-location binding after covariates were accounted for. This model is described in Table 4.

Table 4. Hierarchical linear regression modeling of object-location binding and carotenoids.

Step & Variable		β	ΔR^2
Step 1	Age	0.290 **	0.142 **
	IQ	−0.197 *	
Step 2	Serum Lutein	0.159 *	0.166 *

*$p < 0.05$, ** $p < 0.01$.

4. Discussion

The present work examined the relationship between carotenoids in the macula, diet, and serum, and their relationship with hippocampal-dependent relational memory performance. Given previous literature indicating that lutein disproportionately accumulates in neural tissue, including the hippocampus, we anticipated that serum and macular lutein concentrations, in particular, would be related to relational memory. Herein, the results indicated that higher serum, but not macular, lutein concentrations were positively associated with greater relational memory performance on a spatial reconstruction task. Although dietary lutein + zeaxanthin and both dietary and serum beta-carotene were correlated with performance, these relationships did not persist after covariate adjustment. Taken together, these findings provide additional evidence that serum carotenoid status may impact memory performance among adults who are overweight or obese.

There are many potential roles lutein may play in the brain. One proposed mechanism, which is relevant in our sample, is that lutein can modulate inflammatory and oxidative stress pathways [13,14]. Participants with overweight or obesity are more susceptible to oxidative and inflammatory stress due to the higher levels of chronic inflammation associated with excess adipose tissue [27,28]. Inflammation and oxidative stress can be mitigated by fruit and vegetable intake, foods that are often rich in carotenoids [27]. Inflammation is detrimental to hippocampal function, specifically by inhibiting long term potentiation, the molecular mechanism for memory formation [29]. In the retina, lutein is protective against age-related macular degeneration by reducing oxidative stress [13,14]. Thus, we hypothesize that lutein may play a similar role in the hippocampus.

Previous studies have shown that elevated serum lutein concentrations are positively associated with multiple realms of cognition, particularly in brain regions where lutein is known to deposit. A recent MRI study found that the parahippocampal cortex mediates the relationships between serum lutein concentrations and fluid intelligence [20]. Additionally, serum lutein concentrations were positively related to delayed recall memory task performance, controlled oral word association tests, and the Weshler adults intelligence scale-III similarities subtest [11]. Vishwanathan et al. did not observe similar relationships with delayed recall and serum lutein + zeaxanthin in their sample [19]. Our results in comparison to these studies indicate that lutein, in particular, may impact cognition in participants with overweight or obesity. When lutein is assessed as a combined metric (dietary, MPOD, serum L + Z), significant associations have not been found. Our results, however, display a strong association with lutein assessed independently of zeaxanthin.

Contrary to our *a priori* hypothesis, macular xanthophylls, assessed by MPOD, were not related to relational memory performance. This was surprising given previous work linking MPOD to relational memory in children, and in older adults which has shown positive associations between other memory forms (e.g., delayed recall) and MPOD [19]. However, our finding may differ from the previous literature due to the weight status of our participants. It is thought that increased adipose tissue, owing to its capacity to store carotenoids, may limit carotenoid availability for other tissues (e.g., retina), which may contribute to this disparity. Though the current body of literature has displayed positive relationships between MPOD and relational memory or executive function, these samples were mainly healthy weight [7,18,26,30,31]. Nevertheless, our team has previously shown that MPOD is associated with intellectual abilities among persons who are overweight or obese [17]. Therefore, it is possible that the relationship between MPOD and cognitive function among persons with overweight or obesity may be domain-dependent. While previous adult studies have displayed relationships with memory function and MPOD, the memory assessments used primarily assessed item-memory and therefore may not have depended on the hippocampus to the extent the spatial reconstruction task does [18,26,30,31]. Additionally, we may have failed to observe the relationship between MPOD and memory due to limitations in the technique we used to assess MPOD. Previous work has shown that, while macular pigmentation is densest at the foveal pit at two sites, we were unable to assess the complete spatial distribution of xanthophylls in the macula, thus, limiting our ability to link macular lutein to memory performance. Additional research studies examining the impact of the spatial profiles of the xanthophylls in the macula on memory function are necessary to characterize the impact of macular lutein to hippocampal function comprehensively. Nevertheless, this work suggests that serum lutein is unique among carotenoids in its relationship with memory. Future studies should assess both blood and macular carotenoids to further elucidate this relationship in a larger and more diverse population.

While we found significant correlations between memory performance and serum lutein concentrations, there are some limitations worth considering. While spatial reconstruction tasks have been shown to elicit the hippocampus, hippocampal-dependent cognition can also be assessed via other tasks/paradigms that could inform the specific memory processes that benefit from lutein. Second, relative to a previously studied Midwest sample, our serum lutein concentrations were lower when compared to the population average (0.28 μmol/L $\pm$ 0.13 versus 0.129 μmol/L $\pm$ 0.06) [32]. This, however, may be explained by the higher BMI of our sample. Finally, this study design was cross-sectional, providing no insights into the causal mechanisms that may underlie the carotenoid and relational memory relationship. Intervention studies are necessary to understand whether improvement in lutein status does, in fact, positively influence hippocampal-dependent relational memory performance.

Nutrients **2019**, *11*, 768

5. Conclusions

This study aimed to understand the relationship between dietary, serum, and macular carotenoids, and relational memory. Our results revealed that serum lutein was significantly related to two metrics of relational memory performance even after adjusting for significant covariates. While this study was correlational, it lays the groundwork for subsequent research in this area. Further intervention studies where carotenoids are assessed in the macula and serum must be conducted to better understand this relationship, particularly in participants with overweight or obesity.

Author Contributions: N.A.K., C.N.C, and K.M.H analyzed the data. C.N.C., C.G.E., and S.V.T. were involved in data collection. N.A.B., H.D.H., and N.A.K. conceived, designed, and acquired funding for the study. N.A.B was responsible for blood sample collection. J.W.E.J. supervised serum carotenoid analyses. N.J.C. and K.M.H. designed the spatial reconstruction task. All authors contributed to the writing of the manuscript draft.

Funding: This work was supported by funds provided by the Department of Kinesiology and Community Health at the University of Illinois and the USDA National Institute of Food and Agriculture, Hatch Project 1009249. Partial support was also provided by the Hass Avocado Board.

Conflicts of Interest: The authors declare no conflict of interest.

References

1. Hales, C.M.; Carroll, M.D.; Fryar, C.D.; Ogden, C.L. Prevalence of Obesity Among Adults and Youth: United States, 2015–2016. *NCHS Data Brief* **2017**, *288*, 1–8.
2. Burkhalter, T.M.; Hillman, C.H. A Narrative Review of Physical Activity, Nutrition, and Obesity to Cognition and Scholastic Performance across the Human Lifespan. *Adv. Nutr.* **2011**, *2*, 201S–206S. [CrossRef] [PubMed]
3. Kanoski, S.E.; Davidson, T.L. Western diet consumption and cognitive impairment: Links to hippocampal dysfunction and obesity. *Physiol. Behav.* **2011**, *103*, 59–68. [CrossRef] [PubMed]
4. Khan, N.A.; Baym, C.L.; Monti, J.M.; Raine, L.B.; Drollette, E.S.; Scudder, M.R.; Moore, R.D.; Kramer, A.F.; Hillman, C.H.; Cohen, N.J. Central adiposity is negatively associated with hippocampal-dependent relational memory among overweight and obese children. *J. Pediatr.* **2015**, *166*, 302–308. [CrossRef]
5. Dupret, D.; Revest, J.-M.; Koehl, M.; Ichas, F.; De Giorgi, F.; Costet, P.; Abrous, D.N.; Piazza, P.V. Spatial Relational Memory Requires Hippocampal Adult Neurogenesis. *PLoS ONE* **2008**, *3*, e1959. [CrossRef]
6. Monti, J.M.; Cooke, G.E.; Watson, P.D.; Voss, M.W.; Kramer, A.F.; Cohen, N.J. Relating Hippocampus to Relational Memory Processing across Domains and Delays. *J. Cogn. Neurosci.* **2014**, *27*, 234–245. [CrossRef]
7. Hassevoort, K.M.; Khazoum, S.E.; Walker, J.A.; Barnett, S.M.; Raine, L.B.; Hammond, B.R.; Renzi-Hammond, L.M.; Kramer, A.F.; Khan, N.A.; Hillman, C.H.; et al. Macular Carotenoids, Aerobic Fitness, and Central Adiposity Are Associated Differentially with Hippocampal-Dependent Relational Memory in Preadolescent Children. *J. Pediatr.* **2017**, *183*, 108–114. [CrossRef]
8. Baym, C.L.; Khan, N.A.; Monti, J.M.; Raine, L.B.; Drollette, E.S.; Moore, R.D.; Scudder, M.R.; Kramer, A.F.; Hillman, C.H.; Cohen, N.J. Dietary lipids are differentially associated with hippocampal-dependent relational memory in prepubescent children. *Am. J. Clin. Nutr.* **2014**, *99*, 1026–1032. [CrossRef]
9. Abdel-Aal, E.-S.; Akhtar, H.; Zaheer, K.; Ali, R. Dietary Sources of Lutein and Zeaxanthin Carotenoids and Their Role in Eye Health. *Nutrients* **2013**, *5*, 1169–1185. [CrossRef] [PubMed]
10. Vishwanathan, R.; Kuchan, M.J.; Sen, S.; Johnson, E.J. Lutein and Preterm Infants With Decreased Concentrations of Brain Carotenoids. *J. Pediatr. Gastroenterol. Nutr.* **2014**, *59*, 659–665. [CrossRef]
11. Johnson, E.J.; Vishwanathan, R.; Johnson, M.A.; Hausman, D.B.; Davey, A.; Scott, T.M.; Green, R.C.; Miller, L.S.; Gearing, M.; Woodard, J.; et al. Relationship between Serum and Brain Carotenoids, α-Tocopherol, and Retinol Concentrations and Cognitive Performance in the Oldest Old from the Georgia Centenarian Study. *J. Aging Res.* **2013**, *2013*, 951786. [CrossRef]
12. Erdman, J.; Smith, J.; Kuchan, M.; Mohn, E.; Johnson, E.; Rubakhin, S.; Wang, L.; Sweedler, J.; Neuringer, M. Lutein and Brain Function. *Foods* **2015**, *4*, 547–564. [CrossRef]
13. Seddon, J.M.; Ajani, U.A.; Sperduto, R.D.; Hiller, R.; Blair, N.; Burton, T.C.; Farber, M.D.; Gragoudas, E.S.; Haller, J.; Miller, D.T.; et al. Dietary Carotenoids, Vitamins A, C, and E, and Advanced Age-Related Macular Degeneration. *JAMA* **1994**, *272*, 1413–1420. [CrossRef]

14. Kamoshita, M.; Toda, E.; Osada, H.; Narimatsu, T.; Kobayashi, S.; Tsubota, K.; Ozawa, Y. Lutein acts via multiple antioxidant pathways in the photo-stressed retina. *Sci. Rep.* **2016**, *6*, 30226. [CrossRef]

15. Beatty, S.; Nolan, J.; Kavanagh, H.; O'Donovan, O. Macular pigment optical density and its relationship with serum and dietary levels of lutein and zeaxanthin. *Arch. Biochem. Biophys.* **2004**, *430*, 70–76. [CrossRef] [PubMed]

16. Vishwanathan, R.; Neuringer, M.; Snodderly, D.M.; Schalch, W.; Johnson, E.J. Macular lutein and zeaxanthin are related to brain lutein and zeaxanthin in primates. *Nutr. Neurosci.* **2013**, *16*, 21–29. [CrossRef] [PubMed]

17. Khan, N.A.; Walk, A.M.; Edwards, C.G.; Jones, A.R.; Cannavale, C.N.; Thompson, S.V.; Reeser, G.E.; Holscher, H.D. Macular xanthophylls are related to intellectual ability among adults with overweight and obesity. *Nutrients* **2018**, *10*, 396. [CrossRef] [PubMed]

18. Renzi, L.M.; Dengler, M.J.; Puente, A.; Miller, L.S.; Hammond, B.R. Relationships between macular pigment optical density and cognitive function in unimpaired and mildly cognitively impaired older adults. *Neurobiol. Aging* **2014**, *35*, 1695–1699. [CrossRef] [PubMed]

19. Vishwanathan, R.; Iannaccone, A.; Scott, T.M.; Kritchevsky, S.B.; Jennings, B.J.; Carboni, G.; Forma, G.; Satterfield, S.; Harris, T.; Johnson, K.C.; et al. Macular pigment optical density is related to cognitive function in older people. *Age Ageing* **2014**, *43*, 271–275. [CrossRef]

20. Zamroziewicz, M.K.; Paul, E.J.; Zwilling, C.E.; Johnson, E.J.; Kuchan, M.J.; Cohen, N.J.; Barbey, A.K. Parahippocampal Cortex Mediates the Relationship between Lutein and Crystallized Intelligence in Healthy, Older Adults. *Front. Aging Neurosci.* **2016**, *8*, 297. [CrossRef]

21. Horecka, K.M.; Dulas, M.R.; Schwarb, H.; Lucas, H.D.; Duff, M.; Cohen, N.J. Reconstructing relational information. *Hippocampus* **2018**, *28*, 164–177. [CrossRef]

22. Naugle, R.I.; Chelune, G.J.; Tucker, G.D. Validity of the Kaufman Brief Intelligence Test. *Psychol. Assess.* **1993**, *5*, 182–186. [CrossRef]

23. Wang, J.-J.; Kaufman, A.S. Changes in Fluid and Crystallized Intelligence Across the 20- to 90-Year Age Range on the K-Bit. *J. Psychoeduc. Assess.* **1993**, *11*, 29–37. [CrossRef]

24. Jeon, S.; Neuringer, M.; Kuchan, M.J.; Erdman, J.W. Relationships of carotenoid-related gene expression and serum cholesterol and lipoprotein levels to retina and brain lutein deposition in infant rhesus macaques following 6 months of breastfeeding or formula feeding. *Arch. Biochem. Biophys.* **2018**, *654*, 97–104. [CrossRef]

25. Wooten, B.R.; Hammond, B.R., Jr.; Land, R.I.; Snodderly, D.M. A Practical Method for Measuring Macular Pigment Optical Density. *Invest. Ophthalmol. Vis. Sci.* **1999**, *40*, 2481–2489.

26. Renzi-Hammond, L.; Bovier, E.; Fletcher, L.; Miller, L.; Mewborn, C.; Lindbergh, C.; Baxter, J.; Hammond, B. Effects of a Lutein and Zeaxanthin Intervention on Cognitive Function: A Randomized, Double-Masked, Placebo-Controlled Trial of Younger Healthy Adults. *Nutrients* **2017**, *9*, 1246. [CrossRef] [PubMed]

27. Kaulmann, A.; Bohn, T. Carotenoids, inflammation, and oxidative stress—Implications of cellular signaling pathways and relation to chronic disease prevention. *Nutr. Res.* **2014**, *34*, 907–929. [CrossRef] [PubMed]

28. Dandona, P. Inflammation: The link between insulin resistance, obesity and diabetes. *Trends Immunol.* **2004**, *25*, 4–7. [CrossRef]

29. Dantzer, R.; O'Connor, J.C.; Freund, G.G.; Johnson, R.W.; Kelley, K.W. From inflammation to sickness and depression: When the immune system subjugates the brain. *Nat. Rev. Neurosci.* **2008**, *9*, 46–56. [CrossRef] [PubMed]

30. Lindbergh, C.A.; Renzi-Hammond, L.M.; Hammond, B.R.; Terry, D.P.; Mewborn, C.M.; Puente, A.N.; Miller, L.S. Lutein and Zeaxanthin Influence Brain Function in Older Adults: A Randomized Controlled Trial. *J. Int. Neuropsychol. Soc.* **2018**, *24*, 77–90. [CrossRef]

31. Johnson, E.J. Role of lutein and zeaxanthin in visual and cognitive function throughout the lifespan. *Nutr. Rev.* **2014**, *72*, 605–612. [CrossRef] [PubMed]

32. Curran-Celentano, J.; Hammond, B.R.; Ciulla, T.A.; Cooper, D.A.; Pratt, L.M.; Danis, R.B. Relation between dietary intake, serum concentrations, and retinal concentrations of lutein and zeaxanthin in adults in a Midwest population. *Am. J. Clin. Nutr.* **2001**, *74*, 796–802. [CrossRef] [PubMed]

nutrients

MDPI

Article

Low Serum Carotenoids Are Associated with Self-Reported Cognitive Dysfunction and Inflammatory Markers in Breast Cancer Survivors

Krystle E. Zuniga [1],* and Nancy E. Moran [2]

[1] Nutrition and Foods, Family and Consumer Sciences, Texas State University, 601 University Drive, San Marcos, TX 78666, USA
[2] United States Department of Agriculture/Agricultural Research Service Children's Nutrition Research Center, Department of Pediatrics, Baylor College of Medicine, Houston, TX 77030, USA; nancy.moran@bcm.edu
* Correspondence: k_z17@txstate.edu; Tel.: +1-512-245-3786

Received: 13 July 2018; Accepted: 8 August 2018; Published: 17 August 2018

Abstract: Background: Dietary carotenoids may exert anti-inflammatory activities to reduce inflammation-driven cognitive impairments during cancer and cancer treatment. Our objective was to explore if cognitive function in breast cancer survivors (BCS) differs by serum carotenoid concentrations, and if blood carotenoids concentrations are associated with reduced systemic inflammation. **Methods:** Objective cognitive function and perceived cognitive impairment of 29 BCS and 38 controls were assessed cross-sectionally with the National Institutes of Health Toolbox Cognition Battery and The Functional Assessment of Cancer Therapy-Cognitive Function Questionnaire, respectively. Serum carotenoid and inflammatory marker (sTNF-RII, IL-6, IL-1ra, CRP) concentrations were measured. **Results:** Low-carotenoid BCS had more cognitive complaints compared to the low-carotenoid controls (Mdiff = −43.0, $p < 0.001$) and high-carotenoid controls (Mdiff = −44.5, $p < 0.001$). However, the cognitive complaints of high-carotenoid BCS were intermediate to and not different than the low-carotenoid BCS, or low- or high-carotenoid controls. BCS performed similarly to controls on all objective cognitive measures. Multiple linear regression, controlling for age and body mass index (BMI), demonstrated an inverse association between serum carotenoid concentrations and pro-inflammatory sTNFR-II ($\beta = 0.404$, $p = 0.005$) and IL-6 concentrations ($\beta = -0.35$, $p = 0.001$), but not IL-1ra or CRP. **Conclusions:** Higher serum carotenoid concentrations may convey cognitive and anti-inflammatory benefits in BCS. Future research should identify dietary components and patterns that support cognitive health in cancer survivors.

Keywords: cancer-related cognitive impairment; cognition; carotenoid; memory; inflammation

1. Introduction

With over 252,000 new breast cancer cases estimated in 2017 and a 90% five-year survival rate, the population of over 3.5 million U.S. breast cancer survivors (BCS) will continue to increase [1,2]. BCS live with many treatment-related side effects including cognitive dysfunction [3]. There is evidence that up to 75% of breast cancer patients experience cognitive decline during cancer treatment and that these impairments can persist up to twenty years following treatment completion [4,5]. However, there is marked variability in the incidence, duration, and severity of cancer-related cognitive impairments, suggesting that risk is modulated by treatment type, demographics, lifestyle, or genetics [4]. Recent advances in the field of nutritional neuroscience suggest that diet may play a role in preventing aging-associated cognitive decline; however, little to no research has focused on the mechanisms by which dietary factors reduce or prevent cancer-related cognitive decline.

Cancer-related cognitive impairments are hypothesized to arise as a result of cancer- and treatment-related inflammation. Cytokine elevation occurs in breast cancer patients before and during chemotherapy [6] and persists post-treatment [7–9]. These cytokines may be stimulated by the tumor, treatment, or psychological factors like fatigue or depression [10]. In breast cancer patients, elevated cytokines have been associated with reduced cognitive performance, changes in perceived cognitive function, and smaller hippocampal volume [10]. Oxidative stress and inflammation adversely affect neurogenesis and have a role in the pathogenesis of cognitive dysfunction [10], making these important target mechanisms to investigate with regard to the cognitive side effects of cancer treatment.

Nutrition has an important role in brain structure and function; for example, dietary components maintain neural tissue and membrane structure, provide substrates for signaling molecule synthesis, fuel metabolism, and some have antioxidant and anti-inflammatory actions [11]. Evidence suggests a correlation between fruit and vegetable intake and cognitive function [12,13], and fruits and vegetables are components of dietary patterns associated with reduced risks of cognitive impairment and neurodegenerative diseases [14–16]. In a prospective study of over 1000 breast cancer patients, both fruit and vegetable intake were associated with better verbal fluency scores post-diagnosis [17]. We previously discovered a positive relationship between fruit and vegetable intake and a measure of executive function (the ability to focus on relevant aspects of the environment) among BCS and age-matched controls [18]; however, the mechanisms underlying this relationship remain unknown. While fruits and vegetables contain an abundance of nutrients and bioactives, carotenoids, which are particularly concentrated in fruits and vegetables [19], have recently gained attention for potential cognitive benefits across the lifespan [20–22]. In addition to some carotenoids serving as vitamin A precursors, carotenoids confer many physiological activities, such as acting as anti-inflammatory or antioxidant agents, and may exert benefits for cognitive health, as some carotenoids may have potent activity in scavenging free radicals, protecting lipid membranes from peroxidation and reducing cellular inflammation [19,23,24]. A number of recent studies have suggested a particular benefit of dietary carotenoid intake for attenuating the symptoms of aging-related cognitive decline [25–28].

Some inflammatory and oxidative processes associated with cognitive decline in aging are believed to be similar to those that occur in cancer-related cognitive impairment; therefore, we hypothesize that potential anti-inflammatory and neuroprotective actions of carotenoids may elicit cognitive benefits in BCS. It is of interest to determine if the benefits of carotenoids to cognition and antioxidant status observed among a healthy population would be seen in other populations at risk of cognitive decline such as BCS. The objective of this research is to explore if cognitive function in BCS differs depending on their serum carotenoid concentrations and to explore reduction of inflammation as a potential mechanism by which carotenoids are associated with cognitive function in BCS.

2. Materials and Methods

Recruitment: Breast cancer survivors who completed primary treatment (chemotherapy, radiation therapy, or both) within the past 60 months and age-matched controls with no history of a cancer diagnosis were recruited from the Central Texas area. Potential participants were recruited via local oncology clinics, support groups, print media (flyers), social networking sites, websites, and listserv announcements. After expressing initial interest, individuals were contacted by phone and provided with a full study description and screened for eligibility. Inclusion criteria included: female; breast cancer survivor ≤60 months from last treatment, 30–70 years old; no history of stroke, heart attack, or transient ischemic attack; not currently pregnant; could speak, read, and write English; could attend all testing sessions; not blind or legally blind; nonsmoker. The exclusion criteria for this study included: current use of computer-based brain training games (e.g., Lumosity®, BrainHQ®). Eligible participants were scheduled for two testing appointments. Of the 83 total contacts, 70 consented and were eligible for testing (30 cancer survivors, 40 controls). Three women withdrew after consenting and did not complete testing owing to being no longer interested or no longer eligible (1 cancer, 2 control); thus, data are presented from sixty-seven women that consented and completed testing (38 control; 29 cancer). Participants were remunerated for their participation. All study procedures and recruitment materials were approved by the Texas State University Institutional Review Board and were performed in accordance with the ethical standards as laid down in the 1964 Declaration of Helsinki and its later amendments. Written informed consent was obtained from all subjects. This trial was registered at ClinicalTrials.gov as NCT02591316.

Demographics: Participants self-reported medical history, marital status, age, race, ethnicity, occupation, income, and education level. Additionally, BCS self-reported breast cancer specific diagnosis and treatment history. Height and weight were measured to calculate body mass index (BMI).

Dietary Intakes: Participants completed a modified version of the 2005 Block Food Frequency Questionnaire [29] reporting on intakes over the past 3 months delivered via NutritionQuest's online Data-on-Demand System. The questionnaire included 110 food items with details on portion sizes.

Neuropsychological assessment: The computer-based National Institutes of Health Toolbox Neurological and Behavioral Function Cognitive Function Battery (NIHTB-CB) was used to assess multiple domains of cognitive function. The NIHTB-CB has high reliability and validity and has been widely used in various research applications to assess cognitive performance [30]. Episodic memory and working memory were tested by the picture sequence and list sorting working memory tasks, respectively. Picture Vocabulary and Oral Reading Recognition tested language abilities. Fully-adjusted scores for each participant were provided by the NIHTB-CB and are normalized to reference groups by age, ethnicity, gender, and education.

Self-reported cognitive dysfunction: Perceived cognitive function was measured with the Functional Assessment of Cancer Therapy-Cognitive Function (FACT-Cog), a 37-item, 5-pt Likert scale questionnaire [31]. Possible scores range from 0–148, with higher FACT-Cog scores indicating fewer cognitive complaints. In this study, the Cronbach alpha was 0.98 for the total score, with subscale scores ranging from 0.91 to 0.96.

Inflammatory markers: Participants were instructed to fast for 10 h prior to blood collection. Fasted blood samples were collected via venipuncture into silica-coated BD Vacutainer serum collection tubes. Samples were allowed to clot for 30 min, centrifuged for 15 min at $2000 \times g$, and the resultant serum was aliquoted and stored at $-80\,^{\circ}$C for subsequent analysis. Serum C-reactive protein (CRP), soluble tumor necrosis factor-alpha receptor type II (sTNF-RII), and interleukin-6 (IL-6) were quantified using their respective Quantikine ELISA Immunoassay kit (R&D Systems, Minneapolis, MN) per the manufacturer's instructions. Interleukin-1 receptor antagonist (IL-1ra) was analyzed with Biosource IL-1ra Cytoscreen kit (Biosource Europe S.A., Nivelles, Belgium) per the manufacturer's instructions.

Serum carotenoids: All sample analyses were performed under yellow lighting to protect carotenoids. Carotenoids were extracted from 300 uL thawed serum based on a previously described method with several modifications [32]. Apo-8'-carotenal (75 ng, #10810, Sigma-Aldrich, St. Louis, MO, USA) was added as an internal standard to duplicate samples from each subject. Dried extracts were reconstituted in 30 μL 80% methyl *tert*-butyl ether (MtBE): 20% methanol, held at 4 °C in a refrigerated autosampler, and 20 μL were injected onto an Ultimate 3000 UHPLC system coupled with a photodiode array detector (ThermoFisher, Waltham, MA, USA). Analytes were separated on a C30 Carotenoid Column (250 × 3 mm, 3 μm particle size) (ThermoFisher, Waltham, MA, USA) cooled to 18 °C, over 42 min at 0.4 mL/min by a gradient elution method, which was a modification of a previously published method [33]. The method utilized 3 HPLC-grade mobile phases: A: Methanol, B: MtBE, C: 1.5% Ammonium Acetate in water (*w/v*). The initial conditions were 98% A, 0% B, 2% C, which reached to 60% A, 38% B, 2% C at 12.5 min, then 48% A, 50% B, and 2% C at 20 min, then 38% A, 60% B, and 2% C at 30 min, which was held until 32 min, then returned to initial conditions (98% A, 0% B, 2% C) by 35 min and was held at initial conditions until 42 min. Analytes were detected by photodiode array detector at 472 nm (lycopene, beta-apo-8'-carotenal), 290 nm (phytoene), 450 nm (alpha- and beta-carotene, lutein & zeaxanthin, and beta-cryptoxanthin), and 330 nm (phytofluene). The data were analyzed using Chromeleon 7 software (ThermoFisher, Waltham, MA, USA). Analyte signals were quantitated by external calibration curves of authentic analytical standards for (*E/Z*)-phytoene, lycopene, beta-carotene, alpha-carotene, lutein, zeaxanthin, and beta-cryptoxanthin (Sigma-Aldrich, St. Louis, MO, USA), phytofluene (Carotenature, Lupsingen, Switzerland), and results were adjusted to the internal standard to correct for extraction efficiency. Total carotenoid concentration was the sum of all carotenoids analyzed.

Plasma Cholesterol: Samples were analyzed by commercial lab (LabCorp, Houston, TX, USA) using the lipid panel test (#303756) for frozen, EDTA-collected plasma. Triglycerides and total and HDL cholesterol were measured directly by enzymatic colorimetric assay on a clinical chemistry analyzer (COBAS 8000 c701, Roche Diagnostics International Ltd., Rotkreuz, Zug Switzerland) and VLDL and LDL cholesterol were calculated using standard clinical algorithms.

Statistical Analysis: The primary objective of this study was to determine if cognitive function in BCS differed by serum total carotenoid status. We also confirmed that serum carotenoid concentrations were positively associated with reported fruit and vegetable intake. As secondary outcomes, we explored whether serum carotenoids were associated with markers of inflammation. Characteristics of the BCS and controls were compared using independent t-tests for continuous variables and X^2-tests for categorical variables. Differences in serum carotenoids between groups were tested with the Mann–Whitney U-test. Univariate analysis of covariance (ANCOVA) was used to determine the difference in subjective cognitive function between BCS and controls after controlling for age. In order to compare cognitive function between BCS and controls with either low or high serum carotenoid concentrations, BCS and controls were each split into two groups: (1) BCS or control with serum carotenoid concentrations lower than, and including, the median; and (2) BCS or control with serum carotenoid concentrations above the median. Then, univariate ANCOVA, including age as a covariate, was used to compare cognitive function as a function of group (low carotenoid BCS, high carotenoid BCS, low carotenoid control, high carotenoid control), and post hoc analysis with *t*-tests was performed with a Bonferroni adjustment for the six comparisons, yielding an $\alpha = 0.008$. The significance level was set at $\alpha = 0.05$ for all other statistical analyses. Associations between serum carotenoids and inflammatory markers were determined using multiple linear regression, controlling for the a priori-identified covariates of age and BMI. All statistical analyses were performed in SPSS (v.24) (IBM, Armonk, NY, USA). When the assumption of normality was violated, values were log-transformed to the base 10 ($\log_{10}$) to provide increased normality.

3. Results

Participant Characteristics: There were no significant differences in age, education, race, ethnicity, socioeconomic status, or BMI between BCS and healthy controls (Table 1). Most BCS had early stage breast cancer at diagnosis (75.9% at stage II or lower) and were an average of 1.6 years post-completion of cancer treatment (M = 18.6 months, SD = 16.3). All had undergone surgery, 65.5% had received radiation therapy, 72.4% had received chemotherapy, and 65.5% were currently receiving hormonal therapy.

Table 1. Participant Characteristics.

	Breast Cancer Survivors	Controls	*p*
Age, y (Mean ± SD)	50.1 ± 10.1	50.8 ± 10.0	0.783
Income, *n* (%)			
≥$60,000	16 (61.5)	29 (78.4)	0.145
Race, *n* (%)			
White	20 (69.0)	33 (86.8)	0.075
Black	3 (10.3)	0 (0.0)	
Asian	2 (6.9)	1 (2.6)	
More than One Race	4 (13.8)	4 (10.5)	
Ethnicity, *n* (%)			
Hispanic or Latino	6 (20.7)	9 (23.7)	0.771
Education, *n* (%)			
≥ 4 year College Degree	20 (69.0)	32 (84.2)	0.138
Body Mass Index (BMI), kg/m^2 (Mean ± SD)	29.7 ± 6.3	27.3 ± 7.6	0.168
Stage at Diagnosis, *n* (%)			
Ductal Carcinoma in Situ (DCIS)	3 (10.3)	-	
Stage I	8 (27.6)	-	
Stage II	11 (37.9)	-	
Stage III	6 (20.7)	-	
Unknown	1 (3.4)	-	
Treatment, *n* (%)			
Chemotherapy Only	10 (34.5)	-	
Radiation Only	8 (27.6)	-	
Chemotherapy + Radiation	11 (37.9)	-	
Current Hormone Therapy	19 (65.5)	-	
Surgery	29 (100)	-	
Time since Treatment-Months (Mean ± SD)	18.6 ± 16.3	-	
Total Cholesterol (mg/dL) [1]	167 (130–189)	166 (146–189)	0.463
Serum Carotenoid Concentrations (nmol/L) [1]			
Alpha-carotene	41.1 (19.7–64.5)	53.3 (35.8–122.7)	0.121
Beta-carotene	163.0 (79.5–298.8)	217.5 (122.6–367.2)	0.260
Lycopene	298.1 (225.8–369.8)	312.4 (213.6–411.3)	0.968
Lutein & Zeaxanthin	163.2 (92.9–272.9)	214.7 (145.7–308.3)	0.199
Beta-cryptoxanthin	68.7 (45.3–115.8)	78.2 (49.6–154.9)	0.354
Phytofluene	65.2 (50.7–92.4)	62.1 (48.7–100.2)	0.958
Phytoene	68.6 (50.3–82.4)	53.6 (43.1–75.6)	0.083
Total Carotenoids	933.3 (663.2–1120.5)	1052.3 (782.2–1356.1)	0.314
Average Daily Intakes (Mean ± SD)			
Fruit (cups)	1.0 ± 0.8	1.1 ± 0.9	0.568
Vegetables (cups)	2.8 ± 1.7	2.6 ± 1.6	0.761
Total Carotenoid Intake (mg)	20.8 ± 13.5	18.7 ± 10.3	0.514

[1] median and interquartile range (25–75th percentile).

Dietary and serum carotenoids: Serum and dietary carotenoids were not significantly different between BCS and controls (Table 1). Fruit and vegetable intake were similar between BCS and age-matched controls (Table 1). After controlling for BMI and total cholesterol, reported fruit and vegetable intake (total servings per day) was positively correlated with total serum carotenoid concentrations ($r = 0.434$, $p < 0.001$) (data not shown).

Objective and Subjective Cognitive Function: BCS performed similarly to controls on all objective cognitive measures (Table 2). However, BCS reported significantly more cognitive complaints than controls (Table 3). After adjustment for age, there was a statistically significant difference in perceived cognitive function between the groups across all subscales and total score of the FACT-Cog, indicating more self-reported cognitive difficulties in BCS (Table 3).

Table 2. Objective Cognitive Function [1].

	Breast Cancer Survivors	Controls	p
List-Sorting Working Memory	108.8 (17.5)	105.3 (11.2)	0.353
Picture Vocabulary	108.2 (16.2)	115.2 (14.5)	0.072
Picture Sequence Memory	100.5 (18.9)	103.1 (14.7)	0.536
Oral Reading Recognition	117.4 (14.3)	116.7 (16.4)	0.864

Mean (SD).[1] fully-adjusted, standardized scores.

Table 3. Self-Reported Cognitive Dysfunction.

	Breast Cancer Survivors	Controls	p
Total FACT-Cog Score [1]	88.70 (5.27)	119.17 (4.60)	<0.001
Perceived Cognitive Impairments	45.43 (3.01)	61.66 (2.63)	<0.001
Impact of Perceived Cognitive Impairments on Quality of Life	10.06 (0.83)	13.60 (0.73)	0.002
Comments from Others	12.94 (0.53)	15.57 (0.46)	<0.001
Perceived Cognitive Abilities	20.28 (1.44)	28.34 (1.26)	<0.001

Data are adjusted mean (SE). ANCOVA with age as a covariate. [1] Functional Assessment of Cancer Therapy-Cognitive Function

Serum carotenoids and cognitive complaints: Serum carotenoid concentrations by group are described in Table 4. ANCOVA was used to determine the differences in FACT-Cog scores between serum carotenoid status groups. After adjustment for age, there was a statistically significant difference in FACT-Cog scores between the groups, $F (3, 60) = 9.498$, $p < 0.001$, partial $\eta2 = 0.322$. The low-carotenoid BCS had significantly lower FACT-Cog scores (more cognitive complaints) compared to the low-carotenoid controls (Mdiff = -43.0, 95% CI $[-68.8, -17.1]$, $p < 0.001$) and high-carotenoid controls (Mdiff = -44.54, 95% CI $[-71.1, -18.0]$, $p < 0.001$) (Figure 1). However, FACT-Cog scores in high-carotenoid BCS were not significantly different than those of the low-carotenoid BCS, or low- or high-carotenoid controls.

Table 4. Serum Carotenoid Concentrations.

Serum Carotenoid Concentrations (nmol/L).	Low Carotenoid BCS	High Carotenoid BCS	Low Carotenoid Control	High Carotenoid Control
Alpha-carotene	27.3 (13.0)	153.8 (269.9)	43.3 (33.6)	90.3 (150.8)
Beta-carotene	103.9 (58.7) [a]	461.1 (427.4) [b]	140.7 (78.5) [a]	425.4 (215.7) [b]
Lycopene	275.0 (74.4) [a]	344.1 (99.4) [ab]	241.4 (80.8) [a]	417.3 (198.4) [b]
Lutein & Zeaxanthin	117.4 (44.5) [a]	366.3 (263.6) [b]	174.0 (79.5) [ac]	299.8 (121.7) [bc]
Beta-cryptoxanthin	54.4 (34.5) [a]	135.9 (133.5) [ab]	69.1 (58.1) [ab]	149.4 (105.3) [b]
Phytofluene	55.5 (34.6) [ab]	113.8 (91.8) [a]	53.9 (22.5) [b]	108.4 (63.4) [a]
Phytoene	56.5 (23.1) [ab]	87.5 (40.1) [a]	47.9 (9.9) [b]	79.7 (39.0) [a]
Total Carotenoids	689.9 (185.0) [a]	1662.4 (1065.7) [b]	770.3 (179.9) [a]	1619.6 (604.2) [b]

Mean (SD); Different letters indicate significant differences between groups ($p < 0.05$).

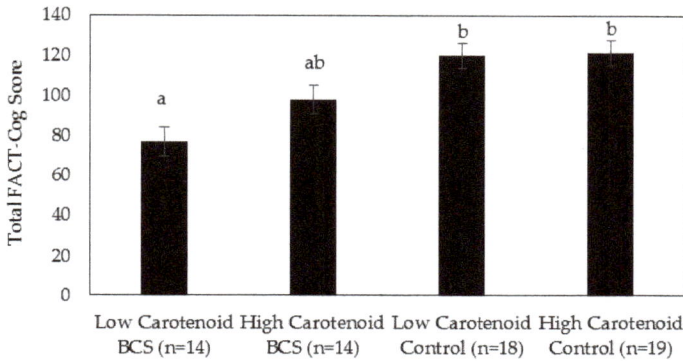

Figure 1. Perceived Cognitive Dysfunction by Group Carotenoid Status. Data are adjusted mean (standard error). ANCOVA with age as covariate. Different letters indicate significant differences between groups ($p < 0.05$).

Inflammatory Markers: Serum CRP, sTNF-RII, IL-6, and IL-1ra levels were not significantly different between BCS and controls (Table 5). In multiple linear regression models controlling for age and BMI, serum carotenoids were significant predictors of lower IL-6 ($\beta = -0.353$, $p = 0.001$) and sTNFR-II ($\beta = -0.404$, $p = 0.005$) concentrations, but not IL-1ra or CRP (Table 6).

Table 5. Inflammatory Markers.

	Breast Cancer Survivors	Controls	*p*
IL-6 (pg/mL)	2.0 (1.1)	1.8 (1.1)	0.431
IL-1ra (pg/mL)	476.8 (91.6)	513.1 (78.4)	0.766
sTNFRII (pg/mL)	3037.41 (128.0)	2717.3 (115.8)	0.073
CRP (ng/mL)	2156.8 (358.2)	3012.3 (296.3)	0.072

Data are mean adjusted (SE). ANCOVA with age, BMI, and moderate to vigorous physical activity as covariates.

Table 6. Regression analyses explaining variability in inflammatory markers.

	IL-6			sTNFR-II			CRP			IL-1ra		
	β	p	Model Adjusted R^2	β	p	Model Adjusted R^2	β	p	Model Adjusted R^2	β	p	Model Adjusted R^2
Age	0.225	0.020	0.483 **	0.284	0.022	0.175 *	−0.064	0.622	0.142 *	0.264	0.046	0.032
BMI	0.433	<0.001		0.036	0.790		0.337	0.024		−0.018	0.900	
Serum Carotenoids	−0.353	0.001		−0.404	0.005		−0.158	0.277		−0.162	0.267	

* $p < 0.01$, ** $p < 0.001$.

4. Discussion

Here we report, for the first time, that BCS with low blood carotenoid concentrations reported more cognitive complaints than healthy controls with no prior cancer diagnosis, and BCS with higher serum carotenoid concentrations reported cognitive complaints not different from both groups of controls. Furthermore, we found that several, but not all, serum markers of inflammation were significantly inversely associated with serum total carotenoid concentrations, suggesting a potential mechanism by which carotenoids may affect cognitive function.

In our sample, objectively assessed cognitive performance was similar between BCS and controls. Indeed, several meta-analyses have concluded that cognitive deficits in BCS are generally subtle [34–36], and may be more detectable by subjective report than by objective laboratory tests. Standard neuropsychological assessments likely command more focus than the tasks of everyday life and may not be representative of the routine cognitive challenges that BCS face [37,38]. Compared to age-matched controls, BCS reported significantly greater levels of perceived cognitive impairment. In non-cancer populations, memory complaints are a predictor of worse physical health and a higher degree of dependency for daily life activities [39,40]. In aging populations, self-reported, perceived memory impairments are associated with an increased risk of cognitive decline and impairment and all-cause dementia, suggesting that cognitive complaints may indicate neurodegenerative changes that precede significant changes in neuropsychological performance assessed with objective laboratory tests [41–43]. A minimal clinically important difference for FACT-Cog has been estimated to be 7–11 points [44] and has been demonstrated to be sensitive to interventions to improve cognitive function in cancer survivors including cognitive rehabilitation [45], physical activity [46,47], and meditation [48]. The low carotenoid BCS had an over 40 point difference on FACT-Cog scores compared to low and high carotenoid controls, which is clinically significant.

The neurotoxicity of cancer and treatment may be mediated, at least in part, through the action of cytokines crossing the blood-brain barrier to elicit oxidative stress and negatively impact neurogenesis [49]. Evidence suggests that several inflammatory markers are elevated in BCS including the cytokines IL-6, IL-8, and IL-10 [7–10]; however, BCS in our sample did not have significantly higher levels of inflammatory markers compared to controls. This may have been partially attributed to the heterogeneity in time since treatment in our small sample size, as it is suggested that some cytokine changes may be acute [8,50].

Carotenoids can accumulate in the brain and may function as reactive oxygen species scavengers, protecting lipid membranes from peroxidation and reducing cellular inflammation [23,51,52]. Evidence indicates an altered redox status in cancer patients [53,54], and other lines of evidence suggest oxidative stress in the brain can negatively affect neurogenesis and promote cognitive dysfunction [55,56]. However, the association between biomarkers of oxidative stress in cancer survivors and cancer-related cognitive dysfunction has not been thoroughly assessed, and future research is needed to determine which oxidative stress markers have clinical and prognostic utility in this sample. The current analyses focused on inflammatory markers that have consistently been identified to be both elevated and correlated with cognitive function in breast cancer patients and survivors. In this study, serum carotenoids were inversely associated with IL-6 and sTNFR-II, which is notable, as elevations in these markers of systemic inflammation have been associated with alterations in brain structure and function in breast cancer patients and survivors. sTNF-RII is a marker of the cytokine TNF-α's pro-inflammatory activity and can be reliably measured in circulation [57]. Higher sTNF-RII levels have been associated with greater memory complaints [50], worse memory performance [58], alterations in brain metabolism [50], and decreased gray matter volume in specific brain regions [59] of BCS. Additionally, a significant correlation between lower hippocampal volume and higher TNF-α has been observed among BCS [60]. In a longitudinal sample of BCS, elevated IL-6 levels were associated with more cognitive complaints [9], and in a sample of breast cancer patients, impaired memory function by radiation therapy was suggested to be partially mediated by elevated IL-6 [61]. The precise mechanism underlying how peripheral inflammatory cytokines impairs cognitive function in BCS is unknown. However, it is hypothesized that peripherally elevated levels of cytokines induce alterations among numerous neural substrates, including neurotransmitters and brain-derived neurotrophic factor [10]. It is mechanistically plausible that carotenoids have a direct effect on systemic inflammation, as carotenoids have been shown in model systems to reduce oxidative stress and downstream inflammatory signaling (reviewed in Reference [62]). In BCS, higher plasma carotenoid concentrations have previously shown to be associated with reduced oxidative stress [63,64]. Therefore, strategies to reduce inflammation, such as dietary changes to increase carotenoid intake, could provide cognitive benefits.

In addition to being bioactive compounds themselves, serum concentrations of carotenoids are a well-recognized biomarker of fruit and vegetable intake [65]. In our sample, serum carotenoid concentrations were positively associated with fruit and vegetable intake. Fruits and vegetables contain an array of antioxidants and anti-inflammatory nutrients and bioactives that may elicit neuroprotection by inhibiting neuroinflammation and neurodegeneration [66,67]. The carotenoids lutein and zeaxanthin have been more strongly associated with cognitive function than other individual carotenoids [22]. Further investigation is warranted to determine which specific compounds contribute to cognitive benefits observed with fruit and vegetable consumption, as many compounds likely interact to elicit cognitive benefits. The limited evidence of interventions with fruits and vegetables or carotenoids for improvement in cognitive function in aging populations does suggest a positive benefit [25–28,66,68,69]; thus, promotion of this dietary strategy may be useful for reducing the short- and/or long-term adverse cognitive effects of cancer treatment.

There are important limitations of this study including the cross-sectional design and small sample size which limits the interpretation of our findings. Our sample of BCS was heterogeneous in cancer stage, time since diagnosis, and treatment variables. The neuropsychological assessment was brief (~40 min), thus, a longer battery or field testing may have identified the subtle cognitive impairments common in BCS. There are several other inflammatory markers that were not assessed; however, CRP, IL-6, sTNF-RII, and IL-1ra were chosen because previous studies have demonstrated these markers to be elevated and correlated with cognitive decline in BCS [50,60,70]. While we cannot directly measure brain oxidative stress or inflammation, evidence indicates systemic inflammation is associated with lower white matter integrity, poorer cognitive performance, dementia, and cognitive decline in aging [71,72]. Future basic research can define the impact of carotenoids on other proposed mechanisms of cognitive dysfunction requiring model systems like oxidative stress, epigenetic changes, impaired neurogenesis, and treatment-induced hormonal alterations [4]. Despite these limitations, this is the first study, to our knowledge, to examine a potential mechanism by which food components could elicit cognitive benefits in cancer survivors. Although previous research has supported the hypothesis that dietary carotenoids have anti-inflammatory activity and may elicit cognitive benefits, these associations have never been explored in individuals with cancer-related cognitive impairment. Only two studies have examined the association between dietary intake and cognitive dysfunction in cancer survivors [17,18], and neither explored a potential mechanism of cognitive benefits from dietary components. Therefore, our findings are novel and provide much needed preliminary information to guide future dietary interventions.

Cancer survivorship research to date has primarily focused on improving survival and reducing risk of recurrence, yet strategies to support survivors' cognitive function are limited. Pharmacotherapies modestly alleviate cognitive deficits [73,74]; however, possible side effects and interactions highlight a need to explore modifiable lifestyle factors. While cognitive behavioral training and physical activity convey functional improvements in cancer survivors [75,76], there are no published dietary strategies. Nevertheless, survivors seek dietary changes to improve symptoms and reduce cancer risk [77]. Our findings suggest an association between serum carotenoids, perceived cognitive function, and markers of systemic inflammation. Future research should explore the mechanisms by which dietary patterns and corresponding food components can reduce memory complaints and support cognitive health in cancer survivors, providing evidence for future dietary interventions to improve BCS' quality of life.

Author Contributions: Conceptualization, K.E.Z. and N.E.M.; Methodology, K.E.Z.; Formal Analysis, K.E.Z.; Investigation, K.E.Z. and N.E.M.; Resources, K.E.Z. and N.E.M.; Data Curation, K.E.Z.; Writing – Original Draft Preparation, K.E.Z. and N.E.M.; Writing – Review & Editing, K.E.Z. and N.E.M.; Visualization, K.E.Z.; Supervision, K.E.Z.; Project Administration, K.E.Z.; Funding Acquisition, K.E.Z. and N.E.M.

Funding: The study was funded by Texas State University Research Enhancement Program (KEZ). The work was also supported in part by the National Institutes of Health National Center for Complementary and Integrative Health and Office of Dietary Supplements (R00 AT008576, NEM), and by The United States Department of Agriculture-Agricultural Research Service under CRIS 3092-51000-056-03S (NEM). The content is solely the responsibility of the authors and does not necessarily represent the official views of the National Institutes of Health or the United States Department of Agriculture.

Conflicts of Interest: The authors declare no conflict of interest.

References

1. Howlader, N.; Noone, A.M.; Krapcho, M.; Miller, D.; Bishop, K.; Kosary, C.L.; Yu, M.; Ruhl, J.; Tatalovich, Z.; Mariotto, A.; et al. SEER Cancer Statistics Review (CSR) 1975–2014. Available online: https://seer.cancer.gov/csr/1975_2015/ (accessed on 16 April 2018).
2. Miller, K.D.; Siegel, R.L.; Lin, C.C.; Mariotto, A.B.; Kramer, J.L.; Rowland, J.H.; Stein, K.D.; Alteri, R.; Jemal, A. Cancer treatment and survivorship statistics, 2016. *CA. Cancer J. Clin.* **2016**, *66*, 271–289. [CrossRef] [PubMed]
3. Palesh, O.; Scheiber, C.; Kesler, S.; Mustian, K.; Koopman, C.; Schapira, L. Management of side effects during and post-treatment in breast cancer survivors. *Breast J.* **2018**, *24*, 164–175. [CrossRef] [PubMed]
4. Janelsins, M.C.; Kesler, S.R.; Ahles, T.A.; Morrow, G.R. Prevalence, mechanisms, and management of cancer-related cognitive impairment. *Int. Rev. Psychiatry* **2014**, *26*, 102–113. [CrossRef] [PubMed]
5. Koppelmans, V.; Breteler, M.M.B.; Boogerd, W.; Seynaeve, C.; Gundy, C.; Schagen, S.B. Neuropsychological performance in survivors of breast cancer more than 20 years after adjuvant chemotherapy. *J. Clin. Oncol.* **2012**, *30*, 1080–1086. [CrossRef] [PubMed]
6. Janelsins, M.C.; Mustian, K.M.; Palesh, O.G.; Mohile, S.G.; Peppone, L.J.; Sprod, L.K.; Heckler, C.E.; Roscoe, J.A.; Katz, A.W.; Williams, J.P.; et al. Differential expression of cytokines in breast cancer patients receiving different chemotherapies: Implications for cognitive impairment research. *Support Care Cancer* **2012**, *20*, 831–839. [CrossRef] [PubMed]
7. Williams, A.M.; Shah, R.; Shayne, M.; Huston, A.J.; Krebs, M.; Murray, N.; Thompson, B.D.; Doyle, K.; Korotkin, J.; van Wijngaarden, E.; et al. Associations between inflammatory markers and cognitive function in breast cancer patients receiving chemotherapy. *J. Neuroimmunol.* **2018**, *314*, 17–23. [CrossRef] [PubMed]
8. Lyon, D.E.; Cohen, R.; Chen, H.; Kelly, D.L.; McCain, N.L.; Starkweather, A.; Ahn, H.; Sturgill, J.; Jackson-Cook, C.K. Relationship of systemic cytokine concentrations to cognitive function over two years in women with early stage breast cancer. *J. Neuroimmunol.* **2016**, *301*, 74–82. [CrossRef] [PubMed]
9. Cheung, Y.T.; Ng, T.; Shwe, M.; Ho, H.K.; Foo, K.M.; Cham, M.T.; Lee, J.A.; Fan, G.; Tan, Y.P.; Yong, W.S.; et al. Association of proinflammatory cytokines and chemotherapy-associated cognitive impairment in breast cancer patients: A multi-centered, prospective, cohort study. *Ann. Oncol.* **2015**, *26*, 1446–1451. [CrossRef] [PubMed]
10. Wang, X.M.; Walitt, B.; Saligan, L.; Tiwari, A.F.; Cheung, C.W.; Zhang, Z.J. Chemobrain: A critical review and causal hypothesis of link between cytokines and epigenetic reprogramming associated with chemotherapy. *Cytokine* **2015**, *72*, 86–96. [CrossRef] [PubMed]
11. Wahl, D.; Cogger, V.C.; Solon-Biet, S.M.; Waern, R.V.; Gokarn, R.; Pulpitel, T.; Cabo, R.; Mattson, M.P.; Raubenheimer, D.; Simpson, S.J.; et al. Nutritional strategies to optimise cognitive function in the aging brain. *Ageing Res. Rev.* **2016**, *31*, 80–92. [CrossRef] [PubMed]
12. Loef, M.; Walach, H. Fruit, vegetables and prevention of cognitive decline or dementia: A systematic review of cohort studies. *J. Nutr. Heal. Aging* **2012**, *16*, 626–630. [CrossRef]
13. Nurk, E.; Refsum, H.; Drevon, C.A.; Tell, G.S.; Nygaard, H.A.; Engedal, K.; Smith, A.D. Cognitive performance among the elderly in relation to the intake of plant foods. The Hordaland Health Study. *Br. J. Nutr.* **2010**, *104*, 1190–1201. [CrossRef] [PubMed]
14. Cheung, B.H.; Ho, I.C.; Chan, R.S.; Sea, M.M.; Woo, J. Current evidence on dietary pattern and cognitive function. *Adv. Food Nutr. Res.* **2014**, *71*, 137–163. [PubMed]

15. Morris, M.C.; Tangney, C.C.; Wang, Y.; Sacks, F.M.; Bennett, D.A.; Aggarwal, N.T. MIND diet associated with reduced incidence of Alzheimer's disease. *Alzheimers Dement.* **2015**, *11*, 1007–1014. [CrossRef] [PubMed]

16. Allès, B.; Samieri, C.; Féart, C.; Jutand, M.A.; Laurin, D.; Barberger-Gateau, P. Dietary patterns: A novel approach to examine the link between nutrition and cognitive function in older individuals. *Nutr. Res. Rev.* **2012**, *25*, 207–222. [CrossRef] [PubMed]

17. Huang, Z.; Shi, Y.; Bao, P.; Cai, H.; Hong, Z.; Ding, D.; Jackson, J.; Shu, X.-O.; Dai, Q. Associations of dietary intake and supplement use with post-therapy cognitive recovery in breast cancer survivors. *Breast Cancer Res. Treat.* **2018**, 1–10. [CrossRef] [PubMed]

18. Zuniga, K.E.; Mackenzie, M.J.; Roberts, S.A.; Raine, L.B.; Hillman, C.H.; Kramer, A.F.; McAuley, E. Relationship between fruit and vegetable intake and interference control in breast cancer survivors. *Eur. J. Nutr.* **2016**, *55*, 1555–1562. [CrossRef] [PubMed]

19. Hammond, B.R., Jr.; Renzi, L.M. Nutrient Information: Carotenoids. *Adv. Nutr.* **2013**, *4*, 474–476. [CrossRef] [PubMed]

20. Akbaraly, N.T.; Faure, H.; Gourlet, V.; Favier, A.; Berr, C. Plasma carotenoid levels and cognitive performance in an elderly population: Results of the EVA study. *J. Gerontol. A Biol. Sci. Med. Sci.* **2007**, *62*, 308–316. [CrossRef] [PubMed]

21. Kesse-Guyot, E.; Andreeva, V.A.; Ducros, V.; Jeandel, C.; Julia, C.; Hercberg, S.; Galan, P. Carotenoid-rich dietary patterns during midlife and subsequent cognitive function. *Br. J. Nutr.* **2014**, *111*, 915–923. [CrossRef] [PubMed]

22. Johnson, E.J. Role of lutein and zeaxanthin in visual and cognitive function throughout the lifespan. *Nutr. Rev.* **2014**, *72*, 605–612. [CrossRef] [PubMed]

23. Mohn, E.S.; Erdman, J.W.; Kuchan, M.J.; Neuringer, M.; Johnson, E.J. Lutein accumulates in subcellular membranes of brain regions in adult rhesus macaques: Relationship to DHA oxidation products. *PLoS ONE* **2017**, *12*, e0186767. [CrossRef] [PubMed]

24. Ruxton, C.H.S.; Derbyshire, E.; Toribio-Mateas, M. Role of fatty acids and micronutrients in healthy ageing: A systematic review of randomised controlled trials set in the context of European dietary surveys of older adults. *J. Hum. Nutr. Diet.* **2016**, *29*, 308–324. [CrossRef] [PubMed]

25. Lindbergh, C.A.; Renzi-Hammond, L.M.; Hammond, B.R.; Terry, D.P.; Mewborn, C.M.; Puente, A.N.; Miller, L.S. Lutein and zeaxanthin influence brain function in older adults: A randomized controlled trial. *J. Int. Neuropsychol. Soc.* **2018**, *24*, 1–14. [CrossRef] [PubMed]

26. Hammond, B.R.; Miller, L.S.; Bello, M.O.; Lindbergh, C.A.; Mewborn, C.; Renzi-Hammond, L.M.; Renzi-Hammond, L.M. Effects of lutein/zeaxanthin supplementation on the cognitive function of community dwelling older adults: A randomized, double-masked, placebo-controlled trial. *Front. Aging Neurosci.* **2017**, *9*, 254. [CrossRef] [PubMed]

27. Power, R.; Coen, R.F.; Beatty, S.; Mulcahy, R.; Moran, R.; Stack, J.; Howard, A.N.; Nolan, J.M. Supplemental retinal carotenoids enhance memory in healthy individuals with Low levels of macular pigment in a randomized, double-blind, placebo-controlled clinical trial. *J. Alzheimers Dis.* **2018**, *61*, 947–961. [CrossRef] [PubMed]

28. Grodstein, F.; Kang, J.H.; Glynn, R.J.; Cook, N.R.; Gaziano, J.M. A randomized trial of beta carotene supplementation and cognitive function in men: The Physicians' Health Study II. *Arch. Intern. Med.* **2007**, *167*, 2184. [CrossRef] [PubMed]

29. Block, G.; Wakimoto, P.; Jensen, C.; Mandel, S.; Green, R.R. Validation of a food frequency questionnaire for Hispanics. *Prev. Chronic Dis.* **2006**, *3*, A77. [PubMed]

30. Heaton, R.K.; Akshoomoff, N.; Tulsky, D.; Mungas, D.; Weintraub, S.; Dikmen, S.; Beaumont, J.; Casaletto, K.B.; Conway, K.; Slotkin, J.; et al. Reliability and validity of composite scores from the NIH toolbox cognition battery in adults. *J. Int. Neuropsychol. Soc.* **2014**, *20*, 588–598. [CrossRef] [PubMed]

31. Cella, D.F.; Tulsky, D.S.; Gray, G.; Sarafian, B.; Linn, E.; Bonomi, A.; Silberman, M.; Yellen, S.B.; Winicour, P.; Brannon, J.; et al. The Functional Assessment of Cancer Therapy scale: Development and validation of the general measure. *J. Clin. Oncol.* **1993**, *11*, 570–579. [CrossRef] [PubMed]

32. Campbell, J.K.; Engelmann, N.J.; Lila, M.A.; Erdman, J.W. Phytoene, phytofluene, and lycopene from tomato powder differentially accumulate in tissues of male Fisher 344 rats. *Nutr. Res.* **2007**, *27*, 794–801. [CrossRef] [PubMed]

33. Lu, C.H.; Choi, J.H.; Engelmann Moran, N.; Jin, Y.S.; Erdman, J.W. Laboratory-scale production of 13C-labeled lycopene and phytoene by bioengineered Escherichia coli. *J. Agric. Food Chem.* **2011**, *59*, 9996–10005. [CrossRef] [PubMed]

34. Jim, H.S.; Phillips, K.M.; Chait, S.; Faul, L.A.; Popa, M.A.; Lee, Y.H.; Hussin, M.G.; Jacobsen, P.B.; Small, B.J. Meta-analysis of cognitive functioning in breast cancer survivors previously treated with standard-dose chemotherapy. *J. Clin. Oncol.* **2012**, *30*, 3578–3587. [CrossRef] [PubMed]

35. Ono, M.; Ogilvie, J.M.; Wilson, J.S.; Green, H.J.; Chambers, S.K.; Ownsworth, T.; Shum, D.H.K. A meta-analysis of cognitive impairment and decline associated with adjuvant chemotherapy in women with breast cancer. *Front. Oncol.* **2015**, *5*, 59. [CrossRef] [PubMed]

36. Bernstein, L.J.; McCreath, G.A.; Komeylian, Z.; Rich, J.B. Cognitive impairment in breast cancer survivors treated with chemotherapy depends on control group type and cognitive domains assessed: A multilevel meta-analysis. *Neurosci. Biobehav. Rev.* **2017**, *83*, 417–428. [CrossRef] [PubMed]

37. Hutchinson, A.D.; Hosking, J.R.; Kichenadasse, G.; Mattiske, J.K.; Wilson, C. Objective and subjective cognitive impairment following chemotherapy for cancer: A systematic review. *Cancer Treat. Rev.* **2012**, *38*, 926–934. [CrossRef] [PubMed]

38. Pullens, M.J.; De Vries, J.; Roukema, J.A. Subjective cognitive dysfunction in breast cancer patients: A systematic review. *Psychooncology* **2010**, *19*, 1127–1138. [CrossRef] [PubMed]

39. Clarnette, R.M.; Almeida, O.P.; Forstl, H.; Paton, A.; Martins, R.N. Clinical characteristics of individuals with subjective memory loss in Western Australia: Results from a cross-sectional survey. *Int. J. Geriatr. Psychiatry* **2001**, *16*, 168–174. [CrossRef]

40. Montejo, P.; Montenegro, M.; Fernandez, M.A.; Maestu, F. Subjective memory complaints in the elderly: Prevalence and influence of temporal orientation, depression and quality of life in a population-based study in the city of Madrid. *Aging Ment. Health* **2011**, *15*, 85–96. [CrossRef] [PubMed]

41. Reid, L.; MacLullich, A.M. Subjective memory complaints and cognitive impairment in older people. *Dement. Geriatr. Cogn. Disord.* **2006**, *22*, 471–485. [CrossRef] [PubMed]

42. Wang, L.; van Belle, G.; Crane, P.K.; Kukull, W.A.; Bowen, J.D.; McCormick, W.C.; Larson, E.B. Subjective memory deterioration and future dementia in people aged 65 and older. *J. Am. Geriatr. Soc.* **2004**, *52*, 2045–2051. [CrossRef] [PubMed]

43. Saykin, A.J.; Wishart, H.A.; Rabin, L.A.; Santulli, R.B.; Flashman, L.A.; West, J.D.; McHugh, T.L.; Mamourian, A.C. Older adults with cognitive complaints show brain atrophy similar to that of amnestic MCI. *Neurology* **2006**, *67*, 834–842. [CrossRef] [PubMed]

44. Cheung, Y.T.; Foo, Y.L.; Shwe, M.; Tan, Y.P.; Fan, G.; Yong, W.S.; Madhukumar, P.; Ooi, W.S.; Chay, W.Y.; Dent, R.A.; et al. Minimal clinically important difference (MCID) for the functional assessment of cancer therapy: Cognitive function (FACT-Cog) in breast cancer patients. *J. Clin. Epidemiol.* **2014**, *67*, 811–820. [CrossRef] [PubMed]

45. Bray, V.J.; Dhillon, H.M.; Bell, M.L.; Kabourakis, M.; Fiero, M.H.; Yip, D.; Boyle, F.; Price, M.A.; Vardy, J.L. Evaluation of a web-based cognitive rehabilitation program in cancer survivors reporting cognitive symptoms after chemotherapy. *J. Clin. Oncol.* **2017**, *35*, 217–225. [CrossRef] [PubMed]

46. Hartman, S.J.; Nelson, S.H.; Myers, E.; Natarajan, L.; Sears, D.D.; Palmer, B.W.; Weiner, L.S.; Parker, B.A.; Patterson, R.E. Randomized controlled trial of increasing physical activity on objectively measured and self-reported cognitive functioning among breast cancer survivors: The memory and motion study. *Cancer* **2018**, *124*, 192–202. [CrossRef] [PubMed]

47. Derry, H.M.; Jaremka, L.M.; Bennett, J.M.; Peng, J.; Andridge, R.; Shapiro, C.; Malarkey, W.B.; Emery, C.F.; Layman, R.; Mrozek, E.; et al. Yoga and self-reported cognitive problems in breast cancer survivors: A randomized controlled trial. *Psychooncology* **2015**, *24*, 958–966. [CrossRef] [PubMed]

48. Milbury, K.; Chaoul, A.; Biegler, K.; Wangyal, T.; Spelman, A.; Meyers, C.A.; Arun, B.; Palmer, J.L.; Taylor, J.; Cohen, L. Tibetan sound meditation for cognitive dysfunction: Results of a randomized controlled pilot trial. *Psychooncology* **2013**, *22*, 2354–2363. [CrossRef] [PubMed]

49. Lacourt, T.E.; Heijnen, J.C. Mechanisms of neurotoxic symptoms as a result of breast cancer and its treatment: Considerations on the contribution of stress, inflammation, and cellular bioenergetics. *Curr. Breast Cancer Rep.* **2017**, *9*, 70–81. [CrossRef] [PubMed]

50. Ganz, P.A.; Bower, J.E.; Kwan, L.; Castellon, S.A.; Silverman, D.H.; Geist, C.; Breen, E.C.; Irwin, M.R.; Cole, S.W. Does tumor necrosis factor-alpha (TNF-a) play a role in post-chemotherapy cerebral dysfunction? *Brain Behav. Immun.* **2013**, *30*. [CrossRef] [PubMed]

51. Johnson, E.J.; Vishwanathan, R.; Johnson, M.A.; Hausman, D.B.; Davey, A.; Scott, T.M.; Green, R.C.; Miller, L.S.; Gearing, M.; Woodard, J.; et al. Relationship between serum and brain carotenoids, α -tocopherol, and retinol concentrations and cognitive performance in the oldest old from the georgia centenarian study. *J. Aging Res.* **2013**, *2013*, 951786. [CrossRef] [PubMed]

52. Snoddenly, D.M. Evidence for protection against age-related macular degeneration by carotenoids and antioxidant vitamins. *Am. J. Clin. Nutr.* **1995**, *62*, 1448S–1461S. [CrossRef] [PubMed]

53. Chen, Y.; Jungsuwadee, P.; Vore, M.; Butterfield, D.A.; St Clair, D.K. Collateral damage in cancer chemotherapy: Oxidative stress in nontargeted tissues. *Mol. Interv.* **2007**, *7*, 147–156. [CrossRef] [PubMed]

54. Vera-Ramirez, L.; Ramirez-Tortosa, Mc.; Perez-Lopez, P.; Granados-Principal, S.; Battino, M.; Quiles, J.L. Long-term effects of systemic cancer treatment on DNA oxidative damage: The potential for targeted therapies. *Cancer Lett.* **2012**, *327*, 134–141. [CrossRef] [PubMed]

55. Yuan, T.F.; Gu, S.; Shan, C.; Marchado, S.; Arias-Carrión, O. Oxidative Stress and Adult Neurogenesis. *Stem Cell Rev. Rep.* **2015**, *11*, 706–709. [CrossRef] [PubMed]

56. Lam, V.; Hackett, M.; Takechi, R. Antioxidants and Dementia Risk: Consideration through a Cerebrovascular Perspective. *Nutrients* **2016**, *8*. [CrossRef] [PubMed]

57. Aderka, D. The potential biological and clinical significance of the soluble tumor necrosis factor receptors. *Cytokine Growth Factor Rev.* **1996**, *7*, 231–240. [CrossRef]

58. Patel, S.K.; Wong, A.L.; Wong, F.L.; Breen, E.C.; Hurria, A.; Smith, M.; Kinjo, C.; Paz, I.B.; Kruper, L.; Somlo, G.; et al. Inflammatory biomarkers, comorbidity, and neurocognition in women With newly diagnosed breast cancer. *J. Natl. Cancer Inst.* **2015**, *107*, djv131. [CrossRef] [PubMed]

59. Jenkins, V.; Thwaites, R.; Cercignani, M.; Sacre, S.; Harrison, N.; Whiteley-Jones, H.; Mullen, L.; Chamberlain, G.; Davies, K.; Zammit, C.; et al. A feasibility study exploring the role of pre-operative assessment when examining the mechanism of "chemo-brain" in breast cancer patients. *Springerplus* **2016**, *5*, 390. [CrossRef] [PubMed]

60. Kesler, S.; Janelsins, M.; Koovakkattu, D.; Palesh, O.; Mustian, K.; Morrow, G.; Dhabhar, F.S. Reduced hippocampal volume and verbal memory performance associated with interleukin-6 and tumor necrosis factor-alpha levels in chemotherapy-treated breast cancer survivors. *Brain Behav. Immun.* **2013**, *30*, S109–S116. [CrossRef] [PubMed]

61. Shibayama, O.; Yoshiuchi, K.; Inagaki, M.; Matsuoka, Y.; Yoshikawa, E.; Sugawara, Y.; Akechi, T.; Wada, N.; Imoto, S.; Murakami, K.; et al. Association between adjuvant regional radiotherapy and cognitive function in breast cancer patients treated with conservation therapy. *Cancer Med.* **2014**, *3*, 702–709. [CrossRef] [PubMed]

62. Kaulmann, A.; Bohn, T. Carotenoids, inflammation, and oxidative stress—implications of cellular signaling pathways and relation to chronic disease prevention. *Nutr. Res.* **2014**, *34*, 907–929. [CrossRef] [PubMed]

63. Thomson, C.A.; Stendell-Hollis, N.R.; Rock, C.L.; Cussler, E.C.; Flatt, S.W.; Pierce, J.P. Plasma and dietary carotenoids are associated with reduced oxidative stress in women previously treated for breast cancer. *Cancer Epidemiol. Biomarkers Prev.* **2007**, *16*, 2008–2015. [CrossRef] [PubMed]

64. Butalla, A.C.; Crane, T.E.; Patil, B.; Wertheim, B.C.; Thompson, P.; Thomson, C.A. Effects of a carrot juice intervention on plasma carotenoids, oxidative stress, and inflammation in overweight breast cancer survivors. *Nutr. Cancer* **2012**, *64*, 331–341. [CrossRef] [PubMed]

65. Institute of Medicine, Panel on Micrnutrients. *DRI: Dietary Reference Intakes for Vitamin A, Vitamin K, Arsenic, Boron, Chromium, Copper, Iodine, Iron, Manganese, Molybdenum, Nickel, Silicon, Vanadium, and Zinc*; National Academies Press: Washington, DC, USA, 2002.

66. Polidori, M.C.; Praticó, D.; Mangialasche, F.; Mariani, E.; Aust, O.; Anlasik, T.; Mang, N.; Pientka, L.; Stahl, W.; Sies, H.; et al. High fruit and vegetable intake is positively correlated with antioxidant status and cognitive performance in healthy subjects. *J. Alzheimers Dis.* **2009**, *17*, 921–927. [CrossRef] [PubMed]

67. Liu, R.H. Health-promoting components of fruits and vegetables in the diet. *Adv. Nutr.* **2013**, *4*, 384S–392S. [CrossRef] [PubMed]

68. Lamport, D.J.; Saunders, C.; Butler, L.T.; Spencer, J.P. Fruits, vegetables, 100% juices, and cognitive function. *Nutr. Rev.* **2014**, *72*, 774–789. [CrossRef] [PubMed]

69. Miller, M.G.; Thangthaeng, N.; Poulose, S.M.; Shukitt-Hale, B. Role of fruits, nuts, and vegetables in maintaining cognitive health. *Exp. Gerontol.* **2017**, *94*, 24–28. [CrossRef] [PubMed]

70. Pomykala, K.L.; Ganz, P.A.; Bower, J.E.; Kwan, L.; Castellon, S.A.; Mallam, S.; Cheng, I.; Ahn, R.; Breen, E.C.; Irwin, M.R.; et al. The association between pro-inflammatory cytokines, regional cerebral metabolism, and cognitive complaints following adjuvant chemotherapy for breast cancer. *Brain Imaging Behav.* **2013**, *7*, 511–523. [CrossRef] [PubMed]

71. Simen, A.A.; Bordner, K.A.; Martin, M.P.; Moy, L.A.; Barry, L.C. Cognitive dysfunction with aging and the role of inflammation. *Ther. Adv. Chronic Dis.* **2011**, *2*, 175–195. [CrossRef] [PubMed]

72. Sartori, A.; Vance, D. The impact of inflammation on cognitive function in older adults: Implications for health care practice and research. *J. Neurosci. Nurs.* **2012**, *44*, 206–217. [CrossRef] [PubMed]

73. Fardell, J.E.; Vardy, J.; Johnston, I.N.; Winocur, G. Chemotherapy and cognitive impairment: Treatment options. *Clin. Pharmacol. Ther.* **2011**, *90*, 366–376. [CrossRef] [PubMed]

74. Wefel, J.S.; Kesler, S.R.; Noll, K.R.; Schagen, S.B. Clinical characteristics, pathophysiology, and management of noncentral nervous system cancer-related cognitive impairment in adults. *CA Cancer J. Clin.* **2015**, *65*, 123–138. [CrossRef] [PubMed]

75. Oh, P.J.; Kim, J. The effects of nonpharmacologic interventions on cognitive function in patients with cancer: A meta-analysis. *Oncol. Nurs. Forum* **2016**, *43*, E205–E217. [CrossRef] [PubMed]

76. Treanor, C.J.; McMenamin, U.C.; O'Neill, R.F.; Cardwell, C.R.; Clarke, M.J.; Cantwell, M.; Donnelly, M. Non-pharmacological interventions for cognitive impairment due to systemic cancer treatment. *Cochrane Database Syst. Rev.* **2016**, CD011325. [CrossRef] [PubMed]

77. Demark-Wahnefried, W.; Aziz, N.M.; Rowland, J.H.; Pinto, B.M. Riding the crest of the teachable moment: Promoting long-term health after the diagnosis of cancer. *J. Clin. Oncol.* **2005**, *23*, 5814–5830. [CrossRef] [PubMed]

![nutrients logo] **nutrients**

MDPI

Article

Astaxanthin Prevents Alcoholic Fatty Liver Disease by Modulating Mouse Gut Microbiota

Huilin Liu [1], Meihong Liu [2,3], Xueqi Fu [1], Ziqi Zhang [4], Lingyu Zhu [4], Xin Zheng [4] and Jingsheng Liu [2,3,*]

[1] School of Life Sciences, Jilin University, Changchun 130012, China; lhl14@mails.jlu.edu.cn (H.L.); fxq@jlu.edu.cn (X.F.)
[2] College of Food Science and Engineering, Jilin Agricultural University, Changchun 130118, China; liumh@jlau.edu.cn
[3] National Engineering Laboratory for Wheat and Corn Deep Processing, Changchun 130118, China
[4] College of Animal Science and Technology, Jilin Agricultural University, Changchun 130118, China; zhangziqi@jlau.edu.cn (Z.Z.); zhulingyu@jlau.edu.cn (L.Z.); zhengxin@jlau.edu.cn (X.Z.)
* Correspondence: liujs1007@vip.sina.com; Tel.: +86-0431-84533505

Received: 16 August 2018; Accepted: 10 September 2018; Published: 13 September 2018

Abstract: The development and progression of alcoholic fatty liver disease (AFLD) is influenced by the intestinal microbiota. Astaxanthin, a type of oxygenated carotenoid with strong antioxidant and anti-inflammatory properties, has been proven to relieve liver injury. However, the relationship between the gut microbiota regulation effect of astaxanthin and AFLD improvement remains unclear. The effects of astaxanthin on the AFLD phenotype, overall structure, and composition of gut microbiota were assessed in ethanol-fed C57BL/6J mice. The results showed that astaxanthin treatment significantly relieves inflammation and decreases excessive lipid accumulation and serum markers of liver injury. Furthermore, astaxanthin was shown to significantly decrease species from the phyla Bacteroidetes and Proteobacteria and the genera *Butyricimonas*, *Bilophila*, and *Parabacteroides*, as well as increase species from Verrucomicrobia and *Akkermansia* compared with the Et (ethanol)group. Thirteen phylotypes related to inflammation as well as correlated with metabolic parameters were significantly altered by ethanol, and then notably reversed by astaxanthin. Additionally, astaxanthin altered 18 and 128 KEGG (Kyoto Encyclopedia of Genes and Genomes) pathways involved in lipid metabolism and xenobiotic biodegradation and metabolism at levels 2 and 3, respectively. These findings suggest that *Aakkermansia* may be a potential target for the astaxanthin-induced alleviation of AFLD and may be a potential treatment for bacterial disorders induced by AFLD.

Keywords: astaxanthin; *Akkermansia*; alcoholic fatty liver disease; inflammation; gut microbiota

1. Introduction

According to the "Global Status Report on Alcohol and Health 2014" released by the World Health Organization (WHO), the use of alcohol has reached a detrimental level, resulting in more than 3 million deaths every year and accounting for 5.9% of all deaths worldwide. As the metabolism of alcohol in the organism mainly depends on the liver, the long-term consumption and over-consumption of alcohol leads to liver damage and triggers alcoholic fatty liver disease (AFLD) [1], thereafter causing health issues ranging from fatty liver to alcoholic hepatitis and even cirrhosis [2,3]. The first two phases can be reversible through alcohol abstinence and lifestyle intervention. Thus, the detection of potential functional ingredients for AFLD prevention is significant.

Astaxanthin is one of the major xanthophyll carotenoids in marine organisms, and is found in shrimp, crabs, fish, algae, yeast, and feathers of birds [4,5]. As astaxanthin cannot be synthesized in humans, its uptake fully depends on dietary sources. Astaxanthin-associated protection against aging as well as cardiovascular and cancerous diseases is attributed to its great antioxidative and anti-inflammatory activity [4]. In particular, it has been suggested that astaxanthin is protective against various types of liver damage [6,7], such as non-alcoholic fatty liver disease (NAFLD) [8–10] and liver fibrosis [11]; however, further investigation is required to determine the effect of astaxanthin on AFLD protection.

Previous studies have confirmed that gut microbiota imbalance is related to a variety of chronic diseases, such as obesity, cardiovascular disease, and cancer [12–14]. The relationship between gut microbiota and AFLD has become a topic of increasing concern to researchers. In healthy humans, the intestinal microbiota remains in symbiotic balance, but alcohol intake induces the modification of its composition, thereby inhibiting dominant intestinal bacteria and promoting a small amount of pathogenic bacteria overgrowth [15], resulting in the impairment of intestinal physiological function. The long-term intake of alcohol breaks the barrier function of the intestine and promotes the growth of intestinal pathogenic bacteria and harmful metabolites (such as lipopolysaccharides). These health-threatening components lead to the development of AFLD by invading into other tissues and organs through blood circulation [16,17]. *Akkermansia muciniphila* (*A. muciniphila*), which can degrade mucin and maintain intestinal barrier integrity, colonizes in the human gut mucus layer and is negatively associated with certain diseases like obesity, diabetes, inflammation, and metabolic disorders [18]. Long-term dietary intervention can restore the gut microbiota and improve intestinal physiological function [19]. The concentration of *A. muciniphila* can be increased through the ingestion of *Bifidobacteria*, fructo-oligosaccharides, short-chain carbohydrates, metformin, rhubarb extract, or specific antibiotics [18]. Bacterial imbalance is an important factor in inducing disease in AFLD patients. An increase in probiotics species and the inhibition of pathogenic bacteria may improve AFLD status.

Astaxanthin has been reported to prevent liver damage [6,20], but there has been little research on the effect of astaxanthin on AFLD. Studying intestinal microbiota regulation in AFLD may provide new insight into the pathogenesis and therapeutic target of AFLD. In the present study, we investigated whether astaxanthin can protect against alcohol-induced liver injury and its association with the intestinal microecology. We also determined the main influence of bacteria during the astaxanthin intervention.

2. Materials and Methods

2.1. Animal Experimentation

Male C57BL/6J mice (22 ± 2 g, six weeks old) were purchased from the Beijing Vital River Laboratory Animal Technology Co., Ltd. (Beijing, China) and housed individually in cages with a 12-h light/dark cycle at 23 ± 2 °C with full access to chow diet and water. All experiment protocols were approved by the Institutional Animal Care and Use Committee at the Jilin Institute of Traditional Chinese Medicine (approval number: SYXK (JI) 2015-0009).

The alcoholic liver disease (AFLD) mouse model was established using the modified Lieber–DeCarli liquid diet [21]. The 60 mice were randomly separated into five groups, each consisting of 12 mice. The first group was fed with a normal standard growth diet while the other four groups were fed high-fat liquid diets (35% fat, 18% protein, 47% carbohydrates, provided by TROPHIC Animal Feed High-Tech Co., Ltd., Nantong, China) for an acclimation period of two weeks acclimation. Then the four high-fat diet groups were fed with either only the high-fat liquid diet (Con), or combined with the astaxanthin (AST group, 50 mg/kg bw), ethanol-containing (Et group, 5% ethanol v/v, accounted for 36% of the total caloric intake), or ethanol plus astaxanthin (EtAST group) treatments for 12 weeks. Compositions of the diets are shown in Supplementary Table S1. The astaxanthin was purchased from Sigma-Aldrich (St. Louis, MO, USA; purity ≥97%, SML0982), isolated from *Blakeslea trispora* and dissolved in corn oil for utilization in this experiment. The dose selection of astaxanthin was chosen in accordance with previous research [22], while the amount of ethanol included increased over two weeks to reach a final concentration of 5% (v/v). Body weight gain and food intake were assessed once a week. The pair-fed control group (Con) was included in this model.

After 12 weeks of treatment, the fasted mice were euthanized and blood samples were collected. The feces in the colon were also harvested into 2-mL sterile tubes to assess the gut microbiota. The tissues were immediately removed and weighed, and the liver coefficient (liver weight/body weight) was calculated. The left lobe of the liver was instantly fixed in 10% buffered formalin for histology analysis with the rest of the tissues frozen in liquid nitrogen and stored at −80 °C until further use.

2.2. Biochemical Analyses

The serum was obtained from the collected blood by centrifugation at 5000 rpm for 10 min at 4 °C and stored at −80 °C. Plasma triacylglycerol (TG), total cholesterol (TC), high-density lipoprotein (HDL), low-density lipoprotein (LDL), aspartate aminotransferase (AST), and alanine aminotransferase (ALT) were measured by enzymatic colorimetric assays using commercial detection kits (Nanjing Jiancheng of Bioengineering Institute, Nanjing, China).

2.3. Hepatic Triglyceride Staining

The left lobe of the liver tissue was separated and rapidly fixed in 4% neutral buffered formalin solution for 24 h and then processed for paraffin embedding. Five-micrometer-thick paraffin sections were stained with hematoxylin and eosin (H&E) and oil-red solution. The liver steatosis status was examined under a light microscope (Olympus, Tokyo, Japan), and photographed at 200× magnification.

2.4. Quantification of Genes Expression in Liver Tissue

Total RNA was extracted and purified in the liver tissue using TRIzol reagent (TAKARA, Beijing, China. RNAiso Plus, Code No. 9108). The purity, concentration, and quality of RNA were measured. High-quality RNA was converted to cDNA using PrimeScript™ RT reagent Kit with gDNA Eraser (TAKARA, Beijing China. Code No. RR047A). The SYBR fluorescent dye method (TAKARA, Beijing China. Code No. RR420A) and Agilent Stratagene Mx3000P Real-Time PCR System (Santa Clara, CA, USA) were used to detect the gene expression. The primers used are shown in Supplementary Table S2. The data were calculated using the $2^{-\Delta\Delta Ct}$ relative quantification method and normalized to β-actin.

2.5. Fecal DNA Extraction

Total bacterial genomic DNA was extracted from the fecal material in the colon using Fast DNA SPIN extraction kits (MP Biomedicals, Santa Ana, CA, USA), following the manufacturer's instructions. The DNA yield was measured for quantity using a NanoDrop ND-1000 spectrophotometer (Thermo Fisher Scientific, Waltham, MA, USA) and the quality was analyzed by 0.8% agarose gel electrophoresis; the DNA was then stored at −20 °C until further analysis.

2.6. Amplification and Sequencing of the 16S rRNA Genes

The V3–V4 region of bacterial 16S rRNA genes was subjected to PCR amplification using the forward primer 5′-ACTCCTACGGGAGGCAGCA-3′ and the reverse primer 5′-GGACTACH VGGGTWTCTAAT-3′. The specific system was a 25-μL reaction including 5 μL of Q5 High-Fidelity DNA Polymerase (New England Biolabs (Beijing) Ltd., Beijing, China). PCR amplicons were purified with Agencourt AMPure Beads (Beckman Coulter, Indianapolis, IN, USA) and quantified using the PicoGreen dsDNA Assay Kit (Invitrogen, Carlsbad, CA, USA). The amplicons were pooled and normalized, and then paired-end 2 × 300 bp sequencing was performed using the Illumina

MiSeq platform with the MiSeq Reagent Kit v3 at Shanghai Personal Biotechnology Co., Ltd. (Shanghai, China).

Sequencing data for the 16S rRNA sequences have been deposited in the SRA database under GenBank accession NO. SRP148082.

2.7. Bioinformatics and Statistical Analysis

Sequence data were processed using Quantitative Insights into Microbial Ecology (QIIME, v1.8.0), as previously described [23]. The low-quality sequences, which had lengths of <150 bp and average Phred scores of <20, and contained ambiguous bases and mononucleotide repeats of >8 bp, were filtered with the following criteria. Paired-end reads were assembled using FLASH. The remaining high-quality sequences were clustered into operational taxonomic units (OTUs) at 97% sequence identity by UCLUST (Edgar 2010) [24]. OTU taxonomic classification was conducted by BLAST and the OTUs containing more than 99.999% of total sequences across all samples were reserved.

All results are presented as means $\pm$ standard deviation. Data were analyzed with SPSS 19.0 using one-way analysis of variance (ANOVA), and, when appropriate, using a two-tailed Student's *t*-test between different groups. Differences among groups were evaluated for significance with the comparable variances, followed by Tukey's and least significant difference (LSD) tests. $p < 0.05$ was considered statistically significant. GraphPad Prism 7.0 software (GraphPad Software, La Jolla, CA, USA) was used for graph-making.

Sequence data analysis was mainly performed using QIIME (v1.8.0, University of Colorado, Denver, CO, USA) and R packages (v3.2.0, Bell Labs Technology Showcase, Murray Hill, NJ, USA). The alpha diversity indices, Chao1 richness estimator, and the Shannon diversity index were calculated using the OTU table in QIIME. Beta diversity analysis was performed using UniFrac distance metrics and visualized by principle coordinate analysis (PCoA), and the unweighted pair-group method with arithmetic means (UPGMA) hierarchical clustering. The significance of microbiota structure differentiation among groups was assessed by PERMANOVA (permutational multivariate analysis of variance) and ANOSIM (analysis of similarities) using the R package "vegan". Taxa abundances at different taxonomies were statistically compared among groups by Metastats and visualized as box plots. Microbial functions were predicted by PICRUSt (phylogenetic investigation of communities by reconstruction of unobserved states), based on high-quality sequences.

3. Results

3.1. Astaxanthin Protects Mice from High-Fat Diet and Ethanol-Induced Liver Lesions

We investigated whether different diets affect the growth status of mice. The five groups have similar body weights at baseline which were still not significantly difference after 12 weeks of intervention (Figure 1A). On the other hand, compared with the Control (Con) group, the liver indices of mice in the ethanol group were significantly increased, and markedly reversed in the EtAST (ethanol plus astaxanthin treatments) group, which was not different from that of the Con group (Figure 1B). The light microscopy images of the liver slices, stained by Oil Red O, showed that lipid droplets increased in size and number in the Et group compared with those in the Con group. Astaxanthin intervention significantly relieved fat accumulation in the liver induced by ethanol. H&E staining showed the structure of hepatic lobules and neatly arranged normal liver cells in the normal diet (ND) group, which showed no obvious differences among the Con and AST groups. Ethanol supplementation markedly increased the amount of hepatic steatosis and the number of necrotic cells, which appeared in a large number of fat vacuoles within liver cells and significantly enlarged them. Hepatic steatosis caused by ethanol was reversed by astaxanthin intervention (Figure 1C).

There were no significant changes in the serum markers involved in lipid dysmetabolism and liver injury in normal and astaxanthin-supplied groups compared with those in the Con group.

However, the levels of ALT and AST were markedly increased in the Et group compared to the Con group, indicating the existence of liver damage, which affected cell membrane permeability and promoted ALT and AST overflow, suggesting that mice suffered significant alcoholic liver injuries. Moreover, 50 mg kg^{-1} astaxanthin treatment significantly decreased the levels of ALT, AST, TG, and LDL compared to the Et diet. However, there were no obvious differences in the levels when compared with the ND group. These results suggest that astaxanthin has the ability to alleviate lipid dysmetabolism and alcohol-induced liver injury (Figure 1D).

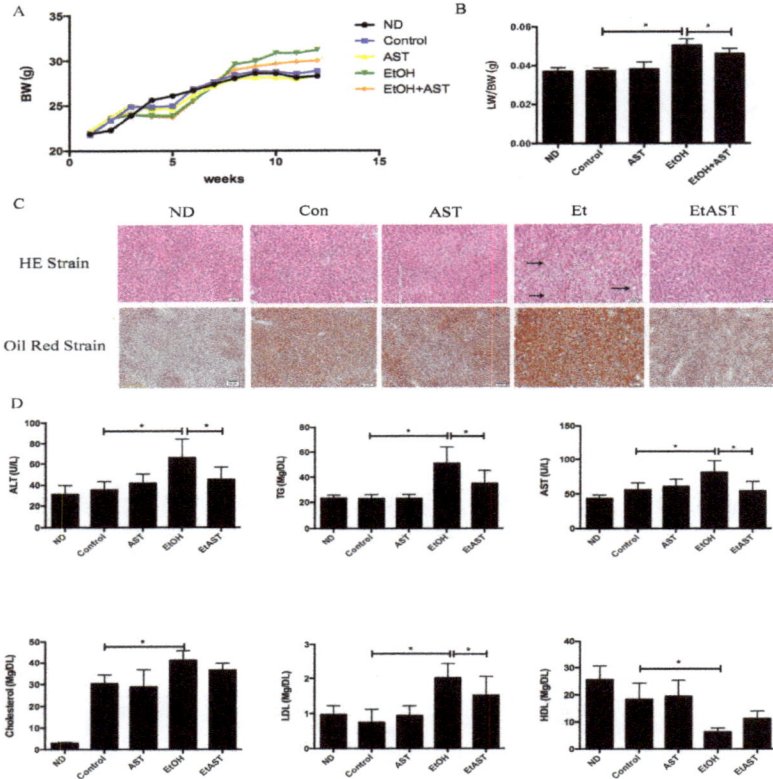

Figure 1. Effect of astaxanthin on body weight, pathological morphology, and serum markers involved in liver injury in ethanol diet-fed mice. C57BL/6J mice were fed the Lieber–DeCarli liquid diet containing 5% ethanol for 12 weeks ad libitum with or without 50 mg kg^{-1} of astaxanthin. Body weight was measured once a week (**A**). The liver index was represented by calculating liver weight/body weight (**B**). The statues of hepatic steatosis were checked by hematoxylin and eosin (H&E) staining and hepatic lipid accumulation was detected by Oil red O staining. The symbol → plays a indicate role to identify the fat vacuoles (**C**). The serum markers of liver injury were determined by the enzyme activities of alanine aminotransferase (ALT) and aspartate aminotransferase (AST). Lipid accumulation in the liver was reflected by the levels of hepatic markers, including plasma triacylglycerol (TG), low-density lipoprotein (LDL), high-density lipoprotein (HDL), and total cholesterol (TC) (**D**). ND: normal diet; Con: high-fat diet with 35% of total calories from fat; AST: high-fat diet + astaxanthin; Et: Lieber–DeCarli liquid ethanol diet with 35% of total calories from fat; EtAST: Lieber–DeCarli liquid ethanol diet with 35% of total calories from fat + astaxanthin. All values represent means ± SD. * $p < 0.05$ represents significant differences in each group compared with the ethanol (Et) group by ANOVA analysis.

3.2. Astaxanthin Can Relieve Liver Injury Through the Regulation of Inflammatory Genes Expression in Mice

To explore whether astaxanthin can reverse the development of AFLD that is associated with inflammatory responses, liver inflammatory gene expression in liver was measured. A high-fat diet did not promote inflammatory gene expression. However, as expected, the consumption of ethanol in a high-fat diet significantly induced the mRNA expression of interleukin-1 alpha (IL-1α), macrophage inflammatory protein 2 (MIP-2), interleukin-6 (IL-6), and tumor necrosis factor-alpha (TNF-α). However, these effects were markedly reversed by astaxanthin supplementation (Figure 2).

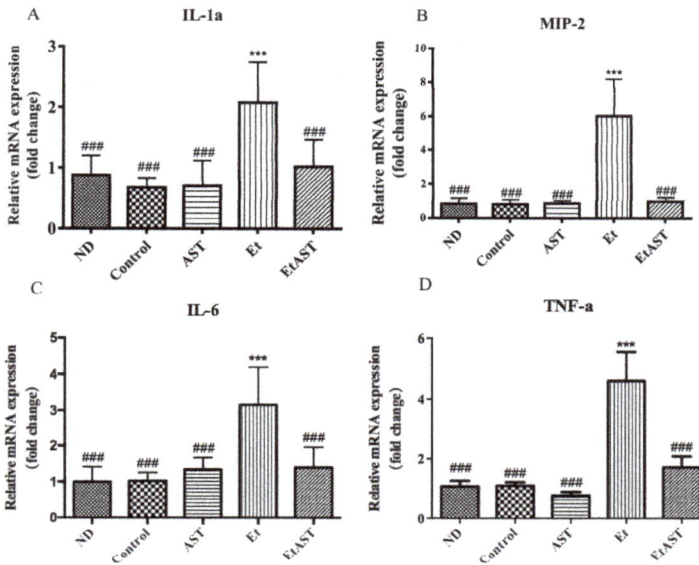

Figure 2. Effect of astaxanthin on inflammatory genes expression in alcoholic fatty liver disease (AFLD) mice. At the end of the experiment, total RNA was extracted from liver tissues. The mRNA expressions of interleukin-1 alpha (IL-1α) (**A**), macrophage inflammatory protein 2 (MIP-2) (**B**), interleukin-6 (IL-6) (**C**), and tumor necrosis factor-alpha (TNF-α) (**D**) were normalized to that of β-actin. Data are presented as means $\pm$ SD, $n = 6$. *** $p < 0.001$ compared with the control (Con) group, and ### $p < 0.001$ compared with the Et group.

3.3. Astaxanthin Alters the Profiles of Gut Microbiota in Ethanol-Fed Mice

The microbiota can influence the development and progression of AFLD. High-throughput 16S rRNA gene sequencing produced a total of 332,981 good-quality sequences from 15 samples (with 21,532 $\pm$ 231 sequences per sample) (Table S3). Rarefaction curves and rank abundance curves have shown that the most gut microbes in samples were captured based on the current sequencing depth and the data can be used for further analysis (Figure S1).

Next, we analyzed the gut bacterial community in mice affected by high-fat and/or ethanol exposure with or without astaxanthin. Alpha-diversity analysis indicated that the values of Chao1, abundance-based coverage estimator (ACE), and Shannon were significantly increased in the Con and Et groups compared to the ND group. After treatment with 50 mg kg^{-1} astaxanthin, the values were not influenced by the high-fat diet alone but markedly decreased when ethanol was added, and even improved relative to the normal group (Figure 3A–C). Furthermore, to investigate the similarities in gut microbial community structure among different samples, certain analyses were conducted. The PCoA plot indicated that the structure of gut microbiota in the Et group was statistically different from the ND along the PC1 axis (59.16% and 42.3% of overall variation based on weighted and unweighted UniFrac, respectively), and a significant structural shift was also shown for most of the astaxanthin-supplemented mice compared with those in the Et group (Figure 3D–F). As expected, the results of the unweighted pair group method with arithmetic mean (UPGMA) based on weighted and unweighted Unifrac also showed overt changes in the composition of gut microbiota in the EtAST group compared with the Et group (Figure S2), which is in line with the PCoA results, indicating that the microbial structure was disturbed by ethanol feeding, but remedied and returned to normal status by astaxanthin administration.

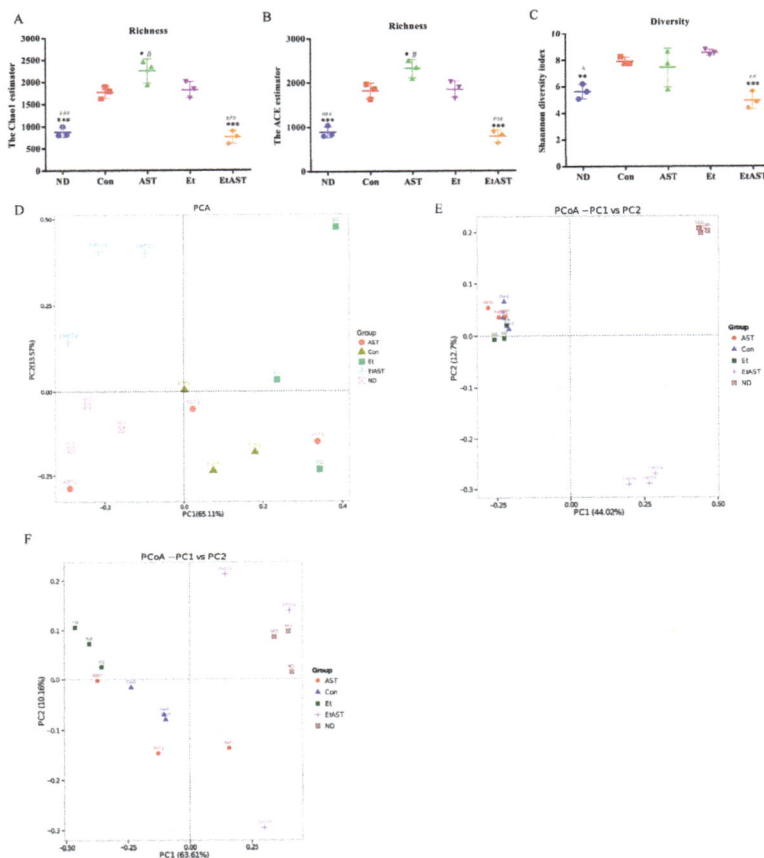

Figure 3. Astaxanthin treatment ameliorates the microbiota profiles affected by the ethanol diet. Bacterial genomic DNA was extracted from the feces collected at the end of week 12, and 16S rDNA sequence analysis was performed. Alpha diversity analysis, the Chao1 estimator (**A**), the ACE estimator (**B**), and the Shannon diversity index (**C**) were used for evaluation. To investigate the similarity of gut microbial community structure among different samples, we conducted beta-diversity analysis and PCA analysis (**D**), PCoA score plot based on unweighted UniFrac (**E**), and weighted UniFrac (**F**). ND: normal diet; Con: high-fat diet with 35% of total calories from fat; AST: high-fat diet + astaxanthin; Et: Lieber–DeCarli liquid ethanol diet with 35% of total calories from fat; EtAST: Lieber–DeCarli liquid ethanol diet with 35% of total calories from fat + astaxanthin. # represents $p < 0.05$, ## represents $p < 0.01$, and ### represents $p < 0.001$ compared with the Con group, * represents $p < 0.05$, ** represents $p < 0.01$, and *** represents $p < 0.001$ compared with the Et group.

3.4. Astaxanthin Regulates the Gut Microbiota Composition in Ethanol Feeding Mice

We detected nine bacterial phyla and 60 genera among the mice. The structure and composition of gut microbiota were significantly influenced by the high-fat plus ethanol diet. The results of the top 20 most abundant OTUs at all taxonomic levels in samples, as inferred by GraPhlAn, showed that *Firmicutes, Bacteroidetes, Proteobacteria,* and *Verrucomicrobia* were the most abundant phyla, and *Akkermansia, Bacteroides, Prevotella,* and *Paraprevotella* were the most abundant genera among the OTUs (Figure S3A). Additionally, the relative bacterial abundance in groups was reflected by the cladogram and linear discriminant analysis (LDA) score (Figure S3B,C). The taxonomic profiles indicated that the proportions of *Bacteroidetes* and *Proteobacteria* increased significantly, and the abundance of *Verrucomicrobia* decreased markedly in the Con group, especially in the Et group when compared with those of the ND group, while astaxanthin was shown to significantly reverse the tendency of this bacterial abundance, similar to the ND group (Figure 4A,B). Specifically, *Cyanobacteria* was completely depleted in both the Et and EtAST groups (Figure 4B). At the genus level, bacterial taxa displayed obvious changes in the heat maps, which were affected by different diets. The genera abundance, including *Akkermansia, Bacillus, Adlercreutzia, Lactococcus, Bacteroides, Butyricimonas, Parabacteroides,* and *Bilophila,* was significantly switched by ethanol feeding compared with the normal diet and partially reversed by the addition of astaxanthin (Figure S4). Astaxanthin treatment significantly decreased the *Butyricimonas, Bilophila,* and *Parabacteroides* concentrations relative to the Et group, and the abundance of *Akkermansia* decreased markedly in the Et group (3%) and recovered dramatically in the EtAST group (34%), while the abundance was similar to that of the ND group (38%, Figure 4D and Figure S5).

Next, we identified changes in the strain-specific key phenotypes which were affected by astaxanthin in the AFLD mice. The results showed that the abundance levels of 27 OTUs were markedly changed by ethanol supplementation (15 increased and 12 decreased OTUs) compared with those of the Con group, while astaxanthin intervention significantly altered 31 OTUs including enhanced or reduced abundance in 2 and 29 OTUs, respectively (EtAST group vs. Et group). Particularly, among the 43 OTUs, 13 were significantly increased or decreased by ethanol, and afterwards notably reversed by astaxanthin (Figure 4E), which including bacteria belonging to *Akkermansia_muciniphila,* species from *Butyricimonas, S24-7, Oscillospira, Clostridiales,* and *Bilophila.*

3.5. Associations of the Bacterial Abundance Altered by Astaxanthin with the AFLD Phenotype

To assess the relationships between OTUs and metabolic parameters altered by astaxanthin, Spearman's correlation coefficient was employed. Among the 43 OTUs that were altered in abundance by ethanol or astaxanthin shown in Figure 5, 33 OTUs were markedly correlated with at least one of the following metabolic parameters: AST, ALT, TG, LDL, liver weight/body weight (LW/BW). Thirty-eight of these OTUs were positively correlated with the abnormal parameters, and five OTUs were negatively correlated with abnormal parameters. The abundance levels of 20 OTUs were markedly changed by ethanol supplementation (15 increased and five decreased OTUs) compared with those of the Con group, while astaxanthin intervention significantly altered 27 OTUs, including enhanced and reduced abundances in two and 25 OTUs, respectively, compared with those of the Et group. Notably, 13 OTUs were significantly increased or decreased by ethanol and then markedly reversed by astaxanthin (Figure 5).

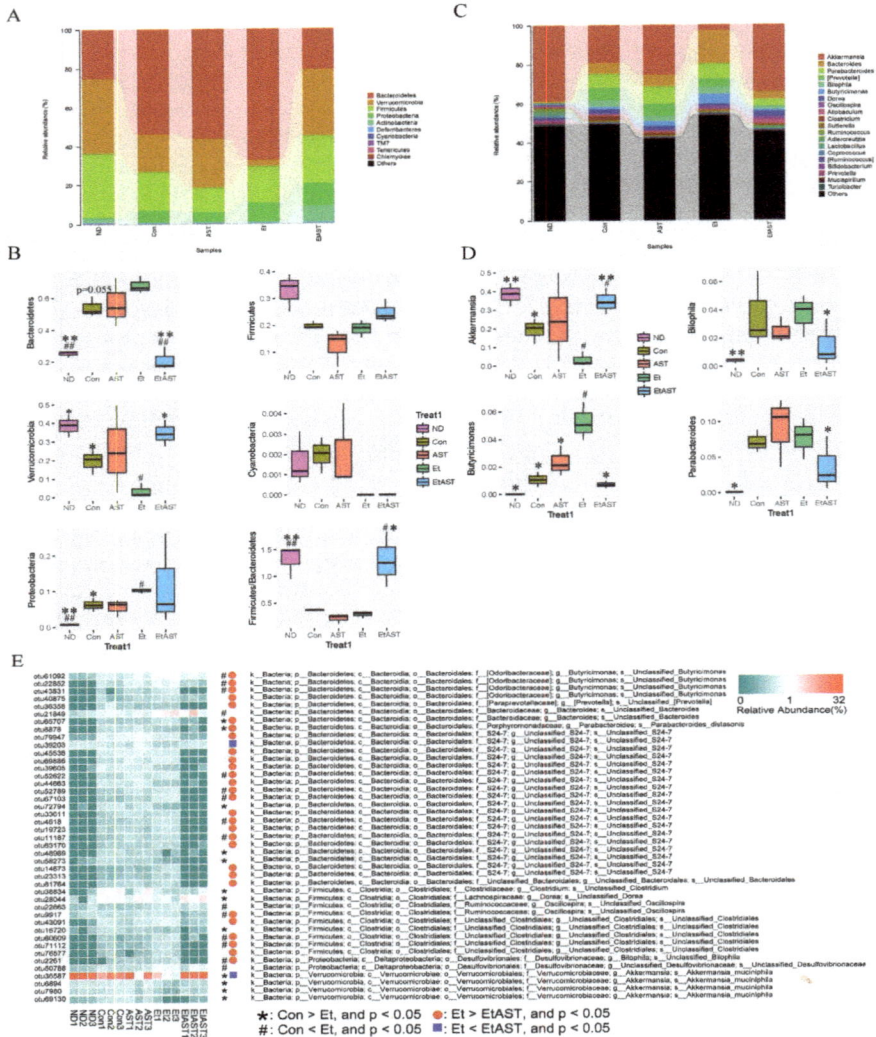

Figure 4. Bacterial community analysis and comparison. The composition and abundance distributions of each group at the phylum (**A**) and genus (**C**) levels were shown using QIIME software. At the phylum (**B**) and genus (**D**) levels, pairwise comparisons, conducted to determine the sequence amounts between two groups, were presented as pair-wise comparisons using Metastats analysis. # represents $p < 0.05$ and ## represents $p < 0.01$ compared with the Con group; * represents $p < 0.05$ and ** represents $p < 0.01$ compared with the Et group. A heat map of 43 operational taxonomic units (OTUs) which were altered in abundance by ethanol or astaxanthin is shown, based on the redundancy analysis (RDA) model. OTUs with a relative abundance greater than 0.1% in at least in one group were selected and used to analyze these differences. The red and green colors indicate the relative abundances of OTUs that were more or less abundant. The symbols represent the OTUs whose abundance were reduced and increased in the Et group relative to the Con group, while the circles in red and rectangles in blue represent the OTUs whose abundances were reduced and increased in the EtAST group relative to the Et group (**E**).

OTU						Taxonomy
otu61092					# ●	k__Bacteria; p__Bacteroidetes; c__Bacteroidia; o__Bacteroidales; f__[Odoribacteraceae]; g__Butyricimonas; s__Unclassified_Butyricimonas
otu22852					# ●	k__Bacteria; p__Bacteroidetes; c__Bacteroidia; o__Bacteroidales; f__[Odoribacteraceae]; g__Butyricimonas; s__Unclassified_Butyricimonas
otu43831					# ●	k__Bacteria; p__Bacteroidetes; c__Bacteroidia; o__Bacteroidales; f__[Odoribacteraceae]; g__Butyricimonas; s__Unclassified_Butyricimonas
otu40875					●	k__Bacteria; p__Bacteroidetes; c__Bacteroidia; o__Bacteroidales; f__[Odoribacteraceae]; g__Butyricimonas; s__Unclassified_Butyricimonas
otu36358					●	k__Bacteria; p__Bacteroidetes; c__Bacteroidia; o__Bacteroidales; f__[Paraprevotellaceae]; g__[Prevotella]; s__Unclassified_[Prevotella]
otu21849					#	k__Bacteria; p__Bacteroidetes; c__Bacteroidia; o__Bacteroidales; f__Bacteroidaceae; g__Bacteroides; s__Unclassified_Bacteroides
otu65707					● ●	k__Bacteria; p__Bacteroidetes; c__Bacteroidia; o__Bacteroidales; f__Bacteroidaceae; g__Bacteroides; s__Unclassified_Bacteroides
otu79947					●	k__Bacteria; p__Bacteroidetes; c__Bacteroidia; o__Bacteroidales; f__S24-7; g__Unclassified_S24-7; s__Unclassified_S24-7
otu39203					■	k__Bacteria; p__Bacteroidetes; c__Bacteroidia; o__Bacteroidales; f__S24-7; g__Unclassified_S24-7; s__Unclassified_S24-7
otu69886					●	k__Bacteria; p__Bacteroidetes; c__Bacteroidia; o__Bacteroidales; f__S24-7; g__Unclassified_S24-7; s__Unclassified_S24-7
otu52622					# ●	k__Bacteria; p__Bacteroidetes; c__Bacteroidia; o__Bacteroidales; f__S24-7; g__Unclassified_S24-7; s__Unclassified_S24-7
otu52789					# ●	k__Bacteria; p__Bacteroidetes; c__Bacteroidia; o__Bacteroidales; f__S24-7; g__Unclassified_S24-7; s__Unclassified_S24-7
otu67103					# ●	k__Bacteria; p__Bacteroidetes; c__Bacteroidia; o__Bacteroidales; f__S24-7; g__Unclassified_S24-7; s__Unclassified_S24-7
otu33611					●	k__Bacteria; p__Bacteroidetes; c__Bacteroidia; o__Bacteroidales; f__S24-7; g__Unclassified_S24-7; s__Unclassified_S24-7
otu4618					# ●	k__Bacteria; p__Bacteroidetes; c__Bacteroidia; o__Bacteroidales; f__S24-7; g__Unclassified_S24-7; s__Unclassified_S24-7
otu19723					●	k__Bacteria; p__Bacteroidetes; c__Bacteroidia; o__Bacteroidales; f__S24-7; g__Unclassified_S24-7; s__Unclassified_S24-7
otu11187					# ●	k__Bacteria; p__Bacteroidetes; c__Bacteroidia; o__Bacteroidales; f__S24-7; g__Unclassified_S24-7; s__Unclassified_S24-7
otu63170					●	k__Bacteria; p__Bacteroidetes; c__Bacteroidia; o__Bacteroidales; f__S24-7; g__Unclassified_S24-7; s__Unclassified_S24-7
otu14673					●	k__Bacteria; p__Bacteroidetes; c__Bacteroidia; o__Bacteroidales; f__S24-7; g__Unclassified_S24-7; s__Unclassified_S24-7
otu23313					●	k__Bacteria; p__Bacteroidetes; c__Bacteroidia; o__Bacteroidales; f__S24-7; g__Unclassified_S24-7; s__Unclassified_S24-7
otu81764					●	k__Bacteria; p__Bacteroidetes; c__Bacteroidia; o__Bacteroidales; f__Unclassified_Bacteroidales; g__Unclassified_Bacteroidales; s__Unclassified_Bacteroidales
otu22663					#	k__Bacteria; p__Firmicutes; c__Clostridia; o__Clostridiales; f__Ruminococcaceae; g__Oscillospira; s__Unclassified_Oscillospira
otu9917					# ●	k__Bacteria; p__Firmicutes; c__Clostridia; o__Clostridiales; f__Ruminococcaceae; g__Oscillospira; s__Unclassified_Oscillospira
otu43091					●	k__Bacteria; p__Firmicutes; c__Clostridia; o__Clostridiales; f__Unclassified_Clostridiales; g__Unclassified_Clostridiales; s__Unclassified_Clostridiales
otu60609					# ●	k__Bacteria; p__Firmicutes; c__Clostridia; o__Clostridiales; f__Unclassified_Clostridiales; g__Unclassified_Clostridiales; s__Unclassified_Clostridiales
otu71112					# ●	k__Bacteria; p__Firmicutes; c__Clostridia; o__Clostridiales; f__Unclassified_Clostridiales; g__Unclassified_Clostridiales; s__Unclassified_Clostridiales
otu76577					●	k__Bacteria; p__Firmicutes; c__Clostridia; o__Clostridiales; f__Unclassified_Clostridiales; g__Unclassified_Clostridiales; s__Unclassified_Clostridiales
otu2261					# ●	k__Bacteria; p__Proteobacteria; c__Deltaproteobacteria; o__Desulfovibrionales; f__Desulfovibrionaceae; g__Bilophila; s__Unclassified_Bilophila
otu50788					●	k__Bacteria; p__Proteobacteria; c__Deltaproteobacteria; o__Desulfovibrionales; f__Desulfovibrionaceae; g__Unclassified_Desulfovibrionaceae; s__Unclassified_Desulfovibrionaceae
otu38587					● ■	k__Bacteria; p__Verrucomicrobia; c__Verrucomicrobiae; o__Verrucomicrobiales; f__Verrucomicrobiaceae; g__Akkermansia; s__Akkermansia_muciniphila
otu6894					●	k__Bacteria; p__Verrucomicrobia; c__Verrucomicrobiae; o__Verrucomicrobiales; f__Verrucomicrobiaceae; g__Akkermansia; s__Akkermansia_muciniphila
otu7980					●	k__Bacteria; p__Verrucomicrobia; c__Verrucomicrobiae; o__Verrucomicrobiales; f__Verrucomicrobiaceae; g__Akkermansia; s__Akkermansia_muciniphila
otu69130					●	k__Bacteria; p__Verrucomicrobia; c__Verrucomicrobiae; o__Verrucomicrobiales; f__Verrucomicrobiaceae; g__Akkermansia; s__Akkermansia_muciniphila

Columns: ALT, AST, TG, LDL, LW/BW

Spearman rho: −0.5 0 0.5

★ : Con > Et, and p < 0.05   ● : Et > EtAST, and p < 0.05
: Con < Et, and p < 0.05   ■ : Et < EtAST, and p < 0.05

Figure 5. Heat map of 33 OTUs which were significantly associated with the AFLD disease phenotype altered by ethanol or astaxanthin as determined by Spearman's correlation coefficient. These 33 OTUs were selected from the 43 OTUs which had significant changes after ethanol or astaxanthin treatment. The red and blue colors indicate the relative abundance of OTUs that were more or less abundant. The symbols * and #, shown on the right-hand side of the map, represent the OTUs whose abundances were reduced and increased in the Et group relative to those of the Con group, while the red circles and blue rectangles represent the OTUs whose abundances were reduced and increased in the EtAST group relative to the Et group. The symbol * in the cells of the heat map represents the significant correlation between the corresponding metabolic parameters and OTU abundance.

3.6. Predicted Metabolic Functions of the Metagenome in Gut Microbiota

The PICRUSt analysis based on metagenomes was used to predict the metabolic function of gut microbiota that were influenced by astaxanthin in AFLD mice. The results revealed that 18 and 128 KEGG (Kyoto Encyclopedia of Genes and Genomes) pathways were changed in the EtAST group at levels 2 (Figure 6B) and 3 (Figure S6), respectively, among which seven were increased and 11 were decreased compared to the Et group at level 2. In particular, we found several interesting changes among the 128 altered KEGG pathways at level 3. Firstly, the biosynthesis processes of bacteria, such as nucleotide metabolism (level 2), was increased in the Et group compared to that in the Con group, including pyrimidine metabolism, energy metabolism, DNA replication proteins, and cytoskeleton proteins (level 3), while astaxanthin intervention significantly restrained these pathways. In addition, the metagenome of the EtAST group was enriched in the pathways related to lipid metabolism, including glycerophospholipid metabolism; arachidonic acid metabolism; the biosynthesis of unsaturated fatty acids; fatty acid elongation in mitochondria; fatty acid metabolism; the synthesis and degradation of ketone bodies; xenobiotic biodegradation and metabolism, including aminobenzoate degradation, atrazine degradation, caprolactam degradation, drug metabolism, cytochrome P450 and fluorobenzoate degradation; and amino acid metabolism, including tryptophan metabolism, lysine degradation, tyrosine metabolism, and phenylalanine metabolism.

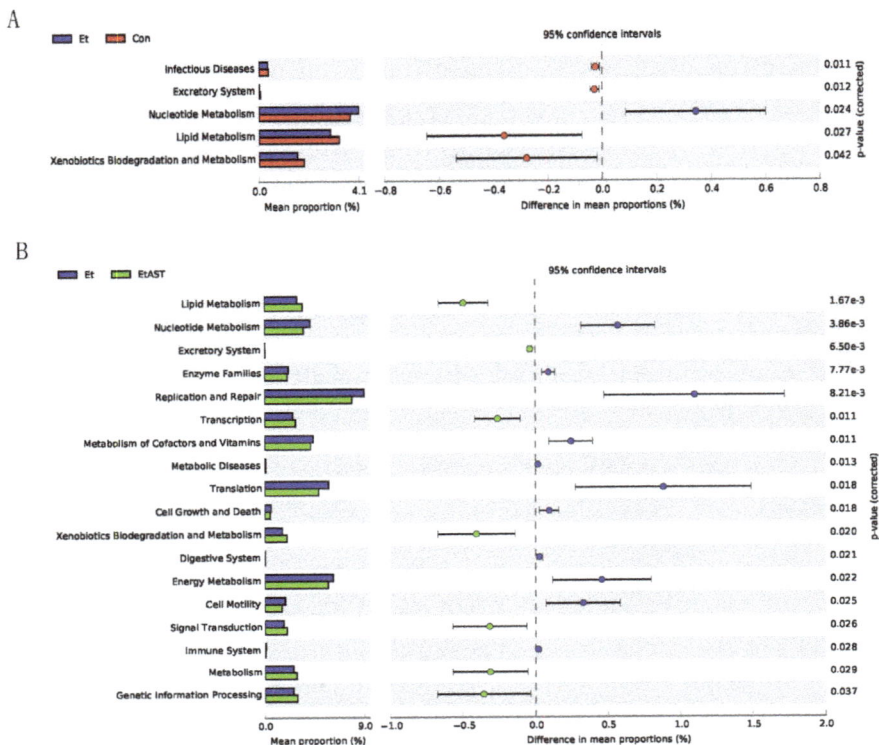

Figure 6. Predicted functions for the altered metagenome of gut microbiota in each group shown with KEGG (Kyoto Encyclopedia of Genes and Genomes) pathways. These data were obtained by PICRUSt. (**A**) A total of five markedly altered KEGG pathways at level 2 in the Et group compared with that in the Con group. (**B**) A total of 18 significantly changed the pathways through astaxanthin supplementation in alcoholic fatty liver disease (AFLD) mice.

4. Discussion

Increasing evidence indicates that the gut microbiota could significantly affect the emergence and development of AFLD [25,26]. Astaxanthin can prevent liver injury by suppressing inflammation, fibrosis [27], and fat accumulation [10,28]. However, previously, there was little evidence suggesting an effect of astaxanthin on AFLD protection. Here, we demonstrated that astaxanthin intervention can mitigate ethanol-induced hepatic steatosis through the reconstruction of the gut microbiota structure and a subsequent increased abundance of *Akkermansia*.

Organ dysfunction is mainly associated with prolonged excessive drinking, which leads to tissue injury-related conditions, such as AFLD, with initial symptoms like fatty liver, and may be responsible for the development of alcoholic hepatitis, liver fibrosis, and cirrhosis. Proper dietary habits can reverse the pathological state of fatty liver in the early stages [29]. Astaxanthin has been proven to protect against liver injury in mice suffering from NAFLD [27] and liver fibrosis in mice [7] due to its anti-inflammation and antioxidant ability. In addition, Zheng et al. has demonstrated that astaxanthin can protect against maternal ethanol-induced embryonic developmental retardation in C57BL/6J mice [22]. In the present study, we found that astaxanthin can protect against ethanol-induced liver injury in mice by alleviating lipid accumulation, inflammatory cell infiltration, and necrosis in the liver. Furthermore, we explored whether astaxanthin could reverse the AFLD development associated with inflammatory responses, and we detected inflammatory gene expression. Various inflammatory mediators and cytokines can be generated after the activation of inflammatory signaling, particularly the major inflammatory markers TNF-α and IL-6 [30]. IL-1α and MIP-2 are important pro-inflammatory cytokines [31,32] that are significantly induced by ethanol. Astaxanthin significantly decreased these mRNA expression level and partly protected the liver from inflammation and injury.

Increasing evidence indicates that the gut microbiota is responsible for the pathogenesis and development of liver disease. A large microbial ecosystem exists in the human gastrointestinal tract, which is associated closely with health and disease control [33]. The gut–liver circulation pathway plays a critical role in alcohol metabolism and is strongly regulated by the gut microbiota. It has been proven that dietary perturbations have dominant effects on the gut microbiota [34]. We further identified bacterial groups that are significantly affected by ethanol and astaxanthin supplementation. Phyla and genera imbalance are associated with health disorders. In this model of alcohol-induced liver injury, we found that several *Bacteroidetes*, including the genera *Bacteroides*, *Butyricimonas*, and *Parabacteroides*, were significantly increased. Consistent with our results, Llopis et al. reported that after five weeks' consumption of a diet including alcohol, *Bacteroides* was significantly more represented in severe alcoholic hepatitis mice relative to the control group [35]. Similarly, other studies have reported that the relative abundance of *Bacteroidetes* increases in alcohol-fed mice [36–38]. The phylum of *Bacteroidetes* is composed of three major classes of Gram-negative bacteria, normally resident in the intestines, mouth, upper respiratory tract, and genital tracts of humans and animals, which have been described as having both beneficial and detrimental features. *Bacteroidetes* can lead to endogenous infections due to a micro-ecological imbalance. In particular, *Bacteroides fragilis* can produce polysaccharide A to relieve colitis in animals [39]. However, it can also produce a toxin, which triggers a pro-carcinogenic effect, to induce colon tumorigenesis [40]. In addition, *Proteobacteria*, as pro-inflammatory intestinal microbes, can multiply in the gut in response to an imbalanced microbial composition and are associated with disease occurrence and development [41]. During this experiment, astaxanthin intervention was found to significantly reverse the ethanol-induced increases in *Bacteroidetes* and *Proteobacteria*, restoring their proportions to the levels of the ND group. This indicates that the protective effect of astaxanthin is likely associated with its anti-inflammatory activity.

Cyanobacteria was completely suppressed in mice after treatment with ethanol as compared to the ethanol-free treatment. Although the abundance of *Cyanobacteria* is extremely low in the microbiota, we focused on this phylum because *Cyanobacteria* is completely inhibited in the gut when the diet contains ethanol, either with or without astaxanthin. This phenomenon indicated that astaxanthin did not affect the action of alcohol.

Accumulating reports have focused on the beneficial effects of *Akkermansia* on host metabolism. *Akkermansia* is a dominant genus in *Verrucomicrobia* and can degrade intestinal mucin [42], increasing mucus thickness and enhancing gut barrier function, which correlates inversely with the incidence of inflammation [43] and metabolic syndrome [18]. Lack of *Akkermansia* has been determined to be an early marker of alcohol-induced gut dysbiosis [37]. Moreover, ethanol exposure reduces the abundance of *Akkermansia* in both mice and humans, and the status of AFLD can be improved by oral supplementation of the genus, directly demonstrating the protective effect of this bacterium in AFLD [44]. As a probiotic, the abundance of *Akkermansia* can be affected by the administration of specific dietary components [45]. In the current research, the abundance of *Akkermansia* in ethanol-exposed AFLD mice dramatically reduced, and astaxanthin intervention obviously recovered its abundance, to even higher than that of the ND group. Our data suggest that mice with AFLD might benefit from astaxanthin supplementation increasing *Akkermansia*. Studies have proved that *Akkermansia* can play a protective role in ethanol-induced liver injury as a result of its function of improving the gut barrier [46], and the present research requires the further investigation of astaxanthin barrier function in the future.

The different bacterial species in the same genus may reflect different responses by the same treatment. Thus, it is indispensable to identify changes in the microbiota at the species level. In the present study, among the 13 OTUs altered by ethanol and reversed by astaxanthin intervention, the proportions of *Akkermansia muciniphila* (OTU36578) were significantly decreased in the Et group, while they were enriched in the EtAST and ND groups. *Akkermansia muciniphila* is a species of the *Akkermansia* genus, and extensive research has shown that it can improve the status of obesity, diabetes, and inflammation [47]. On the other hand, the species from *Butyricimonas* (OTU 61092, OTU22852, and OTU43831), S24-7 (OTU52622, OTU52789, OTU67103, OTU4618, and OTU11187), *Oscillospira* (OTU9917), *Clostridiales* (OTU60609 and OTU71112), and *Bilophila* (OTU2261) were significantly increased by ethanol diet and markedly reversed by astaxanthin supplementation. It has been reported that the abundance of *Oscillospira* and *Clostridales* increased in inflammatory response and is associated with the barrier injury of intestinal mucosa [48,49]. Stanislawski et al. demonstrated that *Bilophila* was positively correlated with the fat fraction of the liver [50]. Our research found that astaxanthin treatment significantly decreased the levels of ALT, AST, TG, LDL, and the index of LW/BW, which were increased by ethanol diet. Among the 13 OTUs altered by ethanol and reversed by astaxanthin supplementation, Spearman's correlation analysis suggested that *Akkermansia muciniphila*, enriched in the EtAST group, was negatively associated with the AFLD phenotype. On the other hand, the 12 enriched OTUs in the Et group were positively associated with the AFLD phenotype. These results indicate that astaxanthin relieves the AFLD phenotype through its ability to enrich the bacteria responsible for intestinal integrity and anti-inflammation.

The typical characteristic of AFLD is excessive hepatic lipid accumulation. Therefore, lipid metabolic regulation plays an important role in the pathogenesis of metabolic disorders. In this study, the predicted metabolic function of the gut microbiota showed that the lipid metabolism pathway was increased after astaxanthin treatment, while further research proved that *Akkermansia* acts as a probiotic to promote lipid metabolism and avoid lipid excessive accumulation [51]. The results indicate that the inhibition by astaxanthin of excessive lipid accumulation in the liver may be associated with gut bacteria that promote lipid metabolism. In addition, astaxanthin increased the xenobiotic biodegradation and metabolism pathways. Xenobiotics are a kind of foreign chemical in living systems, which, after entering an organism, may induce adverse or even very serious consequences [52]. The liver is the major organ of metabolism, and astaxanthin supplementation can increase the abundance of bacteria that promote hepatic xenobiotic biodegradation and metabolism, resulting in detoxification and liver injury protection. However, the current findings need to be verified in larger samples.

Interestingly, astaxanthin was more effective in ameliorating alcohol-induced liver injury rather than that related to HFD. There is evidence that astaxanthin can protect against liver damage conditions like NAFLD and AFLD through multiple mechanisms, including antioxidant and anti-inflammatory effects [10,22,53]. In this research, mice in the Con group fed a high-fat diet did not show obvious inflammation, as measured by the liver index, histopathology, and liver gene expression. However, ethanol interference induced a marked inflammatory response which promoted astaxanthin to exert its anti-inflammatory effects to protect the liver from injury. This can also be explained from another aspect; specifically, astaxanthin will not disturb the metabolic balance under non-inflammatory conditions [54], which is consistent with the result that most responses in the AST group were similar to those in the Con group in our research. Future research should determine the content of astaxanthin in the mouse (liver and serum) and its feces.

5. Conclusions

Our research explored the protective effects of astaxanthin on AFLD injury mice and its regulation of gut microbiota. Astaxanthin has the ability to reverse the ethanol-induced liver weight ratio increase, hepatic inflammation, and lipid dysmetabolism. Furthermore, the overall structure and composition of the gut microbiota altered by ethanol feeding can be balanced by astaxanthin, particularly recovering the beneficial bacterium *Akkermansia*. Our results indicate that *Akkermansia* may be a potential target for the astaxanthin-related alleviation of AFLD, which provides further evidence for its molecular mechanism, and suggests that it could be used for the treatment of bacterial disorders induced by AFLD.

Supplementary Materials: The following are available online at http://www.mdpi.com/2072-6643/10/9/1298/s1, Table S1: Compositions of the diets (DOC); Table S2: Primer sequences used for q-PCR reactions (DOC); Table S3: The OTUs obtained from high-throughput 16S rRNA gene sequencing (XLS); Figure S1: Rationality of sequencing data was evaluated by rarefaction and rank abundance curve (DOC 359); Figure S2: Cluster analysis using the UPGMA method based on unweighted Unifrac and weighted Unifrac analysis (DOC); Figure S3: The overall classification levels based on GraphlAn and LEfSe analyses were used to visualize the taxonomy compositions and bacterial abundance; Figure S4: Heatmap of the abundance of bacteria at the genus level at different groups (DOC); Figure S5: Astaxanthin protect ethanol-induced *Akkermansia muciniphila* depletion (DOC); Figure S6: Predicted functions for the altered metagenome of gut microbiota (PDF).

Author Contributions: H.L., X.F., X.Z. and J.L. conceived and designed the experiments; H.L., Z.Z. and L.Z. performed the experiments; H.L. and M.L. analyzed the data; X.F., X.Z. and J.L. contributed reagents/materials/analysis tools; H.L. and M.L. wrote the paper. All authors reviewed the manuscript.

Funding: This research received no external funding.

Acknowledgments: The sequencing service was provided by Personal Biotechnology Co., Ltd. of Shanghai, China. The program was supported by the National Natural Science Foundation of China–General program (grant number: 31672511).

Conflicts of Interest: The authors declare no conflict of interest.

Abbreviations

AST	astaxanthin
AFLD	alcoholic fatty liver disease
IL-1α	interleukin-1 alpha
MIP-2	macrophage inflammatory protein 2
ND	normal diet
Et	ethanol
ALT	alanine aminotransferase
TG	triacylglycerol
LDL	low-density lipoprotein
LW	liver weight
BW	body weight
Con	control
PCoA	principle coordinate analysis
OTUs	operational taxonomic units
KEGG	Kyoto Encyclopedia of Genes and Genomes

References

1. O'Shea, R.S.; Dasarathy, S.; McCullough, A.J. Alcoholic liver disease. *Hepatology* **2010**, *51*, 307–328. [CrossRef] [PubMed]
2. Menon, K.V.; Gores, G.J.; Shah, V.H. Pathogenesis, diagnosis, and treatment of alcoholic liver disease. *Mayo Clin. Proc.* **2001**, *76*, 1021–1029. [CrossRef] [PubMed]
3. MacSween, R.N.; Burt, A.D. Histologic spectrum of alcoholic liver disease. *Semin. Liver Dis.* **1986**, *6*, 221–232. [CrossRef] [PubMed]
4. Hussein, G.; Sankawa, U.; Goto, H.; Matsumoto, K.; Watanabe, H. Astaxanthin, a carotenoid with potential in human health and nutrition. *J. Nat. Prod.* **2006**, *69*, 443–449. [CrossRef] [PubMed]
5. Yuan, J.P.; Peng, J.; Yin, K.; Wang, J.H. Potential health-promoting effects of astaxanthin: A high-value carotenoid mostly from microalgae. *Mol. Nutr. Food Res.* **2011**, *55*, 150–165. [CrossRef] [PubMed]
6. Sila, A.; Kamoun, Z.; Ghlissi, Z.; Makni, M.; Nasri, M.; Sahnoun, Z.; Nedjar-Arroume, N.; Bougatef, A. Ability of natural astaxanthin from shrimp by-products to attenuate liver oxidative stress in diabetic rats. *Pharmacol. Rep.* **2015**, *67*, 310–316. [CrossRef] [PubMed]

7. Shen, M.; Chen, K.; Lu, J.; Cheng, P.; Xu, L.; Dai, W.; Wang, F.; He, L.; Zhang, Y.; Chengfen, W.; et al. Protective effect of astaxanthin on liver fibrosis through modulation of TGF-beta1 expression and autophagy. *Mediat. Inflamm.* **2014**, 954502. [CrossRef]

8. Kobori, M.; Takahashi, Y.; Sakurai, M.; Ni, Y.; Chen, G.; Nagashimada, M.; Kaneko, S.; Ota, T. Hepatic Transcriptome Profiles of Mice with Diet-Induced Nonalcoholic Steatohepatitis Treated with Astaxanthin and Vitamin E. *Int. J. Mol. Sci.* **2017**, *18*, 593. [CrossRef] [PubMed]

9. Takemoto, M.; Yamaga, M.; Furuichi, Y.; Yokote, K. Astaxanthin Improves Nonalcoholic Fatty Liver Disease in Werner Syndrome with Diabetes Mellitus. *J. Am. Geriatr. Soc.* **2015**, *63*, 1271–1273. [CrossRef] [PubMed]

10. Jia, Y.; Wu, C.; Kim, J.; Kim, B.; Lee, S.J. Astaxanthin reduces hepatic lipid accumulations in high-fat-fed C57BL/6J mice via activation of peroxisome proliferator-activated receptor (PPAR) alpha and inhibition of PPAR gamma and Akt. *J. Nutr. Biochem.* **2016**, *28*, 9–18. [CrossRef] [PubMed]

11. Ni, Y.; Nagashimada, M.; Zhuge, F.; Zhan, L.; Nagata, N.; Tsutsui, A.; Nakanuma, Y.; Kaneko, S.; Ota, T. Astaxanthin prevents and reverses diet-induced insulin resistance and steatohepatitis in mice: A comparison with vitamin E. *Sci. Rep.* **2015**, *5*, 17192. [CrossRef] [PubMed]

12. Battson, M.L.; Lee, D.M.; Weir, T.L.; Gentile, C.L. The gut microbiota as a novel regulator of cardiovascular function and disease. *J. Nutr. Biochem.* **2015**, *56*, 1–15. [CrossRef] [PubMed]

13. Koh, J.C.; Loo, W.M.; Goh, K.L.; Sugano, K.; Chan, W.K.; Chiu, W.Y.; Choi, M.G.; Gonlachanvit, S.; Lee, W.J.; Lee, W.J.; et al. Asian consensus on the relationship between obesity and gastrointestinal and liver diseases. *J. Gastroenterol. Hepatol.* **2016**, *31*, 1405–1413. [CrossRef] [PubMed]

14. Dao, M.C.; Clement, K. Gut microbiota and obesity: Concepts relevant to clinical care. *Eur. J. Intern. Med.* **2018**, *48*, 18–24. [CrossRef] [PubMed]

15. Chen, P.; Torralba, M.; Tan, J.; Embree, M.; Zengler, K.; Starkel, P.; van Pijkeren, J.P.; DePew, J.; Loomba, R.; Ho, S.B.; et al. FoutsandB. Schnabl. Supplementation of saturated long-chain fatty acids maintains intestinal eubiosis and reduces ethanol-induced liver injury in mice. *Gastroenterology* **2015**, *148*, 203–214. [CrossRef] [PubMed]

16. Leclercq, S.; De Saeger, C.; Delzenne, N.; de Timary, P.; Starkel, P. Role of inflammatory pathways, blood mononuclear cells, and gut-derived bacterial products in alcohol dependence. *Biol. Psychiatry* **2014**, *76*, 725–733. [CrossRef] [PubMed]

17. Lippai, D.; Bala, S.; Catalano, D.; Kodys, K.; Szabo, G. Micro-RNA-155 deficiency prevents alcohol-induced serum endotoxin increase and small bowel inflammation in mice. *Alcohol. Clin. Exp. Res.* **2014**, *38*, 2217–2224. [CrossRef] [PubMed]

18. Dao, M.C.; Everard, A.; Aron-Wisnewsky, J.; Sokolovska, N.; Prifti, E.; Verger, E.O.; Kayser, B.D.; Levenez, F.; Chilloux, J.; Hoyles, L. *Akkermansia muciniphila* and improved metabolic health during a dietary intervention in obesity: Relationship with gut microbiome richness and ecology. *Gut* **2016**, *65*, 426–436. [CrossRef] [PubMed]

19. Requena, T.; Martinez-Cuesta, M.C.; Pelaez, C. Diet and microbiota linked in health and disease. *Food Funct.* **2018**, *9*, 688–704. [CrossRef] [PubMed]

20. Chen, J.T.; Kotani, K. Astaxanthin as a Potential Protector of Liver Function: A. Review. *J. Clin. Med. Res.* **2016**, *8*, 701–704. [CrossRef] [PubMed]

21. Lieber, C.S.; DeCarli, L.M.; Sorrell, M.F. Experimental methods of ethanol administration. *Hepatology* **1989**, *10*, 501–510. [CrossRef] [PubMed]

22. Zheng, D.; Li, Y.; He, L.; Tang, Y.; Li, X.; Shen, Q.; Yin, D.; Peng, Y. The protective effect of astaxanthin on fetal alcohol spectrum disorder in mice. *Neuropharmacology* **2014**, *84*, 13–18. [CrossRef] [PubMed]

23. Caporaso, J.G.; Kuczynski, J.; Stombaugh, J.; Bittinger, K.; Bushman, F.D.; Costello, E.K.; Fierer, N.; Pena, A.G.; Goodrich, J.K.; Gordon, J.I.; et al. QIIME allows analysis of high-throughput community sequencing data. *Nat. Methods* **2010**, *7*, 335–336. [CrossRef] [PubMed]

24. Edgar, R.C. Search and clustering orders of magnitude faster than BLAST. *Bioinformatics* **2010**, *26*, 2460–2461. [CrossRef] [PubMed]

25. Neuman, M.G.; French, S.W.; Zakhari, S.; Malnick, S.; Seitz, H.K.; Cohen, L.B.; Salaspuro, M.; Voinea-Griffin, A.; Barasch, A.; Kirpich, I.A.; et al. Alcohol, microbiome, life style influence alcohol and non-alcoholic organ damage. *Exp. Mol. Pathol.* **2017**, *102*, 162–180. [CrossRef] [PubMed]

26. Scarpellini, E.; Forlino, M.; Lupo, M.; Rasetti, C.; Fava, G.; Abenavoli, L.; De Santis, A. Gut Microbiota and Alcoholic Liver Disease. *Rev. Recent Clin. Trials* **2016**, *11*, 213–219. [CrossRef] [PubMed]

27. Kim, B.; Farruggia, C.; Ku, C.S.; Pham, T.X.; Yang, Y.; Bae, M.; Wegner, C.J.; Farrell, N.J.; Harness, E.; Park, Y.K.; et al. Astaxanthin inhibits inflammation and fibrosis in the liver and adipose tissue of mouse models of diet-induced obesity and nonalcoholic steatohepatitis. *J. Nutr. Biochem.* **2017**, *43*, 27–35. [CrossRef] [PubMed]

28. Bhuvaneswari, S.; Arunkumar, E.; Viswanathan, P.; Anuradha, C.V. Astaxanthin restricts weight gain, promotes insulin sensitivity and curtails fatty liver disease in mice fed a obesity-promoting diet. *Process Biochem.* **2010**, *45*, 1406–1414. [CrossRef]

29. Bae, M.; Park, Y.K.; Lee, J.Y. Food components with antifibrotic activity and implications in prevention of liver disease. *J. Nutr. Biochem.* **2017**, *55*, 1–11. [CrossRef] [PubMed]

30. Fard, M.T.; Arulselvan, P.; Karthivashan, G.; Adam, S.K.; Fakurazi, S. Bioactive extract from Moringa oleifera inhibits the pro-inflammatory mediators in lipopolysaccharide stimulated macrophages. *Pharmacogn. Mag.* **2015**, *11*, 556–563.

31. Malik, A.; Kanneganti, T.D. Function and regulation of IL-1alpha in inflammatory diseases and cancer. *Immunol. Rev.* **2018**, *281*, 124–137. [CrossRef] [PubMed]

32. Qin, C.C.; Liu, Y.N.; Hu, Y.; Yang, Y.; Chen, Z. Macrophage inflammatory protein-2 as mediator of inflammation in acute liver injury. *World J. Gastroenterol.* **2017**, *23*, 3043–3052. [CrossRef] [PubMed]

33. Lynch, S.V.; Pedersen, O. The Human Intestinal Microbiome in Health and Disease. *N. Engl. J. Med.* **2016**, *375*, 2369–2379. [CrossRef] [PubMed]

34. Carmody, R.N.; Gerber, G.K.; Jr Luevano, J.M.; Gatti, D.M.; Somes, L.; Svenson, K.L.; Turnbaugh, P.J. Diet dominates host genotype in shaping the murine gut microbiota. *Cell Host Microbe* **2015**, *17*, 72–84. [CrossRef] [PubMed]

35. Llopis, M.; Cassard, A.M.; Wrzosek, L.; Boschat, L.; Bruneau, A.; Ferrere, G.; Puchois, V.; Martin, J.C.; Lepage, P.; Le Roy, T.; et al. Intestinal microbiota contributes to individual susceptibility to alcoholic liver disease. *Gut* **2016**, *65*, 830–839. [CrossRef] [PubMed]

36. Yan, A.W.; Fouts, D.E.; Brandl, J.; Starkel, P.; Torralba, M.; Schott, E.; Tsukamoto, H.; Nelson, K.E.; Brenner, D.A.; Schnabl, B. Enteric dysbiosis associated with a mouse model of alcoholic liver disease. *Hepatology* **2011**, *53*, 96–105. [CrossRef] [PubMed]

37. Lowe, P.P.; Gyongyosi, B.; Satishchandran, A.; Iracheta-Vellve, A.; Ambade, A.; Kodys, K.; Catalano, D.; Ward, D.V.; Szabo, G. Alcohol-related changes in the intestinal microbiome influence neutrophil infiltration, inflammation and steatosis in early alcoholic hepatitis in mice. *PLoS ONE* **2017**, *12*, e0174544.

38. Neyrinck, A.M.; Etxeberria, U.; Taminiau, B.; Daube, G.; Van Hul, M.; Everard, A.; Cani, P.D.; Bindels, L.B.; Delzenne, N.M. Rhubarb extract prevents hepatic inflammation induced by acute alcohol intake, an effect related to the modulation of the gut microbiota. *Mol. Nutr. Food Res.* **2017**, *61*, 1–12. [CrossRef] [PubMed]

39. Surana, N.K.; Kasper, D.L. The yin yang of bacterial polysaccharides: Lessons learned from B. fragilis PSA. *Immunol. Rev.* **2012**, *245*, 13–26. [CrossRef] [PubMed]

40. Chung, L.; Thiele, O.E.; Geis, A.L.; Chan, J.L.; Fu, K.; DeStefano, S.C.E.; Dejea, C.M.; Fathi, P.; Chen, J.; Finard, B.B.; et al. Bacteroides fragilis Toxin Coordinates a Pro-carcinogenic Inflammatory Cascade via Targeting of Colonic Epithelial Cells. *Cell Host Microbe* **2018**, *23*, 203–214.e5. [CrossRef] [PubMed]

41. Shin, N.R.; Whon, T.W.; Bae, J.W. Proteobacteria: Microbial signature of dysbiosis in gut microbiota. *Trends Biotechnol.* **2015**, *33*, 496–503. [CrossRef] [PubMed]

42. Derrien, M.; Vaughan, E.E.; Plugge, C.M.; de Vos, W.M. *Akkermansia muciniphila* gen. nov., sp. nov., a human intestinal mucin-degrading bacterium. *Int. J. Syst. Evol. Microbiol.* **2004**, *54*, 1469–1476. [CrossRef] [PubMed]

43. Li, J.; Lin, S.; Vanhoutte, P.M.; Woo, C.W.; Xu, A. Akkermansia Muciniphila Protects Against Atherosclerosis by Preventing Metabolic Endotoxemia-Induced Inflammation in Apoe$^{-/-}$ Mice. *Circulation* **2016**, *133*, 2434–2446. [CrossRef] [PubMed]

44. Grander, C.; Adolph, T.E.; Wieser, V.; Lowe, P.; Wrzosek, L.; Gyongyosi, B.; Ward, D.V.; Grabherr, F.; Gerner, R.R.; Pfister, A.; et al. Recovery of ethanol-induced *Akkermansia muciniphila* depletion ameliorates alcoholic liver disease. *Gut* **2018**, *67*, 891–901. [CrossRef] [PubMed]

45. Derrien, M.; Belzer, C.; de Vos, W.M. *Akkermansia muciniphila* and its role in regulating host functions. *Microb. Pathog.* **2017**, *106*, 171–181. [CrossRef] [PubMed]

46. Wu, W.; Lv, L.; Shi, D.; Ye, J.; Fang, D.; Guo, F.; Li, Y.; He, X.; Li, L. Protective Effect of *Akkermansia muciniphila* against Immune-Mediated Liver Injury in a Mouse Model. *Front. Microbiol.* **2017**, *8*, 1804. [CrossRef] [PubMed]

47. Ottman, N.; Geerlings, S.Y.; Aalvink, S.; de Vos, W.M.; Belzer, C. Action and function of *Akkermansia muciniphila* in microbiome ecology, health and disease. *Best Pract. Res. Clin. Gastroenterol.* **2017**, *31*, 637–642. [CrossRef] [PubMed]

48. Power, K.A.; Lu, J.T.; Monk, J.M.; Lepp, D.; Wu, W.; Zhang, C.; Liu, R.; Tsao, R.; Robinson, L.E.; Wood, G.A.; et al. Purified rutin and rutin-rich asparagus attenuates disease severity and tissue damage following dextran sodium sulfate-induced colitis. *Mol. Nutr. Food Res.* **2016**, *60*, 2396–2412. [CrossRef] [PubMed]

49. Mastrocola, R.; Ferrocino, I.; Liberto, E.; Chiazza, F.; Cento, A.S.; Collotta, D.; Querio, G.; Nigro, D.; Bitonto, V.; Cutrin, J.C.; et al. Fructose liquid and solid formulations differently affect gut integrity, microbiota composition and related liver toxicity: A comparative in vivo study. *J. Nutr. Biochem.* **2018**, *55*, 185–199. [CrossRef] [PubMed]

50. Stanislawski, M.A.; Lozupone, C.A.; Wagner, B.D.; Eggesbo, M.; Sontag, M.K.; Nusbacher, N.M.; Martinez, M.; Dabelea, D. Gut microbiota in adolescents and the association with fatty liver: The EPOCH study. *Pediatr. Res.* **2018**. [CrossRef] [PubMed]

51. Everard, A.; Belzer, C.; Geurts, L.; Ouwerkerk, J.P.; Druart, C.; Bindels, L.B.; Cani, P.D. Cross-talk between Akkermansia muciniphila and intestinal epithelium controls diet-induced obesity. *Proc. Natl. Acad. Sci. USA* **2013**, *110*, 9066–9071. [CrossRef] [PubMed]

52. Macpherson, A.J.; Heikenwalder, M.; Ganal-Vonarburg, S.C. The Liver at the Nexus of Host-Microbial Interactions. *Cell Host Microbe* **2016**, *20*, 561–571. [CrossRef] [PubMed]

53. Kaulmann, A.; Bohn, T. Carotenoids, inflammation, and oxidative stress—Implications of cellular signaling pathways and relation to chronic disease prevention. *Nutr. Res.* **2014**, *34*, 907–929. [CrossRef] [PubMed]

54. Speranza, L.; Pesce, M.; Patruno, A.; Franceschelli, S.; de Lutiis, M.A.; Grilli, A.; Felaco, M. Astaxanthin Treatment Reduced Oxidative Induced Pro-Inflammatory Cytokines Secretion in U937: SHP-1 as a Novel Biological Target. *Mar. Drugs* **2012**, *10*, 890–899. [CrossRef] [PubMed]

nutrients

MDPI

Article

A Lipophilic Fucoxanthin-Rich *Phaeodactylum tricornutum* Extract Ameliorates Effects of Diet-Induced Obesity in C57BL/6J Mice

Andrea Gille [1], Bojan Stojnic [2], Felix Derwenskus [3,4], Andreas Trautmann [5],
Ulrike Schmid-Staiger [4], Clemens Posten [5], Karlis Briviba [1], Andreu Palou [2,6,7],
M. Luisa Bonet [2,6,7,*] and Joan Ribot [2,6,7]

[1] Max Rubner-Institut, Federal Research Institute of Nutrition and Food, Department of Physiology and Biochemistry of Nutrition, 76131 Karlsruhe, Germany; andrea.gille@mri.bund.de (A.G.); karlis.briviba@mri.bund.de (K.B.)
[2] Laboratory of Molecular Biology, Nutrition and Biotechnology, Universitat de les Illes Balears, 07122 Palma de Mallorca, Spain; bojan.stojnic@uib.es (B.S.); andreu.palou@uib.es (A.P.); joan.ribot@uib.es (J.R.)
[3] Institute of Interfacial Process Engineering and Plasma Technology IGVP, University of Stuttgart, 70569 Stuttgart, Germany; felix.derwenskus@igb.fraunhofer.de
[4] Fraunhofer Institute for Interfacial Engineering and Biotechnology IGB, 70569 Stuttgart, Germany; ulrike.schmid-staiger@igb.fraunhofer.de
[5] Karlsruhe Institute of Technology (KIT), Institute of Process Engineering in Life Sciences III Bioprocess Engineering, 76131 Karlsruhe, Germany; andi.t@gmx.de (A.T.); clemens.posten@kit.edu (C.P.)
[6] CIBER de Fisiopatología de la Obesidad y Nutrición (CIBEROBN), 07122 Palma de Mallorca, Spain
[7] Institut d'Investigació Sanitària Illes Balears (IdISBa), 07120 Palma de Mallorca, Spain
[*] Correspondence: luisabonet@uib.es; Tel.: +34-971172734; Fax: +34-971173426

Received: 6 March 2019; Accepted: 4 April 2019; Published: 6 April 2019

Abstract: *Phaeodactylum tricornutum* (*P. tricornutum*) comprise several lipophilic constituents with proposed anti-obesity and anti-diabetic properties. We investigated the effect of an ethanolic *P. tricornutum* extract (PTE) on energy metabolism in obesity-prone mice fed a high fat diet (HFD). Six- to eight-week-old male C57BL/6J mice were switched to HFD and, at the same time, received orally placebo or PTE (100 mg or 300 mg/kg body weight/day). Body weight, body composition, and food intake were monitored. After 26 days, blood and tissue samples were collected for biochemical, morphological, and gene expression analyses. PTE-supplemented mice accumulated fucoxanthin metabolites in adipose tissues and attained lower body weight gain, body fat content, weight of white adipose tissue (WAT) depots, and inguinal WAT adipocyte size than controls, independent of decreased food intake. PTE supplementation was associated with lower expression of *Mest* (a marker of fat tissue expandability) in WAT depots, lower gene expression related to lipid uptake and turnover in visceral WAT, increased expression of genes key to fatty acid oxidation and thermogenesis (*Cpt1*, *Ucp1*) in subcutaneous WAT, and signs of thermogenic activation including enhanced UCP1 protein in interscapular brown adipose tissue. In conclusion, these data show the potential of PTE to ameliorate HFD-induced obesity in vivo.

Keywords: *Phaeodactylum tricornutum*; microalgae; fucoxanthin; eicosapentanoic acid; obesity; browning; brown adipose tissue

1. Introduction

Microalgae constitute a sustainable source of a multitude of nutrients with interesting properties such as proteins, ω-3 fatty acids, carotenoids, vitamins, and minerals [1], leading to an increasing market of microalgae-containing nutraceuticals and food products with important clinical and economic implications [2]. The marine diatom microalga *Phaeodactylum tricornutum* has potential for use in animal feed and human nutrition especially because it contains polyunsaturated fatty acids (PUFAs) and phytochemicals (e.g., polyphenols and carotenoids) [3,4]. *P. tricornutum* is particularly enriched in the ω-3 PUFA eicosapentanoic acid (EPA) [5,6] and the carotenoid fucoxanthin [6,7], which likely mediate the physiological and nutritional value of this microalga. Beneficial health effects such as anti-inflammatory [8–12], anti-obesity, and anti-diabetic effects [13–18] have been reported in cell and in vivo studies for these two compounds, mostly derived from fish oil (EPA) and edible macroalgae (fucoxanthin). The anti-obesity effects of ω-3 long-chain PUFA comprise decreased lipogenesis and the enhancement of fatty acid oxidation in liver and adipose tissues [13,19]. Fucoxanthin anti-obesity activity has been attributed to the stimulation of thermogenesis by increasing the expression of uncoupling protein 1 (UCP1) in adipose tissues [16,17,20] as well as to effects on intestinal lipid absorption and lipid metabolism [17,20–23]. UCP1 is a mitochondrial inner membrane protein, typically expressed in brown adipose tissue (BAT) and inducible in white adipose tissue (WAT) through a process known as WAT browning or beigeing [24], whose activity allows the dissipation of substrate-derived energy as heat.

Despite its interesting composition, few studies to date have addressed the anti-obesity properties of *P. tricornutum* in vivo [25,26]. In these studies, supplementation of the diet with *P. tricornutum* lipid extract [25] or *P. tricornutum* powder [26] ameliorated body weight and body fat gain of mice on a high fat diet (HFD) independently of decreases in food intake. In the study of Kang et al., supplementation was also shown to ameliorate HFD-induced metabolic derangements, such as hyperglycemia, hyperlipidemia, and insulin resistance, and to exert antioxidant effects in the liver [25]. In the study by Kim et al., evidence was provided that *P. tricornutum* powder may activate the AMP-activated protein kinase (AMPK) pathway in the liver [26]. However, these previous reports did not address changes in cellular and metabolic features of adipose tissues as potential contributors to the anti-obesity activity of *P. tricornutum* supplementation.

We here aimed to investigate the ability of a lipophilic ethanol extract of *P. tricornutum* (PTE) to oppose the development of obesity in obesity-prone (C57BL/6J) mice fed an obesogenic HFD, with focus on effects in adipose tissues. Therefore, body weight gain, adipose depots weight, adipocyte size distribution, and expression in adipose tissues of selected genes related to lipid and energy metabolism were analyzed, together with parameters related to glucose control.

2. Materials and Methods

2.1. Materials

Chemicals were purchased from Merck (Darmstadt, Germany), Sigma-Aldrich (Taufkirchen, Germany), and VWR (Bruchsal, Germany) or from Carl Roth (Karlsruhe, Germany), unless otherwise noted.

2.2. Microalgae Cultivation, Processing, and Preparation of Ethanolic Extract

The *P. tricornutum* strain UTEX 640 (SAG 1090-1b) was obtained from the culture collection of Algae (SAG) from the University of Goettingen (Germany) and was cultivated under controlled and axenic conditions, as described previously [27]. The biomass was harvested by centrifugation, the supernatant was discarded, and the remaining pellets were stored at −20 °C until cell disruption. The biomass of several cultivations was combined and lyophilized, and it was protected from light in a Christ Alpha 1–2 LD freeze drier (Osterode a. Harz, Germany). This was followed by cell disruption using the tissue homogenizer Precellys 24 from Bertin Technologies (Frankfurt/Main, Germany). The resulting *P. tricornutum* powder was applied to pressurized liquid extraction (ASE 350, Thermo-Fisher Scientific, Waltham, MA, USA) in accordance with the method described earlier by Derwenskus et al. using ethanol as extraction solvent [6]. The obtained extract was aliquoted, the ethanol was evaporated under a stream of nitrogen, and the extract was stored at −80 °C until used for animal experiments. In order to apply PTE to the mice, the dried extract was resolved in olive oil:water (2:1, *v:v*) to achieve a concentration of 0.1 mg/µL or 0.3 mg/µL and homogenized in an ultrasonic bath for 5 min.

2.3. Animal Experiment

The study was approved by the Bioethical Committee of the University of the Balearic Islands (UIB, Ref. CEEA 43/07/15). International standards for the use and care of laboratory animals were followed. C75Bl/6J mice were originally obtained from Charles River Laboratories (Barcelona, Spain) and expanded at the UIB animal house. The animals were housed in standard cages (without running wheel) at 22 °C with a 12-h light/dark cycle and ad libitum access to food (chow) and water. One week prior to the start of the experiment, six- to eight-week-old male C75BL/6J mice were divided into three groups with six animals per group (three animals per cage) and switched from chow to a defined low fat diet (3.8 kcal/g, 10% energy as fat, Research Diets D12450J, New Brunswick, NJ, USA). The diet was then changed to a defined HFD containing 4.7 kcal/g and 45% energy as fat (Research Diets D12451). At the same time as the HFD, the animals received daily, orally with the aid of a pipette, the vehicle (olive oil:water, 2:1, *v:v*) (placebo group) or PTE at a dose of 100 mg/kg body weight (bw)/day (PTE100) or 300 mg/kg bw/day (PTE300). Body weight and food intake were regularly monitored. Food intake was estimated on a per-cage basis, from the actual amount of food consumed by the animals and its caloric equivalence. Body composition was analyzed using an Echo MRI body composition analyzer (EchoMRI, LLC, Houston, TX, USA). At day 22, animals were starved for 6 h (from 06:00 a.m. to 12:00 p.m.) after which tail blood was collected for the measurement of circulating parameters. After 26 days, the animals were euthanized. Blood, liver, and adipose tissues (including interscapular BAT and epididymal, inguinal, and retroperitoneal WAT) were dissected and stored at −80 °C until used for analysis. Samples of the liver, BAT, and inguinal WAT were fixed for histology.

2.4. Circulating Parameters

Blood glucose was determined using an Accu-Chek Aviva system (Roche Diagnostics, Risch, Switzerland). Commercial kits for measurement of serum insulin (Mercodia, Uppsala, Sweden), non-esterified fatty acids (NEFA; Wako Chemicals GmbH, Neuss, Germany) and triacylglycerides (TAG; Sigma-Aldrich, St. Louis, MO, USA) were applied following the manufacturer's protocols. The homeostatic model assessment for insulin resistance (HOMA-IR) and the revised quantitative insulin sensitivity check index (R-QUICKI) were calculated as described earlier [28].

2.5. Total Liver Fat Content

The total fat content in the liver was determined by Folch extraction with minor modifications [29]. In brief, 50–80 mg of fresh liver was weighted in a sample tube, followed by adding 500 µL of PBS and homogenization with a sonication probe for 10 s. To each sample, 500 µL methanol were added, mixed thoroughly for 2 min, followed by addition of 1 mL chloroform and mixing for 2 min. The mixture was centrifuged for 3 min at 4000× *g*, and the lower chloroform phase was transferred to a new tube using a glass pipette. The extraction procedure was repeated three times, and the solvent was evaporated under a stream of nitrogen. The tube with the dried residue was weighed, and the total fat content was calculated by subtracting the weight of the empty tube.

2.6. Carotenoid and Fatty Acid Analyses

Fatty acids in PTE were analyzed, as described earlier [6], using gas chromatography and a flame ionization detector. Carotenoids in PTE and in liver, BAT, and epididymal and inguinal WAT samples of animals were analyzed by HPLC coupled to a photodiode array detector, as described previously [6,30].

2.7. Total RNA Isolation and Quantitative Real Time PCR (qPCR) Analysis

Total RNA was extracted from tissues or cells using TRI Reagent (Sigma-Aldrich, St. Louis, MO, USA) following the manufacturer's instructions. Isolated RNA was quantified using NanoDrop ND-1000 spectrophotometer (NanoDrop Technologies Inc., Wilmington, DE, USA) and its integrity confirmed by agarose gel electrophoresis. A 0.25 µg sample of total RNA was reverse-transcribed using reagents from Life Technologies (Carlsbad, CA, USA). The resulting cDNA was subjected to qPCR analysis on a StepOnePlus instrument (Life technologies). *Hprt-1* transcript was used as a reference housekeeping gene. The sequences of the employed primers for qPCR are available on request.

2.8. Histology and Immunohistochemistry

Tissue samples were fixed by immersion in 4% paraformaldehyde in 0.1 M sodium phosphate buffer, pH 7.4, overnight at 4 °C, dehydrated in a graded series of ethanol, cleared in xylene, and embedded in paraffin blocks for light microscopy. Five-micrometer-thick sections of tissues were cut with a microtome, mounted on slides, and stained with hematoxylin/eosin. Morphometric analysis of inguinal WAT sections was performed by digital acquisition of adipose tissue areas using AxioVision 40V 4.6.3.0 software and a Zeiss Axioskop 2 microscope equipped with an AxioCam ICc3 digital camera (Carl Zeiss S.A., Barcelona, Spain). Distributions of adipocyte size were obtained from individual data of cell sizes. Immunohistochemical detection of UCP1, mitofusin-2 (MFN2), and galectin-3 in fixed tissue sections was performed, essentially as previously described [31], using polyclonal antibodies against UCP1 (Catalog number GTX112784, GeneTex, Irvine, CA, USA), Mfn2 (Catalog number HPA030554, Sigma-Aldrich, St. Louis, MO, USA), and galectin-3 (MAC-2; Catalog number CL8942AP, Cedarlane, Burlington, Ontario, Canada).

2.9. Immunoblotting

Total protein was isolated from tissues using TRI Reagent (Sigma-Aldrich, St. Louis, MO, USA) following the manufacturer's instructions. Protein concentration was determined with Pierce™ BCA Protein Quantification Assay kit (Thermo-Fisher Scientific, Waltham, MA, USA). Ten micrograms of protein was loaded and separated in a precast 12% gel (Bio-Rad, Hercules, CA, USA) and transferred onto a 0.2 µm nitrocellulose membrane using a Trans-Blot Turbo semi-dry transfer apparatus (Bio-Rad, Hercules, CA, USA). Membranes were blocked for 1 h at room temperature with Odyssey Blocking Buffer (Li-Cor, Lincoln, NE, USA) and incubated overnight at 4 °C with gentle shaking with primary antibodies (1:1000 in Tris Buffered Saline-Tween 20, TBS-T) against UCP1 and MFN2 (same sources as in 2.8). Membranes were then incubated with the corresponding secondary IRDye antibodies (1:10000 in TBS-T, 1 h at room temperature), and the signal was detected using an Odyssey near-infrared scanner (Li-Cor, Lincoln, NE, USA).

2.10. Cell Culture Experiment

3T3-L1 preadipocytes obtained from ATCC (American Type Culture Collection, Manassas, VA, USA) were grown and differentiated in six-well culture plates, using commercial media from Zen-Bio Inc (Research Triangle Park, NC, USA) and following the manufacturer standard protocol. Preadipocytes were routinely cultured in Preadipocyte medium at 37 °C and 5% CO_2. For differentiation into adipocytes, the cells were allowed to reach confluence, and two days later (defined as day zero) the preadipocyte medium was replaced by Differentiation medium. On day three, differentiation medium was replaced by Adipocyte Maintenance medium. Cells were cultured until day eight, with medium replacement every two days. Adipogenic differentiation of the cells was regularly monitored through phase contrast microscopical examination. On day seven, when more than 95% of the cells showed intracellular lipid accumulation, cells were treated with PTE (100 mg/L), fucoxanthin (5 µM; Sigma-Aldrich, St. Louis, MO, USA) or vehicle (ethanol; 10 µL) for 24 h prior to harvesting. In parallel plates intracellular lipid content was quantified by Oil Red O (ORO) staining, as previously described [32], evidencing no differences between treatments (results not shown). The treatments applied had no cytotoxic effects in 3T3-L1 preadipocytes as assessed by the lactate dehydrogenase release assay (results not shown).

2.11. Statistical Analysis

Data are presented as mean ± standard deviation (SD) or as mean ± standard error of the mean (SEM). Comparisons between three groups (placebo, PTE100, and PTE300) were assessed by a non-parametric Kruskal–Wallis test. To compare between two groups, the non-parametric Mann–Whitney U test was used. In both cases, threshold of significance was set at $p < 0.05$. IBM SPSS Statistics for Windows, version 23.0 (IBM Corp., Armonk, NY, USA) was used for the analyses.

3. Results

3.1. Phaeodactylum tricornutum Extract (PTE) Characterization

The extract contained 15 mg dry matter (dm)/mL. Table 1 shows the fatty acid and carotenoid composition of PTE. Palmitoleic acid, palmitic acid, and EPA represented the main fatty acids. The PTE contained relatively high amounts of the carotenoid fucoxanthin and lower amounts of zeaxanthin and β-carotene.

Table 1. Fatty acid and carotenoid spectra of the ethanolic *Phaeodactylum tricornutum* extract (PTE).

Constituent	Concentration (µg/mg dm)
Fatty acids:	
Myristic acid	1.82 ± 0.10
Myristoleic acid	0.30 ± 0.02
Palmitic acid	7.73 ± 0.74
Palmitoleic acid	15.24 ± 0.44
cis-Oleic acid	1.52 ± 0.06
trans-Oleic acid	1.02 ± 0.05
α-Linoleic acid	0.71 ± 0.04
γ-Linoleic acid	0.30 ± 0.01
Eicosatrinoic acid	1.25 ± 0.05
Eicosapentanoic acid	7.32 ± 0.40
Carotenoids:	
Fucoxanthin	23.54 ± 0.60
Zeaxanthin	0.30 ± 0.05
β-Carotene	0.12 ± 0.04

dm—dry matter. Data are mean ± SD of three independent measurements.

3.2. PTE Supplementation Led to an Accumulation of Fucoxanthin Metabolites in Adipose Tissues

Animals received through PTE supplementation a daily dose of ~2.4 mg fucoxanthin/kg bw (PTE100) or ~7.1 mg fucoxanthin/kg bw (PTE300). Fucoxanthin is rapidly metabolized to fucoxanthinol through deacetylation during intestinal digestion, so that little is absorbed intact, and fucoxanthinol is further dehydrogenated/isomerized to amarouciaxanthin A in the liver and other tissues [33–35]. Traces of fucoxanthin metabolites, potentially the sum of fucoxanthinol and amarouciaxanthin A [30], were found in the interscapular BAT of PTE300-supplemented mice and in the epididymal and inguinal WAT depots of PTE100-supplemented mice, whereas supplementation with PTE300 resulted in a consistent accumulation up to 1.33 ± 0.72 µg fucoxanthinol/g epididymal WAT and 1.48 ± 0.76 µg fucoxanthinol/g inguinal WAT. Levels of fucoxanthin metabolites in the liver of all experimental groups and the adipose tissues of the control (HFD-fed, placebo-treated) group were below detection. Lack of detection of fucoxanthin metabolites in the liver of PTE-supplemented mice might be consistent with previous findings that fucoxanthin metabolites have a shorter half-life in the liver than in adipose tissues [36].

3.3. PTE Supplementation Partly Counteracted High Fat Diet (HFD)-Induced Obesity

In general, PTE supplementation was well accepted by the animals, and there were no apparent effects on spontaneous physical activity or adverse health effects. The average initial body weight was 27.7 ± 0.5 g and did not differ significantly among the three experimental groups. HFD feeding led to a gradual increase of body weight in the three groups, which was already evident after two days. The PTE-treated mice gained less body weight than controls upon HFD feeding, an effect that reached statistical significance for PTE300 (Figure 1A). At the end of the experiment (day 26), control, PTE100, and PTE300 mice had gained 7.0 ± 0.5, 6.0 ± 1.0, and 4.5 ± 0.5 g; final body weights were 35.1 ± 0.6, 33.5 ± 1.4, and 33 ± 0.7 g, respectively. Differences in body weight gain were not because of the differences in energy intake, which was similar in the three experimental groups throughout the entire duration of the HFD challenge (Figure 1B). The PTE300 group also revealed less total body fat mass than controls in body composition analyses performed after 5, 14, and 22 days on the HFD (Figure 1C). Body weight lost upon a 6 h fast, used as an indicator of energy expenditure, was maximal in the PTE300 group (Figure 1D). In keeping with these results, at the end of the study the mass of the epididymal and inguinal WAT depots was significantly lower (by 24% and 17%, respectively) in the PTE300 group compared with the control group (Figure 1E). There were no differences between groups in retroperitoneal WAT mass and BAT mass (Figure 1E). The weight of the epididymal depot expressed as percentage of body weight, which is commonly used as an adiposity index in mice [37], was significantly lower in the PTE300 group compared to the control group (Figure 1F). Furthermore, mRNA expression levels of mesoderm-specific transcript homolog protein (*Mest*), used as a marker of WAT expansion [38,39], were markedly decreased in inguinal WAT of PTE-supplemented mice (Figure 1G). Histological analysis of the liver was largely normal and did not reveal obvious hepatosteatosis in any of the experimental groups, possibly because of the relatively short period of HFD feeding applied (Supplementary Figure S1A). Biochemical analysis showed a tendency for lower total liver lipid content in the PTE-supplemented mice (Supplementary Figure S1B).

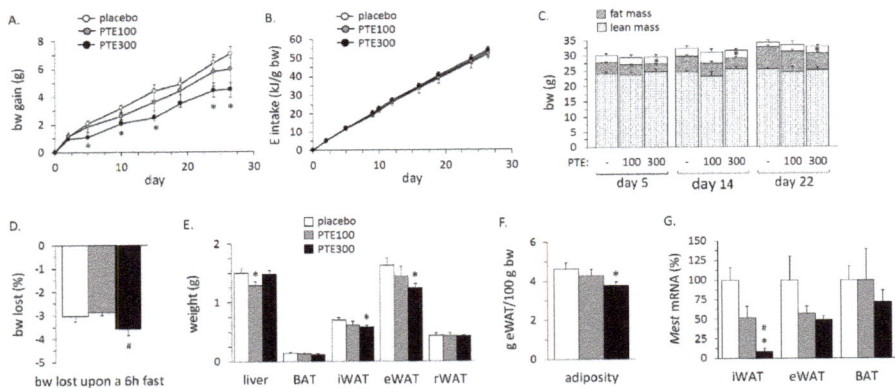

Figure 1. *Phaeodactylum tricornutum* ethanolic extract (PTE) ameliorates fat deposition in C57BL/6J mice fed with a high fat diet (HFD). Evolution of body weight (bw) gain (**A**), cumulative energy intake from food (**B**), and body composition (**C**) from day 1 to 26 of dietary challenge. Body weight lost upon a 6 h fast, on day 22 (**D**). Liver, interscapular brown adipose tissue (BAT), and inguinal, epididymal, and retroperitoneal white adipose tissue (iWAT, eWAT, and rWAT) weights (**E**), adiposity as eWAT weight as percent body weight (**F**), and mRNA expression levels of *Mest* in iWAT, eWAT, and BAT (**G**) at the end of the experiment. HFD-fed mice received daily an oral dose of PTE (100 mg or 300 mg/kg bw) or placebo (olive oil:water, 2:1, *v:v*) for 26 days. Data are mean ± SEM of 5–6 male mice/group. To compare between two groups, the non-parametric Mann–Whitney U test was used: *, different ($p < 0.05$) from vehicle; and #, different ($p < 0.05$) between doses.

Microscopical examination evidenced that inguinal WAT adipocytes were smaller in PTE-supplemented mice than in control mice (see representative microphotographs in Figure 2A). Detailed morphometric analysis confirmed a shift of adipocyte population distribution toward an increased percentage of small adipocytes and a lower percentage of large adipocytes in PTE-supplemented mice compared to controls (Figure 2B). The Kolmogorov–Smirnov test indicated that the difference in distributions of cell size between control and PTE-supplemented mice was statistically significant ($p < 0.001$ for both PTE doses) (Figure 2B). Microscopical examination of inguinal WAT sections also revealed a sporadic occurrence of crown-like structures (CLSs) in two out of five control mice and three out of five PTE100 mice examined, and an even more consistent occurrence of CLS was found in the PTE300-supplemented group, for which CLSs were detected in all five animals examined (Supplementary Figure S2). These CLSs were positive for immunostaining against the macrophage marker galectin-3 (MAC-2).

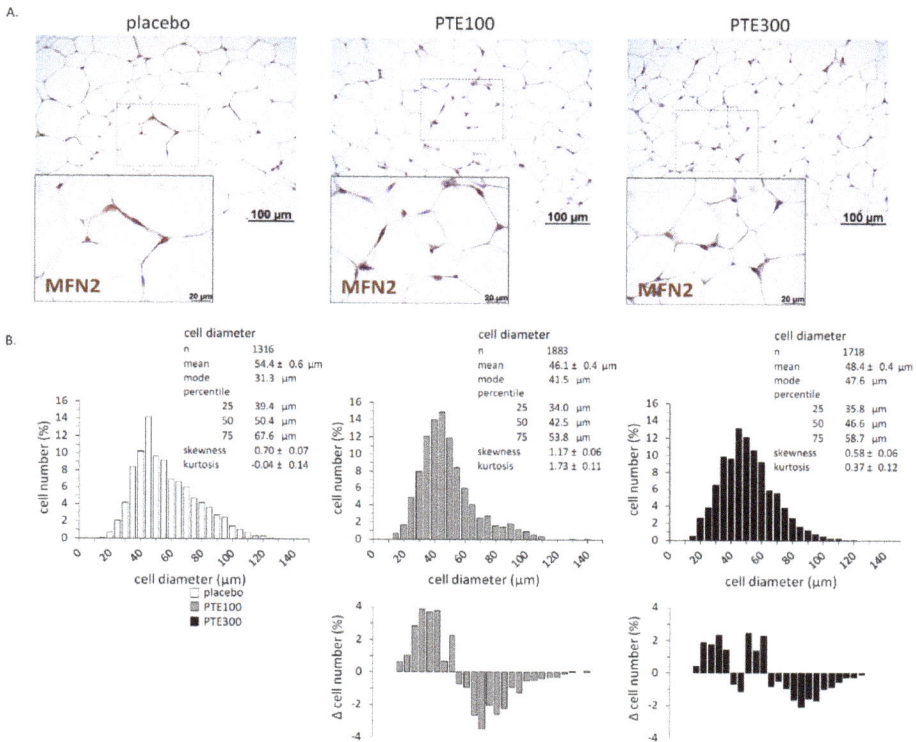

Figure 2. *Phaeodactylum tricornutum* ethanolic extract (PTE) decreases adipocyte size in inguinal white adipose tissue (iWAT) of C57BL/6J mice fed with a high fat diet (HFD). Representative microphotographs illustrating adipocyte size and Mitofusin (MFN) 2 immunostaining (**A**), and distribution of adipocytes size (**B**) in iWAT at the end of the experiment. HFD-fed mice received daily an oral dose of PTE (100 mg or 300 mg/kg body weight) or placebo (olive oil:water, 2:1, *v:v*) for 26 days. Five to six animals per group and between 200 and 300 cells per animal were included in the analysis of distribution of adipocytes size. The area of individual adipocytes was measured using a quantitative morphometric method at 20× magnification with the assistance of Axio Vision software. Adipocyte size distribution was statistically different ($p < 0.001$) between the control and the PTE groups, according to the Kolmogorov–Smirnov test. The bottom panels in (**B**) correspond to the difference in frequency for each adipocyte size interval between the PTE-supplemented group (PTE100 or PTE300) and the control (vehicle receiving) group.

Table 2 shows parameters related to glucose control and insulin sensitivity determined after a short fasting on day 22 of HFD challenge. There was a tendency for decreased HOMA-IR index ($p = 0.068$; Mann–Whitney U test) and increased R-QUICKI index ($p = 0.100$; Mann–Whitney U test) in the PTE100-supplemented group, due to lower levels of fasting glucose as well as (though non-significantly) insulin in blood of animals in this group compared to controls. These results suggested improved glucose control and insulin sensitivity on HFD in the PTE100 mice. Such trends were absent in the PTE300-supplemented mice, which had fasting blood glucose levels significantly higher than control and PTE100-supplemented mice, and fasting insulin, NEFA, HOMA-IR, and R-QUICKI indexes were very similar to those of control mice. Fed blood glucose levels at the end of the experiment did not significantly differ between groups (control, 180 ± 8.1; PTE100, 186 ± 9.8; PTE300, 191 ± 10.7 mg/dL).

Table 2. Plasma analyses and insulin resistance/sensitivity indexes in animals.

	Placebo	PTE100	PTE300
Glucose (mg/dL)	161 ± 8.2	134 ± 5.8 *	179 ± 5.5 *#
Insulin (mU/L)	42.1 ± 7.8	27.7 ± 1.9	33.5 ± 2.4
NEFA (mEq/L)	0.556 ± 0.046	0.714 ± 0.083	0.661 ± 0.113
HOMA-IR	17.0 ± 3.5	9.1 ± 0.71	14.7 ± 0.94 #
R-QUICKI	0.283 ± 0.005	0.295 ± 0.005	0.281 ± 0.005

Data are mean ± SEM; n = 5–6; Parameters were obtained from blood collected at day 22 after a 6 h fast. To compare between two groups, the non-parametric Mann–Whitney U test was used: *, different ($p < 0.05$) from vehicle; and #, different ($p < 0.05$) between doses; NEFA—non esterified fatty acids; HOMA-IR—homeostatic model assessment for insulin resistance; R-QUICKI—revised quantitative insulin sensitivity check index; PTE100—100 mg PTE/kg body weight/day; and PTE300—300 mg PTE/kg body weight/day.

3.4. PTE Supplementation Affected Transcriptional Control of Lipid Metabolism in White Adipose Tissue (WAT) Depots and Favored Browning of Subcutaneous WAT

Gene expression of key proteins related to different aspects of fatty acid and energy metabolism was compared in visceral (epididymal) and subcutaneous (inguinal) WAT depots of control and PTE-supplemented mice under HFD. In the epididymal WAT of PTE-supplemented mice, mRNA levels of lipolysis-related genes *Lipe*, encoding hormone sensitive lipase, and *Plin1*, encoding perilipin 1, were significantly down-regulated ($p = 0.039$ and $p = 0.033$, respectively; Kruskal–Wallis test), and there were trends to down-regulation for the lipogenesis-related gene *Srebf1* and the fatty acid uptake-related gene *Cd36* as well ($p = 0.099$ and $p = 0.078$, respectively; Kruskal–Wallis test) (Figure 3A). Further, mRNA levels in the epididymal fat depot of *Lpl*, coding for lipoprotein lipase that enables utilization of fatty acids from circulating triacylglycerols, were significantly down-regulated in the PTE300 mice relative to controls.

In subcutaneous (inguinal) WAT, expression of these same genes was unaffected by PTE supplementation (Figure 3B). However, *Cpt1* was 4 times and *Ucp1* at least 11 times upregulated in inguinal WAT of PTE-supplemented mice, indicating an increased capacity for fatty acid oxidation and thermogenesis (Figure 3B) ($p = 0.085$ and $p = 0.022$, respectively; Kruskal–Wallis test). Further, at the protein level, UCP1 could not be detected in inguinal WAT in any control mice, but it was detected by immunoblotting in one-sixth PTE100 and two-fifths PTE300 mice analyzed (results not shown). Moreover, immunostaining of inguinal WAT sections for MFN2—an outer mitochondrial membrane protein whose activity has been linked to an enhancement of oxygen consumption and substrate oxidation [40]—was more intense in PTE-supplemented mice than in controls (see the brown color in the periphery of adipocytes in Figure 2A).

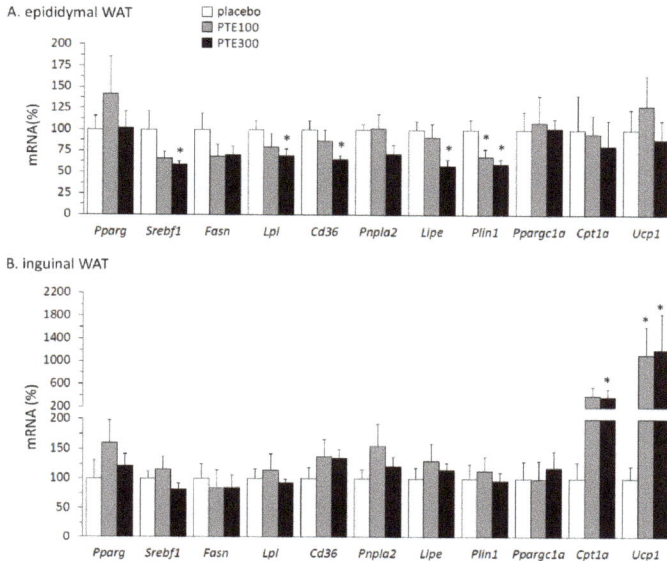

Figure 3. *Phaeodactylum tricornutum* ethanolic extract (PTE) down-regulates fatty acid uptake and lipid turnover capacities in epididymal white adipose tissue (eWAT) and increases oxidative/thermogenic capacity in inguinal WAT (iWAT) of C57BL/6J mice fed with a high fat diet (HFD). mRNA levels of selected genes as indicated were analyzed in eWAT (**A**) and iWAT (**B**) at the end of the experiment. HFD-fed mice received daily an oral dose of PTE (100 mg or 300mg/kg body weight) or placebo (olive oil:water, 2:1, *v:v*) for 26 days. Data are the mean ± SEM of 5–6 male mice/group and are expressed relative to the mean value of the vehicle group, which was set to 100. To compare between two groups, the non-parametric Mann–Whitney U test was used: *, different ($p < 0.05$) from vehicle.

3.5. PTE Supplementation Favored Brown Adipose Tissue (BAT) Activation

PTE supplementation led to BAT activation as indicated by the smaller size of brown adipocytes and their enrichment in UCP1 protein immunostaining (see the representative microphotographs in Figure 4A). This was confirmed by immunoblotting analysis of UCP1 and MFN2, showing dose-dependent increased levels of both proteins in BAT of PTE-supplemented animals as compared to controls (Figure 4B). Moreover, PTE-supplemented mice showed an increased gene expression in BAT of *Cd36* ($p = 0.046$; Kruskal–Wallis test) that was especially evident in the PTE100 group, which also showed increased mRNA levels of *Ppargc1a* in BAT (Figure 4C). PTE supplementation had no effect on the expression of *Ucp1* or *Cpt1* at the mRNA level, and resulted in a downregulated expression in BAT of the lipolytic genes *Lipe* and *Pnpla2* ($p = 0.029$ and $p = 0.057$, respectively; Kruskal–Wallis test), encoding hormone sensitive lipase and adipose triglyceride lipase. mRNA levels of the lipogenic genes *Fasn* and *Srebf1* were also down-regulated in BAT of PTE-supplemented mice ($p = 0.049$ and $p = 0.074$, respectively; Kruskal–Wallis test) (Figure 4C).

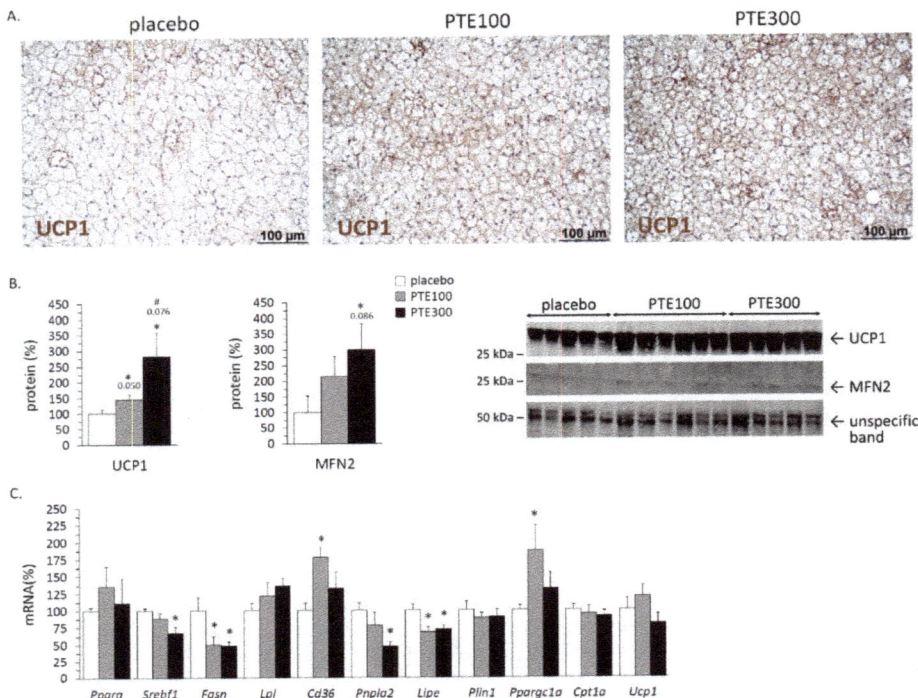

Figure 4. *Phaeodactylum tricornutum* ethanolic extract (PTE) activates interscapular brown adipose tissue (BAT) in C57BL/6J mice fed with a high fat diet (HFD). Representative microphotographs illustrating BAT activation and Uncoupling protein (UCP) 1 immunostaining (**A**), UCP1 and Mitofusin (MFN) 2 protein levels as determined by immunoblotting in BAT (**B**), and mRNA levels of selected genes in BAT (**C**) at the end of the experiment. HFD-fed mice received daily an oral dose of PTE (100 mg or 300 mg/kg body weight) or placebo (olive oil:water, 2:1, *v:v*) for 26 days. Data are the mean ± SEM of 5–6 male mice/group and are expressed relative to the mean value of the vehicle group, which was set to 100. To compare between two groups the non-parametric Mann–Whitney U test was used: *, different from vehicle; and #, different between doses. Threshold of statistical significance was set at mboxemphp < 0.05; in (**B**), *p* values < 0.1 are also indicated.

3.6. PTE and Fucoxanthin had Both Overlapping and Distinct Effects on Gene Expression of Lipid Metabolism-Related Genes in Mature 3T3-L1 Adipocytes

Expression levels of a series of genes related to lipid metabolism and thermogenesis were compared in mature 3T3-L1 adipocytes exposed to the vehicle (control cells) or to either 100 mg PTE/L, contributing ~3.6 µM fucoxanthin, or a similar dose of pure fucoxanthin (5 µM) for 24 h (Figure 5). *Cpt1a* mRNA levels were strongly, relative to levels in control cells, similarly induced by both PTE and fucoxanthin exposure. *Cd36* mRNA levels were induced following exposure to PTE, but not fucoxanthin, whereas *Fasn* mRNA levels were decreased following exposure to fucoxanthin, but not PTE. *Ucp1* mRNA could not be detected in mature 3T3-L1 adipocytes irrespective of treatment.

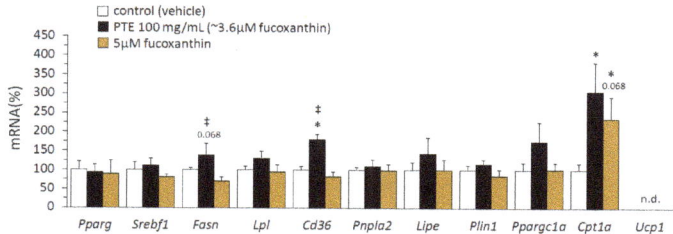

Figure 5. *Phaeodactylum tricornutum* ethanolic extract (PTE) and fucoxanthin effects on gene expression in mature 3T3-L1 adipocytes are not equivalent. mRNA levels of selected genes in mature 3T3-L1 adipocytes are shown. 3T3-L1 preadipocytes were grown and differentiated following a standard protocol. On day 7, cultures were treated with PTE (100 mg/L), fucoxanthin (5 μM; Sigma-Aldrich), or vehicle (ethanol 0.5%) for 24 h. Data are the mean ± SEM of two independent experiments made in triplicate and are expressed relative to the mean value of the vehicle group, which was set to 100. To compare between two groups the non-parametric Mann–Whitney test was used: *, different from vehicle; and ‡, different from fucoxanthin. Threshold of statistical significance was set at $p < 0.05$, p values < 0.1 are also indicated.

4. Discussion

The current obesity pandemic [41] is boosting research on the use of plant and algae-based products including extracts or isolated components in obesity prevention and therapy [42,43]. Adipose tissues are active players in energy metabolism and a target of anti-obesity strategies. WAT is the main storage site of excess energy taken up from food, and both WAT and BAT are plastic tissues where substrate (mainly fatty acid) oxidation and thermogenesis can be activated, through pharma or food compounds, to oppose body fat accrual and preserve metabolic health [44–47]. In this work, we provide evidence that an ethanolic extract of the microalga *Phaeodactylum tricornutum* ameliorates the development of diet-induced obesity and insulin resistance in mice independent of decreased food intake. *P. tricornutum* supplementation was linked to increased energy expenditure, as indicated by increased body weight loss upon fasting, molecular and histological signs of BAT activation, and molecular signs of enhanced mitochondrial oxidative metabolism in subcutaneous WAT. Thus, the results point to thermogenesis and metabolic activation in adipose tissues as one mechanism for the anti-obesity effects of PTE, even if (as a limitation of the study) indirect calorimetry measurements of energy expenditure are lacking. Other mechanisms that could be involved are decreased dietary lipid absorption, since energy excreted in feces was not measured, and increased spontaneous physical activity, since this parameter was not continuously monitored.

An effect of lipophilic constituents of *P. tricornutum* on obesity-related metabolic changes was suggested by previous findings. In particular, studies have reported anti-obesity effects of isolated fucoxanthin or fucoxanthin-containing extracts of edible seaweeds (macroalgae) such as *Undaria pinnatifida* or *Laminaria japonica* in genetic and dietary rodent models of obesity, which have been ascribed to metabolic effects in tissues including, notably, the adipose tissues [16,17,20,23]. *P. tricornutum* is 10 times richer in fucoxanthin than macroalgae [7]. Further, in previous studies we found a dose-dependent accumulation of fucoxanthin metabolites in adipose tissues of mice that were fed diets containing 5% to 25% *P. tricornutum* biomass [30,48]. This scenario prompted us to assay the anti-obesity activity of an ethanolic extract of *P. tricornutum* (PTE) with focus on its impact on energy and lipid metabolism in white and brown adipose tissues.

Mice on HFD receiving PTE supplementation accumulated fucoxanthin metabolites in adipose tissues, as expected [30,48], and displayed lower body weight gain, body fat content, and weight of WAT depots than control mice receiving placebo. Whereas effects on macroscopic biometric parameters were observed mainly at the high PTE300 dose, it is noteworthy that favorable effects of supplementation on adipocyte size and distribution (e.g., decreased mean adipocyte size, increased proportion of small adipocytes, and decreased proportion of large adipocytes) were already evident in inguinal WAT at the low PTE100 dose. Smaller adipocytes in obesity have been linked to a better metabolic profile, both in humans [49] and rodent models [50]. Decreased *Mest* mRNA levels found in WAT depots of PTE-supplemented mice are also in keeping with PTE opposing the development of obesity, since *Mest* expression is a known predictive marker of WAT expansion sensitive to dietary anti-obesity interventions [38,39]. Two previous papers evidencing anti-obesity effects of *P. tricornutum* employed high or very high doses as lipid extract (0.7% in diet, corresponding to ~800 mg extract/kg bw per day or ~255 mg fucoxanthin/kg bw per day) [25] or dry powder (15% and 30% in diet, corresponding to ~14 and 36 g dry powder/kg bw per day or ~72 and 155 mg fucoxanthin/kg bw per day) [26], which were well over the doses used in the present work (100 and 300 mg extract/kg bw per day or 2.4 and 7.1 mg fucoxanthin/kg bw per day). Further, these previous studies did not address cellular or molecular effects of supplementation on adipose tissues.

Both visceral (epididymal) and subcutaneous (inguinal) WAT were reduced in mass in the PTE–supplemented animals compared to controls on the HFD, and molecular results suggest that metabolic effects in both depots may contribute to the local depot weight reduction and the whole-body anti-obesity effect of supplementation. On the one hand, gene expression data are suggestive of decreased lipid uptake and turnover in the visceral (epididymal) fat depot of PTE-supplemented mice, where gene expressions related to fatty acid uptake (*Lpl* and *Cd36*), fatty acid mobilization from intracellular lipid stores (*Pnpla2*, *Lipe*, and *Plin1*), and fatty acid and triacylglycerol synthesis (*Srebf1* and *Fasn*) were simultaneously found to be down-regulated. A decreased lipid turnover in visceral fat might be of interest in the context of obesity, as it has been suggested that increased lipid turnover in visceral WAT (and/or decreased turnover in subcutaneous WAT) may result in metabolic complications of overweight or obesity [51].

On the other hand, results herein point to an effect of PTE supplementation favoring the browning of subcutaneous (inguinal) WAT. Thus, although we lacked to detect the appearance of adipocytes with the typical brown adipocyte multilocular distribution of intracellular fat, molecular signs of an enhanced capacity for oxidative metabolism and thermogenesis were present in the inguinal WAT of PTE-supplemented mice. These included an increased expression of *Ucp1* (and probably UCP1), which is the hallmark of WAT browning, and also *Cpt1a* and MFN2. The *Cpt1a* protein product is traditionally considered the rate-limiting enzyme for long-chain fatty acid uptake and beta-oxidation by mitochondria [52]. MFN2 is a protein involved in mitochondria dynamics that favors mitochondria fusion and whose activity enhances mitochondrial oxidative metabolism in cells [40].

Not only WAT browning but also increased phagocytosis of large adipocytes could contribute to the decreased inguinal WAT mass and the favorable changes in adipocyte size distribution observed in PTE-supplemented mice relative to controls on the HFD. This is suggested by the increased occurrence of CLS in WAT of PTE-supplemented mice, especially at the high PTE300 dose. CLS corresponds to dead adipocytes that are being cleared by surrounding macrophages [53]. The biological significance of these structures is not straightforward. Massive macrophage infiltration in WAT (especially visceral WAT) in obesity has been related to pro-inflammatory cytokine production, chronic inflammation, and systemic insulin resistance [54], yet there is emerging evidence for beneficial functions of WAT macrophages during diet-induced obesity, as the clearance of dead adipocytes may promote adipocyte turnover [55,56]. Further studies are required to discard a pro-inflammatory potential of PTE, yet we note that the increased presence of CLS in inguinal WAT of PTE-supplemented mice did not associate with an aggravation of insulin resistance on the HFD. On the contrary, PTE100 supplementation resulted in lower fasting blood glucose levels—in keeping with reported antihyperglycemic effects of fucoxanthin [12]—and the insulin resistance index (HOMA-IR) of PTE-supplemented mice was tendentially lower (PTE100) or indistinguishable (PTE300) from that of control mice. Further, anti-inflammatory effects have been reported for *P. tricornutum* extracts in cell studies in human blood mononuclear cells and murine macrophages [57] and for fucoxanthin metabolites in adipocytes [12,58], and fucoxanthin supplementation is shown to improve skeletal muscle insulin responsiveness in mice with genetic diabesity [59]. However, our results do suggest that lower doses of PTE might exert better effects than higher doses on glucose control and insulin sensitivity.

Observed effects in BAT most likely contribute to the ability of PTE supplementation to counteract the development of diet-induced obesity. Activation of BAT in the supplemented mice was very clear from the tissue microphotographs and the results of both BAT UCP1 immunohistochemical staining and immunoblotting. Activation of BAT was also consistent with the observed up-regulation of other genes and proteins known to be required for BAT thermogenesis, namely MFN2, *Cd36*, and *Ppargc1a*. MFN2 expression is shown to play a major role in BAT metabolism by physically coupling the mitochondria with lipid droplets and maintaining mitochondrial oxidative capacity [60]. *Cd36* encodes a transport protein present at both the plasma and the mitochondrial membrane that mediates fatty acid uptake in the cell and the mitochondria, and it is known to play an essential role in BAT thermogenesis [61,62]. *Ppargc1a* encodes PGC1α, a transcriptional coactivator first identified for its stimulatory role of BAT thermogenesis [63]. To be highlighted is the fact that PTE supplementation exerted opposite effects in visceral WAT and BAT regarding gene expression of proteins for cellular fatty acid provision and uptake (*Lpl* and *Cd36*), decreasing it in the epididymal depot and increasing it in BAT, while having no effects on these genes in the subcutaneous WAT depot. Overall, these results suggest that PTE supplementation in the context of HFD may favor channeling of dietary fatty acids away from visceral fat depots and toward ignition in thermogenic fat tissues, mainly BAT and also subcutaneous WAT.

In previous animal studies, supplementation with fucoxanthin or fucoxanthin-rich seaweed extracts led to inconsistent results regarding UCP1 expression in adipose tissues. Different authors observed either up-regulation [20] or lack of effect on UCP1 expression in BAT [16,64], yet in the latter reports BAT mass normalized to body weight increased following supplementation (contrary to WAT mass, which was decreased). Therefore, a contribution of increased BAT activity to observed anti-obesity effects cannot be discarded. An induction of UCP1 expression in visceral (gonadal) WAT following fucoxanthin supplementation has been reported and highlighted [16,17,20]; however, other reports failed to detect UCP1 induction in WAT of supplemented animals [64,65]. Further, visceral WAT has a minor tendency to turn to a BAT-like phenotype than subcutaneous (inguinal) WAT [66], yet most previous reports did not compare UCP1 induction in visceral and subcutaneous fat depots. As an exception, Wu et al. assessed gene expression related to mitochondriogenesis and thermogenesis in inguinal WAT, gonadal (epididymal) WAT, and BAT of mice fed obesogenic diets supplemented or not with fucoxanthin [64]. They confirmed an increase in the animals' metabolic rate following fucoxanthin supplementation, and—different from our results in PTE-supplemented mice—they found little evidence of up-regulation of thermogenic genes in BAT and a similar up-regulation of many of such genes in visceral and subcutaneous WAT (though results for *Ucp1* did not reach statistical significance) [64].

Overall, while most animal studies to date point to anti-obesity properties of fucoxanthin, it would appear that parameters such as the time length of supplementation and the formulation of the fucoxanthin source (isolated vs. extract) influence the exact regulatory and metabolic mechanisms involved. In fact, effects of PTE and purified fucoxanthin on gene expression in mature 3T3-L1 adipocytes were not fully equivalent, suggesting that components other than fucoxanthin contribute to PTE effects in cultured adipocytes and likely also in supplemented animals in vivo. For different components of PTE including fucoxanthin, metabolic effects have been related to the induction of the PGC1α network in adipose tissues [64,67] and the activation of AMPK in tissues such as liver and muscle [68]. These two molecules are important players in the control of energy metabolism. While further mechanistic studies are warranted, we note that BAT activation brought about by PTE supplementation involved the induction of the PGC1α gene (Figure 4), but it did not affect levels of phosphorylated (active) AMPK nor the ratio phosphoAMPK/totalAMPK in the tissue (results not shown).

5. Conclusions

In summary, this work demonstrates that a *Phaeodactylum tricornutum* extract ameliorates the development of diet-induced obesity in a well-established rodent model, and it links this effect to the stimulation of oxidative metabolism in BAT and WAT depots, notably the induction of BAT recruitment and the browning of subcutaneous WAT. Knowledge on the anti-obesity action of *P. tricornutum* and its mechanisms may pave the way for novel uses of this microalga in the functional food and nutraceutical arena.

Supplementary Materials: The following are available online at http://www.mdpi.com/2072-6643/11/4/796/s1, Figure S1: Effects of PTE supplementation to C57BL/6J mice fed with a high fat diet on liver histological appearance and lipid content, Figure S2: Effects of PTE supplementation to C57BL/6J mice fed with a high fat diet on the appearance of crown-like structures in inguinal white adipose tissue.

Author Contributions: Authors' contributions were as follows: M.L.B., J.R., A.G., and K.B. conceptualized and designed the experiments. A.P. contributed funding acquisition. A.G. and A.T. cultivated and harvested *P. tricornutum* biomass, supervised by C.P., F.D., and U.S.-S. prepared *P. tricornutum* extract and analyzed fatty acids. A.G. analyzed carotenoids. B.S. acquired data. A.G., J.R., and M.L.B. planed and performed the cell culture and animal experiments, analyzed the data, and wrote the manuscript. All authors critically read and approved the final manuscript.

Funding: This work was funded in part by the Spanish Government (Agencia Estatal de Investigación, MINECO/FEDER, EU), grant AGL2015-67019-P (to AP). The work by AG was supported by a grant (7533-10-5/91/2) from the Ministry of Science, Research and the Arts of Baden-Württemberg (MWK). AG also acknowledges generous support by the bioeconomy graduate program BBW ForWerts, supported by the MWK. AG received a travel grant of the DAAD (German Academic Exchange Service). BS is the recipient of a "La Caixa" Foundation pre-doctoral contract at the Universitat de les Illes Balears.

Acknowledgments: We gratefully thank the employees of the animal care unit of the Universitat de les Illes Balears as well as Enzo Ceresi and Benjamin Peters for their contribution and excellent technical assistance during the performance of the experiments. The Universitat de les Illes Balears group is a member of the European COST-Action EUROCAROTEN (CA15136; EU Framework Programme Horizon 2020), and the Spanish Network of Excellence CaRed (BIO2017-90877-REDT; Agencia Estatal de Investigación, MINECO/FEDER, EU). CIBER de Fisiopatología de la Obesidad y Nutrición (CIBEROBN) is an initiative of the ISCIII (Spanish Government).

Conflicts of Interest: The authors have no conflict of interest to declare.

References

1. Becker, W. Microalgae in human and animal nutrition. In *Handbook of Microalgal Culture: Biotechnology and Applied Phycology*; Richmond, A., Ed.; Blackwell Science: Ames, IA, USA, 2007; pp. 312–351.

2. Caporgno, M.P.; Mathys, A. Trends in microalgae incorporation into innovative food products with potential health benefits. *Front. Nutr.* **2018**, *5*, 58. [CrossRef]

3. Ryckebosch, E.; Muylaert, K.; Eeckhout, M.; Ruyssen, T.; Foubert, I. Influence of drying and storage on lipid and carotenoid stability of the microalga phaeodactylum tricornutum. *J. Agric. Food Chem.* **2011**, *59*, 11063–11069. [CrossRef] [PubMed]

4. Sorensen, M.; Berge, G.M.; Reitan, K.I.; Ruyter, B. Microalga phaeodactylum tricornutum in feed for atlantic salmon (salmo salar)—Effect on nutrient digestibility, growth and utilization of feed. *Aquaculture* **2016**, *460*, 116–123. [CrossRef]

5. Meiser, A.; Schmid-Staiger, U.; Trösch, W. Optimization of eicosapentaenoic acid production by phaeodactylum tricornutum in the flat panel airlift (fpa) reactor. *J. Appl. Phycol.* **2004**, *16*, 215–225. [CrossRef]

6. Derwenskus, F.; Metz, F.; Gille, A.; Schmid-Staiger, U.; Briviba, K.; Schließmann, U.; Hirth, T. Pressurized extraction of unsaturated fatty acids and carotenoids from wet chlorella vulgaris and phaeodactylum tricornutum biomass using subcritical liquids. *GCB Bioenergy* **2019**, *11*, 335–344. [CrossRef]

7. Kim, S.M.; Jung, Y.J.; Kwon, O.N.; Cha, K.H.; Um, B.H.; Chung, D.; Pan, C.H. A potential commercial source of fucoxanthin extracted from the microalga phaeodactylum tricornutum. *Appl. Biochem. Biotechnol.* **2012**, *166*, 1843–1855. [CrossRef] [PubMed]

8.   Vedin, I.; Cederholm, T.; Freund Levi, Y.; Basun, H.; Garlind, A.; Faxen Irving, G.; Jonhagen, M.E.; Vessby, B.; Wahlund, L.O.; Palmblad, J. Effects of docosahexaenoic acid-rich *n*-3 fatty acid supplementation on cytokine release from blood mononuclear leukocytes: The omegad study. *Am. J. Clin. Nutr.* **2008**, *87*, 1616–1622. [CrossRef]

9.   Serini, S.; Bizzarro, A.; Piccioni, E.; Fasano, E.; Rossi, C.; Lauria, A.; Cittadini, A.R.; Masullo, C.; Calviello, G. Epa and dha differentially affect in vitro inflammatory cytokine release by peripheral blood mononuclear cells from alzheimer's patients. *Curr. Alzheimer Res.* **2012**, *9*, 913–923. [CrossRef]

10. Heo, S.J.; Yoon, W.J.; Kim, K.N.; Ahn, G.N.; Kang, S.M.; Kang, D.H.; Affan, A.; Oh, C.; Jung, W.K.; Jeon, Y.J. Evaluation of anti-inflammatory effect of fucoxanthin isolated from brown algae in lipopolysaccharide-stimulated raw 264.7 macrophages. *Food Chem. Toxicol.* **2010**, *48*, 2045–2051. [CrossRef] [PubMed]

11. Kim, K.N.; Heo, S.J.; Yoon, W.J.; Kang, S.M.; Ahn, G.; Yi, T.H.; Jeon, Y.J. Fucoxanthin inhibits the inflammatory response by suppressing the activation of nf-kappab and mapks in lipopolysaccharide-induced raw 264.7 macrophages. *Eur. J. Pharmacol.* **2010**, *649*, 369–375. [CrossRef] [PubMed]

12. Hosokawa, M.; Miyashita, T.; Nishikawa, S.; Emi, S.; Tsukui, T.; Beppu, F.; Okada, T.; Miyashita, K. Fucoxanthin regulates adipocytokine mrna expression in white adipose tissue of diabetic/obese kk-ay mice. *Arch. Biochem. Biophys.* **2010**, *504*, 17–25. [CrossRef]

13. Flachs, P.; Horakova, O.; Brauner, P.; Rossmeisl, M.; Pecina, P.; Franssen-van Hal, N.; Ruzickova, J.; Sponarova, J.; Drahota, Z.; Vlcek, C.; et al. Polyunsaturated fatty acids of marine origin upregulate mitochondrial biogenesis and induce beta-oxidation in white fat. *Diabetologia* **2005**, *48*, 2365–2375. [CrossRef]

14. Flachs, P.; Rossmeisl, M.; Kopecky, J. The effect of *n*-3 fatty acids on glucose homeostasis and insulin sensitivity. *Physiol. Res.* **2014**, *63* (Suppl. 1), S93–S118.

15. Djousse, L.; Gaziano, J.M.; Buring, J.E.; Lee, I.M. Dietary omega-3 fatty acids and fish consumption and risk of type 2 diabetes. *Am. J. Clin. Nutr.* **2011**, *93*, 143–150. [CrossRef]

16. Maeda, H.; Hosokawa, M.; Sashima, T.; Funayama, K.; Miyashita, K. Fucoxanthin from edible seaweed, undaria pinnatifida, shows antiobesity effect through ucp1 expression in white adipose tissues. *Biochem. Biophys. Res. Commun.* **2005**, *332*, 392–397. [CrossRef]

17. Jeon, S.M.; Kim, H.J.; Woo, M.N.; Lee, M.K.; Shin, Y.C.; Park, Y.B.; Choi, M.S. Fucoxanthin-rich seaweed extract suppresses body weight gain and improves lipid metabolism in high-fat-fed c57bl/6j mice. *Biotechnol. J.* **2010**, *5*, 961–969. [CrossRef] [PubMed]

18. Maeda, H. Nutraceutical effects of fucoxanthin for obesity and diabetes therapy: A review. *J. Oleo Sci.* **2015**, *64*, 125–132. [CrossRef]

19. Jump, D.B. *n*-3 polyunsaturated fatty acid regulation of hepatic gene transcription. *Curr. Opin. Lipidol.* **2008**, *19*, 242–247. [CrossRef] [PubMed]

20. Woo, M.N.; Jeon, S.M.; Shin, Y.C.; Lee, M.K.; Kang, M.A.; Choi, M.S. Anti-obese property of fucoxanthin is partly mediated by altering lipid-regulating enzymes and uncoupling proteins of visceral adipose tissue in mice. *Mol. Nutr. Food Res.* **2009**, *53*, 1603–1611. [CrossRef] [PubMed]

21. Matsumoto, M.; Hosokawa, M.; Matsukawa, N.; Hagio, M.; Shinoki, A.; Nishimukai, M.; Miyashita, K.; Yajima, T.; Hara, H. Suppressive effects of the marine carotenoids, fucoxanthin and fucoxanthinol on triglyceride absorption in lymph duct-cannulated rats. *Eur. J. Nutr.* **2010**, *49*, 243–249. [CrossRef]

22. Woo, M.N.; Jeon, S.M.; Kim, H.J.; Lee, M.K.; Shin, S.K.; Shin, Y.C.; Park, Y.B.; Choi, M.S. Fucoxanthin supplementation improves plasma and hepatic lipid metabolism and blood glucose concentration in high-fat fed c57bl/6n mice. *Chem. Biol. Interact.* **2010**, *186*, 316–322. [CrossRef] [PubMed]

23. Jang, W.S.; Choung, S.Y. Antiobesity effects of the ethanol extract of *Laminaria japonica* Areshoung in high-fat-diet-induced obese rat. *Evid. Based Complement. Altern. Med.* **2013**, *2013*, 492807. [CrossRef] [PubMed]

24. Jankovic, A.; Otasevic, V.; Stancic, A.; Buzadzic, B.; Korac, A.; Korac, B. Physiological regulation and metabolic role of browning in white adipose tissue. *Horm. Mol. Biol. Clin. Investig.* **2017**, *31*. [CrossRef] [PubMed]

25. Kang, M.; Kim, S.M.; Jeong, S.; Choi, H.; Jang, Y.; Kim, J. Antioxidant effect of phaeodactylum tricornutum in mice fed high-fat diet. *Food Sci. Biotechnol.* **2013**, *22*, 107–113. [CrossRef]

26. Kim, J.H.; Kim, S.M.; Cha, K.H.; Mok, I.; Koo, S.Y.; Pan, C.; Lee, J.K. Evaluation of the anti-obesity effect of the microalga phaeodactylum tricornutum. *Appl. Biol. Chem.* **2016**, *59*, 283–290. [CrossRef]

27. Gille, A.; Hollenbach, R.; Trautmann, A.; Posten, C.; Briviba, K. Effect of sonication on bioaccessibility and cellular uptake of carotenoids from preparations of photoautotrophic phaeodactylum tricornutum. *Food Res. Int.* **2019**, *118*, 40–48. [CrossRef]

28. Reynes, B.; Serrano, A.; Petrov, P.; Ribot, J.; Chetrit, C.; Martinez-Puig, D.; Bonet, M.L.; Palou, A. Anti-obesity and insulin-sensitising effects of a glycosaminoglycan mix. *J. Funct. Foods* **2016**, *26*, 350–362. [CrossRef]

29. Folch, J.; Lees, M.; Sloane Stanley, G.H. A simple method for the isolation and purification of total lipides from animal tissues. *J. Biol. Chem.* **1957**, *226*, 497–509. [PubMed]

30. Gille, A.; Neumann, U.; Sandrine, L.; Bischoff, S.C.; Briviba, K. Microalgae as a potential source of carotenoids: Comparative results of an in vitro digestion method and a feeding experiment with c57bl/6j mice. *J. Funct. Foods* **2018**, *49*, 285–294. [CrossRef]

31. Petrov, P.D.; Ribot, J.; Palou, A.; Bonet, M.L. Improved metabolic regulation is associated with retinoblastoma protein gene haploinsufficiency in mice. *Am. J. Physiol. Endocrinol. Metab.* **2015**, *308*, E172–E183. [CrossRef] [PubMed]

32. Tacherfiout, M.; Petrov, P.D.; Mattonai, M.; Ribechini, E.; Ribot, J.; Bonet, M.L.; Khettal, B. Antihyperlipidemic effect of a rhamnus alaternus leaf extract in triton-induced hyperlipidemic rats and human hepg2 cells. *Biomed. Pharmacother. Biomed. Pharmacother.* **2018**, *101*, 501–509. [CrossRef] [PubMed]

33. Sugawara, T.; Baskaran, V.; Tsuzuki, W.; Nagao, A. Brown algae fucoxanthin is hydrolyzed to fucoxanthinol during absorption by caco-2 human intestinal cells and mice. *J. Nutr.* **2002**, *132*, 946–951. [CrossRef] [PubMed]

34. Asai, A.; Sugawara, T.; Ono, H.; Nagao, A. Biotransformation of fucoxanthinol into amarouciaxanthin a in mice and hepg2 cells: Formation and cytotoxicity of fucoxanthin metabolites. *Drug Metab. Dispos.* **2004**, *32*, 205–211. [CrossRef] [PubMed]

35. Hashimoto, T.; Ozaki, Y.; Taminato, M.; Das, S.K.; Mizuno, M.; Yoshimura, K.; Maoka, T.; Kanazawa, K. The distribution and accumulation of fucoxanthin and its metabolites after oral administration in mice. *Br. J. Nutr.* **2009**, *102*, 242–248. [CrossRef]

36. Yonekura, L.; Kobayashi, M.; Terasaki, M.; Nagao, A. Keto-carotenoids are the major metabolites of dietary lutein and fucoxanthin in mouse tissues. *J. Nutr.* **2010**, *140*, 1824–1831. [CrossRef]

37. Eisen, E.J.; Leatherwood, J.M. Predicting percent fat in mice. *Growth* **1981**, *45*, 100–107.

38. Takahashi, M.; Kamei, Y.; Ezaki, O. Mest/peg1 imprinted gene enlarges adipocytes and is a marker of adipocyte size. *Am. J. Physiol. Endocrinol. Metab.* **2005**, *288*, E117–E124. [CrossRef]

39. Voigt, A.; Ribot, J.; Sabater, A.G.; Palou, A.; Bonet, M.L.; Klaus, S. Identification of mest/peg1 gene expression as a predictive biomarker of adipose tissue expansion sensitive to dietary anti-obesity interventions. *Genes Nutr.* **2015**, *10*, 477. [CrossRef]

40. Zorzano, A.; Hernandez-Alvarez, M.I.; Sebastian, D.; Munoz, J.P. Mitofusin 2 as a driver that controls energy metabolism and insulin signaling. *Antioxid. Redox Signal.* **2015**, *22*, 1020–1031. [CrossRef]

41. Bray, G.A.; Kim, K.K.; Wilding, J.P.H.; World Obesity, F. Obesity: A chronic relapsing progressive disease process. A position statement of the world obesity federation. *Obes. Rev.* **2017**, *18*, 715–723. [CrossRef]

42. Jung, H.S.; Lim, Y.; Kim, E.K. Therapeutic phytogenic compounds for obesity and diabetes. *Int. J. Mol. Sci.* **2014**, *15*, 21505–21537. [CrossRef]

43. Wan-Loy, C.; Siew-Moi, P. Marine algae as a potential source for anti-obesity agents. *Mar. Drugs* **2016**, *14*, 222. [CrossRef]

44. Hursel, R.; Westerterp-Plantenga, M.S. Thermogenic ingredients and body weight regulation. *Int. J. Obes. (Lond.)* **2010**, *34*, 659–669. [CrossRef]

45. Bonet, M.L.; Oliver, P.; Palou, A. Pharmacological and nutritional agents promoting browning of white adipose tissue. *Biochim. Biophys. Acta* **2013**, *1831*, 969–985. [CrossRef]

46. Bartelt, A.; Heeren, J. Adipose tissue browning and metabolic health. *Nat. Rev. Endocrinol.* **2014**, *10*, 24–36. [CrossRef]

47. Bonet, M.L.; Mercader, J.; Palou, A. A nutritional perspective on ucp1-dependent thermogenesis. *Biochimie* **2017**, *134*, 99–117. [CrossRef]

48. Neumann, U.; Derwenskus, F.; Gille, A.; Louis, S.; Schmid-Staiger, U.; Briviba, K.; Bischoff, S.C. Bioavailability and safety of nutrients from the microalgae chlorella vulgaris, nannochloropsis oceanica and phaeodactylum tricornutum in c57bl/6 mice. *Nutrients* **2018**, *10*, 965. [CrossRef]

49. Hoffstedt, J.; Arner, E.; Wahrenberg, H.; Andersson, D.P.; Qvisth, V.; Lofgren, P.; Ryden, M.; Thorne, A.; Wiren, M.; Palmer, M.; et al. Regional impact of adipose tissue morphology on the metabolic profile in morbid obesity. *Diabetologia* **2010**, *53*, 2496–2503. [CrossRef]

50. Kim, J.Y.; van de Wall, E.; Laplante, M.; Azzara, A.; Trujillo, M.E.; Hofmann, S.M.; Schraw, T.; Durand, J.L.; Li, H.; Li, G.; et al. Obesity-associated improvements in metabolic profile through expansion of adipose tissue. *J. Clin. Investig.* **2007**, *117*, 2621–2637. [CrossRef]

51. Spalding, K.L.; Bernard, S.; Naslund, E.; Salehpour, M.; Possnert, G.; Appelsved, L.; Fu, K.Y.; Alkass, K.; Druid, H.; Thorell, A.; et al. Impact of fat mass and distribution on lipid turnover in human adipose tissue. *Nat. Commun.* **2017**, *8*, 15253. [CrossRef]

52. Zammit, V.A. Carnitine palmitoyltransferase 1: Central to cell function. *IUBMB Life* **2008**, *60*, 347–354. [CrossRef] [PubMed]

53. Cinti, S.; Mitchell, G.; Barbatelli, G.; Murano, I.; Ceresi, E.; Faloia, E.; Wang, S.; Fortier, M.; Greenberg, A.S.; Obin, M.S. Adipocyte death defines macrophage localization and function in adipose tissue of obese mice and humans. *J. Lipid Res.* **2005**, *46*, 2347–2355. [CrossRef]

54. Olefsky, J.M.; Glass, C.K. Macrophages, inflammation, and insulin resistance. *Annu. Rev. Physiol.* **2010**, *72*, 219–246. [CrossRef]

55. Fitzgibbons, T.P.; Czech, M.P. Emerging evidence for beneficial macrophage functions in atherosclerosis and obesity-induced insulin resistance. *J. Mol. Med.* **2016**, *94*, 267–275. [CrossRef]

56. Coats, B.R.; Schoenfelt, K.Q.; Barbosa-Lorenzi, V.C.; Peris, E.; Cui, C.; Hoffman, A.; Zhou, G.; Fernandez, S.; Zhai, L.; Hall, B.A.; et al. Metabolically activated adipose tissue macrophages perform detrimental and beneficial functions during diet-induced obesity. *Cell Rep.* **2017**, *20*, 3149–3161. [CrossRef] [PubMed]

57. Neumann, U.; Louis, S.; Gille, A.; Derwenskus, F.; Schmid-Staiger, U.; Briviba, K.; Bischoff, S.C. Anti-inflammatory effects of phaeodactylum tricornutum extracts on human blood mononuclear cells and murine macrophages. *J. Appl. Phycol.* **2018**, *30*, 2837–2846. [CrossRef]

58. Maeda, H.; Kanno, S.; Kodate, M.; Hosokawa, M.; Miyashita, K. Fucoxanthinol, metabolite of fucoxanthin, improves obesity-induced inflammation in adipocyte cells. *Mar. Drugs* **2015**, *13*, 4799–4813. [CrossRef] [PubMed]

59. Nishikawa, S.; Hosokawa, M.; Miyashita, K. Fucoxanthin promotes translocation and induction of glucose transporter 4 in skeletal muscles of diabetic/obese kk-a(y) mice. *Phytomed. Int. J. Phytother. Phytopharmacol.* **2012**, *19*, 389–394. [CrossRef]

60. Boutant, M.; Kulkarni, S.S.; Joffraud, M.; Ratajczak, J.; Valera-Alberni, M.; Combe, R.; Zorzano, A.; Canto, C. Mfn2 is critical for brown adipose tissue thermogenic function. *EMBO J.* **2017**, *36*, 1543–1558. [CrossRef]

61. Putri, M.; Syamsunarno, M.R.; Iso, T.; Yamaguchi, A.; Hanaoka, H.; Sunaga, H.; Koitabashi, N.; Matsui, H.; Yamazaki, C.; Kameo, S.; et al. Cd36 is indispensable for thermogenesis under conditions of fasting and cold stress. *Biochem. Biophys. Res. Commun.* **2015**, *457*, 520–525. [CrossRef]

62. Anderson, C.M.; Kazantzis, M.; Wang, J.; Venkatraman, S.; Goncalves, R.L.; Quinlan, C.L.; Ng, R.; Jastroch, M.; Benjamin, D.I.; Nie, B.; et al. Dependence of brown adipose tissue function on cd36-mediated coenzyme q uptake. *Cell Rep.* **2015**, *10*, 505–515. [CrossRef]

63. Puigserver, P.; Wu, Z.; Park, C.W.; Graves, R.; Wright, M.; Spiegelman, B.M. A cold-inducible coactivator of nuclear receptors linked to adaptive thermogenesis. *Cell* **1998**, *92*, 829–839. [CrossRef]

64. Wu, M.T.; Chou, H.N.; Huang, C.J. Dietary fucoxanthin increases metabolic rate and upregulated mrna expressions of the pgc-1alpha network, mitochondrial biogenesis and fusion genes in white adipose tissues of mice. *Mar. Drugs* **2014**, *12*, 964–982. [CrossRef] [PubMed]

65. Maeda, H.; Hosokawa, M.; Sashima, T.; Murakami-Funayama, K.; Miyashita, K. Anti-obesity and anti-diabetic effects of fucoxanthin on diet-induced obesity conditions in a murine model. *Mol. Med. Rep.* **2009**, *2*, 897–902. [CrossRef] [PubMed]

66. Seale, P.; Conroe, H.M.; Estall, J.; Kajimura, S.; Frontini, A.; Ishibashi, J.; Cohen, P.; Cinti, S.; Spiegelman, B.M. Prdm16 determines the thermogenic program of subcutaneous white adipose tissue in mice. *J. Clin. Investig.* **2011**, *121*, 96–105. [CrossRef] [PubMed]

67. Pahlavani, M.; Razafimanjato, F.; Ramalingam, L.; Kalupahana, N.S.; Moussa, H.; Scoggin, S.; Moustaid-Moussa, N. Eicosapentaenoic acid regulates brown adipose tissue metabolism in high-fat-fed mice and in clonal brown adipocytes. *J. Nutr. Biochem.* **2017**, *39*, 101–109. [CrossRef]

68. Zhang, Y.; Xu, W.; Huang, X.; Zhao, Y.; Ren, Q.; Hong, Z.Y.; Huang, M.; Xing, X. Fucoxanthin ameliorates hyperglycemia, hyperlipidemia and insulin resistance in diabetic mice partially through irs-1/pi3k/akt and ampk pathways. *J. Funct. Foods* **2018**, *48*, 515–524. [CrossRef]

nutrients

MDPI

Review

β-carotene in Obesity Research: Technical Considerations and Current Status of the Field

Johana Coronel [1], Ivan Pinos [2] and Jaume Amengual [1,2,*]

[1]   Department of Food Sciences and Human Nutrition, University of Illinois Urbana Champaign, Urbana, IL 61801, USA; acoronel@illinois.edu
[2]   Division of Nutritional Sciences, University of Illinois Urbana Champaign, Urbana, IL 61801, USA; ivanp2@illinois.edu
*   Correspondence: jaume6@illinois.edu

Received: 16 February 2019; Accepted: 6 April 2019; Published: 13 April 2019

Abstract: Over the past decades, obesity has become a rising health problem as the accessibility to high calorie, low nutritional value food has increased. Research shows that some bioactive components in fruits and vegetables, such as carotenoids, could contribute to the prevention and treatment of obesity. Some of these carotenoids are responsible for vitamin A production, a hormone-like vitamin with pleiotropic effects in mammals. Among these effects, vitamin A is a potent regulator of adipose tissue development, and is therefore important for obesity. This review focuses on the role of the provitamin A carotenoid β-carotene in human health, emphasizing the mechanisms by which this compound and its derivatives regulate adipocyte biology. It also discusses the physiological relevance of carotenoid accumulation, the implication of the carotenoid-cleaving enzymes, and the technical difficulties and considerations researchers must take when working with these bioactive molecules. Thanks to the broad spectrum of functions carotenoids have in modern nutrition and health, it is necessary to understand their benefits regarding to metabolic diseases such as obesity in order to evaluate their applicability to the medical and pharmaceutical fields.

Keywords: Vitamin A; adipocyte; β-carotene oxygenase 1

1. Introduction

Metabolic diseases are a growing cause of morbidity and mortality, becoming a heavy economic burden for patients and healthcare systems worldwide. More than 2.1 billion people worldwide were overweight or obese in 2014, and current predictions estimate that obesity will affect almost half of the world's adult population by 2030 [1]. According to the US Centers for Disease Control and Prevention, the prevalence of obesity in the US was 39.8% in 2015–2016 [2], and the medical costs associated with obesity were approximately $2 trillion in 2014 [1]. While obesity per se is the direct cause of only a few disorders, such as bone and joint-related disease [3], obese individuals are more susceptible to suffer metabolic alterations leading to type 2 diabetes, high blood pressure, and heart disease. Obesity is also associated with depression and certain cancers, overall affecting longevity and life quality [4,5].

From a simplified point of view, obesity is a consequence of disproportionate energy intake in combination with a reduction in energy expenditure, which leads to a positive energy balance in the organism [6,7]. The hallmark of obesity is the excessive accumulation of triglycerides in adipocytes, the main cellular component of the white adipose tissue [8]. Even though multifactorial aspects such as genetic or epigenetic predisposition can cause obesity [9], acquired behaviors such as eating habits and reduced physical activity are the main contributing factors responsible for the development of this disease [10,11]. For example, the ingestion of only 5% more calories than those expended could result in the accumulation of approximately 5 kg of adipose tissue in just one year [12]. An adequate dietary intervention is the foundation of weight loss therapy, as it is easier for most obese people to achieve a negative energy balance by decreasing food intake than just by increasing physical activity [13–15]. Therefore, healthy eating is a great strategy for reducing obesity, and the key signature of a healthy diet is the high consumption of a plant-based diet [16].

While fruits and vegetables promote weight loss thanks to their macronutrient composition (e.g., elevated fiber, water, and complex carbohydrates), we cannot dismiss their content in micronutrients with bioactive properties, some of which have documented effects on energy metabolism [17,18]. Among these micronutrients, carotenoids appear as potential candidates to prevent and treat obesity [19]. Carotenoids are a diverse group of compounds responsible for most of the yellow, orange and red colors in fruits and vegetables. Over the years, researchers have attributed carotenoids multiple biological functions, and they are widely considered to be some of the most important bioactive compounds in our food [20]. Among the six most abundant carotenoids in plasma, β-carotene, α-carotene, and β-cryptoxanthin are provitamin A carotenoids. Two other carotenoids, lutein, and zeaxanthin, have an important role in vision as they largely accumulate in the human eye [21]. Lycopene, the last carotenoid on this list, plays a crucial role in preventing some types of cancer [22].

Several comprehensive reviews have recently been published discussing the effect of carotenoids in metabolic diseases [23,24], and so we will describe them briefly here. The focus of this review article is to combine the technical considerations and limitations on carotenoid research with a special emphasis on obesity research. For this, we will explore the mechanistic insights obtained from in vitro, cellular, and animal models, as well as observational and interventional studies in human subjects.

2. Vitamin A Sources in Mammals—Provitamin A Carotenoids

Carotenoids are pigments synthesized mostly by photosynthetic organisms to function as light-harvesting scavengers during photosynthesis [25]. Chemically, carotenoids contain forty carbons, usually organized in a single tetraterpenoid chain with conjugated double bonds that are responsible for their coloration. Most carotenoids in our diet are cyclic on both ends, forming an ionone ring. Depending on the absence or presence of oxygen groups, carotenoids are classified as carotenes or xanthophylls, respectively. As such, β, β-carotene (β-carotene) (cyclic, two β ionone rings) and lycopene (acyclic) are two of the most common carotenes in nature, while lutein, zeaxanthin, and β-cryptoxanthin are the most abundant xanthophylls. There are approximately 650 carotenoids in nature, 50 are abundant in the human diet, and only 20 are significantly present in human plasma. Fruits and vegetables are the primary sources of dietary carotenoids, but some animal products such as eggs and salmon also contain these pigments in significant amounts [26].

Scientists have attributed many functions to carotenoids, most of which are positive on human health [20,27], although some notable exceptions have sparked intense debate in the scientific community [28–31]. Despite these controversies, carotenoids have an indisputable role in mammals, as some of them serve as a vitamin A precursor [32]. Only a handful of carotenoids have documented provitamin A activity, as they contain at least one unsubstituted β-ionone ring, required to produce vitamin A. Three of these carotenoids are carotenes: α,β-carotene (α-carotene), γ,β-carotene (γ-carotene) and β-carotene; and only one is a xanthophyll: β-cryptoxanthin. β-carotene, β-cryptoxanthin, and α-carotene are the most abundant provitamin A carotenoids in our diet (Table 1).

Table 1. List of food sources abundant on pro-vitamin A carotenoids (red) and retinyl esters (blue). RAE, Retinol activity equivalent. Source; USDA Food composition database. Data expressed as µg/100 g food. Pro-vitamin A carotenoids in meat sources are very low or not present, while vegetables do not contain retinyl esters.

Source	β-Carotene	α-Carotene	β-Cryptoxanthin	Vitamin A, RAE
Peppers	42,891	6931	-	3863
Carrots	33,954	14,251	-	3423
Paprika	26,162	595	6186	2463
Pepper (spices)	21,840	-	6252	2081
Grape leaves	16,194	629	9	1376
Chili powder	15,000	2090	3490	1483
Sweet potatoes	12,498	47	-	1043
Pumpkin	6940	4795	-	778
Lettuce	5226	-	-	435
Squash	4226	834	3471	532
Seaweed	4872	-	-	406
Fish oil, cod liver	-	-	-	30,000
Liver (beef, other meats)	-	-	-	28,318
Lamb, liver	-	-	-	19,872
Duck, liver	-	-	-	11,984
Turkey, liver	-	-	-	10,751
Chicken, liver	-	-	-	4374

1 µg RAE = 1 µg retinol, 12 µg beta-carotene, 24 µg alpha-carotene, or 24 µg beta-cryptoxanthin. RAE conversion values obtained from [33].

It is widely believed that β-carotene is the preferential source of vitamin A in mammals, since its two β-ionone rings can render two vitamin A molecules. This logical statement, however, has not been carefully characterized to date, as many other factors, such as the differential absorption and transport, cell type specificity, tissue storage or enzymatic accessibility, could define which provitamin A carotenoid is the preferred source for vitamin A production in humans. For example, some researchers postulate that β-cryptoxanthin could be a better source of vitamin A than β-carotene, presumably because of the presence of a hydroxyl group on its structure, which affects its solubility and absorption [34,35]. Previous studies have also shown that, unlike β-carotene, β-cryptoxanthin conversion to vitamin A involves a multi-step enzymatic process [36], introducing an extra layer of complexity that involves enzymatic activities and substrate partitioning. Furthermore, genetic variations localized in both the promoter and/or coding region transporters and enzymes involved in carotenoid metabolism are increasingly gaining attention, as their implications for human health could have a tremendous impact in the way we understand carotenoid metabolism [37–40].

Carotenoids are not the only source of vitamin A in animals. After carotenoids are ingested and converted to vitamin A, this fat-soluble vitamin is stored as retinyl esters in different organs, such as the liver and the adipose tissue. Therefore, the ingestion of certain organs such as the liver from other animals could be a significant source of vitamin A for humans (see Table 1). Vitamin A is also present in milk, providing this crucial nutrient to the newborn [41]. Since vitamin A deficiency is still a public health problem in more than half of the countries worldwide, fortification efforts, such as the addition of retinyl esters, are currently part of a preventive strategy carried out by many governments [42]

Another strategy for reducing vitamin A deficiency consists of the genetic modification of certain crops poor in carotenoids, but highly consumed in countries exposed to vitamin A deficiency. A notable example of these strategies is the development of crops fortified with β-carotene such as golden rice [43] or cassava [44,45]. The biggest advantage of this second approach is that β-carotene is present in the food matrix and absorbed in the intestine by a protein-mediated process [46,47]. In the absence of enough vitamin A in the body, two intestinal proteins are dramatically upregulated to promote carotenoid uptake and conversion to vitamin A [48]. On the contrary, many subjects have reported vitamin A toxicity due to vitamin A ingestion with a wide variety of physiological manifestations such as liver damage, bone abnormalities, headaches, skin desquamation or death [49,50].

Taken together, the scientific consensus in the field considers provitamin A carotenoids as the preferred source of vitamin A for humans, and β-carotene as the main precursor of vitamin A [20,24,48].

3. β–Carotene Oxygenase 1 as Solely Responsible for Vitamin A Production in Mammals

β–carotene conversion to vitamin A was first postulated by Moore in 1930 after observing that vitamin A-deficient animals recovered health when fed β-carotene [51]. In 1965, two research groups showed that β-carotene was cleaved when incubated in the presence of rat intestine homogenates, and this cleavage occurred at the central 15-15′ double bond, yielding two molecules of vitamin A aldehyde (retinal). They proposed that the enzyme behaves like an oxygenase, naming it β-carotene 15-15′ dioxygenase (later named β-carotene oxygenase 1, BCO1), as it seemed to involve a dioxygenase reaction mechanism [52,53]. Recently, two laboratories confirmed these finding by using human recombinant BCO1 and its homolog in insects (NinaB) [54,55].

In 2000, von Lintig and Vogt cloned and identified for the first time an enzyme with β-carotene dioxygenase activity using *Drosophila melanogaster*, showing that this protein is a non-heme, iron-dependent soluble (cytosolic) enzyme [56]. In mammals, BCO1 is present in most tissues, but its expression is higher in the intestine and the liver indicating that β-carotene is largely converted to vitamin A in these shortly after its ingestion. However, since β-carotene is present in human plasma, and BCO1 has a broad expression pattern, it is plausible that peripheral tissues such as the adipose tissue can cleave β-carotene to form vitamin A locally [57]. This review discusses the implications of this pathway below.

In 2001, von Lintig's group cloned and characterized a second carotenoid cleaving enzyme named β-carotene oxygenase 2 (BCO2) [58]. These findings finally provided a mechanistic explanation by which asymmetric apocarotenoids are present in various animal tissues [59], a finding that, together with the cloning of BCO1, shook up the carotenoid field and facilitated many of the groundbreaking discoveries that would come after.

4. β-Carotene and Obesity; Key Findings and Technical Limitations

In recent years, many research groups have studied the effects of carotenoids and apocarotenoids on energy metabolism. Due to its importance in nutrition and health, this field has been extensively reviewed in the past [23,24]. In this section, we will highlight the main findings related to β-carotene and its derivatives (retinoids) on energy metabolism, adipocyte function, and adipose tissue biology in various experimental models. We will also focus on some of the technical approaches and limitations of these studies, which we divided into cell culture, animal, and human studies.

4.1. Cell Culture Studies—β-carotene and Adipogenesis

Cell culture studies represent a good strategy for elucidating the mechanism of action of bioactive compounds. These techniques allow the manipulation of genetic and environmental factors such as dose and time-dependent effects, unbiased screening of bioactive compounds, and tissue-specific targeted effects. However, cell culture studies using carotenoids are technically challenging due to their hydrophobicity, which often hampers their bioavailability in aqueous phases as cell culture media. Some carotenoids, such as β-carotene, are in relatively elevated concentrations in human plasma (Figure 1), but the direct dissolution of β-carotene into the media, for example at 2 μM using a common solvent such as dimethyl sulfoxide will result in its irreversible precipitation. Another strategy, the addition of strong detergents, such as Tween, is one of the most acceptable techniques used to promote the formation of β-carotene micelles [60–62]. This approach will lead to increased solubility in the media, but could cause toxicity and unexpected side effects, as these detergents produce holes in cell membranes. Less harmful approaches, such as cyclodextrin, serum, or purified lipoproteins, have been used in the past as an alternative [63], but in all cases, the stability of β-carotene must be considered, especially in the absence of antioxidants in solution [64].

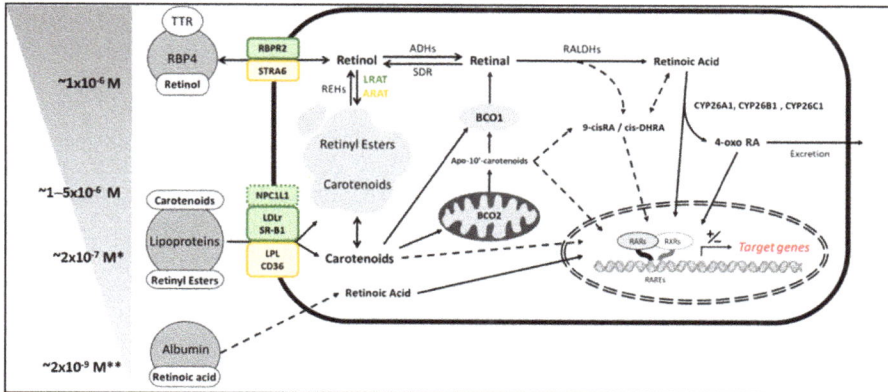

Figure 1. Schematic representation of carotenoid and vitamin A uptake and metabolism. **Left:** extracellular sources of provitamin A carotenoids (named carotenoids) and vitamin A and the relative concentration. * indicates that these sources can vary more than one order of magnitude depending on the fasting vs. fed conditions. Data show fasting values. **Right:** main proteins and conversion pathways involved in the uptake, cleavage/conversion and catabolism of carotenoids and vitamin A. For the purpose of this review, we only represented the proteins present in adipocytes and the hepatocytes in yellow and green, respectively. Dotted arrows are pathways not fully established. TTR, transthyretin; RBP4, retinol-binding protein 4; RBPR2, RBP4-receptor 2; STRA6, stimulated retinoic acid gene 6; NPC1L1, Niemann-Pick C1-Like 1; LDLr, Low-density lipoprotein receptor; SR-B1, scavenger receptor class B type 1; LPL, lipoprotein lipase; CD36, cluster of differentiation 36; REHs; Retinyl ester hydrolases; LRAT, lecithin:retinol acyl transferase, ARAT, acyl:retinol acyl transferase; ADHs, aldehyde hydrogenases; SDR, short-chain dehydrogenase/reductase; RALDH, retinaldehyde hydrogenases; 9-cisRA, 9-cis-retinoic acid; cis-DHRA, 9-cis-13,14-dihydroretinoic acid; CYP26s, cytochrome P450s; RARs, retinoic acid receptors; RXRs, retinoid X receptors; RAREs, retinoic acid-response element.

Another hindrance to carotenoid research is the identification of the bioactive compound responsible for the cellular effects. As we mentioned above, most cell types express BCO1 and BCO2, which cleave carotenoids generating bioactive molecules with properties that often resemble those observed or described by the parent compound. For example, the exposure of either lycopene [65] or its BCO2-mediated cleavage product apo-10′-lycopenoic acid [66] attenuates the expression of pro-inflammatory cytokines in cultured adipocytes. Similarly to the transcriptionally active form of vitamin A retinoic acid, apo-10′-lycopenoic acid seems to activate the retinoic acid receptors (RARs) (see next section for more details), providing a mechanistic explanation of the effects observed in cell culture. Whether all BCO2- derived apocarotenoids can act as RAR ligands is not clear. Some studies show that some apocarotenoids do not bind RARs [67], while other studies suggest that some of them can act as RAR antagonists [68,69] (Figure 1).

Altogether, the utilization of carotenoids in cell culture experiments can lead to misinterpretation of the results, as technical issues such as solubility and stability can arise. In addition, the presence and enzymatic activity of BCO1 and BCO2 will also affect the interpretation of the results. These issues must be addressed before interpreting the conclusions of an experiment using carotenoids in cell culture, a practice that often does not occur.

4.2. Cell Culture Studies—Retinoic Acid and Adipogenesis

The cleavage products of β-carotene by the action of BCO1, commonly named vitamin A, have a defined role in vision and gene expression, being involved in vital processes such as cellular differentiation, embryo development and immune system maturation [24,70–73]. Among the different forms of vitamin A, retinoic acid is the main regulator of gene expression by directly binding specific members of the nuclear receptor superfamily: RARs and the retinoid X receptors (RXRs). All-trans-retinoic acid, the predominant isomer in cells, has the highest binding affinity for RARs, while the 9-cis-retinoic acid isomer can bind both the RARs and the RXRs (See Figure 1). These nuclear receptors form different homo and heterodimer combinations (RXR:RAR, RXR:RXR, RXR:other), and modulate the transcription of over 650 genes. There are three different RAR and three RXR isoforms named alpha, beta and gamma (recently reviewed in [74]). Additionally, all-trans-retinoic acid, but not 9-cis-retinoic acid, can directly bind the PPARδ, an obligate RXR heterodimer [75]. Additionally, the endogenous retinoid 9-cis-13,14-dihydroretinoic acid (cis-DHRA) has been recently characterized as a novel RXR agonist [76]. While the precursor of this compound has not been established yet [77], its specificity for RXRs indicates that it could activate different pathways other than those activated by retinoic acid.

The role of all-trans-retinoic acid (referred as retinoic acid from now on) in adipocyte biology was established many years ago through seminal work from Lazar's group using cell culture models. They showed that retinoic acid inhibits adipocyte differentiation by blocking the expression of key adipogenic transcription factors [78]. Adding retinoic acid to fully differentiated adipocytes favors lipolysis instead, promoting adipocyte "browning" characterized by induction of UCP1 and brown/beige adipocyte markers [79,80].

β-carotene, the main provitamin A carotenoid, exerts similar effects to those observed by retinoic acid in mature brown adipocytes [81] and other cell types prone to accumulate lipids such as macrophages [62], indicating that β-carotene is efficiently cleaved by BCO1 to produce retinal, which is rapidly converted to its transcriptionally active form retinoic acid.

Due to the physiological relevance of white adipose tissue, and since this organ efficiently stores both β-carotene and vitamin A, observations by von Lintig's group showed that β-carotene could also serve as a physiological source of retinoic acid in adipocytes. We observed that BCO1 expression is upregulated during adipocyte differentiation, which would facilitate the conversion of β-carotene to vitamin A in adipocytes as they become fully differentiated and capable of storing intracellular lipids. We also observed that exogenous β-carotene, but not retinol, triggered retinoic acid production in cultured adipocytes, indicating that β-carotene is the preferred source of retinoic acid in mature adipocytes over preformed vitamin A (retinol). Lastly, and in accordance with follow-up animal studies (see below), we observed that β-carotene-exposed adipocytes presented reduced lipolysis accompanied by a downregulation of PPARγ [82].

In summary, the use of retinoic acid to study the effects of β-carotene as a precursor of vitamin A will allow the researcher to bypass the difficulty of dissolving β-carotene in the media (β-carotene solubility in water is approximately 0.7 nM, while retinoic acid is 16 nM, values calculated using ALOGP software). Additionally, carotenoid oxygenases are not uniformly expressed in all cell types, and some carotenoids such as β-carotene can be cleaved by both BCO1 and BCO2, which could lead to confounding results. Lastly, while the concentration or retinoic acid to obtain significant effects in cell culture is generally considered pharmacological or even toxic, scientists should take into consideration that these compounds do not cross the lipid bilayer easily when added to the media [36].

4.3. Animal Models in Carotenoid Research

The use of animal models to resemble human physiology is a very common practice in biomedical research. Humans accumulate large amounts of carotenoids in tissues and plasma, but also cleave them to produce vitamin A and other apocarotenoids via BCO1 and BCO2. Similarly, some animal models, such as ferrets, gerbils and non-human primates, accumulate and cleave carotenoids. These models are expensive and the possibility to study certain diseases is very limited as not many genetically modified organisms are commercially available [83,84]. Rodents are the most common animal model in biomedical research, but do not accumulate carotenoids when fed at physiological doses. The cause of these interspecific differences could be explained by the catalytical properties of the carotenoid cleaving enzymes; while human or primate BCO1 and BCO2 show little enzyme activity in vitro, murine BCO1 and BCO2 are very active on cleaving carotenoids [85,86].

Based on these observations, scientists should be careful when choosing the adequate animal model to study carotenoid metabolism. For example, ferrets and gerbils are good animal models to study carotenoid absorption and biodistribution, while rodents are recommended if the goal is to examine the properties of the carotenoid cleavage products. Unfortunately, these recommendations are not always followed, and while some experts in the field have exposed multiple times the caveats of rodents to study intact carotenoids, many scientists choose to supplement carotenoids at supraphysiological concentrations, or even by injecting these dietary compounds intraperitoneally or intravenously without having a clear scientific justification.

It was not until over a decade ago when Johannes von Lintig and Adrian Wyss' teams joined forces and developed two knockout mouse models lacking BCO1 and BCO2, respectively. In 2007, $Bco1^{-/-}$ mice were characterized, and later, in 2011, $Bco2^{-/-}$ mice followed. The most striking characteristic of both mouse models was their ability to accumulate carotenoids as it occurs in humans; $Bco1^{-/-}$ accumulate β-carotene [87], while $Bco2^{-/-}$ mice accumulate xanthophylls such as lutein, zeaxanthin [88], and β-cryptoxanthin [36], as well as the carotene lycopene [89]. While BCO1 is a cytosolic enzyme, BCO2 is present in the inner mitochondrial membrane where it protects this organelle against carotenoid accumulation [88,90,91]. Altogether, these experiments demonstrated that despite their similar catalytical properties, BCO1 and BCO2 have differential substrate specificity and subcellular localization, facilitating the compartmentalization of carotenoid metabolism. Taking β-carotene as an example, this carotenoid is cleaved to vitamin A in the cytosol (where BCO1 is present), but in case of an excess of intracellular β-carotene, BCO2 will prevent its accumulation in the mitochondria [88,90].

This is not the case for β-cryptoxanthin, another provitamin A carotenoid. While this carotenoid is the only xanthophyll with provitamin A activity, the exact mechanism of how vitamin A formation occurred was not clear. β-cryptoxanthin is one of the most abundant carotenoids in our diet (Table 1) and some authors consider this xanthophyll a better provitamin A carotenoid than β-carotene [34,35]. By using a combination of in vitro enzyme activity assays, cell culture experiments, and BCO1 and BCO2 deficient mice, we demonstrated that β-cryptoxanthin suffers a sequential cleavage to apo-10′-carotenal and retinal by BCO2 and BCO1, respectively [36]. These studies have recently been expanded, further confirming these findings [92] and pointing out the complexity of carotenoid metabolism and vitamin A production.

Therefore, $Bco1^{-/-}$ and $Bco2^{-/-}$ mice seem to be a suitable model for studying carotenoid accumulation in humans, as they accumulate carotenoids as people do. However, these mouse models have a few limitations that researchers cannot forget. For example, as they are full-body knockouts, these mice do not cleave carotenoids at all. Additionally, feeding xanthophylls such as lutein or zeaxanthin to $Bco2^{-/-}$ mice results in the accumulation of oxidized forms of these carotenoids, with little presence of the parent compound in tissues or sera [57,93]. This key difference should be considered if the research goal is to extrapolate the metabolism of xanthophylls to humans, in which the formation of oxidized carotenoids is minimal compared to rodents [57,94].

4.4. Animal Models—β-carotene and Obesity

Dietary interventions to study the effect of carotenoids in animal models generally describe their positive effects on the prevention and treatment of body weight gain, either with provitamin A carotenoids [95] or not [96–99]. While the effects of provitamin A carotenoids on obesity can be somehow explained by the conversion of these molecules to vitamin A (Figure 1), the mechanism by which the rest of carotenoids could reduce obesity remains largely elusive. Some authors hypothesize that these effects are mediated by apocarotenoid metabolites, while others suggest they are derived from the effects of the parent compound(s). Additionally, as mentioned before, some authors seem to ignore the importance of the carotenoid concentration in diet, which could potentially result in toxic side effects or alterations in food intake or nutrient absorption.

In 2011, as part of a joint effort including eight laboratories, we studied the effect of β-carotene on obesity using wild-type and $Bco1^{-/-}$ deficient mice. We observed that physiological amounts of dietary β-carotene reduced adipose tissue size in wild-type mice. These effects were associated with a global downregulation of adipogenic genes, most of which had a PPARγ-responsive element, similarly to our observations in cell culture [82]. On the contrary, $Bco1^{-/-}$ mice did not show any significant changes upon β-carotene supplementation, despite $Bco1^{-/-}$ mice accumulating large amounts of β-carotene in the adipose tissue. We also detected for the first time the accumulation of apo-10′-carotenol in $Bco1^{-/-}$ mice, as the BCO2 asymmetric derivative of β-carotene. Yet, no significant changes were observed neither at the level of mRNA nor adipose tissue size or histological analyses. Overall, these results demonstrated for the first time that β-carotene reduces obesity only when it is converted to vitamin A, and that the accumulation of apo-10′-carotenol or β-carotene itself do not affect gene expression or obesity in vivo [57].

Since BCO2 can cleave β-carotene to produce apo-10′-carotenal, which accumulates as its reduced form apo-10′-carotenol, we next aimed to quantify the contribution of this enzyme in β-carotene metabolism. For this purpose, we used BCO1- and BCO2-deficient mice to generate BCO1/BCO2 double knockout mice. We observed that despite BCO2 can produce apo-10′-carotenoids, this pathway is marginal and BCO1 is the main β-carotene cleaving enzyme in mammals. Additionally, we described that apo-10′-carotenol can be stored, transported, and taken up by cells similarly to retinol. This observation is particularly important, as it shows that if BCO2 could easily convert β-carotene in cells, its cleavage product apo-10′-carotenol would interfere with vitamin A transporters/enzymes, potentially affecting key processes such as vision [36].

In summary, the use of animal models to study carotenoid physiology is a common practice that should be undertaken carefully. While some animal, such as ferrets and gerbils, are great models for mimicking carotenoid metabolism in humans, rodents can lead to confounding results, as they completely metabolize carotenoids when ingested. Currently, nearly 1,000 research articles have utilized rodents in carotenoid research, and only a few used BCO1 or BCO2-deficient mice to study carotenoid accumulation. The consideration of the adequate model to understand the function of these compounds, either as a whole or as an apocarotenoid metabolite(s), is crucial for the interpretation of the data.

4.5. Animal Models—Vitamin A and Obesity

The effect of retinoic acid in obesity has been well documented over the years by different groups, and using different experimental methods. Most of these studies have focused on the treatment of adult mice, showing that retinoic acid exposure affects adipogenesis in animals [100], promotes adipose tissue browning [80,101], and favors an overall induction of fatty acid oxidation in adipocytes [102]. These observations were similar in other metabolic tissues, such as liver and muscle, both in cell culture [103,104] and animal models [105,106].

While the role of vitamin A in adipocyte differentiation was previously established in cell culture [78], and some reports implicate this compound in adult mice [100], we designed an experiment in which vitamin A was supplemented at physiological doses during adipose tissue development in vivo. To this end, we supplemented newborn rats with three times the normal content of retinyl esters in milk and determined if vitamin A exposure during adipogenesis could affect adipocyte differentiation and obesity later in life. This intervention was carried out during the entire lactation period (21 days) before weaning, which is the period when rats develop adipose tissue [107]. One set of rats was sacrificed after lactation, showing for the first time that vitamin A supplementation during adipogenesis promotes adipocyte proliferation in newborn animals, defined by an increased number of small adipocytes and an increased number of proliferating cell nuclear antigen (PCNA)-positive cells. A second set of rats were supplemented with retinyl esters following the same methodology, but subjected to a chow or high-fat diet after weaning time for 16 weeks. Two different cohorts of rats, with 5 animals/group and cohort, showed that rats receiving three times more retinyl esters during lactation and were fed high-fat diet, presented an increased obesity index and elevated leptin production in the white adipose tissue [108].

In our opinion, these findings clearly relate to those from Lazar's group, where cultured preadipocytes exposed to retinoic acid showed reduced differentiation [78], which would translate to adipocyte proliferation. Our study showed that rats supplemented with retinyl esters as retinoic acid precursor during adipose tissue development increased adipocyte proliferation, confirming the role of retinoic acid and the importance of the exposure window. If too much vitamin A is administered during the proliferation stage of adipose tissue, this could lead to obesity later in life, while if retinoic acid is supplemented to mature adipocytes it will lead to weight loss [109].

4.6. Human Studies—β-Carotene and Obesity

It is common knowledge that diets rich in fruits and vegetables are associated with improved health and longevity and, since carotenoids are abundant in plants, many studies correlated the presence of these pigments with a healthy status in people (recently reviewed in [110–112]). Nevertheless, these observations could be merely associations based on the chemical properties of carotenoids, i.e., carotenoids are decreased in obesity because these compounds accumulate in the adipose tissue as this tissue works as a "trap" for lipophilic molecules. Osth's work nicely supports this hypothesis by measuring β-carotene in the adipose tissue in lean and obese individuals. They showed that while obese people had lower β-carotene content in the adipose tissue when normalized to triglyceride tissue levels, these individuals had similar β-carotene content as lean control patients [113].

Many studies have tried to establish causation between elevated carotenoids in plasma and improved health status by using supplementation strategies, and while some studies provided positive results [114,115], the negative outcomes of two major clinical trials including β-carotene set back the possibility of β-carotene dietary carotenoid supplementation in humans. Both the alpha-tocopherol, β-carotene cancer prevention study (ATBC) and the β-carotene and retinol efficacy trial (CARET) showed that β-carotene supplementation promoted cancer progression in subjects exposed to asbestos and smoking [28,116]. Some authors have hypothesized that these adverse effects could be mediated by the anti-oxidant properties of β-carotene [117], supported by convincing data for some antioxidants such as vitamin E promoting cancer [118,119], although this nutrient was only present in the ATBC study, and not in the CARET study. Despite the efforts from many research groups, the relationship between cancer progression and β-carotene supplementation still remain elusive. It is possible that the elevated dose of β-carotene (20 and 30mg β-carotene/day) given during the ATBC and CARET studies to subjects already at risk of suffering cancer was the cause of these adverse effects [20]. Another possibility is that the purity of the β-carotene used in these trials was not adequate, as β-carotene is unstable when stored for a long period, or when it is exposed to elevated temperatures, light and/or oxygen ([120–122] and Figure 2).

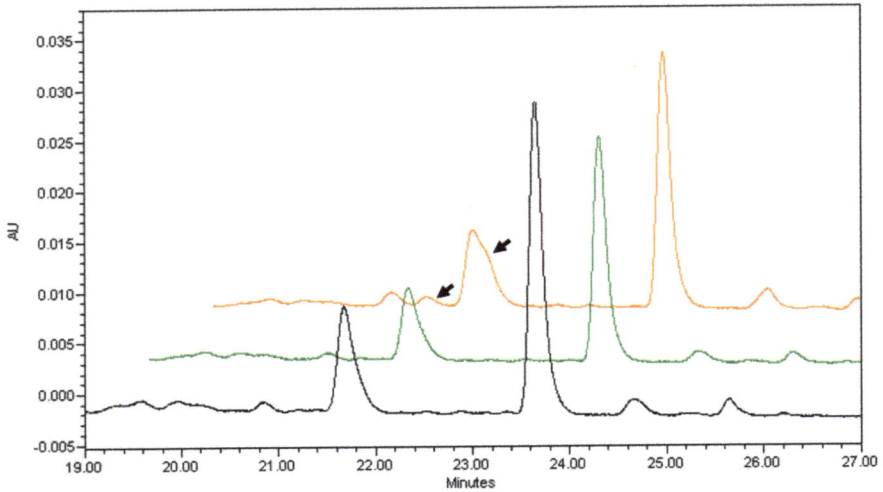

Figure 2. Purity and isomer variation between three commercial sources of β-carotene. HPLC chromatograms detected at 452 nm were obtained using YMC C30 column in three different commercially available β-carotene sources. The largest peak shows all-trans-β-carotene. Smaller peaks correspond to different carotenoid isomers (probably cis forms) in each commercial source. Arrows show the presence of two peaks only present in the orange chromatogram (probably β-carotene cis forms). AU, arbitrary units.

At the end of 2018, a group of researchers analyzed the health outcomes of over 29,000 participants of the ATBC study and correlated the baseline β-carotene plasma levels with overall and cause-specific mortality. Their study concluded that elevated β-carotene concentration is associated with lower cardiovascular disease, stroke, and cancer. This positive effect of β-carotene was dose-dependent, and the correction of their statistical models with the consumption of fruits and vegetables did not alter the outcome of this observation [123]. This study is particularly important, as it examines the effect of plasma β-carotene in a very large population [124], and agrees with a recent meta-analysis study showing that elevated β-carotene correlates with a lower incidence of metabolic syndrome [23]. Regarding obesity, β-carotene in plasma inversely correlate with body weight, as obese individuals have reduced levels compared to normal weight people (reviewed in [19]). Accordingly, Canas' group showed that the supplementation of obese children with a carotenoid mixture showed a significant increase in plasma concentration of various carotenoids including β-carotene, accompanied by a decreased body mass index score, waist-to-height ratio, and subcutaneous adipose tissue content [125].

Nevertheless, β-carotene is the most abundant carotenoid in plasma, and the main precursor of vitamin A, that, as we previously reviewed, it is a crucial nutrient involved in the regulation of lipid metabolism in many organs, and is especially important in the adipose tissue, where it controls adipocyte differentiation, proliferation, and lipid metabolism. These observations are in accordance with the mounting evidence that points out that obesity [126] and other lipid-mediated disorders such as insulin resistance [23], fatty liver disease [117], or coronary artery disease [127] are associated with a lower retinoic acid status in human. Since β-carotene is the main precursor of vitamin A and, therefore, retinoic acid, the hypothesis that elevated β-carotene promotes retinoic acid signaling cannot be ignored.

On the flip side of the coin, patients suffering from anorexia nervosa present elevated plasma carotenoid levels that can, in some cases, lead to hypercarotenemia and changes on skin color [128,129]. Scientists believed that a combination of an elevated intake of low-calorie foods rich on carotenoids such as carrots and spinach, a reduced metabolic conversion to vitamin A, and a reduced carotenoid deposition in adipose tissue could lead to hypercarotenemia [130,131]. In agreement with this hypothesis, patients suffering cachexia do not show signs of hypercarotenemia, although they present a reduced adipose tissue size, [132]. More research on this topic will be necessary to elucidate these interesting observations, as carotenoids could play a role on anorexia nervosa [132], or as carotenoids could worsen this disorder as some of them have leaning effects as they favor fatty acid oxidation via retinoic acid signaling.

4.7. Human Studies—Synthetic Retinoids and Obesity

Considering the positive effects of retinoic acid on lipid metabolism in experimental models, and the indications that elevated retinoic acid status in human is associated with positive health outcomes, we would imagine that retinoic acid treatment in humans could reduce obesity and therefore work as an antiobesity agent. However retinoic acid is highly toxic for humans [133], and the use of less toxic, synthetic retinoids such as fenretinide could suppose a strategy to treat obese patients. Fenretinide is an anticancer drug currently employed in some clinical trials, alone (identifier NCT00546455) or in combination with other molecules [134]. The use of fenretinide as an antiobesity agent is supported by promising data in cell culture and animal studies describing its effects on preventing and treating obesity and some associated disorders [135–138]. We should be cautious, however, as some of the side effects of fenretinide and its metabolites include the reduction of vitamin A delivery to tissues, presumably through the binding of this molecule to the vitamin A transporter retinol-binding protein 4 [82,139–142].

Other synthetic retinoids, which are commonly used in the treatment of acne, have known effects on lipid metabolism. For example, oral treatment with Accutane (13- cis- retinoic acid) increases total triglyceride and LDL-cholesterol levels in some patients [143,144], but the effect on obesity has not been systematically studied. These effects could be mediated by alterations in hepatic lipid metabolism, as retinoic acid upregulates lipogenesis by inducing the expression of lipogenic genes [145,146], and its long-term exposure reduces lipolysis in certain cell types [147]. However, we observed that retinoic acid induces fatty acid oxidation in muscle cells [103], human hepatoma cells [104], and our recent data show that retinoic acid decreases triglyceride and cholesterol secretion in cultured hepatocytes (under preparation).

5. Conclusions

Mechanistic studies developed in cell culture and animal models show that β-carotene, the main precursor of vitamin A and retinoic acid, decreases obesity by promoting fatty acid oxidation in adipocytes and other tissues. Observational and interventional studies corroborate these findings and provide a mechanistic explanation by which the consumption of fruits and vegetables, rich in provitamin A carotenoids, prevent the development of obesity.

In our opinion, the implementation of policies and strategies promoting the consumption of foods rich in provitamin A carotenoids could contribute to prevent the development of obesity and other metabolic diseases such as atherosclerosis and diabetes. Additionally, we also propose to introduce the quantification of retinoic acid and β-carotene into clinical practice as novel markers of health status. This is based on recent data obtained from clinical and preclinical models supporting the notion that low levels of these two compounds are tightly linked to the development of metabolic diseases.

Funding: This research was funded by the American Heart Association (16SDG27550012), the U.S. Department of Agriculture (Multi-State grant project W4002), and the National Institute of Health (HL147252).

Acknowledgments: J.C. is a recipient of Philip L. and Juanita Fitzer Francis Fellowship in Health and Wellness We also thank John Erdman and Molly Black for their assistance with HPLC, and Nuria Granados and Benjamin Abraham for their suggestions to the manuscript.

Conflicts of Interest: The authors declare no conflict of interests.

References

1. Tremmel, M.; Gerdtham, U.G.; Nilsson, P.M.; Saha, S. Economic Burden of Obesity: A Systematic Literature Review. *Int. J. Environ. Res. Public Health* **2017**, *14*, 435. [CrossRef] [PubMed]
2. Hales, C.M.; Carroll, M.D.; Fryar, C.D.; Ogden, C.L. *Prevalence of Obesity Among Adults and Youth: United States, 2015-2016*; NCHS Data Brief No. 288; Centers for Disease Control and Prevention, National Center for Health Statistics: Hyattsville, MD, USA, 2017; pp. 1–8.
3. Briggs, A.M.; Woolf, A.D.; Dreinhofer, K.; Homb, N.; Hoy, D.G.; Kopansky-Giles, D.; Akesson, K.; March, L. Reducing the global burden of musculoskeletal conditions. *Bull. World Health Organ.* **2018**, *96*, 366–368. [CrossRef] [PubMed]
4. Hruby, A.; Hu, F.B. The Epidemiology of Obesity: A Big Picture. *Pharmacoeconomics* **2015**, *33*, 673–689. [CrossRef]
5. Kimmons, J.E.; Blanck, H.M.; Tohill, B.C.; Zhang, J.; Khan, L.K. Associations between body mass index and the prevalence of low micronutrient levels among US adults. *MedGenMed* **2006**, *8*, 59. [PubMed]
6. Church, T.; Martin, C.K. The Obesity Epidemic: A Consequence of Reduced Energy Expenditure and the Uncoupling of Energy Intake? *Obesity* **2018**, *26*, 14–16. [CrossRef]
7. Hill, J.O.; Wyatt, H.R.; Peters, J.C. Energy balance and obesity. *Circulation* **2012**, *126*, 126–132. [CrossRef] [PubMed]
8. Cinti, S. The adipose organ at a glance. *Dis. Model. Mech.* **2012**, *5*, 588–594. [CrossRef] [PubMed]
9. Jirtle, R.L.; Skinner, M.K. Environmental epigenomics and disease susceptibility. *Nat. Rev. Genet.* **2007**, *8*, 253–262. [CrossRef]
10. Cohen, P.; Spiegelman, B.M. Cell biology of fat storage. *Mol. Biol. Cell* **2016**, *27*, 2523–2527. [CrossRef]
11. Spiegelman, B.M.; Flier, J.S. Obesity and the regulation of energy balance. *Cell* **2001**, *104*, 531–543. [CrossRef]
12. Klein, S.; Wadden, T.; Sugerman, H.J. AGA technical review on obesity. *Gastroenterology* **2002**, *123*, 882–932. [CrossRef]
13. Hall, K.D.; Sacks, G.; Chandramohan, D.; Chow, C.C.; Wang, Y.C.; Gortmaker, S.L.; Swinburn, B.A. Quantification of the effect of energy imbalance on bodyweight. *Lancet* **2011**, *378*, 826–837. [CrossRef]
14. Makris, A.; Foster, G.D. Dietary approaches to the treatment of obesity. *Psychiatr. Clin. N. Am.* **2011**, *34*, 813–827. [CrossRef] [PubMed]
15. Westerterp, K.R. Physical activity, food intake, and body weight regulation: Insights from doubly labeled water studies. *Nutr. Rev.* **2010**, *68*, 148–154. [CrossRef] [PubMed]
16. Tian, Y.; Su, L.; Wang, J.; Duan, X.; Jiang, X. Fruit and vegetable consumption and risk of the metabolic syndrome: A meta-analysis. *Public Health Nutr.* **2018**, *21*, 756–765. [CrossRef] [PubMed]
17. Maria, A.G.; Graziano, R.; Nicolantonio, D.O. Carotenoids: Potential allies of cardiovascular health? *Food Nutr. Res.* **2015**, *59*, 26762. [CrossRef] [PubMed]
18. Williams, D.J.; Edwards, D.; Hamernig, I.; Jian, L.; James, A.P.; Johnson, S.K.; Tapsell, L.C. Vegetables containing phytochemicals with potential anti-obesity properties: A review. *Food Res. Int.* **2013**, *52*, 323–333. [CrossRef]
19. Bonet, M.L.; Canas, J.A.; Ribot, J.; Palou, A. Carotenoids in Adipose Tissue Biology and Obesity. In *Carotenoids in Nature*; Springer: Cham, Switzerland, 2016; Volume 79, pp. 377–414.
20. Eggersdorfer, M.; Wyss, A. Carotenoids in human nutrition and health. *Arch. Biochem. Biophys.* **2018**, *652*, 18–26. [CrossRef] [PubMed]
21. Landrum, J.T.; Bone, R.A. Lutein, zeaxanthin, and the macular pigment. *Arch. Biochem. Biophys.* **2001**, *385*, 28–40. [CrossRef]
22. Rowles, J.L., 3rd; Ranard, K.M.; Smith, J.W.; An, R.; Erdman, J.W., Jr. Increased dietary and circulating lycopene are associated with reduced prostate cancer risk: A systematic review and meta-analysis. *Prostate Cancer Prostatic Dis.* **2017**, *20*, 361–377. [CrossRef]
23. Beydoun, M.A.; Chen, X.; Jha, K.; Beydoun, H.A.; Zonderman, A.B.; Canas, J.A. Carotenoids, vitamin A, and their association with the metabolic syndrome: A systematic review and meta-analysis. *Nutr. Rev.* **2019**, *77*, 32–45. [CrossRef] [PubMed]

24. Rodriguez-Concepcion, M.; Avalos, J.; Bonet, M.L.; Boronat, A.; Gomez-Gomez, L.; Hornero-Mendez, D.; Limon, M.C.; Melendez-Martinez, A.J.; Olmedilla-Alonso, B.; Palou, A.; et al. A global perspective on carotenoids: Metabolism, biotechnology, and benefits for nutrition and health. *Prog. Lipid Res.* **2018**, *70*, 62–93. [CrossRef]

25. Moise, A.R.; Al-Babili, S.; Wurtzel, E.T. Mechanistic aspects of carotenoid biosynthesis. *Chem. Rev.* **2014**, *114*, 164–193. [CrossRef] [PubMed]

26. Moran, N.E.; Mohn, E.S.; Hason, N.; Erdman, J.W., Jr.; Johnson, E.J. Intrinsic and Extrinsic Factors Impacting Absorption, Metabolism, and Health Effects of Dietary Carotenoids. *Adv. Nutr.* **2018**, *9*, 465–492. [CrossRef] [PubMed]

27. von Lintig, J.; Eggersdorfer, M.; Wyss, A. News and views about carotenoids: Red-hot and true. *Arch. Biochem. Biophys.* **2018**, *657*, 74–77. [CrossRef] [PubMed]

28. Alpha-Tocopherol, Beta Carotene Cancer Prevention Study Group. The effect of vitamin E and β carotene on the incidence of lung cancer and other cancers in male smokers. *N. Engl. J. Med.* **1994**, *330*, 1029–1035. [CrossRef]

29. Christen, W.; Glynn, R.; Sperduto, R.; Chew, E.; Buring, J. Age-related cataract in a randomized trial of β-carotene in women. *Ophthalmic Epidemiol.* **2004**, *11*, 401–412. [CrossRef] [PubMed]

30. Lee, I.M.; Cook, N.R.; Manson, J.E.; Buring, J.E.; Hennekens, C.H. Beta-carotene supplementation and incidence of cancer and cardiovascular disease: The Women's Health Study. *J. Natl. Cancer Inst.* **1999**, *91*, 2102–2106. [CrossRef] [PubMed]

31. van Poppel, G. Epidemiological evidence for β-carotene in prevention of cancer and cardiovascular disease. *Eur. J. Clin. Nutr.* **1996**, *50* (Suppl. 3), S57–S61.

32. Grune, T.; Lietz, G.; Palou, A.; Ross, A.C.; Stahl, W.; Tang, G.; Thurnham, D.; Yin, S.A.; Biesalski, H.K. Beta-carotene is an important vitamin A source for humans. *J. Nutr.* **2010**, *140*, 2268S–2285S. [CrossRef] [PubMed]

33. Institute of Medicine (US) Panel on Micronutrients. *Dietary Reference Intakes for Vitamin A, Vitamin K, Arsenic, Boron, Chromium, Copper, Iodine, Iron, Manganese, Molybdenum, Nickel, Silicon, Vanadium, and Zinc*; National Academy of Medicine: Washington, DC, USA, 2001.

34. Burri, B.J. Beta-cryptoxanthin as a source of vitamin A. *J. Sci. Food Agric.* **2015**, *95*, 1786–1794. [CrossRef] [PubMed]

35. Burri, B.J.; La Frano, M.R.; Zhu, C. Absorption, metabolism, and functions of β-cryptoxanthin. *Nutr. Rev.* **2016**, *74*, 69–82. [CrossRef]

36. Amengual, J.; Widjaja-Adhi, M.A.; Rodriguez-Santiago, S.; Hessel, S.; Golczak, M.; Palczewski, K.; von Lintig, J. Two carotenoid oxygenases contribute to mammalian provitamin A metabolism. *J. Biol. Chem.* **2013**, *288*, 34081–34096. [CrossRef] [PubMed]

37. Farook, V.S.; Reddivari, L.; Mummidi, S.; Puppala, S.; Arya, R.; Lopez-Alvarenga, J.C.; Fowler, S.P.; Chittoor, G.; Resendez, R.G.; Kumar, B.M.; et al. Genetics of serum carotenoid concentrations and their correlation with obesity-related traits in Mexican American children. *Am. J. Clin. Nutr.* **2017**, *106*, 52–58. [CrossRef]

38. Meyers, K.J.; Johnson, E.J.; Bernstein, P.S.; Iyengar, S.K.; Engelman, C.D.; Karki, C.K.; Liu, Z.; Igo, R.P., Jr.; Truitt, B.; Klein, M.L.; et al. Genetic determinants of macular pigments in women of the Carotenoids in Age-Related Eye Disease Study. *Investig. Ophthalmol. Vis. Sci.* **2013**, *54*, 2333–2345. [CrossRef] [PubMed]

39. Widjaja-Adhi, M.A.; Lobo, G.P.; Golczak, M.; Von Lintig, J. A genetic dissection of intestinal fat-soluble vitamin and carotenoid absorption. *Hum. Mol. Genet.* **2015**, *24*, 3206–3219. [CrossRef] [PubMed]

40. Yabuta, S.; Urata, M.; Wai Kun, R.Y.; Masaki, M.; Shidoji, Y. Common SNP rs6564851 in the BCO1 Gene Affects the Circulating Levels of β-Carotene and the Daily Intake of Carotenoids in Healthy Japanese Women. *PLoS ONE* **2016**, *11*, e0168857. [CrossRef]

41. Haskell, M.J.; Brown, K.H. Maternal vitamin A nutriture and the vitamin A content of human milk. *J. Mammary Gland Biol. Neoplasia* **1999**, *4*, 243–257. [CrossRef]

42. Dary, O.; Mora, J.O.; International Vitamin A Consultative Group. Food fortification to reduce vitamin A deficiency: International Vitamin A Consultative Group recommendations. *J. Nutr.* **2002**, *132*, 2927S–2933S. [CrossRef]

43. Ye, X.; Al-Babili, S.; Kloti, A.; Zhang, J.; Lucca, P.; Beyer, P.; Potrykus, I. Engineering the provitamin A (β-carotene) biosynthetic pathway into (carotenoid-free) rice endosperm. *Science* **2000**, *287*, 303–305. [CrossRef]

44. La Frano, M.R.; Woodhouse, L.R.; Burnett, D.J.; Burri, B.J. Biofortified cassava increases β-carotene and vitamin A concentrations in the TAG-rich plasma layer of American women. *Br. J. Nutr.* **2013**, *110*, 310–320. [CrossRef] [PubMed]

45. Welsch, R.; Arango, J.; Bar, C.; Salazar, B.; Al-Babili, S.; Beltran, J.; Chavarriaga, P.; Ceballos, H.; Tohme, J.; Beyer, P. Provitamin A accumulation in cassava (Manihot esculenta) roots driven by a single nucleotide polymorphism in a phytoene synthase gene. *Plant Cell* **2010**, *22*, 3348–3356. [CrossRef] [PubMed]

46. Choi, M.Y.; Romer, A.I.; Hu, M.; Lepourcelet, M.; Mechoor, A.; Yesilaltay, A.; Krieger, M.; Gray, P.A.; Shivdasani, R.A. A dynamic expression survey identifies transcription factors relevant in mouse digestive tract development. *Development* **2006**, *133*, 4119–4129. [CrossRef] [PubMed]

47. Lobo, G.P.; Amengual, J.; Baus, D.; Shivdasani, R.A.; Taylor, D.; von Lintig, J. Genetics and diet regulate vitamin A production via the homeobox transcription factor ISX. *J. Biol. Chem.* **2013**, *288*, 9017–9027. [CrossRef]

48. Lobo, G.P.; Amengual, J.; Palczewski, G.; Babino, D.; von Lintig, J. Mammalian carotenoid-oxygenases: Key players for carotenoid function and homeostasis. *Biochim. Biophys. Acta* **2012**, *1821*, 78–87. [CrossRef]

49. Beste, L.A.; Moseley, R.H.; Saint, S.; Cornia, P.B. CLINICAL PROBLEM-SOLVING. Too Much of a Good Thing. *N. Engl. J. Med.* **2016**, *374*, 873–878. [CrossRef] [PubMed]

50. Geubel, A.P.; De Galocsy, C.; Alves, N.; Rahier, J.; Dive, C. Liver damage caused by therapeutic vitamin A administration: Estimate of dose-related toxicity in 41 cases. *Gastroenterology* **1991**, *100*, 1701–1709. [CrossRef]

51. Moore, T. Vitamin A and carotene: The absence of the liver oil vitamin A from carotene. VI. The conversion of carotene to vitamin A in vivo. *Biochem. J.* **1930**, *24*, 692–702. [CrossRef] [PubMed]

52. Fidge, N.H.; Smith, F.R.; Goodman, D.S. Vitamin A and carotenoids. The enzymic conversion of β-carotene into retinal in hog intestinal mucosa. *Biochem. J.* **1969**, *114*, 689–694. [CrossRef]

53. Olson, J.A.; Hayaishi, O. The enzymatic cleavage of β-carotene into vitamin A by soluble enzymes of rat liver and intestine. *Proc. Natl. Acad. Sci. USA* **1965**, *54*, 1364–1370. [CrossRef] [PubMed]

54. dela Sena, C.; Narayanasamy, S.; Riedl, K.M.; Curley, R.W., Jr.; Schwartz, S.J.; Harrison, E.H. Substrate specificity of purified recombinant human β-carotene 15,15′-oxygenase (BCO1). *J. Biol. Chem.* **2013**, *288*, 37094–37103. [CrossRef] [PubMed]

55. Babino, D.; Golczak, M.; Kiser, P.D.; Wyss, A.; Palczewski, K.; von Lintig, J. The Biochemical Basis of Vitamin A3 Production in Arthropod Vision. *ACS Chem. Biol.* **2016**, *11*, 1049–1057. [CrossRef] [PubMed]

56. von Lintig, J.; Wyss, A. Molecular analysis of vitamin A formation: Cloning and characterization of β-carotene 15,15′-dioxygenases. *Arch. Biochem. Biophys.* **2001**, *385*, 47–52. [CrossRef] [PubMed]

57. Amengual, J.; Gouranton, E.; van Helden, Y.G.; Hessel, S.; Ribot, J.; Kramer, E.; Kiec-Wilk, B.; Razny, U.; Lietz, G.; Wyss, A.; et al. Beta-carotene reduces body adiposity of mice via BCMO1. *PLoS ONE* **2011**, *6*, e20644. [CrossRef]

58. Kiefer, C.; Hessel, S.; Lampert, J.M.; Vogt, K.; Lederer, M.O.; Breithaupt, D.E.; von Lintig, J. Identification and characterization of a mammalian enzyme catalyzing the asymmetric oxidative cleavage of provitamin A. *J. Biol. Chem.* **2001**, *276*, 14110–14116. [CrossRef] [PubMed]

59. Wang, X.D.; Tang, G.W.; Fox, J.G.; Krinsky, N.I.; Russell, R.M. Enzymatic conversion of β-carotene into β-apo-carotenals and retinoids by human, monkey, ferret, and rat tissues. *Arch. Biochem. Biophys.* **1991**, *285*, 8–16. [CrossRef]

60. O'Sullivan, L.; Ryan, L.; O'Brien, N. Comparison of the uptake and secretion of carotene and xanthophyll carotenoids by Caco-2 intestinal cells. *Br. J. Nutr.* **2007**, *98*, 38–44. [CrossRef]

61. O'Sullivan, S.M.; Woods, J.A.; O'Brien, N.M. Use of Tween 40 and Tween 80 to deliver a mixture of phytochemicals to human colonic adenocarcinoma cell (CaCo-2) monolayers. *Br. J. Nutr.* **2004**, *91*, 757–764. [CrossRef] [PubMed]

62. Zolberg Relevy, N.; Bechor, S.; Harari, A.; Ben-Amotz, A.; Kamari, Y.; Harats, D.; Shaish, A. The inhibition of macrophage foam cell formation by 9-cis β-carotene is driven by BCMO1 activity. *PLoS ONE* **2015**, *10*, e0115272. [CrossRef] [PubMed]

63. Gouranton, E.; Yazidi, C.E.; Cardinault, N.; Amiot, M.J.; Borel, P.; Landrier, J.F. Purified low-density lipoprotein and bovine serum albumin efficiency to internalise lycopene into adipocytes. *Food Chem. Toxicol.* **2008**, *46*, 3832–3836. [CrossRef]

64. Scita, G. Stability of β-carotene under different laboratory conditions. *Methods Enzym.* **1992**, *213*, 175–185.

65. Gouranton, E.; Thabuis, C.; Riollet, C.; Malezet-Desmoulins, C.; El Yazidi, C.; Amiot, M.J.; Borel, P.; Landrier, J.F. Lycopene inhibits proinflammatory cytokine and chemokine expression in adipose tissue. *J. Nutr. Biochem.* **2011**, *22*, 642–648. [CrossRef] [PubMed]

66. Gouranton, E.; Aydemir, G.; Reynaud, E.; Marcotorchino, J.; Malezet, C.; Caris-Veyrat, C.; Blomhoff, R.; Landrier, J.F.; Ruhl, R. Apo-10′-lycopenoic acid impacts adipose tissue biology via the retinoic acid receptors. *Biochim. Biophys. Acta* **2011**, *1811*, 1105–1114. [CrossRef]

67. Marsh, R.S.; Yan, Y.; Reed, V.M.; Hruszkewycz, D.; Curley, R.W.; Harrison, E.H. {beta}-Apocarotenoids do not significantly activate retinoic acid receptors {alpha} or {beta}. *Exp. Biol. Med.* **2010**, *235*, 342–348. [CrossRef]

68. Eroglu, A.; Hruszkewycz, D.P.; dela Sena, C.; Narayanasamy, S.; Riedl, K.M.; Kopec, R.E.; Schwartz, S.J.; Curley, R.W., Jr.; Harrison, E.H. Naturally occurring eccentric cleavage products of provitamin A β-carotene function as antagonists of retinoic acid receptors. *J. Biol. Chem.* **2012**, *287*, 15886–15895. [CrossRef] [PubMed]

69. Narayanasamy, S.; Sun, J.; Pavlovicz, R.E.; Eroglu, A.; Rush, C.E.; Sunkel, B.D.; Li, C.; Harrison, E.H.; Curley, R.W., Jr. Synthesis of apo-13- and apo-15-lycopenoids, cleavage products of lycopene that are retinoic acid antagonists. *J. Lipid Res.* **2017**, *58*, 1021–1029. [CrossRef] [PubMed]

70. Shete, V.; Quadro, L. Mammalian metabolism of β-carotene: Gaps in knowledge. *Nutrients* **2013**, *5*, 4849–4868. [CrossRef]

71. Ross, A.C. Vitamin A and retinoic acid in T cell-related immunity. *Am. J. Clin. Nutr.* **2012**, *96*, 1166S–1172S. [CrossRef] [PubMed]

72. von Lintig, J. Provitamin A metabolism and functions in mammalian biology. *Am. J. Clin. Nutr.* **2012**, *96*, 1234S–1244S. [CrossRef]

73. Shete, V.; Costabile, B.K.; Kim, Y.K.; Quadro, L. Low-Density Lipoprotein Receptor Contributes to β-Carotene Uptake in the Maternal Liver. *Nutrients* **2016**, *8*, 765. [CrossRef]

74. Benbrook, D.M.; Chambon, P.; Rochette-Egly, C.; Asson-Batres, M.A. History of retinoic acid receptors. *Subcell. Biochem.* **2014**, *70*, 1–20. [PubMed]

75. Shaw, N.; Elholm, M.; Noy, N. Retinoic acid is a high affinity selective ligand for the peroxisome proliferator-activated receptor β/delta. *J. Biol. Chem.* **2003**, *278*, 41589–41592. [CrossRef]

76. Ruhl, R.; Krzyzosiak, A.; Niewiadomska-Cimicka, A.; Rochel, N.; Szeles, L.; Vaz, B.; Wietrzych-Schindler, M.; Alvarez, S.; Szklenar, M.; Nagy, L.; et al. 9-cis-13,14-Dihydroretinoic Acid Is an Endogenous Retinoid Acting as RXR Ligand in Mice. *PLoS Genet.* **2015**, *11*, e1005213. [CrossRef] [PubMed]

77. Ruhl, R.; Krezel, W.; de Lera, A.R. 9-Cis-13,14-dihydroretinoic acid, a new endogenous mammalian ligand of retinoid X receptor and the active ligand of a potential new vitamin A category: Vitamin A5. *Nutr. Rev.* **2018**, *76*, 929–941. [CrossRef] [PubMed]

78. Schwarz, E.J.; Reginato, M.J.; Shao, D.; Krakow, S.L.; Lazar, M.A. Retinoic acid blocks adipogenesis by inhibiting C/EBPβ-mediated transcription. *Mol. Cell. Biol.* **1997**, *17*, 1552–1561. [CrossRef] [PubMed]

79. Mercader, J.; Ribot, J.; Murano, I.; Felipe, F.; Cinti, S.; Bonet, M.L.; Palou, A. Remodeling of white adipose tissue after retinoic acid administration in mice. *Endocrinology* **2006**, *147*, 5325–5332. [CrossRef]

80. Wang, B.; Fu, X.; Liang, X.; Deavila, J.M.; Wang, Z.; Zhao, L.; Tian, Q.; Zhao, J.; Gomez, N.A.; Trombetta, S.C.; et al. Retinoic acid induces white adipose tissue browning by increasing adipose vascularity and inducing beige adipogenesis of PDGFRalpha(+) adipose progenitors. *Cell Discov.* **2017**, *3*, 17036. [CrossRef]

81. Serra, F.; Bonet, M.L.; Puigserver, P.; Oliver, J.; Palou, A. Stimulation of uncoupling protein 1 expression in brown adipocytes by naturally occurring carotenoids. *Int. J. Obes. Relat. Metab. Disord.* **1999**, *23*, 650–655. [CrossRef] [PubMed]

82. Lobo, G.P.; Amengual, J.; Li, H.N.; Golczak, M.; Bonet, M.L.; Palczewski, K.; von Lintig, J. β,β-carotene decreases peroxisome proliferator receptor gamma activity and reduces lipid storage capacity of adipocytes in a β,β-carotene oxygenase 1-dependent manner. *J. Biol. Chem.* **2010**, *285*, 27891–27899. [CrossRef]

83. Kim, J.; Kim, Y. Animal models in carotenoids research and lung cancer prevention. *Transl. Oncol.* **2011**, *4*, 271–281. [CrossRef]

84. Lee, C.M.; Boileau, A.C.; Boileau, T.W.; Williams, A.W.; Swanson, K.S.; Heintz, K.A.; Erdman, J.W., Jr. Review of animal models in carotenoid research. *J. Nutr.* **1999**, *129*, 2271–2277. [CrossRef] [PubMed]

85. Babino, D.; Palczewski, G.; Widjaja-Adhi, M.A.; Kiser, P.D.; Golczak, M.; von Lintig, J. Characterization of the Role of β-Carotene 9,10-Dioxygenase in Macular Pigment Metabolism. *J. Biol. Chem.* **2015**, *290*, 24844–24857. [CrossRef] [PubMed]

86. Li, B.; Vachali, P.P.; Gorusupudi, A.; Shen, Z.; Sharifzadeh, H.; Besch, B.M.; Nelson, K.; Horvath, M.M.; Frederick, J.M.; Baehr, W.; et al. Inactivity of human β,β-carotene-9′,10′-dioxygenase (BCO2) underlies retinal accumulation of the human macular carotenoid pigment. *Proc. Natl. Acad. Sci. USA* **2014**, *111*, 10173–10178. [CrossRef]

87. Hessel, S.; Eichinger, A.; Isken, A.; Amengual, J.; Hunzelmann, S.; Hoeller, U.; Elste, V.; Hunziker, W.; Goralczyk, R.; Oberhauser, V.; et al. CMO1 deficiency abolishes vitamin A production from β-carotene and alters lipid metabolism in mice. *J. Biol. Chem.* **2007**, *282*, 33553–33561. [CrossRef]

88. Amengual, J.; Lobo, G.P.; Golczak, M.; Li, H.N.; Klimova, T.; Hoppel, C.L.; Wyss, A.; Palczewski, K.; von Lintig, J. A mitochondrial enzyme degrades carotenoids and protects against oxidative stress. *FASEB J.* **2011**, *25*, 948–959. [CrossRef]

89. Ford, N.A.; Clinton, S.K.; von Lintig, J.; Wyss, A.; Erdman, J.W., Jr. Loss of carotene-9′,10′-monooxygenase expression increases serum and tissue lycopene concentrations in lycopene-fed mice. *J. Nutr.* **2010**, *140*, 2134–2138. [CrossRef] [PubMed]

90. Palczewski, G.; Amengual, J.; Hoppel, C.L.; von Lintig, J. Evidence for compartmentalization of mammalian carotenoid metabolism. *FASEB J.* **2014**, *28*, 4457–4469. [CrossRef]

91. Wu, L.; Guo, X.; Lyu, Y.; Clarke, S.L.; Lucas, E.A.; Smith, B.J.; Hildebrand, D.; Wang, W.; Medeiros, D.M.; Shen, X.; et al. Targeted Metabolomics Reveals Abnormal Hepatic Energy Metabolism by Depletion of β-Carotene Oxygenase 2 in Mice. *Sci. Rep.* **2017**, *7*, 14624. [CrossRef] [PubMed]

92. Kelly, M.E.; Ramkumar, S.; Sun, W.; Colon Ortiz, C.; Kiser, P.D.; Golczak, M.; von Lintig, J. The Biochemical Basis of Vitamin A Production from the Asymmetric Carotenoid β-Cryptoxanthin. *ACS Chem. Biol.* **2018**, *13*, 2121–2129. [CrossRef]

93. Yonekura, L.; Kobayashi, M.; Terasaki, M.; Nagao, A. Keto-carotenoids are the major metabolites of dietary lutein and fucoxanthin in mouse tissues. *J. Nutr.* **2010**, *140*, 1824–1831. [CrossRef]

94. Wang, W.; Connor, S.L.; Johnson, E.J.; Klein, M.L.; Hughes, S.; Connor, W.E. Effect of dietary lutein and zeaxanthin on plasma carotenoids and their transport in lipoproteins in age-related macular degeneration. *Am. J. Clin. Nutr.* **2007**, *85*, 762–769. [CrossRef] [PubMed]

95. Takayanagi, K.; Morimoto, S.; Shirakura, Y.; Mukai, K.; Sugiyama, T.; Tokuji, Y.; Ohnishi, M. Mechanism of visceral fat reduction in Tsumura Suzuki obese, diabetes (TSOD) mice orally administered β-cryptoxanthin from Satsuma mandarin oranges (Citrus unshiu Marc). *J. Agric. Food Chem.* **2011**, *59*, 12342–12351. [CrossRef] [PubMed]

96. Fenni, S.; Hammou, H.; Astier, J.; Bonnet, L.; Karkeni, E.; Couturier, C.; Tourniaire, F.; Landrier, J.F. Lycopene and tomato powder supplementation similarly inhibit high-fat diet induced obesity, inflammatory response, and associated metabolic disorders. *Mol. Nutr. Food Res.* **2017**, *61*. [CrossRef] [PubMed]

97. Ikeuchi, M.; Koyama, T.; Takahashi, J.; Yazawa, K. Effects of astaxanthin in obese mice fed a high-fat diet. *Biosci. Biotechnol. Biochem.* **2007**, *71*, 893–899. [CrossRef]

98. Liu, M.; Liu, H.; Xie, J.; Xu, Q.; Pan, C.; Wang, J.; Wu, X.; Sanabil; Zheng, M.; Liu, J. Anti-obesity effects of zeaxanthin on 3T3-L1 preadipocyte and high fat induced obese mice. *Food Funct.* **2017**, *8*, 3327–3338. [CrossRef]

99. Seca, A.M.L.; Pinto, D. Overview on the Antihypertensive and Anti-Obesity Effects of Secondary Metabolites from Seaweeds. *Mar. Drugs* **2018**, *16*, 237. [CrossRef] [PubMed]

100. Berry, D.C.; DeSantis, D.; Soltanian, H.; Croniger, C.M.; Noy, N. Retinoic acid upregulates preadipocyte genes to block adipogenesis and suppress diet-induced obesity. *Diabetes* **2012**, *61*, 1112–1121. [CrossRef]

101. Mercader, J.; Madsen, L.; Felipe, F.; Palou, A.; Kristiansen, K.; Bonet, M.L. All-trans retinoic acid increases oxidative metabolism in mature adipocytes. *Cell. Physiol. Biochem.* **2007**, *20*, 1061–1072. [CrossRef]

102. Bonet, M.L.; Ribot, J.; Felipe, F.; Palou, A. Vitamin A and the regulation of fat reserves. *Cell. Mol. Life Sci.* **2003**, *60*, 1311–1321. [CrossRef]

103. Amengual, J.; Garcia-Carrizo, F.J.; Arreguin, A.; Musinovic, H.; Granados, N.; Palou, A.; Bonet, M.L.; Ribot, J. Retinoic Acid Increases Fatty Acid Oxidation and Irisin Expression in Skeletal Muscle Cells and Impacts Irisin In Vivo. *Cell. Physiol. Biochem.* **2018**, *46*, 187–202. [CrossRef]

104. Amengual, J.; Petrov, P.; Bonet, M.L.; Ribot, J.; Palou, A. Induction of carnitine palmitoyl transferase 1 and fatty acid oxidation by retinoic acid in HepG2 cells. *Int. J. Biochem. Cell Biol.* **2012**, *44*, 2019–2027. [CrossRef]

105. Amengual, J.; Ribot, J.; Bonet, M.L.; Palou, A. Retinoic acid treatment increases lipid oxidation capacity in skeletal muscle of mice. *Obesity* **2008**, *16*, 585–591. [CrossRef] [PubMed]

106. Amengual, J.; Ribot, J.; Bonet, M.L.; Palou, A. Retinoic acid treatment enhances lipid oxidation and inhibits lipid biosynthesis capacities in the liver of mice. *Cell. Physiol. Biochem.* **2010**, *25*, 657–666. [CrossRef] [PubMed]

107. Cryer, A.; Jones, H.M. The early development of white adipose tissue. Effects of litter size on the lipoprotein lipase activity of four adipose-tissue depots, serum immunoreactive insulin and tissue cellularity during the first four weeks of life in the rat. *Biochem. J.* **1979**, *178*, 711–724. [CrossRef]

108. Granados, N.; Amengual, J.; Ribot, J.; Musinovic, H.; Ceresi, E.; von Lintig, J.; Palou, A.; Bonet, M.L. Vitamin A supplementation in early life affects later response to an obesogenic diet in rats. *Int. J. Obes.* **2013**, *37*, 1169–1176. [CrossRef]

109. Bonet, M.L.; Ribot, J.; Palou, A. Lipid metabolism in mammalian tissues and its control by retinoic acid. *Biochim. Biophys. Acta* **2012**, *1821*, 177–189. [CrossRef] [PubMed]

110. Ames, B.N. Prolonging healthy aging: Longevity vitamins and proteins. *Proc. Natl. Acad. Sci. USA* **2018**, *115*, 10836–10844. [CrossRef]

111. Goncalves, A.; Amiot, M.J. Fat-soluble micronutrients and metabolic syndrome. *Curr. Opin. Clin. Nutr. Metab. Care* **2017**, *20*, 492–497. [CrossRef] [PubMed]

112. Mitchell, P.; Liew, G.; Gopinath, B.; Wong, T.Y. Age-related macular degeneration. *Lancet* **2018**, *392*, 1147–1159. [CrossRef]

113. Osth, M.; Ost, A.; Kjolhede, P.; Stralfors, P. The concentration of β-carotene in human adipocytes, but not the whole-body adipocyte stores, is reduced in obesity. *PLoS ONE* **2014**, *9*, e85610. [CrossRef]

114. Asemi, Z.; Alizadeh, S.A.; Ahmad, K.; Goli, M.; Esmaillzadeh, A. Effects of β-carotene fortified synbiotic food on metabolic control of patients with type 2 diabetes mellitus: A double-blind randomized cross-over controlled clinical trial. *Clin. Nutr.* **2016**, *35*, 819–825. [CrossRef] [PubMed]

115. Qiu, Q.; Yao, Y.; Wu, X.; Han, D. Higher dose lutein and a longer supplementation period would be good for visual performance. *Nutrition* **2013**, *29*, 1072–1073. [CrossRef] [PubMed]

116. Omenn, G.S.; Goodman, G.; Thornquist, M.; Grizzle, J.; Rosenstock, L.; Barnhart, S.; Balmes, J.; Cherniack, M.G.; Cullen, M.R.; Glass, A.; et al. The β-carotene and retinol efficacy trial (CARET) for chemoprevention of lung cancer in high risk populations: Smokers and asbestos-exposed workers. *Cancer Res.* **1994**, *54*, 2038s–2043s. [PubMed]

117. Greenwald, P. Beta-carotene and lung cancer: A lesson for future chemoprevention investigations? *J. Natl. Cancer Inst.* **2003**, *95*, E1. [CrossRef] [PubMed]

118. Le Gal, K.; Ibrahim, M.X.; Wiel, C.; Sayin, V.I.; Akula, M.K.; Karlsson, C.; Dalin, M.G.; Akyurek, L.M.; Lindahl, P.; Nilsson, J.; et al. Antioxidants can increase melanoma metastasis in mice. *Sci. Transl. Med.* **2015**, *7*, 308re8. [CrossRef]

119. Sayin, V.I.; Ibrahim, M.X.; Larsson, E.; Nilsson, J.A.; Lindahl, P.; Bergo, M.O. Antioxidants accelerate lung cancer progression in mice. *Sci. Transl. Med.* **2014**, *6*, 221ra5. [CrossRef] [PubMed]

120. Xu, D.; Wang, X.; Jiang, J.; Yuan, F.; Decker, E.A.; Gao, Y. Influence of pH, EDTA, α-tocopherol, and WPI oxidation on the degradation of β-carotene in WPI-stabilized oil-in-water emulsions. *LWT-Food Sci. Technol.* **2013**, *54*, 236–241. [CrossRef]

121. Yi, J.; Fan, Y.; Yokoyama, W.; Zhang, Y.; Zhao, L. Thermal degradation and isomerization of β-carotene in oil-in-water nanoemulsions supplemented with natural antioxidants. *J. Agric. Food Chem.* **2016**, *64*, 1970–1976. [CrossRef]

122. Yi, J.; Li, Y.; Zhong, F.; Yokoyama, W. The physicochemical stability and in vitro bioaccessibility of β-carotene in oil-in-water sodium caseinate emulsions. *Food Hydrocoll.* **2014**, *35*, 19–27. [CrossRef]

123. Huang, J.; Weinstein, S.J.; Yu, K.; Mannisto, S.; Albanes, D. Serum Beta Carotene and Overall and Cause-Specific Mortality. *Circ. Res.* **2018**, *123*, 1339–1349. [CrossRef]

124. Rimm, E.B. The Challenges of Deconstructing Fruits and Vegetables. *Circ. Res.* **2018**, *123*, 1267–1268. [CrossRef] [PubMed]

125. Canas, J.A.; Lochrie, A.; McGowan, A.G.; Hossain, J.; Schettino, C.; Balagopal, P.B. Effects of Mixed Carotenoids on Adipokines and Abdominal Adiposity in Children: A Pilot Study. *J. Clin. Endocrinol. Metab.* **2017**, *102*, 1983–1990. [CrossRef] [PubMed]

126. Trasino, S.E.; Tang, X.H.; Jessurun, J.; Gudas, L.J. Obesity Leads to Tissue, but not Serum Vitamin A Deficiency. *Sci. Rep.* **2015**, *5*, 15893. [CrossRef]

127. Liu, Y.; Chen, H.; Mu, D.; Li, D.; Zhong, Y.; Jiang, N.; Zhang, Y.; Xia, M. Association of Serum Retinoic Acid With Risk of Mortality in Patients With Coronary Artery Disease. *Circ. Res.* **2016**, *119*, 557–563. [CrossRef]

128. Birmingham, C.L. Hypercarotenemia. *N. Engl. J. Med.* **2002**, *347*, 222–223. Available online: https://www.ncbi.nlm.nih.gov/pubmed/12124418 (accessed on 24 March 2019). [PubMed]

129. Pierach, C.A. Hypercarotenemia. *N. Engl. J. Med.* **2002**, *347*, 222–223. Available online: https://www.ncbi.nlm.nih.gov/pubmed/12132109/?ncbi_mmode=std (accessed on 24 March 2019).

130. Tyler, I.; Wiseman, M.C.; Crawford, R.I.; Birmingham, C.L. Cutaneous manifestations of eating disorders. *J. Cutan. Med. Surg.* **2002**, *6*, 345–353. [CrossRef] [PubMed]

131. Curran-Celentano, J.; Erdman, J.W., Jr.; Nelson, R.A.; Grater, S.J. Alterations in vitamin A and thyroid hormone status in anorexia nervosa and associated disorders. *Am. J. Clin. Nutr.* **1985**, *42*, 1183–1191. [CrossRef]

132. Priyadarshani, A.M.B. Insights of hypercarotenaemia: A brief review. *Clin. Nutr. ESPEN* **2018**, *23*, 19–24. [CrossRef]

133. Penniston, K.L.; Tanumihardjo, S.A. The acute and chronic toxic effects of vitamin A. *Am. J. Clin. Nutr.* **2006**, *83*, 191–201. [CrossRef]

134. Cooper, J.P.; Reynolds, C.P.; Cho, H.; Kang, M.H. Clinical development of fenretinide as an antineoplastic drug: Pharmacology perspectives. *Exp. Biol. Med.* **2017**, *242*, 1178–1184. [CrossRef] [PubMed]

135. Koh, I.U.; Jun, H.S.; Choi, J.S.; Lim, J.H.; Kim, W.H.; Yoon, J.B.; Song, J. Fenretinide ameliorates insulin resistance and fatty liver in obese mice. *Biol. Pharm. Bull.* **2012**, *35*, 369–375. [CrossRef] [PubMed]

136. McIlroy, G.D.; Delibegovic, M.; Owen, C.; Stoney, P.N.; Shearer, K.D.; McCaffery, P.J.; Mody, N. Fenretinide treatment prevents diet-induced obesity in association with major alterations in retinoid homeostatic gene expression in adipose, liver, and hypothalamus. *Diabetes* **2013**, *62*, 825–836. [CrossRef]

137. McIlroy, G.D.; Tammireddy, S.R.; Maskrey, B.H.; Grant, L.; Doherty, M.K.; Watson, D.G.; Delibegovic, M.; Whitfield, P.D.; Mody, N. Fenretinide mediated retinoic acid receptor signalling and inhibition of ceramide biosynthesis regulates adipogenesis, lipid accumulation, mitochondrial function and nutrient stress signalling in adipocytes and adipose tissue. *Biochem. Pharm.* **2016**, *100*, 86–97. [CrossRef]

138. Morrice, N.; McIlroy, G.D.; Tammireddy, S.R.; Reekie, J.; Shearer, K.D.; Doherty, M.K.; Delibegovic, M.; Whitfield, P.D.; Mody, N. Elevated Fibroblast growth factor 21 (FGF21) in obese, insulin resistant states is normalised by the synthetic retinoid Fenretinide in mice. *Sci. Rep.* **2017**, *7*, 43782. [CrossRef] [PubMed]

139. Amengual, J.; Golczak, M.; Palczewski, K.; von Lintig, J. Lecithin:retinol acyltransferase is critical for cellular uptake of vitamin A from serum retinol-binding protein. *J. Biol. Chem.* **2012**, *287*, 24216–24227. [CrossRef] [PubMed]

140. Amengual, J.; Zhang, N.; Kemerer, M.; Maeda, T.; Palczewski, K.; Von Lintig, J. STRA6 is critical for cellular vitamin A uptake and homeostasis. *Hum. Mol. Genet.* **2014**, *23*, 5402–5417. [CrossRef]

141. Poliakov, E.; Gubin, A.; Laird, J.; Gentleman, S.; Salomon, R.G.; Redmond, T.M. The mechanism of fenretinide (4-HPR) inhibition of β-carotene monooxygenase 1. New suspect for the visual side effects of fenretinide. *Adv. Exp. Med. Biol.* **2012**, *723*, 167–174.

142. Poliakov, E.; Samuel, W.; Duncan, T.; Gutierrez, D.B.; Mata, N.L.; Redmond, T.M. Inhibitory effects of fenretinide metabolites N-[4-methoxyphenyl]retinamide (MPR) and 4-oxo-N-(4-hydroxyphenyl)retinamide (3-keto-HPR) on fenretinide molecular targets β-carotene oxygenase 1, stearoyl-CoA desaturase 1 and dihydroceramide Delta4-desaturase 1. *PLoS ONE* **2017**, *12*, e0176487.

143. Beckenbach, L.; Baron, J.M.; Merk, H.F.; Loffler, H.; Amann, P.M. Retinoid treatment of skin diseases. *Eur. J. Dermatol.* **2015**, *25*, 384–391.

144. Lestringant, G.G.; Frossard, P.M.; Agarwal, M.; Galadari, I.H. Variations in lipid and lipoprotein levels during isotretinoin treatment for acne vulgaris with special emphasis on HDL-cholesterol. *Int. J. Dermatol.* **1997**, *36*, 859–862. [CrossRef] [PubMed]

145. Roder, K.; Schweizer, M. Retinoic acid-mediated transcription and maturation of SREBP-1c regulates fatty acid synthase via cis-elements responsible for nutritional regulation. *Biochem. Soc. Trans.* **2007**, *35*, 1211–1214. [CrossRef] [PubMed]

146. Roder, K.; Zhang, L.; Schweizer, M. SREBP-1c mediates the retinoid-dependent increase in fatty acid synthase promoter activity in HepG2. *FEBS Lett.* **2007**, *581*, 2715–2720. [CrossRef] [PubMed]

147. Devereux, D.F.; Taylor, D.D.; Taylor, C.G.; Hollander, D.M. Effects of retinoic acid on lipolytic activity of tumor cells. *Surgery* **1987**, *102*, 277–282.

MDPI

Review

Anti-Obesity Effect of Carotenoids: Direct Impact on Adipose Tissue and Adipose Tissue-Driven Indirect Effects

Lourdes Mounien [1], Franck Tourniaire [1,2] and Jean-Francois Landrier [1,2,*]

[1] Aix Marseille Univ, INSERM, INRA, C2VN, 13385 Marseille, France
[2] CriBioM, criblage biologique Marseille, faculté de Médecine de la Timone, 13256 Marseille, France
* Correspondence: jean-francois.landrier@univ-amu.fr; Tel.: +33-491-324-275

Received: 29 May 2019; Accepted: 7 July 2019; Published: 11 July 2019

Abstract: This review summarizes current knowledge on the biological relevance of carotenoids and some of their metabolites in obesity management. The relationship between carotenoids and obesity is considered in clinical studies and in preclinical studies. Adipose tissue is a key organ in obesity etiology and the main storage site for carotenoids. We thus first describe carotenoid metabolism in adipocyte and adipose tissue and the effects of carotenoids on biological processes in adipose tissue that may be linked to obesity management in in vitro and preclinical studies. It is also now well established that the brain is strongly involved in obesity processes. A section is accordingly devoted to the potential effect of carotenoids on obesity via their direct and/or adipose tissue-driven indirect biological effects on the brain.

Keywords: adipocytes; adipose tissue; brain; carotenoids; obesity

---

1. Obesity, Comorbidities, Adipose Tissue and Brain Dysfunctions

The World Health Organization (WHO) defines obesity and being overweight as abnormal or excessive fat accumulation that presents a risk to health [1]. The risk is mainly related to comorbidities strongly linked to obesity such as metabolic inflammation, insulin resistance, liver steatosis, hypertension, dyslipidemia, certain types of cancer, depression, etc. The WHO states that in 2016, around 39% of the adult population were overweight, and about 13% of the world's adult population were obese [1]. This prefigures a major public health issue in the short term not only in western countries but also in low- and middle-income countries, where an epidemic of obesity and being overweight is emerging.

The excess fat mass that characterizes obesity is produced by an expansion of adipose tissue mediated by hypertrophy and/or hyperplasia of adipocytes [2], which is linked to complex, tightly regulated adipogenesis. This process has been studied in depth, and both the temporal sequences and the transcriptional regulators involved have been identified. Among them, the nuclear receptor peroxisome proliferator-activated receptor gamma (PPARγ) and the CCAAT-enhancer-binding protein (CEBP) families are considered as transcriptional regulators of adipogenesis [3]. Through this mechanism, the adipose tissue can participate in energy homeostasis, allowing the storage of excess energy as triglycerides (lipogenesis) and the release of energy as fatty acids (lipolysis). This balance is tightly regulated, and dysregulation may result in body weight gain or loss.

Adipose tissue is also regarded as an endocrine tissue producing not only free fatty acids but also a wide variety of hormones, cytokines, chemokines and miRNA, together with adipokines and growth factors, acting on many physiological processes. Adipose tissue secretes approximately 50 biologically active proteins acting in an autocrine, paracrine and/or endocrine fashion. Leptin [4] and adiponectin [5] are among those most thoroughly studied. Both adipocytes and cells belonging to the stromal vascular fraction of adipose tissue, especially macrophages, are able to produce and secrete adipokines. Obesity triggers chronic low-grade inflammation associated with abnormal secretion of cytokines [6], chemokines [7], miRNA [8,9], acute phase proteins and other mediators of the immune response together with the activation of inflammatory signaling pathways [6,10]. Adipose tissue is a major contributor to the chronic inflammatory response. The regulation of substances secreted by adipose tissue is multifactorial and is linked to several pathophysiological disorders, including (i) increased levels of circulating free fatty acids, (ii) hypoxia of hypertrophied adipose tissue, (iii) systemic and local oxidative stress, (iv) endoplasmic reticulum stress and/or (v) the production of inflammatory cytokines. All these types of stress converge towards signaling pathways involving c-Jun amino-terminal kinase (JNK) and IκB kinase β (IKKβ) [6,10]. A large part of this inflammatory state is mediated by the increased number of infiltrated macrophages during expansion of adipose tissue [11]. This infiltration has been positively correlated with adiposity, adipocyte size and insulin resistance [12]. Macrophages interfere with adipocyte function through the production of pro-inflammatory cytokines such as tumoral necrosis factor α (TNF-α), interleukin (IL) 1β and IL-6. This can lead to insulin resistance, modified adipokine secretion and an excess of free fatty acid secretion through increased lipolysis and diminished lipogenesis [13], and therefore help install obesity-associated disorders such as insulin resistance.

Besides the effect of inflammatory state on adipocyte and adipose tissue function, it has also been shown that metabolic inflammation is associated with neuro-inflammation. Inflammation at the central level is widely suspected to be involved in obesity aetiologia via modulation of energy homeostasis (both at food intake and energy expenditure level) [14]. The control of energy homeostasis is finely tuned by nervous and endocrine mechanisms that cooperate to balance calorie intake and energy expenditure [15,16]. The central nervous system (CNS) continuously monitors modifications in hormones (insulin, leptin and ghrelin) or metabolic parameters (blood glucose or free fatty acids levels) and elicits adaptive responses, like food intake [15,16]. Among the brain regions involved in this regulation, the hypothalamus plays a pivotal role through specific neuronal networks [15–18]. In particular, leptin is crucial to maintaining both normal body weight and feeding behavior by action in the different regions of the hypothalamus such as arcuate, paraventricular or ventromedial nuclei, and the lateral hypothalamus. More specifically, this peripheral signal is detected by hypothalamic arcuate neurons expressing the anorexigenic peptide proopiomelanocortin (POMC) or the orexigenic peptides neuropeptide Y (NPY)/Agouti-related peptide (AgRP). These neurons project to melanocortin 3 and 4 receptor-expressing neurons located in the hypothalamus and other brain structures [16,19]. These neurons are collectively termed the melanocortin pathway, and regulate feeding behavior, energy expenditure and glucose homeostasis through the activation of the autonomic nervous system and higher brain structures [15–17]. A defect in the communication between brain and peripheral organs can affect fat gain and lead to metabolic syndrome.

Obesity leads to increased inflammatory factors and immune cells in peripheral tissues and in the brain regions that are essential for maintaining energy balance [14]. The production of inflammatory cytokines by adipose tissue and the activation of astrocytes and microglia (the resident immune cells of the brain) in the hypothalamus can interfere with leptin signaling and so contribute to hyperphagia and many other obesity-related diseases [15,16]. In this context, the endocrine function of the adipose tissue is essential to maintain normal weight and regulate energy homeostasis.

Several strategies have been proposed to fight obesity, including pharmacological approaches, limitation of fat and sugar consumption, promotion of physical activity and consumption of fruits and vegetables. Plant-based food is classically associated with weight management not only due to its macronutrient composition, but also to the presence of micronutrients, such as carotenoids. These substances correspond to a large family of C_{40} lipophilic pigments produced by plants, fungi and bacteria [20]. Carotenoids can be divided into two groups according to their chemical structure: carotenes, which are hydrocarbons, and xanthophylls, which also contain oxygen and are therefore less apolar than carotenes (Figure 1). More than 600 different substances have been identified, of which

50 can be found in the human diet, and of which only about 10 are present in significant amounts in human plasma [21]. Carotenoids containing an unsubstituted β-ionone ring are termed provitamin A, as they can be cleaved by animals to release retinal, which can subsequently be converted to retinol [20].

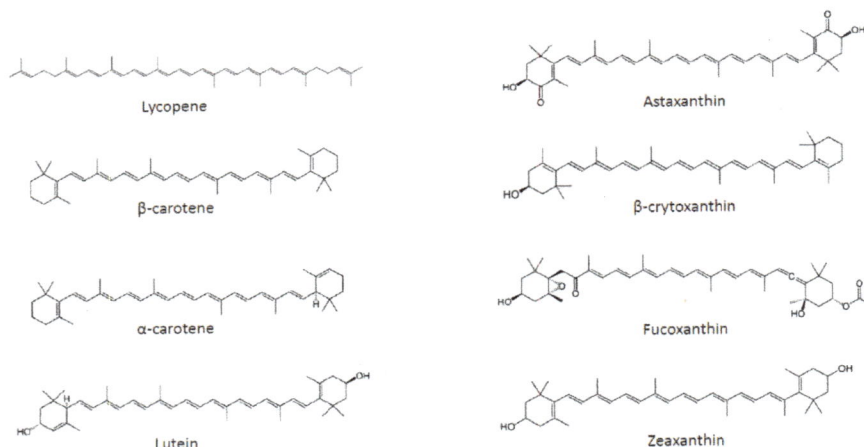

Figure 1. Chemical structures of the main carotenoids.

2. Carotenoids and Obesity in Human Studies

2.1. Observational Studies

Obesity has been associated in many epidemiological and observational studies with low circulating concentrations of carotenoids [22,23]. Strong inverse correlations between body mass index (BMI) and all measured carotenoids in plasma, except lycopene, were highlighted in the CARDIA study [24]. In addition, many obesity-associated disorders, such as low-grade inflammation or insulin resistance, are also strongly inversely associated with serum carotenoid concentrations [25–27].

2.2. Intervention Studies

Several trials have been conducted to study how carotenoids might be used in obesity management. Most of these studies used mixtures of carotenoids and vitamins in a natural matrix, such as fruit juices or plant extracts (reviewed by Bonet et al. [28]), making interpretation of the specific contribution of carotenoids difficult. To our knowledge, only two randomized double-blind placebo-controlled clinical trials have investigated the effect of pure carotenoid or xanthophyll supplementation. Canas et al. [29] reported a decrease in BMI z-score, waist-to-height ratio and subcutaneous adipose tissue in children given a mixture of carotenoids (β-carotene, α-carotene, lutein, zeaxanthin, lycopene, astaxanthin and γ-tocopherol) for 6 months. These beneficial effects were strongly associated with an increase in plasma β-carotene concentration in children [29]. Another study used a mixture of paprika xanthophylls and carotenoids, administered for 12 weeks to healthy overweight volunteers. This supplementation reduced visceral fat area, subcutaneous fat area and total fat area, along with BMI in the treated group compared to a placebo group [30].

3. Carotenoids and/or Metabolites are Involved in Body Weight Management and Limitation of Obesity Comorbidities in Preclinical Studies

Significant research has been devoted to studying the impact of β-carotene on energy metabolism and its outcome on obesity [31]. Its anti-obesity effect was subsequently demonstrated to be linked to a provitamin A effect [32,33], since β-carotene 15, 15′-monooxygenase (BCO) null mice did not display adipose tissue weight modification. This effect was found to be linked to decreased expression of PPARγ in adipose tissue and the involvement of retinoid X receptor (RAR) signaling in this regulation [34].

Astaxanthin prevented obesity in mice fed a high fat diet [35], via the limitation of adipose tissue expansion. Similar anti-obesity effects have been documented in mice fed a high fat and high fructose diet [36], where insulin sensitivity and inflammation were also improved by astaxanthin. Preventive effects of astaxanthin were found for hepatic steatosis [37] and inflammation and fibrosis in the liver in a non alcoholic steato hepatitis NASH and diet induced obesity (DIO) mice model [38].

Anti-adiposity properties have also been reported for β-cryptoxanthin [39], but their mechanism is still unknown. In addition, β-cryptoxanthin reversed liver steatosis and insulin resistance in DIO mice; this effect may be related to the anti-inflammatory effect of this carotenoid in the liver [40].

The potential of fucoxanthin for weight management has been extensively studied and reviewed [41]. This carotenoid limited weight gain and hyperglycemia, and inhibited the expression of several pro-inflammatory cytokines in adipose tissue of KK-a(y) mice [42]. Similar effects have been described in DIO mice, possibly through modulation of lipogenesis, adiponectin production and inflammation in adipose tissue [43], but also via browning of white adipose tissue [41].

Zeaxanthin inhibited obesity in high fat fed mice, presumably by inducing AMP-activated protein kinase (AMPK) activation, and inhibiting lipogenesis in adipose tissue [44].

The anti-obesity effect of lycopene was demonstrated in mice fed a high fat diet, where adiposity was reduced after supplementation. Several comorbidities were concomitantly reduced, such as glucose tolerance, insulin sensitivity and steatosis [45]. We and others have confirmed this beneficial effect of lycopene and/or tomato powder rich in lycopene in a DIO mice model on adiposity, glucose homeostasis, adipose tissue and liver inflammation and steatosis [46–49].

It is also clear that some of the effects of carotenoids (pro-vitamin A or other) are due to the vitamin A effect and are mediated by RAR. Such effects have been extensively reviewed elsewhere [28,31,50] and so will not be detailed here.

Most of these findings not only support the beneficial effect of several carotenoids on obesity management, but also strongly suggest that carotenoids may act on adipocyte/adipose tissue biology to modify several parameters linked to obesity and/or associated comorbidities. This hypothesis is strongly supported by the fact that carotenoids are stored and metabolized and are bioactive in adipocytes and in adipose tissue.

4. Carotenoids and Adipocyte/Adipose Tissue Metabolism

4.1. Carotenoids Are Stored in Adipocytes and Adipose Tissue

It has long been known that carotenoids are notably stored in adipose tissue [51–55]. Lycopene and β-carotene are the predominant carotenoids in human adipose tissue [53,56]. More precisely, Chung et al. identified lycopene as the most abundant carotenoid in adipose tissue (more than 1/2), followed by β-carotene (approx. 1/3 of total carotenoids), lutein + zeaxanthin, β-cryptoxanthin and α-carotene [54].

Total carotenoid concentration appears to be site-specific, with abdomen concentration higher than in the buttocks or thigh [54]. Adipose tissue concentrations of carotenoids are similar in men and women [54]. Interestingly, plasma levels of most carotenoids are inversely correlated to fat mass and to both general and central adiposity [54,57], suggesting that during obesity, carotenoids are sequestered in adipose tissue. Conversely, weight loss is associated with an increase in lutein and zeaxanthin serum concentration [58]. In the case of β-carotene, it is noteworthy that even if its adipose tissue concentration is lower in obese people, the total pool of β-carotene is similar in obese and non-obese when taking into account total fat mass [59].

Factors governing adipose tissue carotenoid uptake, distribution and turnover are poorly understood. However, we recently reported that carotenoid uptake by adipose tissue was independent of the carotenoid's physical and chemical properties [60], suggesting the involvement of putative transporters or facilitators. Consistent with this, we demonstrated the involvement of a cluster of differentiation 36 (CD36) in lycopene and lutein uptake by adipose tissue and adipocytes [61]. We also showed that lycopene was mainly stored in lipid droplets in adipocytes, but was also present in plasma and nuclear membranes [62].

Adipose tissue carotenoid content is usually considered as a good long-term indicator of dietary intake of carotenoids [63]. β-Carotene concentration in adipose tissue increased 5 days after consumption of a large oral dose [64]. Lutein and zeaxanthin levels in adipose tissue significantly increased after spinach and corn supplementation in healthy subjects, with a maximum at 8 weeks of intervention [65]. Finally, tomato-oleoresin supplementation significantly increased lycopene concentration in adipose tissue [66]. Dietary carotenoid intakes were strongly correlated with abdomen adipose tissue concentration (a lower correlation was found for buttock or thigh adipose tissue) for α- and β-carotene, β-cryptoxanthin, cis-lycopene and total carotenoids [54]. However, these correlations vary widely and are strongly sex-related. Notably, El-Sohemy et al. reported correlation in women between intake and concentration in adipose tissues of α-, β-carotene, β-cryptoxanthin and lutein/zeaxanthin (CC 0.25, 0.29, 0.44 and 0.17, respectively), but not in men (CC 0.04, 0.07, 0.23 and 0.06, respectively) [67]. The origin of this discrepancy is presently unknown, but suggests that adipose tissue carotenoid concentration may be affected by factors other than intake, or that carotenoid intake is not appropriately estimated.

Adipose tissue carotenoid content is not only correlated with dietary intake, but also with other tissue concentrations. Thus, lutein adipose tissue content has also been reported to be positively correlated with macular pigment density in men, but not in women [68]. In addition, total carotenoid content in adipose tissue is strongly associated with serum levels [54], except for lycopene and lutein + zeaxanthin. β-Carotene content in adipose tissue is correlated with plasma level, with a correlation coefficient of 0.20 [63,69]. Similarly, breast adipose tissue carotenoid content correlates with plasma levels, except for β-cryptoxanthin [70].

4.2. Carotenoids Are Metabolized in Adipocytes and Adipose Tissue

BCO1, involved in centric cleavage of carotenoids and β-carotene 9′, 10′-dioxygenase (BCO2), involved in eccentric cleavage of carotenoids, are expressed in adipocytes [71], raising the possibility that carotenoid cleavage products, including retinal, derivatives and apocarotenoids, could be found in adipocytes [31,32,72]. In agreement, retinal [73] and free retinol have been identified in the adipocyte fraction of adipose tissue [74]. Several isomers of retinol, including all-trans, 9-cis and 13-cis isomers, were also quantified in white adipose tissue [74–76], together with several isomers of retinoic acid, except for 9-cis retinoic acid [75,77,78]. Adipocytes express BCO1 and BCO2, together with the enzymes necessary for vitamin A metabolism, suggesting that part of the effect of provitamin A carotenoids is mediated via vitamin A production. This topic will not be dealt with here; the reader is referred to the excellent review of Dr. Blaner [50].

Besides these retinoids, β-10′-apocarotenal has been identified in adipose tissue [32]. It is highly probable that other apocarotenoids are produced in adipose tissue, but their function in adipocyte biology needs further research.

4.3. Carotenoids Regulate Gene Expression in Adipocytes and Adipose Tissue

Several molecular mechanisms mediating the effects of carotenoids on gene expression have been described and may be related to the impact of carotenoids on adipocyte biology. In the case of provitamin A carotenoids, leading to retinoic acid synthesis, RARs and retinoid X receptors (RXRs), they constitute specific signaling targets. Two families of receptors mediate the effects of retinoids [79,80]. Three subtypes of each have been described (RARα RARβ, RARγ, RXRα, RXRβ and RXRγ). These receptors work as ligand-dependent transcriptional regulators by binding specific DNA sequences—retinoic acid response element (RARE) or retinoid X response element (RXRE)—found in the promoter region of retinoid target genes either as RAR-RXR or RXR-RXR dimers. RAR and RXR subtypes are found in every cell type. Furthermore, RXRs are dimerization partners for other nuclear receptors such as peroxisome proliferator activated receptors (PPARs), liver X receptor (LXR), farnesoid X receptor (FXR), pregnane X receptor (PXR), RARs, thyroid hormone receptor (TR) and vitamin D receptor (VDR), which are involved in the regulation of a huge number of genes. In addition, several other transcription factors and signaling pathways are modulated by retinoic acid [81], including PPARβ ([82]. Lycopene [83] and apo-10′-lycopenoic acid [84] are also able to activate RAR. Many carotenoids regulate gene expression via ubiquitous signaling pathways such as nuclear factor-kappa B (NF-κB) and mitogen activated proteins (MAP) kinases [85,86], or via transcription factors involved in detoxification such as aryl hydrocarbon receptor (AhR), nuclear factor erythroid-2-related factor 2 (NRF2) or PXR [87,88].

5. Carotenoids and/or Metabolites Impact Adipocyte Biology In Vitro Studies

The impact of some carotenoids has been documented in adipogenesis (Figure 2), which could help obesity management via a limitation of lipid accumulation in adipocytes. Most of the reported effects inhibited adipocyte differentiation [89] by interfering with nuclear receptors such as RAR, RXR or PPAR. β-Carotene inhibited adipogenesis through the production of β-apo-14′-carotenal and repression of PPARα, PPARγ and RXR activation [90], but also through the production of all-trans retinoic acid [34]. Similarly, β-cryptoxanthin suppressed adipogenesis via activation of RAR [91], and astaxanthin inhibited rosiglitazone-induced adipocyte differentiation by antagonizing transcriptional activity of PPARγ [92]. Zeaxanthin [44] and fucoxanthin [93,94] exhibited anti-adipogenic effects via a down-regulation of adipogenic transcription factors C/EBPα and PPARγ, which blunted lipid accumulation. Conversely, lycopene (unpublished personal data) and apo-10′-lycopenoic acid [84] showed no effect on adipogenesis. Besides these effects, there is evidence that some effects of provitamin A carotenoids are mediated through retinol and its metabolite production, which are known to regulate adipogenesis [50].

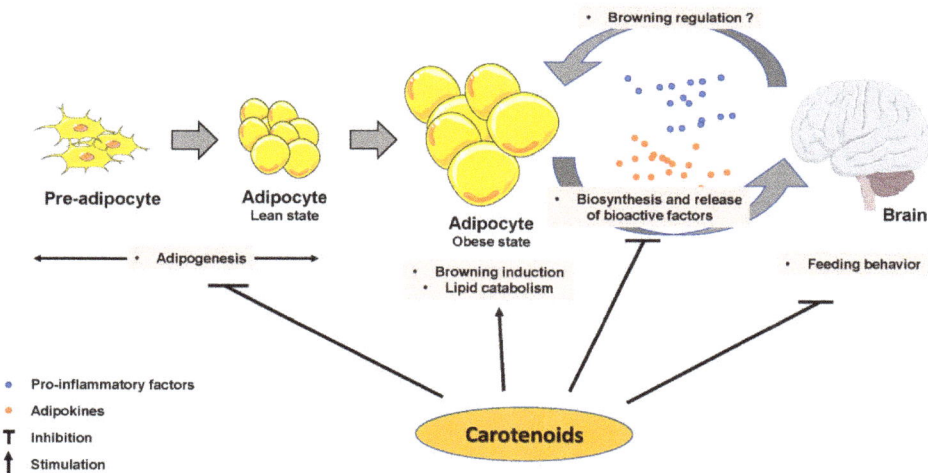

Figure 2. Carotenoids effect on adipose tissue biology parameters, on brain and on adipose tissue–brain crosstalk.

Substances with anti-inflammatory effects are assumed to limit the risk of obesity-associated disorders, including insulin resistance. Such anti-inflammatory effects of β-carotene in 3T3-L1 adipocytes were suggested to arise through limitation of the TNFα-mediated down-regulation of genes linked to adipocyte biology [95]. β-Carotene also counteracted oxidative stress-mediated dysregulation of adiponectin secretion, chemokine expression and NF-κB activation in 3T3-L1 adipocytes [96]. Fucoxanthin blunted TNFα-mediated induction of pro-inflammatory cytokines in adipocytes [42] and in adipocyte/macrophage coculture systems [97]. The most thoroughly studied anti-inflammatory carotenoid is lycopene (all-trans), and we demonstrated its ability to inhibit proinflammatory cytokine and chemokine expression in vitro (in murine and human adipocytes) [98]. These data were also reproduced ex vivo on adipose tissue explants from mice fed a high fat diet (characterized by low-grade inflammation). The molecular mechanism was investigated and the involvement of NF-κB was confirmed. Similar results (inhibition of cytokine and chemokine expression in various in vitro and ex vivo models) were obtained with apo-10′-lycopenoic acid, a metabolite of lycopene [84]. Lycopene also attenuated LPS-mediated induction of TNFα in macrophages via NF-κB and JNK [99], as well as macrophage migration in vitro. Consequently, lycopene decreased macrophage-induced cytokine, acute phase protein and chemokine mRNA in adipocytes. Interestingly, all-trans and 5-cis lycopene, the two main isoforms of lycopene found in vivo, displayed similar effects in terms of inflammation control and glucose uptake in adipocytes [100]. A few studies have shown that retinoids, like carotenoids, have positive effects by decreasing the expression of adipocyte-derived inflammatory mediators such as adipsin [101] and resistin [102]. Our group has also shown that all-trans retinoic acid (ATRA) blunts TNF-α mediated cytokine expression in 3T3-L1 cells [84]. More recently, we demonstrated that ATRA limits the expression of a large range of chemokines in vivo and in vitro. This anti-inflammatory effect of ATRA was associated with a reduction in the phosphorylation levels of IκB and p65, probably mediated by peroxisome proliferator-activated receptor gamma coactivator 1 α (PGC1α) expression [103].

The browning of white adipose tissue has been proposed as a putative mechanism controlling energy homeostasis and insulin sensitivity [104]. Recently, an AMPK-mediated effect on adipocyte browning and mitochondrial biogenesis was demonstrated for zeaxanthin [105] and for lycopene [49]. We reported similar mitochondrial biogenesis, induction of oxidative phosphorylation (OXPHOS) and adipocyte browning in adipocytes incubated with ATRA [106], whereas fucoxanthin and its metabolite fucoxanthinol were inefficient in inducing adipocyte browning [107].

Taken together, these findings suggest that carotenoids impact several adipocyte metabolic pathways, which may in turn explain, at least in part, their anti-obesity effects (Figure 2).

6. Impact of Carotenoids on the Control of Energy Homeostasis by the Brain

As stated above, carotenoids can affect the biology of the adipose tissue and modulate the production of leptin and the inflammatory cytokines [55]. They may consequently have an indirect effect on brain function (Figure 2). However, several food components, including carotenoids, could reach the hypothalamus directly [108], where they could regulate leptin signaling pathways. To support this hypothesis, several carotenoids have been detected in several parts of the adult brain [109,110]. In the study of Johnson et al., lycopene (37 +/− 9 pmol/g), lutein (145 +/− 22 pmol/g), β-carotene (77.6 +/− 10.5 pmol/g) and zeaxanthin (45 +/− 7.5 pmol/g) have been quantified in different structures (cerebellum, frontal, occipital and temporal cortices). More specifically, they could either cross the blood brain barrier or pass through the fenestrated capillaries of circumventricular organs and target the arcuate nucleus neurons. In the context of the central control of feeding behavior, it is important to note that other structures such as the hippocampus play an important role and that they could be targeted by carotenoids as indicated below.

It is presently not clear whether carotenoids act indirectly via adipose tissue or directly on the brain, but several studies suggest involvement of the brain in body weight management under the effect of carotenoids. Continuous intake of lycopene-rich food and intraperitoneal administration of lycopene increased neuronal activity in the paraventricular and ventromedial nuclei, as shown by the immunoreactivity of c-fos, a marker of neuronal activity [111]. This study suggests that lycopene may have some influence on feeding behavior. In support of this hypothesis, the group of Dr. Bishnoi showed that lycopene prevented weight gain and adiposity in mice in a DIO model [45]. Interestingly, this effect was associated with a modulation of hypothalamic anorexigenic and orexigenic gene expression. To date, the direct effect of lycopene on neuronal activity is unclear, and more research is needed to thoroughly understand this mechanism. As stated above, lycopene can impact brain function by limiting peripheral inflammation. In support of this hypothesis, Kuhad et al. showed that chronic treatment with lycopene significantly and dose-dependently attenuated cognitive deficit associated with inflammation in diabetic rats [112].

Interestingly, recent work by Zhao et al. suggests that fucoxanthin may modulate neuroinflammation [113]. In this work, fucoxanthin increased NRF-2 activation in lipopolysaccharide (LPS)-activated microglia. This interesting effect needs to be studied in an in vivo model and especially in brain structures involved in feeding behavior (i.e., hypothalamus or hippocampus). In the same line, a recent paper reported that fucoxanthin treatment reversed LPS-induced defect in body weight and food intake in mice [114]. The authors also showed that fucoxanthin inhibited LPS-induced overexpression of pro-inflammatory cytokines (IL-1β, IL-6 and TNF-α) in the hippocampus and hypothalamus, via the modulation of the AMPK-NF-κB signaling pathway. Interestingly, current studies have shown that the activation of the AMPK pathway is essential to maintaining energy homeostasis, as it is involved in the anorexigenic effect of leptin [115].

7. Conclusions

In vitro and preclinical studies clearly indicate beneficial effects of carotenoid consumption on obesity and associated pathophysiological disorders including metabolic inflammation, insulin resistance and hepatic steatosis. Molecular mechanisms are now better known, although it is not always clear whether carotenoids are active in their native form or after cleavage and metabolization, and adipose tissue appears as a major target of these substances. Nevertheless, recent though limited data suggest that carotenoids or metabolites might also act at the central level, probably by preventing or decreasing obesity-associated neuro-inflammation and comorbidities. Randomized clinical trials using pure carotenoids are urgently needed to support preclinical and observational evidence.

Author Contributions: L.M., F.T. and J.-F.L.; writing—review and editing.

Funding: This research received no external funding.

Conflicts of Interest: The authors declare no conflict of interest.

References

1. The World Health Organization (WHO). *Obesity and Overweight*; WHO Media Center: Geneva, Switzerland, 2012.
2. Arner, E.; Westermark, P.O.; Spalding, K.L.; Britton, T.; Ryden, M.; Frisen, J.; Bernard, S.; Arner, P. Adipocyte turnover: Relevance to human adipose tissue morphology. *Diabetes* **2010**, *59*, 105–109. [CrossRef] [PubMed]
3. Farmer, S.R. Transcriptional control of adipocyte formation. *Cell Metab.* **2006**, *4*, 263–273. [CrossRef] [PubMed]
4. Friedman, J. 20 years of leptin: Leptin at 20: An overview. *J. Endocrinol.* **2014**, *223*, T1–T8. [CrossRef] [PubMed]
5. Ruhl, R.; Landrier, J.F. Dietary regulation of adiponectin by direct and indirect lipid activators of nuclear hormone receptors. *Mol. Nutr. Food Res.* **2015**, *60*, 175–184. [CrossRef] [PubMed]
6. Gregor, M.F.; Hotamisligil, G.S. Inflammatory mechanisms in obesity. *Annu. Rev. Immunol.* **2011**, *29*, 415–445. [CrossRef] [PubMed]
7. Tourniaire, F.; Romier-Crouzet, B.; Lee, J.H.; Marcotorchino, J.; Gouranton, E.; Salles, J.; Malezet, C.; Astier, J.; Darmon, P.; Blouin, E.; et al. Chemokine Expression in Inflamed Adipose Tissue Is Mainly Mediated by NF-kappaB. *PLoS ONE* **2013**, *8*, e66515. [CrossRef] [PubMed]
8. Karkeni, E.; Astier, J.; Tourniaire, F.; El Abed, M.; Romier, B.; Gouranton, E.; Wan, L.; Borel, P.; Salles, J.; Walrand, S.; et al. Obesity-associated Inflammation Induces microRNA-155 Expression in Adipocytes and Adipose Tissue: Outcome on Adipocyte Function. *J. Clin. Endocrinol. Metab.* **2016**, *101*, 1615–1626. [CrossRef] [PubMed]

9. Karkeni, E.; Bonnet, L.; Marcotorchino, J.; Tourniaire, F.; Astier, J.; Ye, J.; Landrier, J.F. Vitamin D limits inflammation-linked microRNA expression in adipocytes in vitro and in vivo: A new mechanism for the regulation of inflammation by vitamin D. *Epigenetics* **2017**. [CrossRef]

10. Olefsky, J.M.; Glass, C.K. Macrophages, inflammation, and insulin resistance. *Annu. Rev. Physiol.* **2010**, *72*, 219–246. [CrossRef]

11. Weisberg, S.P.; McCann, D.; Desai, M.; Rosenbaum, M.; Leibel, R.L.; Ferrante, A.W., Jr. Obesity is associated with macrophage accumulation in adipose tissue. *J. Clin. Investig.* **2003**, *112*, 1796–1808. [CrossRef]

12. Bourlier, V.; Bouloumie, A. Role of macrophage tissue infiltration in obesity and insulin resistance. *Diabetes Metab.* **2009**, *35*, 251–260. [CrossRef] [PubMed]

13. Cornier, M.A.; Dabelea, D.; Hernandez, T.L.; Lindstrom, R.C.; Steig, A.J.; Stob, N.R.; Van Pelt, R.E.; Wang, H.; Eckel, R.H. The metabolic syndrome. *Endocr. Rev.* **2008**, *29*, 777–822. [CrossRef] [PubMed]

14. Guillemot-Legris, O.; Muccioli, G.G. Obesity-Induced Neuroinflammation: Beyond the Hypothalamus. *Trends Neurosci.* **2017**, *40*, 237–253. [CrossRef] [PubMed]

15. Derghal, A.; Djelloul, M.; Trouslard, J.; Mounien, L. The Role of MicroRNA in the Modulation of the Melanocortinergic System. *Front. Neurosci.* **2017**, *11*, 181. [CrossRef] [PubMed]

16. Morton, G.J.; Cummings, D.E.; Baskin, D.G.; Barsh, G.S.; Schwartz, M.W. Central nervous system control of food intake and body weight. *Nature* **2006**, *443*, 289–295. [CrossRef]

17. Berthoud, H.R. Multiple neural systems controlling food intake and body weight. *Neurosci. Biobehav. Rev.* **2002**, *26*, 393–428. [CrossRef]

18. Schneeberger, M.; Gomis, R.; Claret, M. Hypothalamic and brainstem neuronal circuits controlling homeostatic energy balance. *J. Endocrinol.* **2014**, *220*, T25–T46. [CrossRef]

19. Mounien, L.; Bizet, P.; Boutelet, I.; Vaudry, H.; Jegou, S. Expression of melanocortin MC3 and MC4 receptor mRNAs by neuropeptide Y neurons in the rat arcuate nucleus. *Neuroendocrinology* **2005**, *82*, 164–170. [CrossRef]

20. Von Lintig, J. Colors with functions: Elucidating the biochemical and molecular basis of carotenoid metabolism. *Annu. Rev. Nutr.* **2010**, *30*, 35–56. [CrossRef]

21. Paetau, I.; Khachik, F.; Brown, E.D.; Beecher, G.R.; Kramer, T.R.; Chittams, J.; Clevidence, B.A. Chronic ingestion of lycopene-rich tomato juice or lycopene supplements significantly increases plasma concentrations of lycopene and related tomato carotenoids in humans. *Am. J. Clin. Nutr.* **1998**, *68*, 1187–1195. [CrossRef]

22. Kimmons, J.E.; Blanck, H.M.; Tohill, B.C.; Zhang, J.; Khan, L.K. Associations between body mass index and the prevalence of low micronutrient levels among US adults. *MedGenMed* **2006**, *8*, 59. [PubMed]

23. Garcia, O.P.; Long, K.Z.; Rosado, J.L. Impact of micronutrient deficiencies on obesity. *Nutr. Rev.* **2009**, *67*, 559–572. [CrossRef] [PubMed]

24. Andersen, L.F.; Jacobs, D.R., Jr.; Gross, M.D.; Schreiner, P.J.; Dale Williams, O.; Lee, D.H. Longitudinal associations between body mass index and serum carotenoids: The CARDIA study. *Br. J. Nutr.* **2006**, *95*, 358–365. [CrossRef] [PubMed]

25. Calder, P.C.; Ahluwalia, N.; Brouns, F.; Buetler, T.; Clement, K.; Cunningham, K.; Esposito, K.; Jonsson, L.S.; Kolb, H.; Lansink, M.; et al. Dietary factors and low-grade inflammation in relation to overweight and obesity. *Br. J. Nutr.* **2011**, *106*, S5–S78. [CrossRef] [PubMed]

26. Beydoun, M.A.; Shroff, M.R.; Chen, X.; Beydoun, H.A.; Wang, Y.; Zonderman, A.B. Serum antioxidant status is associated with metabolic syndrome among U.S. adults in recent national surveys. *J. Nutr.* **2011**, *141*, 903–913. [CrossRef] [PubMed]

27. Beydoun, M.A.; Chen, X.; Jha, K.; Beydoun, H.A.; Zonderman, A.B.; Canas, J.A. Carotenoids, vitamin A, and their association with the metabolic syndrome: A systematic review and meta-analysis. *Nutr. Rev.* **2019**, *77*, 32–45. [CrossRef]

28. Bonet, M.L.; Canas, J.A.; Ribot, J.; Palou, A. Carotenoids and their conversion products in the control of adipocyte function, adiposity and obesity. *Arch. Biochem. Biophys.* **2015**, *572*, 112–125. [CrossRef]

29. Canas, J.A.; Lochrie, A.; McGowan, A.G.; Hossain, J.; Schettino, C.; Balagopal, P.B. Effects of Mixed Carotenoids on Adipokines and Abdominal Adiposity in Children: A Pilot Study. *J. Clin. Endocrinol. Metab.* **2017**, *102*, 1983–1990. [CrossRef]

30. Kakutani, R.; Hokari, S.; Nishino, A.; Ichihara, T.; Sugimoto, K.; Takaha, T.; Kuriki, T.; Maoka, T. Effect of Oral Paprika Xanthophyll Intake on Abdominal Fat in Healthy Overweight Humans: A Randomized, Double-blind, Placebo-controlled Study. *J. Oleo Sci.* **2018**, *67*, 1149–1162. [CrossRef]

31. Coronel, J.; Pinos, I.; Amengual, J. β-carotene in Obesity Research: Technical Considerations and Current Status of the Field. *Nutrients* **2019**, *11*, 842. [CrossRef]

32. Amengual, J.; Gouranton, E.; van Helden, Y.G.; Hessel, S.; Ribot, J.; Kramer, E.; Kiec-Wilk, B.; Razny, U.; Lietz, G.; Wyss, A.; et al. Beta-Carotene Reduces Body Adiposity of Mice via BCMO1. *PLoS ONE* **2011**, *6*, e20644. [CrossRef]

33. Van Helden, Y.G.; Godschalk, R.W.; von Lintig, J.; Lietz, G.; Landrier, J.F.; Luisa Bonet, M.; van Schooten, F.J.; Keijer, J. Gene expression response of mouse lung, liver and white adipose tissue to beta-carotene supplementation, knockout of Bcmo1 and sex. *Mol. Nutr. Food Res.* **2011**, *55*, 1466–1474. [CrossRef] [PubMed]

34. Lobo, G.P.; Amengual, J.; Li, H.N.; Golczak, M.; Bonet, M.L.; Palczewski, K.; von Lintig, J. Beta,beta-carotene decreases peroxisome proliferator receptor gamma activity and reduces lipid storage capacity of adipocytes in a beta,beta-carotene oxygenase 1-dependent manner. *J. Biol. Chem.* **2010**, *285*, 27891–27899. [CrossRef] [PubMed]

35. Ikeuchi, M.; Koyama, T.; Takahashi, J.; Yazawa, K. Effects of astaxanthin in obese mice fed a high-fat diet. *Biosci. Biotechnol. Biochem.* **2007**, *71*, 893–899. [CrossRef] [PubMed]

36. Arunkumar, E.; Bhuvaneswari, S.; Anuradha, C.V. An intervention study in obese mice with astaxanthin, a marine carotenoid-effects on insulin signaling and pro-inflammatory cytokines. *Food Funct.* **2012**, *3*, 120–126. [CrossRef] [PubMed]

37. Ni, Y.; Nagashimada, M.; Zhuge, F.; Zhan, L.; Nagata, N.; Tsutsui, A.; Nakanuma, Y.; Kaneko, S.; Ota, T. Astaxanthin prevents and reverses diet-induced insulin resistance and steatohepatitis in mice: A comparison with vitamin E. *Sci. Rep.* **2015**, *5*, 17192. [CrossRef] [PubMed]

38. Kim, B.; Farruggia, C.; Ku, C.S.; Pham, T.X.; Yang, Y.; Bae, M.; Wegner, C.J.; Farrell, N.J.; Harness, E.; Park, Y.K.; et al. Astaxanthin inhibits inflammation and fibrosis in the liver and adipose tissue of mouse models of diet-induced obesity and nonalcoholic steatohepatitis. *J. Nutr. Biochem.* **2017**, *43*, 27–35. [CrossRef]

39. Takayanagi, K.; Morimoto, S.; Shirakura, Y.; Mukai, K.; Sugiyama, T.; Tokuji, Y.; Ohnishi, M. Mechanism of visceral fat reduction in Tsumura Suzuki obese, diabetes (TSOD) mice orally administered beta-cryptoxanthin from Satsuma mandarin oranges (Citrus unshiu Marc). *J. Agric. Food Chem.* **2011**, *59*, 12342–12351. [CrossRef]

40. Ni, Y.; Nagashimada, M.; Zhan, L.; Nagata, N.; Kobori, M.; Sugiura, M.; Ogawa, K.; Kaneko, S.; Ota, T. Prevention and reversal of lipotoxicity-induced hepatic insulin resistance and steatohepatitis in mice by an antioxidant carotenoid, beta-cryptoxanthin. *Endocrinology* **2015**, *156*, 987–999. [CrossRef]

41. Maeda, H. Nutraceutical effects of fucoxanthin for obesity and diabetes therapy: A review. *J. Oleo Sci.* **2015**, *64*, 125–132. [CrossRef]

42. Hosokawa, M.; Miyashita, T.; Nishikawa, S.; Emi, S.; Tsukui, T.; Beppu, F.; Okada, T.; Miyashita, K. Fucoxanthin regulates adipocytokine mRNA expression in white adipose tissue of diabetic/obese KK-Ay mice. *Arch. Biochem. Biophys.* **2010**, *504*, 17–25. [CrossRef] [PubMed]

43. Grasa-Lopez, A.; Miliar-Garcia, A.; Quevedo-Corona, L.; Paniagua-Castro, N.; Escalona-Cardoso, G.; Reyes-Maldonado, E.; Jaramillo-Flores, M.E. *Undaria pinnatifida* and Fucoxanthin Ameliorate Lipogenesis and Markers of Both Inflammation and Cardiovascular Dysfunction in an Animal Model of Diet-Induced Obesity. *Mar. Drugs* **2016**, *14*, 148. [CrossRef] [PubMed]

44. Liu, M.; Liu, H.; Xie, J.; Xu, Q.; Pan, C.; Wang, J.; Wu, X.; Zheng, M.; Liu, J. Anti-obesity effects of zeaxanthin on 3T3-L1 preadipocyte and high fat induced obese mice. *Food Funct.* **2017**, *8*, 3327–3338. [CrossRef] [PubMed]

45. Singh, D.P.; Khare, P.; Zhu, J.; Kondepudi, K.K.; Singh, J.; Baboota, R.K.; Boparai, R.K.; Khardori, R.; Chopra, K.; Bishnoi, M. A novel cobiotic-based preventive approach against high-fat diet-induced adiposity, nonalcoholic fatty liver and gut derangement in mice. *Int. J. Obes.* **2016**, *40*, 487–496. [CrossRef] [PubMed]

46. Fenni, S.; Hammou, H.; Astier, J.; Bonnet, L.; Karkeni, E.; Couturier, C.; Tourniaire, F.; Landrier, J.F. Lycopene and tomato powder supplementation similarly inhibit high-fat diet induced obesity, inflammatory response, and associated metabolic disorders. *Mol. Nutr. Food Res.* **2017**, *61*. [CrossRef] [PubMed]

47. Li, C.C.; Liu, C.; Fu, M.; Hu, K.Q.; Aizawa, K.; Takahashi, S.; Hiroyuki, S.; Cheng, J.; von Lintig, J.; Wang, X.D. Tomato Powder Inhibits Hepatic Steatosis and Inflammation Potentially Through Restoring SIRT1 Activity and Adiponectin Function Independent of Carotenoid Cleavage Enzymes in Mice. *Mol. Nutr. Food Res.* **2018**, *62*, e1700738. [CrossRef] [PubMed]

48. Wang, J.; Zou, Q.; Suo, Y.; Tan, X.; Yuan, T.; Liu, Z.; Liu, X. Lycopene ameliorates systemic inflammation-induced synaptic dysfunction via improving insulin resistance and mitochondrial dysfunction in the liver-brain axis. *Food Funct.* **2019**, *10*, 2125–2137. [CrossRef]

49. Wang, J.; Suo, Y.; Zhang, J.; Zou, Q.; Tan, X.; Yuan, T.; Liu, Z.; Liu, X. Lycopene supplementation attenuates western diet-induced body weight gain through increasing the expressions of thermogenic/mitochondrial functional genes and improving insulin resistance in the adipose tissue of obese mice. *J. Nutr. Biochem.* **2019**, *69*, 63–72. [CrossRef]

50. Blaner, W.S. Vitamin A signaling and homeostasis in obesity, diabetes, and metabolic disorders. *Pharmacol. Ther.* **2019**. [CrossRef]

51. Peirce, A.W. Carotene and vitamin A in human fat. *Med. J. Aust.* **1954**, *41*, 589.

52. Virtanen, S.M.; van't Veer, P.; Kok, F.; Kardinaal, A.F.; Aro, A. Predictors of adipose tissue carotenoid and retinol levels in nine countries: The EURAMIC Study. *Am. J. Epidemiol.* **1996**, *144*, 968–979. [CrossRef] [PubMed]

53. Parker, R.S. Carotenoids in human blood and tissues. *J. Nutr.* **1989**, *119*, 101–104. [CrossRef] [PubMed]

54. Chung, H.Y.; Ferreira, A.L.; Epstein, S.; Paiva, S.A.; Castaneda-Sceppa, C.; Johnson, E.J. Site-specific concentrations of carotenoids in adipose tissue: Relations with dietary and serum carotenoid concentrations in healthy adults. *Am. J. Clin. Nutr.* **2009**, *90*, 533–539. [CrossRef] [PubMed]

55. Landrier, J.F.; Marcotorchino, J.; Tourniaire, F. Lipophilic micronutrients and adipose tissue biology. *Nutrients* **2012**, *4*, 1622–1649. [CrossRef] [PubMed]

56. Parker, R.S. Carotenoid and tocopherol composition of human adipose tissue. *Am. J. Clin. Nutr.* **1988**, *47*, 33–36. [CrossRef] [PubMed]

57. Wallstrom, P.; Wirfalt, E.; Lahmann, P.H.; Gullberg, B.; Janzon, L.; Berglund, G. Serum concentrations of beta-carotene and alpha-tocopherol are associated with diet, smoking, and general and central adiposity. *Am. J. Clin. Nutr.* **2001**, *73*, 777–785. [CrossRef] [PubMed]

58. Kirby, M.L.; Beatty, S.; Stack, J.; Harrison, M.; Greene, I.; McBrinn, S.; Carroll, P.; Nolan, J.M. Changes in macular pigment optical density and serum concentrations of lutein and zeaxanthin in response to weight loss. *Br. J. Nutr.* **2011**, *105*, 1036–1046. [CrossRef]

59. Osth, M.; Ost, A.; Kjolhede, P.; Stralfors, P. The concentration of beta-carotene in human adipocytes, but not the whole-body adipocyte stores, is reduced in obesity. *PLoS ONE* **2014**, *9*, e85610. [CrossRef]

60. Sy, C.; Gleize, B.; Dangles, O.; Landrier, J.F.; Veyrat, C.C.; Borel, P. Effects of physicochemical properties of carotenoids on their bioaccessibility, intestinal cell uptake, and blood and tissue concentrations. *Mol. Nutr. Food Res.* **2012**, *56*, 1385–1397. [CrossRef]

61. Moussa, M.; Gouranton, E.; Gleize, B.; Yazidi, C.E.; Niot, I.; Besnard, P.; Borel, P.; Landrier, J.F. CD36 is involved in lycopene and lutein uptake by adipocytes and adipose tissue cultures. *Mol. Nutr. Food Res.* **2011**, *55*, 578–584. [CrossRef]

62. Gouranton, E.; Yazidi, C.E.; Cardinault, N.; Amiot, M.J.; Borel, P.; Landrier, J.F. Purified low-density lipoprotein and bovine serum albumin efficiency to internalise lycopene into adipocytes. *Food Chem. Toxicol.* **2008**, *46*, 3832–3836. [CrossRef] [PubMed]

63. Kardinaal, A.F.; van't Veer, P.; Brants, H.A.; van den Berg, H.; van Schoonhoven, J.; Hermus, R.J. Relations between antioxidant vitamins in adipose tissue, plasma, and diet. *Am. J. Epidemiol.* **1995**, *141*, 440–450. [CrossRef] [PubMed]

64. Johnson, E.J.; Suter, P.M.; Sahyoun, N.; Ribaya-Mercado, J.D.; Russell, R.M. Relation between beta-carotene intake and plasma and adipose tissue concentrations of carotenoids and retinoids. *Am. J. Clin. Nutr.* **1995**, *62*, 598–603. [CrossRef] [PubMed]

65. Johnson, E.J.; Hammond, B.R.; Yeum, K.J.; Qin, J.; Wang, X.D.; Castaneda, C.; Snodderly, D.M.; Russell, R.M. Relation among serum and tissue concentrations of lutein and zeaxanthin and macular pigment density. *Am. J. Clin. Nutr.* **2000**, *71*, 1555–1562. [CrossRef] [PubMed]

66. Walfisch, Y.; Walfisch, S.; Agbaria, R.; Levy, J.; Sharoni, Y. Lycopene in serum, skin and adipose tissues after tomato-oleoresin supplementation in patients undergoing haemorrhoidectomy or peri-anal fistulotomy. *Br. J. Nutr.* **2003**, *90*, 759–766. [CrossRef] [PubMed]

67. El-Sohemy, A.; Baylin, A.; Kabagambe, E.; Ascherio, A.; Spiegelman, D.; Campos, H. Individual carotenoid concentrations in adipose tissue and plasma as biomarkers of dietary intake. *Am. J. Clin. Nutr.* **2002**, *76*, 172–179. [CrossRef]

68. Broekmans, W.M.; Berendschot, T.T.; Klopping-Ketelaars, I.A.; de Vries, A.J.; Goldbohm, R.A.; Tijburg, L.B.; Kardinaal, A.F.; van Poppel, G. Macular pigment density in relation to serum and adipose tissue concentrations of lutein and serum concentrations of zeaxanthin. *Am. J. Clin. Nutr.* **2002**, *76*, 595–603. [CrossRef] [PubMed]

69. Su, L.C.; Bui, M.; Kardinaal, A.; Gomez-Aracena, J.; Martin-Moreno, J.; Martin, B.; Thamm, M.; Simonsen, N.; van't Veer, P.; Kok, F.; et al. Differences between plasma and adipose tissue biomarkers of carotenoids and tocopherols. *Cancer Epidemiol. Prev. Biomark.* **1998**, *7*, 1043–1048.

70. Yeum, K.J.; Booth, S.L.; Roubenoff, R.; Russell, R.M. Plasma carotenoid concentrations are inversely correlated with fat mass in older women. *J. Nutr. Health Aging* **1998**, *2*, 79–83.

71. Hessel, S.; Eichinger, A.; Isken, A.; Amengual, J.; Hunzelmann, S.; Hoeller, U.; Elste, V.; Hunziker, W.; Goralczyk, R.; Oberhauser, V.; et al. CMO1 deficiency abolishes vitamin A production from beta-carotene and alters lipid metabolism in mice. *J. Biol. Chem.* **2007**, *282*, 33553–33561. [CrossRef]

72. Tourniaire, F.; Gouranton, E.; von Lintig, J.; Keijer, J.; Bonet, M.L.; Amengual, J.; Lietz, G.; Landrier, J.F. beta-Carotene conversion products and their effects on adipose tissue. *Genes Nutr.* **2009**, *4*, 179–187. [CrossRef] [PubMed]

73. Ziouzenkova, O.; Orasanu, G.; Sharlach, M.; Akiyama, T.E.; Berger, J.P.; Viereck, J.; Hamilton, J.A.; Tang, G.; Dolnikowski, G.G.; Vogel, S.; et al. Retinaldehyde represses adipogenesis and diet-induced obesity. *Nat. Med.* **2007**, *13*, 695–702. [CrossRef] [PubMed]

74. Tsutsumi, C.; Okuno, M.; Tannous, L.; Piantedosi, R.; Allan, M.; Goodman, D.S.; Blaner, W.S. Retinoids and retinoid-binding protein expression in rat adipocytes. *J. Biol. Chem.* **1992**, *267*, 1805–1810. [PubMed]

75. Kane, M.A. Analysis, occurrence, and function of 9-cis-retinoic acid. *Biochim. Biophys. Acta* **2012**, *1821*, 10–20. [CrossRef] [PubMed]

76. Sima, A.; Manolescu, D.C.; Bhat, P. Retinoids and retinoid-metabolic gene expression in mouse adipose tissues. *Biochem. Cell Biol.* **2011**, *89*, 578–584. [CrossRef] [PubMed]

77. O'Byrne, S.M.; Wongsiriroj, N.; Libien, J.; Vogel, S.; Goldberg, I.J.; Baehr, W.; Palczewski, K.; Blaner, W.S. Retinoid absorption and storage is impaired in mice lacking lecithin: Retinol acyltransferase (LRAT). *J. Biol. Chem.* **2005**, *280*, 35647–35657. [CrossRef] [PubMed]

78. Landrier, J.F.; Kasiri, E.; Karkeni, E.; Mihaly, J.; Beke, G.; Weiss, K.; Lucas, R.; Aydemir, G.; Salles, J.; Walrand, S.; et al. Reduced adiponectin expression after high-fat diet is associated with selective up-regulation of ALDH1A1 and further retinoic acid receptor signaling in adipose tissue. *FASEB J.* **2017**, *31*, 203–211. [CrossRef] [PubMed]

79. Germain, P.; Chambon, P.; Eichele, G.; Evans, R.M.; Lazar, M.A.; Leid, M.; De Lera, A.R.; Lotan, R.; Mangelsdorf, D.J.; Gronemeyer, H. International Union of Pharmacology. LXIII. Retinoid X receptors. *Pharmacol. Rev.* **2006**, *58*, 760–772. [CrossRef] [PubMed]

80. Germain, P.; Chambon, P.; Eichele, G.; Evans, R.M.; Lazar, M.A.; Leid, M.; De Lera, A.R.; Lotan, R.; Mangelsdorf, D.J.; Gronemeyer, H. International Union of Pharmacology. LX. Retinoic acid receptors. *Pharmacol. Rev.* **2006**, *58*, 712–725. [CrossRef] [PubMed]

81. Yasmeen, R.; Jeyakumar, S.M.; Reichert, B.; Yang, F.; Ziouzenkova, O. The contribution of vitamin A to autocrine regulation of fat depots. *Biochim. Biophys. Acta* **2012**, *1821*, 190–197. [CrossRef] [PubMed]

82. Berry, D.C.; Noy, N. All-trans-retinoic acid represses obesity and insulin resistance by activating both peroxisome proliferation-activated receptor beta/delta and retinoic acid receptor. *Mol. Cell. Biol.* **2009**, *29*, 3286–3296. [CrossRef] [PubMed]

83. Aydemir, G.; Carlsen, H.; Blomhoff, R.; Ruhl, R. Lycopene induces retinoic acid receptor transcriptional activation in mice. *Mol. Nutr. Food Res.* **2012**, *56*, 702–712. [CrossRef] [PubMed]

84. Gouranton, E.; Aydemir, G.; Reynaud, E.; Marcotorchino, J.; Malezet, C.; Caris-Veyrat, C.; Blomhoff, R.; Landrier, J.F.; Ruhl, R. Apo-10'-lycopenoic acid impacts adipose tissue biology via the retinoic acid receptors. *Biochim. Biophys. Acta* **2011**, *1811*, 1105–1114. [CrossRef] [PubMed]

85. Rao, A.V.; Ray, M.R.; Rao, L.G. Lycopene. *Adv. Food Nutr. Res.* **2006**, *51*, 99–164. [PubMed]

86. Sharoni, Y.; Linnewiel-Hermoni, K.; Khanin, M.; Salman, H.; Veprik, A.; Danilenko, M.; Levy, J. Carotenoids and apocarotenoids in cellular signaling related to cancer: A review. *Mol. Nutr. Food Res.* **2012**, *56*, 259–269. [CrossRef] [PubMed]

87. Ben-Dor, A.; Steiner, M.; Gheber, L.; Danilenko, M.; Dubi, N.; Linnewiel, K.; Zick, A.; Sharoni, Y.; Levy, J. Carotenoids activate the antioxidant response element transcription system. *Mol. Cancer Ther.* **2005**, *4*, 177–186. [PubMed]

88. Landrier, J.F. *Les Phytomicronutriments*; Lavoisier: Paris, France, 2012.

89. Kawada, T.; Kamei, Y.; Fujita, A.; Hida, Y.; Takahashi, N.; Sugimoto, E.; Fushiki, T. Carotenoids and retinoids as suppressors on adipocyte differentiation via nuclear receptors. *Biofactors* **2000**, *13*, 103–109. [CrossRef]

90. Ziouzenkova, O.; Orasanu, G.; Sukhova, G.; Lau, E.; Berger, J.P.; Tang, G.; Krinsky, N.I.; Dolnikowski, G.G.; Plutzky, J. Asymmetric cleavage of beta-carotene yields a transcriptional repressor of retinoid X receptor and peroxisome proliferator-activated receptor responses. *Mol. Endocrinol.* **2007**, *21*, 77–88. [CrossRef]

91. Shirakura, Y.; Takayanagi, K.; Mukai, K.; Tanabe, H.; Inoue, M. β-Cryptoxanthin suppresses the adipogenesis of 3T3-L1 cells via RAR activation. *J. Nutr. Sci. Vitaminol.* **2011**, *57*, 426–431. [CrossRef]

92. Inoue, M.; Tanabe, H.; Matsumoto, A.; Takagi, M.; Umegaki, K.; Amagaya, S.; Takahashi, J. Astaxanthin functions differently as a selective peroxisome proliferator-activated receptor gamma modulator in adipocytes and macrophages. *Biochem. Pharmacol.* **2012**, *84*, 692–700. [CrossRef]

93. Maeda, H.; Hosokawa, M.; Sashima, T.; Takahashi, N.; Kawada, T.; Miyashita, K. Fucoxanthin and its metabolite, fucoxanthinol, suppress adipocyte differentiation in 3T3-L1 cells. *Int. J. Mol. Med.* **2006**, *18*, 147–152. [CrossRef] [PubMed]

94. Seo, M.J.; Seo, Y.J.; Pan, C.H.; Lee, O.H.; Kim, K.J.; Lee, B.Y. Fucoxanthin Suppresses Lipid Accumulation and ROS Production During Differentiation in 3T3-L1 Adipocytes. *Phytother. Res.* **2016**, *30*, 1802–1808. [CrossRef] [PubMed]

95. Kameji, H.; Mochizuki, K.; Miyoshi, N.; Goda, T. β-Carotene accumulation in 3T3-L1 adipocytes inhibits the elevation of reactive oxygen species and the suppression of genes related to insulin sensitivity induced by tumor necrosis factor-alpha. *Nutrition* **2010**, *26*, 1151–1156. [CrossRef] [PubMed]

96. Cho, S.O.; Kim, M.H.; Kim, H. β-Carotene Inhibits Activation of NF-kappaB, Activator Protein-1, and STAT3 and Regulates Abnormal Expression of Some Adipokines in 3T3-L1 Adipocytes. *J. Cancer Prev.* **2018**, *23*, 37–43. [CrossRef] [PubMed]

97. Maeda, H.; Kanno, S.; Kodate, M.; Hosokawa, M.; Miyashita, K. Fucoxanthinol, Metabolite of Fucoxanthin, Improves Obesity-Induced Inflammation in Adipocyte Cells. *Mar. Drugs* **2015**, *13*, 4799–4813. [CrossRef] [PubMed]

98. Gouranton, E.; Thabuis, C.; Riollet, C.; Malezet-Desmoulins, C.; El Yazidi, C.; Amiot, M.J.; Borel, P.; Landrier, J.F. Lycopene inhibits proinflammatory cytokine and chemokine expression in adipose tissue. *J. Nutr. Biochem.* **2011**, *22*, 642–648. [CrossRef] [PubMed]

99. Marcotorchino, J.; Romier, B.; Gouranton, E.; Riollet, C.; Gleize, B.; Malezet-Desmoulins, C.; Landrier, J.F. Lycopene attenuates LPS-induced TNF-alpha secretion in macrophages and inflammatory markers in adipocytes exposed to macrophage-conditioned media. *Mol. Nutr. Food Res.* **2012**, *56*, 725–732. [CrossRef]

100. Fenni, S.; Astier, J.; Bonnet, L.; Karkeni, E.; Gouranton, E.; Mounien, L.; Couturier, C.; Tourniaire, F.; Bohm, V.; Hammou, H.; et al. (all-E)- and (5Z)-Lycopene Display Similar Biological Effects on Adipocytes. *Mol. Nutr. Food Res.* **2019**, *63*, e1800788. [CrossRef]

101. Antras, J.; Lasnier, F.; Pairault, J. Adipsin gene expression in 3T3-F442A adipocytes is posttranscriptionally down-regulated by retinoic acid. *J. Biol. Chem.* **1991**, *266*, 1157–1161.

102. Felipe, F.; Bonet, M.L.; Ribot, J.; Palou, A. Modulation of resistin expression by retinoic acid and vitamin A status. *Diabetes* **2004**, *53*, 882–889. [CrossRef]

103. Karkeni, E.; Bonnet, L.; Astier, J.; Couturier, C.; Dalifard, J.; Tourniaire, F.; Landrier, J.F. All-trans-retinoic acid represses chemokine expression in adipocytes and adipose tissue by inhibiting NF-kappaB signaling. *J. Nutr. Biochem.* **2017**, *42*, 101–107. [CrossRef] [PubMed]

104. Chondronikola, M.; Volpi, E.; Borsheim, E.; Porter, C.; Annamalai, P.; Enerback, S.; Lidell, M.E.; Saraf, M.K.; Labbe, S.M.; Hurren, N.M.; et al. Brown adipose tissue improves whole-body glucose homeostasis and insulin sensitivity in humans. *Diabetes* **2014**, *63*, 4089–4099. [CrossRef] [PubMed]

105. Liu, M.; Zheng, M.; Cai, D.; Xie, J.; Jin, Z.; Liu, H.; Liu, J. Zeaxanthin promotes mitochondrial biogenesis and adipocyte browning via AMPKalpha1 activation. *Food Funct.* **2019**, *10*, 2221–2233. [CrossRef] [PubMed]

106. Tourniaire, F.; Musinovic, H.; Gouranton, E.; Astier, J.; Marcotorchino, J.; Arreguin, A.; Bernot, D.; Palou, A.; Bonet, M.L.; Ribot, J.; et al. All-trans retinoic acid induces oxidative phosphorylation and mitochondria biogenesis in adipocytes. *J. Lipid Res.* **2016**, *56*, 1100–1109. [CrossRef] [PubMed]

107. Rebello, C.J.; Greenway, F.L.; Johnson, W.D.; Ribnicky, D.; Poulev, A.; Stadler, K.; Coulter, A.A. Fucoxanthin and Its Metabolite Fucoxanthinol Do Not Induce Browning in Human Adipocytes. *J. Agric. Food Chem.* **2017**, *65*, 10915–10924. [CrossRef]

108. Aragones, G.; Ardid-Ruiz, A.; Ibars, M.; Suarez, M.; Blade, C. Modulation of leptin resistance by food compounds. *Mol. Nutr. Food Res.* **2016**, *60*, 1789–1803. [CrossRef] [PubMed]

109. Craft, N.E.; Haitema, T.B.; Garnett, K.M.; Fitch, K.A.; Dorey, C.K. Carotenoid, tocopherol, and retinol concentrations in elderly human brain. *J. Nutr. Health Aging* **2004**, *8*, 156–162.

110. Johnson, E.J.; Vishwanathan, R.; Johnson, M.A.; Hausman, D.B.; Davey, A.; Scott, T.M.; Green, R.C.; Miller, L.S.; Gearing, M.; Woodard, J.; et al. Relationship between Serum and Brain Carotenoids, alpha-Tocopherol, and Retinol Concentrations and Cognitive Performance in the Oldest Old from the Georgia Centenarian Study. *J. Aging Res.* **2013**, *2013*, 951786. [CrossRef]

111. Takayama, K.; Nishiko, E.; Matsumoto, G.; Inakuma, T. Study on the expression of c-Fos protein in the brain of rats after ingestion of food rich in lycopene. *Neurosci. Lett.* **2013**, *536*, 1–5. [CrossRef]

112. Kuhad, A.; Sethi, R.; Chopra, K. Lycopene attenuates diabetes-associated cognitive decline in rats. *Life Sci.* **2008**, *83*, 128–134. [CrossRef]

113. Zhao, D.; Kwon, S.H.; Chun, Y.S.; Gu, M.Y.; Yang, H.O. Anti-Neuroinflammatory Effects of Fucoxanthin via Inhibition of Akt/NF-kappaB and MAPKs/AP-1 Pathways and Activation of PKA/CREB Pathway in Lipopolysaccharide-Activated BV-2 Microglial Cells. *Neurochem. Res.* **2017**, *42*, 667–677. [CrossRef] [PubMed]

114. Jiang, X.; Wang, G.; Lin, Q.; Tang, Z.; Yan, Q.; Yu, X. Fucoxanthin prevents lipopolysaccharide-induced depressive-like behavior in mice via AMPK-NF-kappaB pathway. *Metab. Brain Dis.* **2019**, *34*, 431–442. [CrossRef] [PubMed]

115. Hardie, D.G. AMP-activated protein kinase: Maintaining energy homeostasis at the cellular and whole-body levels. *Annu. Rev. Nutr.* **2014**, *34*, 31–55. [CrossRef] [PubMed]

nutrients

MDPI

Article

β-Cryptoxanthin Reduces Body Fat and Increases Oxidative Stress Response in *Caenorhabditis elegans* Model

Silvia Llopis [1,†], María Jesús Rodrigo [2,†], Nuria González [1], Salvador Genovés [1], Lorenzo Zacarías [2], Daniel Ramón [1] and Patricia Martorell [1,*]

[1] Cell Biology Laboratory, Food Biotechnology Department, Biópolis SL/Archer Daniels Midland, C/Catedrático Agustín Escardino Benlloch 9, Paterna, 46890 Valencia, Spain; silvia.llopis@adm.com (S.L.); nuria.gonzalez@adm.com (N.G.); Salvador.Genoves@adm.com (S.G.); Daniel.RamonVidal@adm.com (D.R.)
[2] Food Biotechnology Department, Instituto de Agroquímica y Tecnología de Alimentos (IATA), Consejo Superior de Investigaciones Científicas (CSIC), C/Catedrático Agustín Escardino 7, Paterna, 46890 Valencia, Spain; mjrodrigo@iata.csic.es (M.J.R.); lzacarias@iata.csic.es (L.Z.)
*   Correspondence: patricia.martorell@adm.com; Tel.: +34-963-160-299
†   These authors contributed equally to this work.

Received: 12 November 2018; Accepted: 15 January 2019; Published: 22 January 2019

Abstract: β-Cryptoxanthin (BCX) is a major dietary pro-vitamin A carotenoid, found mainly in fruits and vegetables. Several studies showed the beneficial effects of BCX on different aspects of human health. In spite of the evidence, the molecular mechanisms of action of BCX need to be further investigated. The *Caenorhabditis elegans* model was used to analyze in vivo the activity of BCX on fat reduction and protection to oxidative stress. Dose-response assays provided evidence of the efficacy of BCX at very low dose (0.025 μg/mL) ($p < 0.001$) on these processes. Moreover, a comparative analysis with other carotenoids, such as lycopene and β-carotene, showed a stronger effect of BCX. Furthermore, a transcriptomic analysis of wild-type nematodes supplemented with BCX revealed upregulation of the energy metabolism, response to stress, and protein homeostasis as the main metabolic targets of this xanthophyll. Collectively, this study provides new in vivo evidence of the potential therapeutic use of BCX in the prevention of diseases related to metabolic syndrome and aging.

Keywords: β-Cryptoxanthin; carotenoids; *Caenorhabditis elegans*; fat reduction; oxidative stress; transcriptomic analysis; metabolic syndrome; aging

1. Introduction

Xanthophyll β-cryptoxanthin (BCX)—also known as 3-hydroxy-β-carotene ((3R)-β,β-caroten-3-ol)—is primarily synthetized in higher plants by non-heme di-iron β-carotene hydroxylases adding a hydroxyl group at the C3 position of the β-carotene ring [1,2]. Xanthophyll β-cryptoxanthin is a major dietary pro-vitamin A carotenoid, the most important pro-vitamin A xanthophyll in the diet and commonly found in human plasma [3–5]. Moreover, several studies suggest that BCX has a relative high bioavailability from common food sources and conversion to retinol can be comparable to that of β-carotene [6,7]. The main dietary sources of BCX are fruits and vegetables; however, only a small number contain significant amounts of BCX; the most common foods rich in this xanthophyll are: sweet red pepper and hot chili peppers, pumpkins, papaya, persimmons, tangerines and sweet oranges, peaches, and sweet corn [6,8]. Among these foods, the citrus tangerines and sweet oranges are considered the main contributors of BCX to the diet in many countries due to their high consumption both as fresh fruit and juice [6,7,9]. In fact, clinical studies indicated that seasonal intake of citrus fruits in many populations significantly increases plasma levels of BCX and its concentration is considered to be a biomarker of citrus fruit consumption [3,10]. Interestingly, the preferential natural accumulating form of BCX in the flesh of ripen citrus fruit is esterified with lauric, myristic, and palmitic acid, and only a minor proportion—ranging from less than 5% up to 20%—remains in free form [11,12].

Besides the well-established function of BCX as a source of vitamin A [3,13], this xanthophyll may exert other beneficial functions in different processes of human health. Evidence indicates that BCX has as potent in vitro antioxidant capacity and may protect human cells against oxidative damage, leading to the suppression of inflammation, acting also as a scavenger of free radicals and preventing the oxidative damage of biomolecules as lipids, proteins, and nucleic acids [14–16]. However, there is scarce information regarding in vivo studies where the antioxidant capacity of BCX has been assessed in targeted tissues at suitable concentration relative to the oxidizing agent [6]. Moreover—related with its antioxidant property—BCX may be also beneficial in the prevention of vascular disease [17–20]. Additionally, other studies propose that a moderated BCX intake may be helpful in reducing the risk of certain cancers [21,22] and in the prevention of age-related cognitive dysfunction in mouse brain [16]. Furthermore, one of the most remarkable biological functions of BCX is its role in bone health [23], since BCX has been proven to play a role in bone homeostasis by promoting osteoclast formation and inhibiting osteoblast action [24,25].

In the case of obesity-related disorders, a relationship between oral intake of BCX with obesity and metabolic syndrome in humans has been described. Thus, some studies established lower serum BCX levels in obese than in non-obese patients, independently of the diet [26,27]. A previous work indicated that a supplementation of BCX in obese subjects improved serum adipocytokine profiles [28]. Other authors described that a continuous oral intake of BCX reduced visceral adipose tissue, body weight, and abdominal circumference in mildly obese males [29,30]. Moreover, oral administration of BCX also repressed body weight and adipocyte hypertrophy in obese model mice [31]. Indeed, some authors concluded that a diet rich in BCX may prevent the development of metabolic syndrome [28,32–34]. However—in spite of these evidences—there are few studies focused on the potential molecular targets of BCX upon obesity and metabolic syndrome. Studies in obese diabetic mice indicate that BCX may act on modulating PPARγ (peroxisome proliferator-activated receptors-gamma type) to reduce adipocyte proliferation and hypertrophy, and its chronic inflammation through the downregulation of genes involved in the cell cycle, chemotaxis, and immune system [31,35]. However—as in many other studies—the mechanism of visceral fat reduction has been mainly addressed in mice upon oral administration of a complex food matrix (powder from pulp of Satsuma mandarin) as the source of BCX, thus the potential activity of other compounds cannot be ruled out [31] and the metabolic targets of BCX need to be further investigated using in vivo models.

Caenorhabditis elegans is an excellent model organism to study obesity. Previous studies have described the possible mechanisms and the genes involved in the pathways that regulate fat metabolism in this nematode [36–38]. Many of the identified genes involved in lipid metabolism have orthologues in humans, sharing also the same control of homeostasis [39]. Furthermore, fat accumulation in lipid droplets are mainly in gut and hypodermal cells, which enables its detection and quantification by different dying techniques. Although there are different methods available to measure lipids in the nematode, each one has advantages and disadvantages. Nile Red is a dye widely used because offers several advantages that are rapid, sensitive, and suitable for live imaging and screening studies [40]. Moreover, it has been used by many authors to identify evolutionarily conserved fat regulatory genes and small molecules that affect fat metabolism when investigated through a variety of independent methodologies [41]. Thereby, several studies have used the nematode *C. elegans* to evaluate potential obesity therapeutics [42–45]. This organism has been also used to validate functional properties of several carotenoids such as increase of longevity and antioxidant activity [46–50], being relevant to study the antioxidant effect of astaxanthin stereoisomers [51]. Regarding the carotenoids conversion to retinoids, two carotenoid cleavage dioxygenase have been identified in *C. elegans* with similarity to β-carotene 15,15-oxygenase (BCO1) (central cleavage of β-carotene) and β-carotene 9′,10′-oxygenase (BCO2) [52]. The BCO1 role in the nematode, as in other animals, has been shown to be essential in the transformation of β-carotene to retinal [53]. However, there is no previous report about the effects of BCX on *C. elegans* and the molecular mechanisms targeted by this carotenoid.

In the present study, the potential fat-reducing effect and antioxidant activity of BCX have been investigated using the in vivo model system of *C. elegans*. As a result of treatment, nematodes reduced the body fat content and were more resistant to an acute oxidative stress. Global transcriptional analysis revealed that BCX influences energy expenditure, response to stress, and protein turnover.

2. Materials and Methods

2.1. Carotenoid Standards and Extracts

Carotenoid β-cryptoxanthin (BCX) (purity ≥ 97%) was supplied by Extrasynthese (Lyon, France) and lycopene (purity ≥ 98%) and β-carotene (purity ≥ 97%) by Sigma-Aldrich (Saint Louis, MO, USA). The carotenoid stock solutions were prepared in chloroform and hexane (HPLC grade, Scharlau, Barcelona, Spain) and purity tested by HPLC-DAD (Waters, Barcelona, Spain) (High Performance Liquid Chromatography-Diode Array Detector) using conditions described in Rodrigo et al. [54]. Aliquots for the experiments were prepared from stocks, completely dried under N2 gas and immediately used. Furthermore, carotenoid extracts rich in BCX were obtained from fresh juice of mature Clementine mandarin (*Citrus clementina*) as described in Rodrigo et al. [54] and individual carotenoid composition was determined by HPLC-DAD immediately before use as indicated elsewhere [54]. Briefly, 2-mL aliquots of fresh juice were weighted in screw-capped polypropylene tubes (50 mL), and 4 mL of methanol were added. The suspension was stirred for 10 min at 4 °C. Tris-HCl (50 mM, pH 7.5) (containing 1 M NaCl) was then added (3 mL) and further stirred for 10 min at 4 °C. Chloroform (5 mL) was added to the mixture, stirred for 5 min at 4 °C and centrifuged at 3000× *g* for 10 min at 4 °C. The hypo-phase was removed, and the aqueous phase re-extracted with chloroform until it was colorless. The pooled chloroform extracts were dried on a rotary evaporator at 40 °C. To test the potential effect of esterified versus free carotenoids extracts on in vivo assays, half of the extracts were saponified in a methanolic solution of KOH (6% *w/v*) overnight at room temperature. Free carotenoids from saponified extracts were recovered from the upper phase after adding 3 mL of MilliQ water and 5 mL of solution A (petroleum ether:diethyl ether, 9:1) to the mixture. Repeated re-extractions by adding 3 mL of solution A were carried out until the hypo-phase was colorless. Non-saponified samples were dissolved in 5 mL of solution A and 3 mL of MilliQ water and proceeded as described above for saponified extracts. The extracts were reduced to dryness by rotary evaporation at 40 °C, dissolved with chloroform and methanol, and the extract was transferred to a 1.5-mL vial, dried under N2 and kept at −20 °C until HPLC analysis or in vivo assays were conducted. All operations were carried out on ice under dim light to prevent photodegradation, isomerization, and structural changes of the carotenoids.

For in vivo assays dried aliquots of carotenoids (commercial standards and Clementine juice extracts) were dissolved in chloroform: ethanol solution (ratio 1:15, *v/v*).

2.2. Quantification of Individual Carotenoids in Extracts

To determine the carotenoid profile and concentration in mandarin juice extracts, aliquots were analyzed by HPLC using a Waters liquid chromatography system equipped with a 600E pump (Waters, Barcelona, Spain) and a model 2998 photodiode array detector (DAD) (Waters, Barcelona, Spain), and Empower software (Waters, Barcelona, Spain). A C30 carotenoid column (250 × 4.6 mm, 5 μm) coupled to a C30 guard column (20 × 4.0 mm, 5 μm) (YMC, Teknokroma, Barcelona, Spain) was used. Samples were prepared for HPLC by dissolving the dried carotenoid extracts in chloroform: methanol: acetone (3:2:1, *v:v:v*). Ternary gradient elution was used for carotenoid separation. The initial solvent composition consisted of 90% methanol, 5% water, and 5% methyl tert-butyl ether (MTBE).

The solvent composition changed in linear fashion to 95% methanol and 5% MTBE at 12 min. During the next 8 min, the solvent composition was changed to 86% methanol and 14% MTBE. After reaching this concentration, the solvent was gradually changed to 75% methanol and 25% MTBE at 30 min. After 20 min, the solvent composition changed linearly, being 50% methanol and 50% MTBE at 50 min and maintained at this proportion until 60 min. The initial conditions were re-established in 5 min and equilibrated for 15 min before the next injection. The flow rate was 1 mL/min, column temperature was set to 25°C, and the injection volume was 20 µL. The PAD was set to scan from 250 to 540 nm, and for each elution a Maxplot chromatogram was obtained, plotting each carotenoid peak at its corresponding maximum absorbance wavelength.

Carotenoids were identified by their retention time, absorption, and fine spectra [54,55]. The carotenoid peaks were integrated at their individual maximal wavelength, and their contents were calculated using the appropriate calibration curves as described elsewhere [54,56]. For the quantification of esterified xanthophylls the calibration curves of the corresponding free carotenoids were used.

2.3. C. elegans Strain and Treatments

Caenorhabditis elegans strain N2, Bristol (wild-type) was provided by the Caenorhabditis Genetics Center (University of Minnesota, Minneapolis, MN, USA). Nematodes were grown and maintained on nematode growth medium (NGM) at 20 °C using *Escherichia coli* OP50 bacteria as food source.

For the evaluation of BCX, worms were grown on NGM as control diet, or NGM supplemented with different doses of BCX (0.005, 0.01, 0.025, 0.05, and 0.1 µg/mL). The carotenoids lycopene and β-carotene were also tested at 0.025 µg/mL. All carotenoids solutions were prepared as described above and added on the agar plate surface. The appropriate amount of chloroform: ethanol (ratio 1:15) was also added on NGM control medium.

In addition, *C. elegans* body-fat reduction was evaluated with NGM plates supplemented with the carotenoid extract from mandarin juice containing BCX at a final concentration of 0.025 µg/mL.

Finally, the activity on *C. elegans* fat-reduction was evaluated in different food matrices supplemented with BCX. Different volumes of each matrix were added to the NGM plates in order to obtain a final dose of 0.025 µg/mL of BCX: dairy fermented product (250 µL/plate), skimmed milk (250 µL/plate), sugar-free soft drink (with or without caffeine) (100 µL/plate) or orange juice (50 µL/plate).

2.4. Bioassimilation Assay in C. elegans

In order to confirm the absorption of BCX by the nematodes during feeding experiments, populations of the *C. elegans* strain N2 Bristol were cultured on NGM or NGM supplemented with BCX (1 µg/mL). At least 40 plates per condition were prepared and worms were recovered with M9 buffer and immediately washed three times to remove *E. coli* OP50 present in the media. Additionally, 2 h of incubation in M9 buffer was performed to facilitate the removal of gut microbiota from the nematodes. The evacuated and washed worm pellets (containing approximately 100 mg per condition) were recovered in 2 mL cryotubes and immediately frozen in liquid nitrogen. The pellets were defrosted on ice, transferred to 1.5-mL tubes with 0.5 mL of acetone (HPLC grade, Scharlab, Barcelona, Spain), and disrupted with micro-pestle. Disrupted pellets were vortexed for 30 s and centrifuged for 3 min at 15,700× *g* at 4 °C. The acetone was recovered, and pellets were re-extracted twice with 2 mL acetone. The pooled acetone extract for each condition was dried under nitrogen stream. The dried residue was dissolved in 250 µL dichloromethane (HPLC grade, Scharlab, Barcelona, Spain), and 250 µL MilliQ water was added and vortexed for 30 s. The organic phases were recovered, and aqueous phase were re-extracted twice with 250 µL of dichloromethane. The pooled organic phases were dried under nitrogen stream and stored at −20 °C until HPLC-PAD analysis. The analytical chromatographic and identification conditions used were the same as indicated in Section 2.2.

2.5. Body Fat Reduction Assays in C. elegans

For Nile Red assays, age-synchronized nematodes were cultured in NGM or NGM supplemented with the different doses of commercial carotenoids, carotenoid extract from mandarin juice, or the different food matrices supplemented with commercial BCX. The NGM plates with 6 µg/mL of Orlistat (Sigma–Aldrich, Madrid, Spain) were used as positive control. The nematode fat content was measured by Nile Red staining as previously described by Martorell and coworkers [45]. Nile Red (Sigma-Aldrich, St. Louis, MO, USA) was added on the surface of NGM agar plates, previously seeded with *E. coli* OP50, to a final concentration of 0.05 µg/mL. Worms were incubated at 20 °C and when they reached young adult stage, nematodes were placed in M9 buffer and fluorescence was measured in a spectrofluorometer FP-6200 (Jasco, Easton, MD, USA) with λ excitation 480 nm and λ emission 571 nm. At least two independent experiments were performed with each sample.

Nematode triglyceride (TG) content was measured using Triglyceride Quantification Kit (Biovision, Mountain View, CA, USA). Age-synchronized N2 nematodes were treated with BCX at a dose of 0.025 µg/mL until nematodes reached young adult stage. Then, worms were collected and washed with PBS buffer. After worm settling, supernatant was removed and 400 µL of TG assay buffer was added to worm pellet. Nematodes were sonicated with a digital sonifier (Branson Ultrasonics Corporation, Danbury, CT, USA) using 4 pulses for 30 s at 10% power. Samples were slowly heated twice at 90 °C for 5 min in a thermomixer (Eppendorf, Hamburg, Germany) to solubilize all TGs in the solution. After brief centrifugation, samples were used for the TG assay following the manufacturer's instructions. Five different biological replicates were included for each condition in four independent experiments. Total protein content was estimated by bicinchoninic acid assay (Pierce™ BCA Protein Assay Kit, Thermo Scientific, Rockford, IL, USA) to normalize TG content following the manufacturer's instructions.

The significance between control and treatment conditions was analyzed by one-way analysis of variance (ANOVA). Statistical comparisons of the different treatments were performed using Tukey's test. All statistical analyses were performed using the GraphPad Prism 4 software (GraphPad Software, Suite, San Diego, CA, USA).

2.6. Oxidative Stress Assays in C. elegans

Populations of age-synchronized worms were obtained by isolating eggs from gravid adults and hatching the eggs in NGM plates (as control media) or NGM plates with the different doses of commercial carotenoids. In order to get a comparison of the antioxidant capacity provided by carotenoids respect to other well-recognized antioxidant compounds, vitamin C (0.1 µg/mL, Sigma-Aldrich, St. Louis, MO, USA) was used as positive control in the different oxidative stress assays. Experiments were performed according to a previously published protocol [57]. At least two independent experiments were performed with each sample.

Viability of worms between nematodes cultured in control and treatment conditions after oxidative stress were evaluated by means of one-way analysis of variance (ANOVA) and the Tukey test for comparative analysis. All the analyses were performed with the GraphPad Prism 4 statistical software (GraphPad Software, Suite, San Diego, CA, USA).

2.7. Transcriptomic Analysis in C. elegans

Gene expression in *C. elegans* N2 strain was analyzed in worm populations treated in NGM or NGM supplemented with BCX at 0.025 µg/mL dose. Synchronized populations were obtained from embryos isolated from gravid adults in the different feeding conditions. Worms were incubated at 20 °C. Then, 5-day adult worms were collected with M9 buffer, washed three times, and collected in 1.5-mL tubes for worm disruption by sonication (3 pulses at 10 W, 20 s/pulse). Total RNA isolation was performed with RNeasy kit (Qiagen, Barcelona, Spain). The RNA samples were processed for hybridization using the GeneChip® *C. elegans* Genome Arrays of Affymetrix (UCIM, University of

Valencia, Valencia, Spain). These chips contain oligonucleotide probe sets designed to asses over 22,500 transcripts from the *C. elegans* genome. Three biological replicates were examined per condition by bioinformatics. Raw data obtained from Affymetrix arrays were background corrected using the RMA (Robust Multi-Array Average) methodology [58]. Signal intensity was standardized across arrays via quantile normalization algorithm. Limma moderated *t*-statistics was used to evaluate differential gene expression between control and treated conditions. To control the false discovery rate, *p*-values were corrected for multiple testing as is described by Benjamini and Hochberg [59]. Finally, gene set analysis was performed for each comparison using logistic regression models [60].

3. Results

3.1. β-Cryptoxanthin Reduces Fat Content in C. elegans N2 in a Dose-Dependent Manner

The effect on BCX on body-fat reduction, expressed as percentage of fluorescence reduction, in nematode populations fed with different concentration (0.005, 0.01, 0.025, 0.05, and 0.1 µg/mL) of the xanthophyll versus control feeding is shown in Figure 1A. All BCX doses assayed produced a significant body fat reduction, and ranged from 13.6% to 29.8% in respect to feeding under control conditions ($p \leq 0.001$). However, the most effective dose was 0.025 µg/mL, with a fluorescence reduction of near 30%. These results indicate that BCX has a significant effect upon fat reduction in *C. elegans* in a dose-dependent manner.

(**A**)

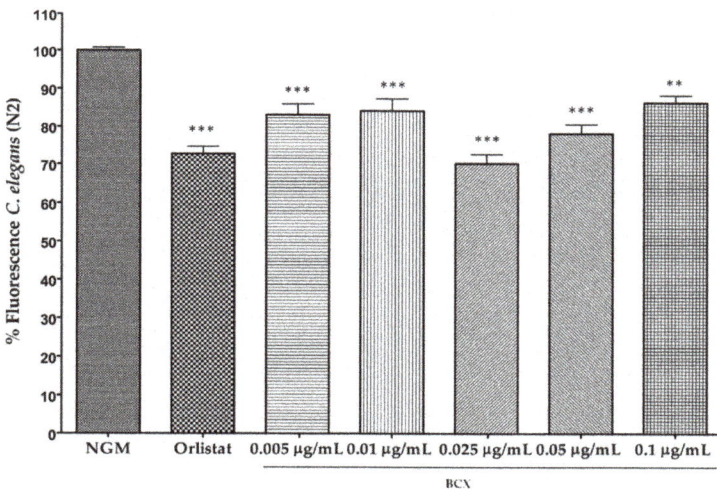

Figure 1. *Cont.*

(B)

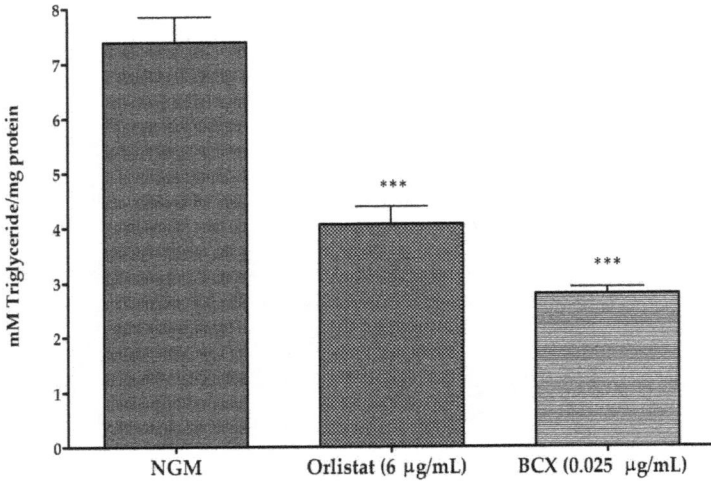

(C)

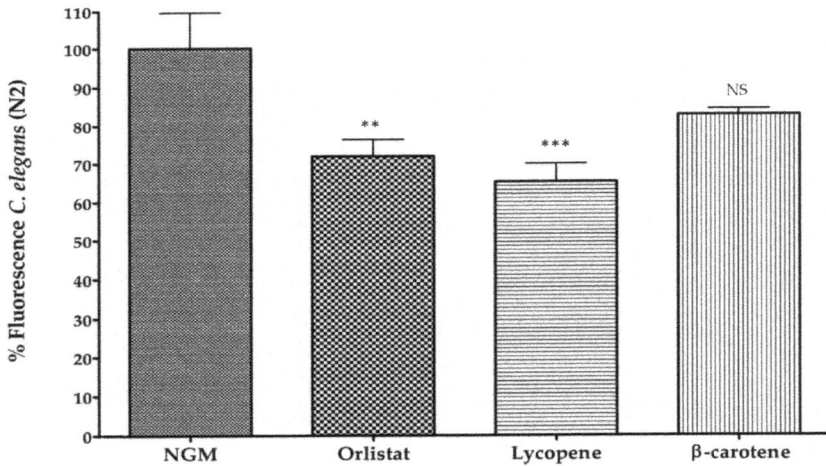

Figure 1. Body fat reduction in *Caenorhabditis elegans* provided by BCX (β-Cryptoxanthin) and other carotenoids. (**A**) Dose-response Nile Red assay with different doses of commercial β-cryptoxanthin. Nematode growth medium (NGM) was used as control feeding condition and Orlistat as positive control. Percentage of fluorescence is the mean of four independent experiments; (**B**) Quantification of triglycerides (TG) content (mM TG/mg protein) in *C. elegans* fed with BCX (0.025 μg/mL), Orlistat, and NGM; (**C**) *C. elegans* were fed with carotenoids lycopene and β-carotene at the dose of 0.025 μg/mL. Data are the average of at least three independent experiments. *** p-value ≤ 0.001, ** p-value ≤ 0.01, NS: not significant.

Triglycerides (TGs) are the main constituents in lipid droplets stored in *C. elegans*. Therefore, TG quantification was performed with worms fed with BCX at the effective dose of 0.025 μg/mL to validate quantification of fluorescence in Nile Red stained worms. Results indicated a significant reduction in total TG in worms fed with BCX ($p \leq 0.001$) compared with the control-fed nematodes (Figure 1B). Worms cultivated under control conditions presented a TG concentration of 7.39 mM/mg protein, while for BCX 0.025 µg/mL had a value of 2.79 mM/mg protein (a reduction of 62.23% of TG content in comparison to control conditions). This reduction was even higher than that of positive control Orlistat, with a TG content of 4.07 mM/mg protein (44.96% reduction over to control-fed nematodes). These results were consistent with the effect observed by Nile Red staining. Therefore, quantification of TG content in nematodes supports the reduction of total fat described in previous works [40,45].

To determine whether the fat reduction was a specific effect of BCX or may be also attributed to other carotenoids with health-related benefits, the fat reducing activity of lycopene and β-carotene was assessed in *C. elegans* (Figure 1C). Nematodes were fed with the concentration of 0.025 µg/mL of each carotenoid, since it was the most effective dose in body fat reduction in the BCX dose-response assays. Results showed a significant reduction in fluorescence produced by lycopene (34.8 %; $p \leq 0.01$) whereas no significant effect was observed in nematodes fed with β-carotene (Figure 1C).

To confirm the intake of BCX by the worms, experiments were performed by feeding nematodes with BCX and using the NGM without supplementation as a control. In these experiments we used a highest dose of BCX (1 µg/mL) to facilitate the detection of BCX in the worm extracts. The pellet corresponding to BCX-supplemented worms showed a cream-yellowish color while that of control populations showed a white-translucent color (Figure 2). The HPLC-PAD analysis showed the presence of BCX only in the extracts derived from worms fed with plates supplemented with BCX, whereas no carotenoids were detected in the extract of nematodes grown in control condition (Figure 2B,C). In the chromatogram from the worms extract grown in BCX supplemented media, a single peak with carotenoid spectrum was detected with a retention time and absorbance spectrum matching that of BCX standard (Figure 2A,B). This result indicates that BCX added to the culture media is uptake by the nematode.

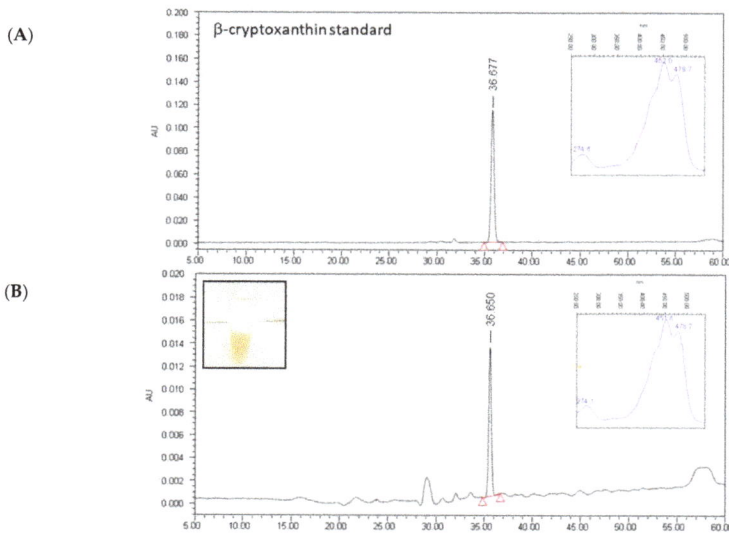

Figure 2. *Cont.*

(C)

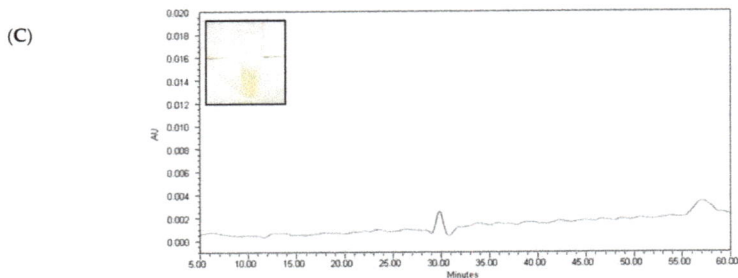

Figure 2. (**A**) BCX uptake by *C. elegans*. HPLC-PAD (High Performance Liquid Chromatography-Photodiode Array Detector) chromatograms of BCX (β-Cryptoxanthin) standard; (**B**) pellet extract from worms previously fed with BCX at 1 μg/mL; and (**C**) from worms grown under control condition (NGM (Nematode growth medium) without supplementation). Insert plots in (**A,B**) show the absorbance spectrum of the carotenoid peak identified in the chromatograms. Insert images in (**B,C**) show the corresponding worm pellets before disruption showing the differences in coloration. AU, arbitrary absorbance units.

One important source of BCX in the human diet are mandarins which is esterified with a variety of fatty acids in a high proportion [8,12]. Analysis of carotenoid composition in the juice extract of Clementine mandarin showed that BCX was the main carotenoid accounting for nearly 60% of the total carotenoid content and being 90% in esterified form (Table 1). As expected, the mandarin extract also contained significant amounts of other β,β-xanthophylls (violaxanthin, zeaxanthin, and antheraxanthin), as well as β-carotene and colorless carotenes (phytoene and phytofluene) (Table 1). To investigate whether carotenoid esterification with fatty acids may have an impact on the reduction of lipid content in the nematode, the reduction in body fat was determined in nematodes fed with a non-saponified carotenoid extract from mandarin juice and compared with that of a saponified extract. Aliquots of free (saponified) or esterified (non-saponified) carotenoid extracts from mandarin juice were prepared to adjust the BCX to a final concentration of 0.025 μg/mL and the effect on worm body fat reduction was determined (Figure 3). Both types of extracts provide a significant reduction of fluorescence respect control medium (30.66% and 31.38%, respectively), without significant differences between them. Therefore, it appears that BCX in both free or esterified form provided a similar reduction in the body fat content of the nematodes.

Table 1. Carotenoid content and composition in saponified and non-saponified juice extracts of Clementine mandarins. The percentage of β-Cryptoxanthin over the total amount of carotenoids is indicated in parenthesis.

Carotenoid	Concentration (μg/mL) in Mandarin Juice Extract [1]	
	Saponified	Non-saponified
Phytoene	0.80 ± 0.05	0.86 ± 0.03
Phytofluene	0.37 ± 0.09	0.45 ± 0.02
ζ-Carotene	0.19 ± 0.04	0.25 ± 0.03
β-Carotene	0.28 ± 0.05	0.35 ± 0.02
β-Cryptoxanthin	4.86 ± 0.23 (61%)	0.30 ± 0.02 (3%)
Zeaxanthin	0.11 ± 0.02	N.D.
Anteraxanthin	0.52 ± 0.01	N.D.
Violaxanthin [2]	0.75 ± 0.01	N.D.
Esters (mono- and diesters)		
β-Crytoxanthin	N.D.	5.35 ± 0.25 (57%)
Zeaxanthin	N.D.	0.20 ± 0.01
Antherxanthin	N.D.	0.70 ± 0.03
Violaxanthin [2]	N.D.	0.92 ± 0.04
Total carotenoids	7.88 ± 0.62	9.38 ± 0.48

[1] Data are mean $\pm$ SD ($n = 3$); [2] Sum of 9-Z- and all-E-isomers; N.D. Not detected.

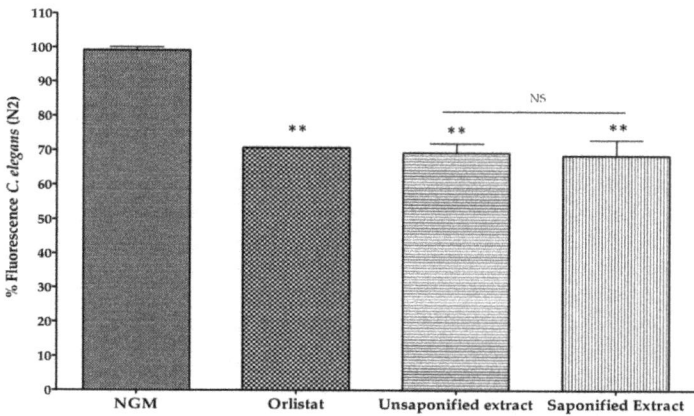

Figure 3. Measurement of fluorescence in Nile Red stained *C. elegans* worms fed with non-saponified or saponified carotenoid extracts from mandarin juice adjusted to contain 0.025 µg/mL of β-Cryptoxanthin (BCX). Nematode growth medium (NGM) was used as control feeding condition and Orlistat (6 µg/mL) as positive control. Percentage of fluorescence is the mean of two independent experiments. ** *p*-value ≤ 0.01, NS: not significant.

To investigate the effect of different food matrices on the BCX-induced body fat reduction, a dairy fermented product, skimmed milk, and sugar free soft drink (with or without caffeine) were tested. Each food matrix was supplemented with the most effective BCX dose (0.025 µg/mL) and the lipid reduction was assayed in the nematode model for obesity. Figure 4 shows that all tested food matrices supplemented with BCX reduced body fat. However, sugar free soft drink with caffeine + BCX was the food matrix with a major fluorescence reduction (28.01%). These data indicate that addition of BCX to the different food matrices preserve its anti-obesity activity.

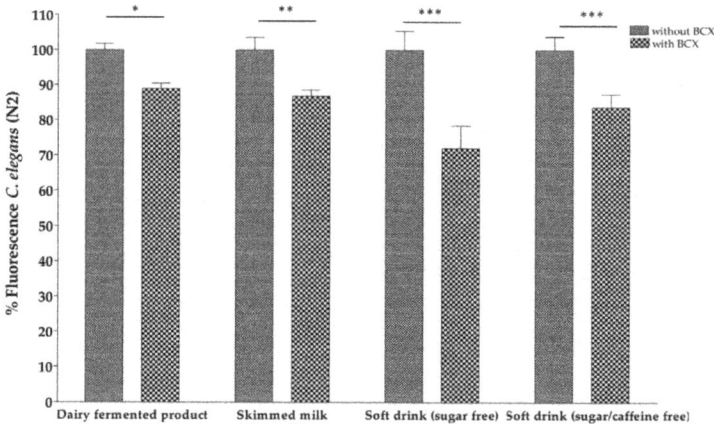

Figure 4. Body fat reduction measured by Red Nile staining in *C. elegans* worms fed with different food matrices supplemented with β-Cryptoxanthin (BCX) at 0.025 µg/mL. Nematode growth medium (NGM) was used as control feeding condition and Orlistat (6 µg/mL) as positive control. Percentage of fluorescence is the mean of two independent experiments. *** Significant *p*-value ≤ 0.001, ** Significant *p*-value ≤ 0.01, * Significant *p*-value ≤ 0.05.

3.2. BCX has Antioxidant Activity in C. elegans N2

Carotenoids are recognized as antioxidants scavenging harmful reactive oxygen species (ROS) generated in the cell [14,61]. In order to check if BCX may also protect *C. elegans* N2 worms from oxidative stress, nematodes were fed with different doses of commercial BCX and subjected to an acute oxidative stress with H_2O_2 [57] (Figure 5A). Although the antioxidant capacity of carotenoids is different to that of the water-soluble molecules, an experiment with worms treated with ascorbic acid (0.1 µg/mL) was performed in order to include in the assays a well-recognized antioxidant as a positive internal control of the assay (not for comparative purposes). The survival percentage of H_2O_2-challenged nematodes was reduced to 33% whereas treatment with vitamin C (0.1 µg/mL) increased survival rate up to 45% (Figure 5A). Among the different BCX doses assayed, 0.025 µg/mL was the dose providing a significant increase in worm survival respect control medium, with survival percentage of 58%, that was even higher than that provided by vitamin C (Figure 5A). No significant effect on worm survival was observed with BCX concentrations lower or higher than 0.025 µg/mL. These results suggest that BCX exerts an antioxidant activity on nematodes after an acute oxidative stress at the same optimum concentration than that observed in fat reduction assays.

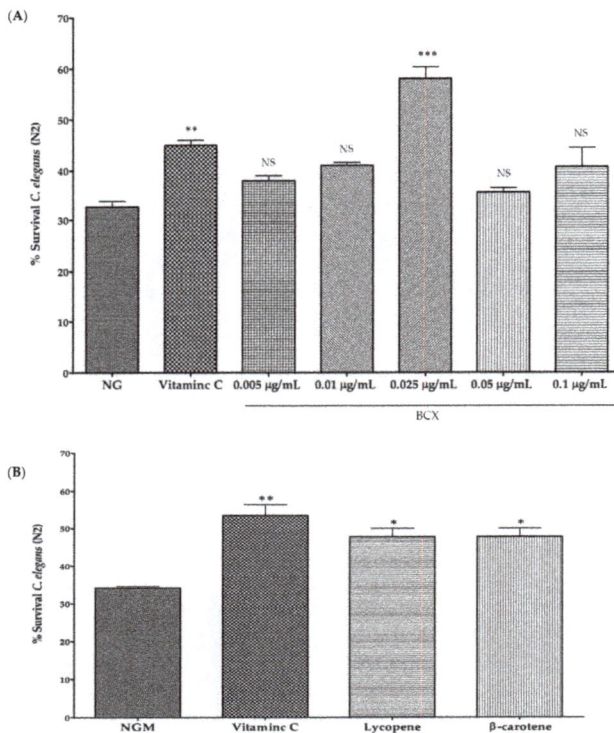

Figure 5. Percentage of *C. elegans* worm's survival after an acute oxidative stress with 2 mM of H_2O_2. Vitamin C (0.1 µg/mL) was used as positive control. (**A**) Worms were fed at different concentrations of commercial β-Cryptoxanthin (BCX). Survival values correspond with the mean of three independent experiments; (**B**) Antioxidant activity of *C. elegans* fed with carotenoids lycopene and β-carotene at the dose of 0.025 µg/mL. In both experiments, Vitamin C (0.1 µg/mL) was used as antioxidant positive control. Survival values correspond with the mean of at least two independent experiments. *** *p*-value ≤ 0.001, ** *p*-value ≤ 0.01, * *p*-value ≤ 0.05, NS: not significant. NGM: nematode growth medium.

Furthermore, to compare the effect of BCX in respect to other carotenoids upon worms' viability after an acute oxidative stress, we evaluated survival capacity in nematodes treated with lycopene and β-carotene at dose of 0.025 μg/mL, that showed the highest antioxidant effect for BCX (Figure 5A). Both carotenoids assayed produced a significant increase in nematode survival in respect to NGM medium in oxidative conditions, with very similar activity (48% of survival) (Figure 5B). Therefore, both carotenoids presented effective protection against an acute oxidative stress in nematodes, but to a lesser extent than BCX.

3.3. BCX Modulates Energy Metabolism, Antioxidant Response, and Protein Homeostasis in C. elegans

To understand the molecular mechanisms underlying the BCX effect in *C. elegans*, the transcriptomic profile of BCX-fed worms was studied using Affymetrix microarrays platform. Five-day adult worms were fed with BCX at 0.025 μg/mL, and the changes in gene expression as well as the study of functional blocks was made by comparing the transcriptomic profiling of BCX–treated worms versus control-fed worms.

First, differential expression at gene level was studied by analyzing the differences between BCX-treated samples and controls using limma moderated *t*-statistic. After feeding BCX, 542 genes were upregulated compared with control-fed nematodes. However, to get a stronger statistical significance, a *p*-value correction was applied, with no significant differences in gene expression. Therefore, a gene set analysis was performed by using two databases, GO (Gene Ontology) Biological Process and KEGG (Kyoto Encyclopedia of Genes and Genomes) Pathways, to facilitate the biological interpretation of microarray data. Thus, in BCX-treated worms, 18 metabolic pathways were found to be upregulated ($p \leq 0.05$) compared with control nematodes (Table 2), while a total of 35 biological processes were upregulated (Table 3).

Table 2. KEGG (metabolic pathways were significantly ($p \leq 0.05$) upregulated in *C. elegans* N2 after feeding with β-Cryptoxanthin (BCX) (0.025 μg/mL) as compared with control feeding conditions.

Upregulated KEGGSs Pathways (BCX vs. Control)			
ID	Name	Size [a]	*p*-Value
04141	Protein processing in endoplasmic reticulum	170	1.26×10^9
00020	Citrate cycle (TCA cycle)	42	0.0005
03040	Spliceosome	123	0.0025
00010	Glycolysis/gluconeogenesis	38	0.0067
03030	DNA replication	39	0.0088
00970	Aminoacyl-tRNA biosynthesis	49	0.0118
03050	Proteasome	45	0.0180
00030	Pentose phosphate pathway	26	0.0180
04144	Endocytosis	109	0.0231
00520	Amino sugar and nucleotide sugar metabolism	32	0.0232
04130	SNARE interactions in vesicular transport	23	0.0232
00240	Pyrimidine metabolism	71	0.0232
00052	Galactose metabolism	13	0.0247
00051	Fructose and mannose metabolism	31	0.0247
00564	Glycerophospholipid metabolism	54	0.0323
04330	Notch signaling pathway	28	0.0393
00190	Oxidative phosphorylation	127	0.0393
04120	Ubiquitin mediated proteolysis	111	0.0393

[a] Number of genes among the group. (KEGG: Kyoto Encyclopedia of Genes and Genomes; TCA: Tricarboxylic Acid Cycle); SNARE: Soluble NSF Attachment Protein Receptor.

Table 3. GO Biological Processes was significantly ($p \leq 0.05$) non-redundantly upregulated in *C. elegans* N2 feeding with β-Cryptoxanthin (BCX) (0.025 µg/mL) as compared with control feeding conditions.

Non-Redundant Upregulated GOs (BCX vs. Control)	
ID	Name
0000910	Cytokinesis
0048598	Embryonic morphogenesis
0006886	Intracellular protein transport
0006400	tRNA modification
0006479	Protein amino acid methylation
0006099	Tricarboxylic acid cycle
0006412	Translation
0019751	Polyol metabolic process
0006261	DNA-dependent DNA replication
0006457	Protein folding
0007346	Regulation of mitotic cell cycle
0030261	Chromosome condensation
0008593	Regulation of Notch signaling pathway
0008284	Positive regulation of cell proliferation
0040028	Regulation of vulval development
0000075	Cell cycle checkpoint
0001708	Cell fate specification
0040019	Positive regulation of embryonic development
0007067	Mitosis
0008356	Asymmetric cell division
0006096	Glycolysis
0040015	Negative regulation of multicellular organism growth
0007283	Spermatogenesis
0051246	Regulation of protein metabolic process
0040039	Inductive cell migration
0033554	Cellular response to stress
0000132	Stablishment of mitotic spindle orientation
0045132	Meiotic chromosome segregation
0045137	Development of primary sexual characteristics
0035194	Posttranscriptional gene silencing by RNA
0007143	Female meiosis
0046580	Negative regulation of Ras protein signal transduction
0000184	Nuclear-transcribed mRNA catabolic process, nonsense-mediated decay
0022613	Ribonucleoprotein complex biogenesis
0051302	Regulation of cell division

GO: Gene Ontology.

Regarding the metabolic pathways in BCX-treated worms, we found an upregulation of metabolic pathways related with energy metabolism (citrate cycle, pentose phosphate pathway, glycolysis/gluconeogenesis, and oxidative phosphorylation). Moreover, other pathways related to protein metabolism/processing were over-expressed under BCX supplementation (spliceosome, proteasome, and ubiquitin mediated proteolysis) (Table 2).

Concerning the non-redundant GO biological processes upregulated by BCX, different processes related to protein metabolism (translation, transport, and folding), as well as energy and carbohydrate metabolism processes (aerobic respiration, tricarboxylic acid cycle, glycolysis, oxidation of organic compounds), Notch signaling pathway and cell division were induced and cellular response to stress was also upregulated under BCX treatment (Table 3). All these processes were consistent with the observed upregulated KEGG pathways.

Finally, only three processes were found to be significantly downregulated in BCX-treated nematodes: potassium transport, sensory perception, and protein signaling pathway (Table 4).

Table 4. GO Biological Processes was significantly ($p \leq 0.05$) non-redundantly downregulated in *C. elegans* N2 feeding with β-Cryptoxanthin (BCX) (0.025 µg/mL) as compared with control feeding conditions.

Non-Redundant Downregulated GOs (BCX vs. Control)	
ID	Name
0006813	Potassium ion transport
0007606	Sensory perception of chemical stimulus
0007186	G-protein coupled receptor protein signaling pathway

GO: Gene Ontology.

4. Discussion

Xanthophyll β-cryptoxanthin is a xanthophyll with significant contribution to the pro-vitamin A activity in the human diet and its bio-accessibility from different food sources is similar or greater than that of α- or β-carotene [3,7]. The occurrence of BCX in food is variable and only a limited number of fruits and vegetables have moderate to high concentrations of BCX, being citrus fruits, especially mandarins and sweet oranges and their derivates, which are the major contributors to its intake in a common Western diet [6,9]. Besides the pro-vitamin A activity, in vitro, in vivo, and epidemiological studies suggest that BCX provides diverse health-related benefits, such as reducing risk of certain cancer [62], antioxidant [63], anti-inflammatory [15], anti-obesity [31], antidiabetic [33], and antiatherosclerosis properties [64]. Moreover, a positive effect on bone-related functions [23] and cognitive function [16] has been also described. However—in spite of the evidence of their beneficial properties in human nutrition and health—information about the mechanisms and metabolic pathways related to these beneficial effects of BCX are largely unknown [6].

Caenorhabditis elegans is a well-studied biological system and is considered an excellent model to study many health-related diseases, such as obesity and aging [37,65,66]. Moreover, other advantages of this model are that many of the pathways related to energy, lipid metabolism, and fat storage are conserved with humans [36]. Several researchers have used *C. elegans* to evaluate different ingredients with anti-obesity and antioxidant properties and to explore their mechanisms of action [40,42,45,67]. In the present study, we have used this nematode to investigate the anti-obesity and antioxidant effects of BCX (Figures 1 and 5) and also to dissect the molecular responses caused by BCX intake in the nematode (Figure 6, Tables 2 and 3). Our study demonstrates that BCX exhibits fat-reducing effect and protects nematodes from acute oxidative stress. Furthermore, BCX displays higher functional activity than other carotenoids as lycopene and β-carotene. This stronger effect of BCX has been previously postulated by other authors, specifically in relation with the improvement of inflammatory markers protecting against rheumatoid arthritis [20,68]. It can be speculated that the differential functional activity among carotenoids may be due to different bioassimilation or uptake. Analysis of carotenoids from worms fed with NGM media supplemented with BCX confirmed the BCX was uptake by the nematodes (Figure 2). The bioassimilation of other carotenoids by *C. elegans* was also demonstrated in feeding experiments with citrus pulp, where the presence of the β,β-xanthophyll violaxanthin, the main carotenoid in sweet orange pulp, was identified in worm pellets [47]. These results suggest that carotenoids are bioassimilated by *C. elegans* when added to the culture media either as a single ingredient or in a food matrix (Figure 2).

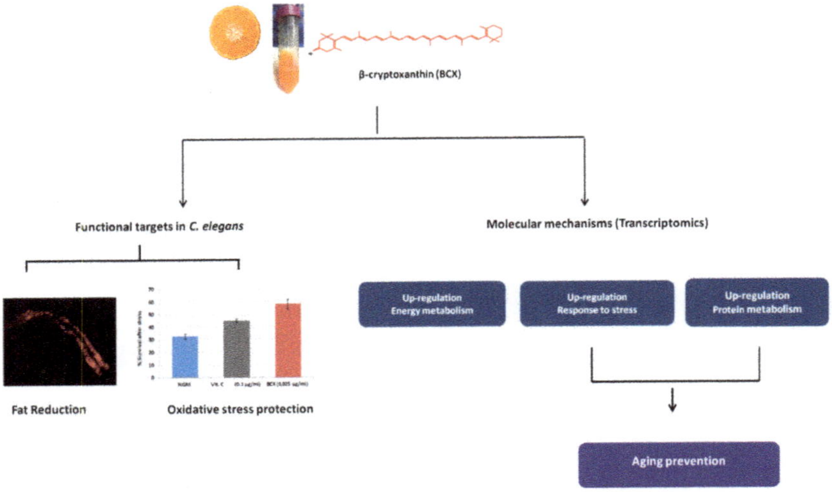

Figure 6. Model of the mechanism of action proposed for β-Cryptoxanthin (BCX) in *C. elegans*. Analysis of phenotype showed the efficacy of BCX on fat reduction and protection upon oxidative stress. The transcriptomic study provided the molecular mechanisms underlying these bioactivities, being the energy metabolism, the response to stress, and the protein metabolism the processes upregulated by BCX. NGM: nematode growth medium. VIt. C: Vitamin C.

To understand the molecular mechanisms underlying the functional activity of BCX, a transcriptomic study was performed in *C. elegans*. Microarray analysis showed an upregulation of biological processes related to energy expenditure. Thus, citrate cycle (TCA cycle) and glycolysis were induced after BCX feeding. Taking into account that the major portion of glucose flux takes place through the TCA cycle [69], our results would suggest that BCX enhances glucose metabolism to ensure ATP demands for cells, being consistent with the anti-diabetic properties proposed for BCX [70]. Moreover, ATP-metabolic process such as oxidative phosphorylation pathway was upregulated in BCX-fed nematodes. Therefore, these findings would explain the fat-reduced phenotype observed in BCX-treated nematodes, and correlate with the mechanism of action exerted by other fat-reducing compounds, which induced the energy expenditure and oxidative stress response [45]. It is worthy to remark that the reduction of body fat accumulation observed in the nematode by supplementation of BCX was maintained in different food matrices (Figure 3). This observation is of particular practical interest since BCX would preserve its beneficial properties when added to different types of food preparation or supplementations, in accordance with the EFSA (European Food Safety Authority) recommendations. All these results, together with the previous evidence showing a reduction of visceral fat in mice, rats, and humans [29,31,71], reinforce the notion that BCX has an anti-obesity function in several biological systems.

The transcriptomic profile of BCX-treated nematodes also showed upregulation of processes related to cellular stress response. The observed transcriptomic response to oxidative stress produced by BCX was confirmed by survival analysis in nematodes treated with BCX and subjected to acute oxidative stress. Results indicate a marked antioxidant effect of BCX in a dose-dependent manner (Figure 5A), in agreement with other authors reporting the anti-inflammatory activity of BCX and its potential use for preventing vascular disease [20,68]. The antioxidant effect was even higher than that induced by other carotenoids with demonstrated antioxidant function as lycopene (Figure 5B) [72]. These results are consistent with those obtained with in vitro systems, where BCX also display potent scavenger activity against several ROS [73]. Under this system, BCX esterification with fatty acids did not affect scavenger activity compared to the free form and it was suggested that this modification may stabilize BCX to allow massive accumulation in plant biosynthetic tissues and also provides thermostability to the xanthophyll structure [11,73]. Moreover, a study in humans showed no differences in plasma response when BCX was given as esterified or non-esterified form, indicating a similar bioavailability of both [74]. This observation is also compatible with our results in the nematode model since no significant differences in the reduction of body fat were found between saponified or non-saponified carotenoid mandarin extracts where nearly 94% of the BCX was naturally accumulated in esterified form (Table 1). In other biological systems, antioxidant properties of BCX has also been observed, as the reduction of DNA oxidative damage in brain of aged mice [16], cultured Caco-2 and HeLa cells [75], or in the liver and the adipose tissues of rats [71]. Overall, these evidences suggest that BCX supplementation contributes to decrease cellular oxidative stress.

Obesity is thought to be a symptom of metabolic syndrome and leading to the production of oxidative stress and ROS [76]. Since our results show that BCX has anti-obesity and antioxidative effects in *C. elegans*, it is tempting to hypothesize that BCX may have a protective effect against development of the metabolic syndrome. Recent epidemiological studies indicate the relationship between the serum antioxidant status and the metabolic syndrome [32,34]. The metabolic syndrome is a complex disorder clustering obesity, diabetes mellitus, and atherosclerotic cardiovascular diseases [76]. It has been suggested that BCX supplementation is able to prevent metabolic syndrome in both mice and humans [28,29,31] and the BCX concentrations in serum are inversely related to indices of oxidative DNA damage and lipid peroxidation [33,77].

Analysis of the microarray data in nematodes fed with BCX also revealed an activation of several pathways related to protein metabolism, such as proteasome, translation, spliceosome, ubiquitin mediated proteolysis, and protein folding. A protein homeostasis process is associated with aging and neurodegeneration [78,79]. A failure in this homeostasis can result in protein misfolding and in the accumulation of insoluble protein fibrils and aggregates, such as amyloids. Thus, the search for compounds or molecules such as BCX that enhance protein turnover could have a beneficial effect against aging and age-related diseases. This suggestion is in accordance with previous report where mice fed with BCX showed a reduction of the risk of age-related cognitive dysfunction and higher learning ability [16].

In summary, the in vivo anti-obesity and antioxidant activity of BCX in the nematode model system has been described in this study. Our results suggest a molecular mechanism based on promoting energy expenditure and cellular response to stress pathways. Moreover, BCX also displays upregulation of biological processes related to protein homeostasis. Therefore, this study provides new evidences for the potential therapeutic use of BCX in the prevention of diseases related to metabolic syndrome and aging. In this respect, further studies would be necessary to confirm the role of BCX on aging-related diseases.

Author Contributions: M.J.R., P.M., and S.G. designed the experiments; M.J.R., and L.Z. obtained the carotenoid extracts and analysis; S.L., and N.G. performed the *C. elegans* assays; P.M., and S.L. run statistical analyses; P.M., and S.L. analyzed the data; P.M., S.L., M.J.R., L.Z. and D.R. wrote the paper; all authors revised and approved the manuscript.

Funding: This study has been funded by the Spanish Ministry of Economy and Competitiveness (MINECO), Consolider Fun-C-Food (CSD 2007-00063 project), and AGL2015-70218. M.J.R. and L.Z. belong to the Spanish Carotenoid Network (CaRed) funded by the Spanish MINECO grant BIO2015-71703-REDT and BIO2017-90877-REDT, and the European Carotenoid Network (EuroCaroten) funded by the European Commission COST action CA15136. We thank to the UCIM at University of Valencia for performing the DNA arrays.

Conflicts of Interest: The authors declare no conflict of interest.

References

1. Rodriguez-Concepcion, M.; Avalos, J.; Bonet, M.L.; Boronat, A.; Gomez-Gomez, L.; Hornero-Mendez, D.; Limon, M.C.; Meléndez-Martínez, A.J.; Olmedilla-Alonso, B.; Palou, A.; et al. A global perspective on carotenoids: Metabolism, biotechnology and benefits for nutrition and health. *Prog. Lipid Res.* **2018**, *70*, 62–93. [CrossRef] [PubMed]

2. Nisar, N.; Li, L.; Lu, S.; Khin, N.C.; Pogson, B.J. Carotenoid metabolism in plants. *Mol. Plant* **2015**, *8*, 68–82. [CrossRef] [PubMed]

3. Burri, B.J. BCX as a source of vitamin A. *J. Sci. Food Agric.* **2015**, *95*, 1786–1794. [CrossRef] [PubMed]

4. Beltrán-de-Miguel, B.; Estévez-Santiago, R.; Olmedilla-Alonso, B. Assessment of dietary vitamin A intake (retinol, α-carotene, β-carotene, β-cryptoxanthin) and its sources in the National Survey of Dietary Intake in Spain (2009–2010). *Int. J. Food Sci. Nutr.* **2015**, *66*, 706–712. [CrossRef]

5. Meléndez-Martínez, A.J.; Mapelli-Brahm, P.; Benítez-González, A.; Stinco, C.M. A comprehensive review on the colorless carotenoids phytoene and phytofluene. *Arch. Biochem. Biophys.* **2015**, *572*, 188–200. [CrossRef] [PubMed]

6. Burri, B.J.; La Frano, M.R.; Zhu, C. Absorption, metabolism, and functions of β-cryptoxanthin. *Nutr. Rev.* **2016**, *74*, 69–82. [CrossRef] [PubMed]

7. Estévez-Santiago, R.; Olmedilla-Alonso, B.; Fernández-Jalao, I. Bioaccessibility of provitamin A carotenoids from fruits: Application of a standardised static in vitro digestion method. *Food Funct.* **2016**, *7*, 1354–1366. [CrossRef] [PubMed]

8. Breithaupt, D.E.; Bamedi, A. Carotenoid esters in vegetables and fruits: A screening with emphasis on beta-cryptoxanthinesters. *J. Agric. Food Chem.* **2001**, *49*, 2064–2070. [CrossRef] [PubMed]

9. Granado, F.; Olmedilla, B.; Blanco, I.; Rojas-Hidalgo, E. Major fruit and vegetable contributors to the main serum carotenoids in the Spanish diet. *Eur. J. Clin. Nutr.* **1996**, *50*, 246–250. [PubMed]

10. Sugiura, M.; Kato, M.; Matsumoto, H.; Nagao, A.; Yano, M. Serum concentration of β-Cryptoxanthin in Japan reflects the frequency of Satsuma mandarin (*Citrus unshiu* Marc.) consumption. *J. Health Sci.* **2002**, *48*, 350–353. [CrossRef]

11. Mercadante, A.Z.; Rodrigues, D.B.; Petry, F.C.; Mariutti, L.R.B. Carotenoid esters in foods—A review and practical directions on analysis and occurrence. *Food Res. Int.* **2017**, *99*, 830–850. [CrossRef] [PubMed]

12. Ma, G.; Zhang, L.; Iida, K.; Madono, Y.; Yungyuen, W.; Yahata, M.; Yamawaki, K.; Kato, M. Identification and quantitative analysis of β-cryptoxanthin and β-citraurin esters in Satsuma mandarin fruit during the ripening process. *Food Chem.* **2017**, *234*, 356–364. [CrossRef] [PubMed]

13. Burri, B.J.; Chang, J.S.; Neidlinger, T.R. β-Cryptoxanthin and α-carotene-rich foods have greater apparent bioavailability than β-carotene-rich foods in Western diets. *Br. J. Nutr.* **2011**, *105*, 212–219. [CrossRef] [PubMed]

14. Stahl, W.; Sies, H. Bioactivity and protective effects of natural carotenoids. *Biochim. Biophys. Acta* **2005**, *1740*, 101–107. [CrossRef] [PubMed]

15. Katsuura, S.; Imamura, T.; Bando, N.; Yamanishi, R. β-carotene and BCX but not lutein evoke redox and immune changes in RAW264 murine macrophages. *Mol. Nutr. Food Res.* **2009**, *53*, 1396–1405. [CrossRef] [PubMed]

16. Unno, K.; Sugiura, M.; Ogawa, K.; Takabayashi, F.; Toda, M.; Sakuma, M.; Maeda, K.; Fujitani, K.; Miyazaki, H.; Yamamoto, H.; et al. Beta-cryptoxanthin, plentiful in Japanese mandarin orange, prevents age-related cognitive dysfunction and oxidative damage in senescence-accelerated mouse brain. *Biol. Pharm. Bull.* **2011**, *34*, 311–317. [CrossRef]

17. Iribarren, C.; Folsom, A.R.; Jacobs, D.R.; Gross, M.D.; Belcher, J.D.; Eckfeldt, J.H. Association of serum vitamin levels, LDL susceptibility to oxidation and autoantibodies aginst MDA-LDL with carotid atherosclerosis: A case-control study. *Artheroscler. Thromb. Vasc. Biol.* **1997**, *17*, 1171–1177. [CrossRef]

18. Dwyer, J.H.; Paul-Labrador, M.J.; Fan, J.; Shircore, A.M.; Merz, C.N.; Dwyer, K.M. Progression of carotid-intima media thickness and plasma antioxidants: The Los Angeles Atherosclerosis Study. *Artheroscler. Thromb. Vasc. Biol.* **2004**, *24*, 231–239. [CrossRef]

19. Nakamura, M.; Sugiura, M.; Aoki, N. High β-carotene and BCX are associated with low pulse wave velocity. *Atherosclerosis* **2006**, *184*, 363–369. [CrossRef]

20. Gammone, M.A. Carotenoids, ROS, and cardiovascular health. In *Reactive Oxyen Species in Biology and Human Health*; Ahmad, S.I., Ed.; CRC Press: Boca Raton, FL, USA, 2016; pp. 325–331, ISBN 9781498735452.

21. Miyazawa, K.; Miyamoto, S.; Suzuki, R.; Yasuu, Y.; Ikeda, R.; Kohno, H.; Yano, M.; Tanaka, T.; Hata, K.; Suzuki, K. Dietary beta-cryptoxanthin inhibits *N*-butyl-*N*-(4-hydroxybutyl)nitrosamine-induced urinary bladder carcinogenesis in male ICR mice. *Oncol. Rep.* **2007**, *17*, 297–304. [CrossRef]

22. Tanaka, T.; Shnimizu, M.; Moriwaki, H. Cancer chemoprevention by carotenoids. *Molecules* **2012**, *17*, 3202–3242. [CrossRef] [PubMed]

23. Yamaguchi, M. Role of carotenoid β-cryptoxanthin in bone homeostasis. *J. Biomed. Sci.* **2012**, *2*, 19–36. [CrossRef] [PubMed]

24. Yamaguchi, M.; Weitzmann, M.N. The bone anabolic carotenoid beta-cryptoxanthin enhances transforming growth factor-beta1-induced SMAD activation in MC3T3 preosteoblasts. *Int. J. Mol. Med.* **2009**, *24*, 671–675. [CrossRef] [PubMed]

25. Sugiura, M.; Nakamura, M.; Ogawa, K.; Ikoma, Y.; Ando, F.; Shimokata, H.; Yano, M. Dietary patterns of antioxidant vitamin and carotenoid intake associated with bone mineral density: Findings from post-menopausal Japanese female subjects. *Osteoporos. Int.* **2011**, *22*, 143–152. [CrossRef] [PubMed]

26. Kabagambe, E.K.; Furtado, J.; Baylin, A.; Campos, H. Some dietary and adipose tissue carotenoids are associated with the risk of nonfatal acute myocardial infarction in Costa Rica. *J. Nutr.* **2005**, *135*, 1763–1769. [CrossRef] [PubMed]

27. Andersen, L.F.; Jacobs, D.R., Jr.; Gross, M.D.; Schreiner, P.J.; Williams, O.D.; Lee, D.H. Longitudinal associations between body mass index and serum carotenoids: The CARDIA study. *Br. J. Nutr.* **2006**, *95*, 358–365. [CrossRef]

28. Iwamoto, M.; Imai, K.; Ohta, H.; Shirouchi, B.; Sato, M. Supplementation of highly concentrated BCX in a satsuma mandarin beverage improves adipocytokine profiles in obese Japanese women. *Lipids Health Dis.* **2012**, *11*, 52–55. [CrossRef]

29. Tsuchida, T.; Minagawa, K.; Mukai, K.; Mizuno, Y.; Mashiko, K.; Minagawa, K. The comparative study of BCX derived from Satsuma mandarin for fat of human body. *Jpn. Pharmacol. Ther.* **2008**, *36*, 247–253.

30. Iwata, A.; Matsubara, S.; Miyazaki, K. Beneficial effects of a beta-cryptoxanthin-containing beverage on body mass index and visceral fat in pre-obese men: Double-blind, placebo-controlled parallel trials. *J. Funct. Foods* **2018**, *41*, 250–257. [CrossRef]

31. Takayanagi, K.; Morimoto, S.; Shirakura, Y.; Mukai, K.; Sugiyama, T.; Tokuji, Y.; Ohnishi, M. Mechanism of visceral fat reduction in Tsumura Suzuki obese, diabetes (TSOD) mice orally administered BCX from Satsuma mandarin oranges (*Citrus unshiu* Marc). *J. Agric. Food Chem.* **2011**, *59*, 12342–12351. [CrossRef]

32. Sugiura, M.; Nakamura, M.; Ogawa, K.; Ikoma, Y.; Matsumoto, H.; Ando, F.; Shimokata, H.; Yano, M. Associations of serum carotenoid concentrations with the metabolic syndrome: Interaction with smoking. *Br. J. Nutr.* **2008**, *100*, 1297–1306. [CrossRef]

33. Ni, Y.; Nagashimada, M.; Zhan, L.; Nagata, N.; Kobori, M.; Sugiura, M.; Ogawa, K.; Kaneko, S.; Ota, T. Prevention and reversal of lipotoxicity-induced hepatic insulin resistance and steatohepatitis in mice by an antioxidant carotenoid, BCX. *Endocrinology* **2015**, *156*, 987–999. [CrossRef] [PubMed]

34. Sugiura, M.; Nakamura, M.; Ogawa, K.; Ikoma, Y.; Yano, M. High serum carotenoids associated with lower risk for the metabolic syndrome and its components among Japanese subjects: Mikkabi cohort study. *Br. J. Nutr.* **2015**, *114*, 1674–1682. [CrossRef] [PubMed]

35. Shirakura, Y.; Takayanagi, K.; Mukai, K.; Tanabe, H.; Inoue, M. BCX suppresses the adipogenesis of 3T3-L1 cells via RAR activation. *J. Nutr. Sci. Vitaminol.* **2011**, *57*, 426–431. [CrossRef] [PubMed]

36. Ashrafi, K.; Chang, F.Y.; Watts, J.L.; Fraser, A.G.; Kamath, R.S.; Ahringer, J.; Ruvkun, G. Genome wide RNAi analysis of *Caenorhabditis elegans* fat regulatory genes. *Nature* **2003**, *421*, 268–272. [CrossRef]

37. Jones, K.T.; Ashrafi, K. *Caenorhabditis elegans* as an emerging model for studying the basic biology of obesity. *Dis. Models Mech.* **2009**, *2*, 224–229. [CrossRef]

38. Zheng, J.; Greenway, F.L. *Caenorhabditis elegans* as a model for obesity research. *Int. J. Obes.* **2012**, *36*, 186–194. [CrossRef]

39. Chiang, S.H.; MacDougald, O.A. Will fatty worms help cure human obesity? *Trends Genet.* **2003**, *19*, 523–525. [CrossRef]

40. Martorell, P.; Llopis, S.; González, N.; Chenoll, E.; López-Carreras, N.; Aleixandre, A.; Chen, Y.; Karoly, E.D.; Ramón, D.; Genovés, S. Probiotic strain *Bifidobacterium animalis* subsp. *lactis* CECT 8145 reduces fat content and modulates lipid metabolism and antioxidant response in *Caenorhabditis elegans*. *J. Agric. Food Chem.* **2016**, *64*, 3462–3472. [CrossRef]

41. Lemieux, G.A.; Ashrafi, K. Insights and challenges in using *C. elegans* for investigation of fat metabolism. *Crit. Rev. Biochem. Mol. Biol.* **2015**, *50*, 69–84. [CrossRef]

42. Kim, H.M.; Do, C.H.; Lee, D.H. Characterization of taurine as anti-obesity agent in *C. elegans*. *J. Biomed. Sci.* **2010**, *17*, S33. [CrossRef] [PubMed]

43. Zheng, J.; Enright, F.; Keenan, M.; Finley, J.; Zhou, J.; Ye, J.; Greenway, F.; Senevirathne, N.; Gissendanner, C.R.; Manaois, R.; et al. Resistant starch, fermented resistant starch, and short-chain fatty acids reduce intestinal fat deposition in *Caenorhabditis elegans*. *J. Agric. Food Chem.* **2010**, *58*, 4744–4748. [CrossRef]

44. Zarse, K.; Bossecker, A.; Müller-Kuhrt, L.; Siems, K.; Hernandez, M.A.; Berendsohn, W.G.; Birringer, M.; Ristow, M. The phytochemical glaucarubinone promotes mitochondrial metabolism, reduces body fat, and extends lifespan of *Caenorhabditis elegans*. *Horm. Metab. Res.* **2011**, *43*, 241–243. [CrossRef]

45. Martorell, P.; Llopis, S.; González, N.; Montón, F.; Ortiz, P.; Genovés, S.; Ramón, D. *Caenorhabditis elegans* as a model to study the effectiveness and metabolic targets of dietary supplements used for obesity treatment: The specific case of a conjugated linoleic acid mixture (Tonalin). *J. Agric. Food Chem.* **2012**, *60*, 11071–11709. [CrossRef] [PubMed]

46. Yazaki, K.; Yoshikoshi, C.; Oshiro, S.; Yanase, S. Supplemental cellular protection by a carotenoid extends lifespan via Ins/IGF-1 signaling in *Caenorhabditis elegans*. *Oxid. Med. Cell. Longev.* **2011**. [CrossRef] [PubMed]

47. Pons, E.; Alquézar, B.; Rodríguez, A.; Martorell, P.; Genovés, S.; Ramón, D.; Rodrigo, M.J.; Zacarías, L.; Peña, L. Metabolic engineering of β-carotene in orange fruit increases its in vivo antioxidant properties. *Plant Biotechnol. J.* **2014**, *12*, 17–27. [CrossRef] [PubMed]

48. Lashmanova, E.; Proshkina, E.; Zhikrivetskaya, S.; Shevchenko, O.; Marusich, E.; Leonov, S.; Melerzanov, A.; Zhavoronkov, A.; Moskalev, A. Fucoxanthin increases lifespan of *Drosophila melanogaster* and *Caenorhabditis elegans*. *Pharmacol. Res.* **2015**, *100*, 228–241. [CrossRef]

49. Chen, W.; Mao, L.; Xing, H.; Xu, L.; Fu, X.; Huang, L.; Huang, D.; Pu, Z.; Li, Q. Lycopene attenuates Aβ1-42 secretion and its toxicity in human cell and *Caenorhabditis elegans* models of Alzheimer disease. *Neurosci. Lett.* **2015**, *608*, 28–33. [CrossRef]

50. You, J.S.; Jeon, S.; Byun, Y.J.; Koo, S.; Choi, S.S. Enhanced biological activity of carotenoids stabilized by phenyl groups. *Food Chem.* **2015**, *177*, 339–345. [CrossRef]

51. Liu, X.; Luo, Q.; Cao, Y.; Goulette, T.; Liu, X.; Xiao, H. Mechanism of different stereoisomeric astaxanthin in resistance to oxidative stress in *Caenorhabditis elegans*. *J. Food Sci.* **2016**, *81*, H2280–H2287. [CrossRef]

52. Harrison, E.H.; Quadro, L. Apocarotenoids: Emerging roles in mammals. *Annu. Rev. Nutr.* **2018**, *38*, 153–172. [CrossRef]

53. Cui, Y.; Freedman, J.H. Cadmium induces retinoic acid signaling by regulating retinoic acid metabolic gene expression. *J. Biol. Chem.* **2009**, *284*, 24925–24932. [CrossRef] [PubMed]

54. Rodrigo, M.J.; Cilla, A.; Barberá, R.; Zacarías, L. Carotenoid bioaccessibility in pulp and fresh juice from carotenoid-rich sweet oranges and mandarins. *Food Funct.* **2015**, *6*, 1950–1959. [CrossRef] [PubMed]

55. Petry, F.C.; Mercadante, A.Z. Composition by LC-MS/MS of new carotenoid esters in mango and citrus. *J. Agric. Food Chem.* **2016**, *64*, 8207–8224. [CrossRef] [PubMed]

56. Carmona, L.; Zacarías, L.; Rodrigo, M.J. Stimulation of coloration and carotenoid biosynthesis during postharvest storage of 'Navelina' orange fruit at 12 °C. *Postharvest Biol. Technol.* **2012**, *74*, 108–117. [CrossRef]

57. Martorell, P.; Forment, J.V.; de Llanos, R.; Montón, F.; Llopis, S.; González, N.; Genovés, S.; Cienfuegos, E.; Monzó, H.; Ramón, D. Use of *Saccharomyces cerevisiae* and *Caenorhabditis elegans* as model organisms to study the effect of cocoa polyphenols in the resistance to oxidative stress. *J. Agric. Food Chem.* **2011**, *59*, 2077–2085. [CrossRef] [PubMed]

58. Irizarry, R.A.; Hobbs, B.; Collin, F.; Beazer-Barclay, Y.D.; Antonellis, K.J.; Scherf, U.; Speed, T.P. Exploration, normalization, and summaries of high density oligonucleotide array probe level data. *Biostatistics* **2003**, *4*, 249–264. [CrossRef]

59. Benjamini, Y.; Hochberg, Y. Controlling the false discovery rate: A practical and powerful approach to multiple testing. *J. R. Stat. Soc. Ser. B (Methodol.)* **1995**, *57*, 289–300. [CrossRef]

60. Montaner, D.; Dopazo, J. Multidimensional gene set analysis of genomic data. *PLoS ONE* **2010**, *5*, e10348. [CrossRef]

61. Barros, R.G.C.; Andrade, J.K.S.; Denadai, M.; Nunes, M.L.; Narain, N. Evaluation of bioactive compounds potential and antioxidant activity in some Brazilian exotic fruit residues. *Food Res. Int.* **2017**, *102*, 84–92. [CrossRef]

62. Wu, C.; Han, L.; Riaz, H.; Wang, S.; Cai, K.; Yang, L. The chemopreventive effect of β-cryptoxanthin from mandarin on human stomach cells (BGC-823). *Food Chem.* **2013**, *136*, 1122–1129. [CrossRef] [PubMed]

63. Eroglu, A.; Schulze, K.J.; Yager, J.; Cole, R.N.; Christian, P.; Nonyane, B.A.S.; Lee, S.E.; Wu, L.S.F.; Khatry, S.; Groopman, J.; et al. Plasma proteins associated with circulating carotenoids in Nepalese school- aged children. *Arch. Biochem. Biophys.* **2018**, *646*, 153–160. [CrossRef] [PubMed]

64. Ciccone, M.M.; Cortese, F.; Gesualdo, M.; Carbonara, S.; Zito, A.; Ricci, G.; De Pascalis, F.; Scicchitano, P.; Riccioni, G. Dietary intake of carotenoids and their antioxidant and anti-inflammatory effects in cardiovascular care. *Mediat. Inflamm.* **2013**. [CrossRef] [PubMed]

65. Kaletsky, R.; Murphy, C.T. The role of insulin/IGF-like signaling in *C. elegans* longevity and aging. *Dis. Models Mech.* **2010**, *3*, 415–419. [CrossRef] [PubMed]

66. Zhou, K.I.; Pincus, Z.; Slack, F.J. Longevity and stress in *Caenorhabditis elegans*. *Aging* **2011**, *3*, 733–753. [CrossRef] [PubMed]

67. Martorell, P.; Llopis, S.; González, N.; Ramón, D.; Serrano, G.; Torrens, A.; Serrano, J.M.; Navarro, M.; Genovés, S. A nutritional supplement containing lactoferrin stimulates the immune system, extends lifespan, and reduces amyloid β peptide toxicity in *Caenorhabditis elegans*. *Food Sci. Nutr.* **2017**, *5*, 255–265. [CrossRef]

68. Gammone, M.A.; Riccioni, G.; D'Orazio, N. Carotenoids: Potential allies of cardiovascular health? *Food Nutr. Res.* **2015**, *59*, 26762.

69. Phelps, M.E.; Barrio, J.R. Correlation of brain amyloid with "aerobic glycolysis": A question of assumptions? *Proc. Natl. Acad. Sci. USA* **2010**, *107*, 17459–17460. [CrossRef]

70. Sugiura, M.; Nakamura, M.; Ikoma, Y.; Yano, M.; Ogawa, K.; Matsumoto, H.; Kato, M.; Ohshima, M.; Nagao, A. The homeostasis model assessment-insulin resistance index is inversely associated with serum carotenoids in non-diabetic subjects. *J. Epidemiol.* **2006**, *16*, 71–78. [CrossRef]

71. Sahin, K.; Orhan, C.; Akdemir, F.; Tuzcu, M.; Sahin, N.; Yilmaz, I.; Juturu, V. BCX ameliorates metabolic risk factors by regulating NF-κB and Nrf2 pathways in insulin resistance induced by high-fat diet in rodents. *Food Chem. Toxicol.* **2017**, *107*, 270–279. [CrossRef]

72. Kaur, P.; Kaur, J. Potentyal role of lycopene as antioxidant and implications for human health and disease. In *Lycopene Food Sources, Potential Role in Human Health and Antioxidant Effects*; Bailey, J.R., Ed.; Nova Science Publishers, Inc.: New York, NY, USA, 2015; pp. 1–38, ISBN 1631179276.

73. Fu, H.; Xie, B.; Fan, G.; Ma, S.; Zhu, X.; Pan, S.Y. Effect of esterification with fatty acid of β-cryptoxanthin on its thermal stability and antioxidant activity by chemiluminescence method. *Food Chem.* **2010**, *122*, 602–609. [CrossRef]

74. Breithaupt, D.E.; Weller, P.; Wolters, M.; Hahn, A. Plasma response to a single dose of dietary BCX esters from papaya (*Carica papaya* L.) or non-esterified BCX in adult human subjects: A comparative study. *Br. J. Nutr.* **2003**, *90*, 795–801. [CrossRef] [PubMed]

75. Lorenzo, Y.; Azqueta, A.; Luna, L.; Bonilla, F.; Domínguez, G.; Collins, A.R. The carotenoid BCX stimulates the repair of DNA oxidation damage in addition to acting as an antioxidant in human cells. *Carcinogenesis* **2009**, *30*, 308–314. [CrossRef] [PubMed]

76. Dandona, P.; Aljada, A.; Chaudhuri, A.; Mohanty, P.; Garg, R. Metabolic syndrome. A comprehensive perspective based on interactions between obesity, diabetes, and inflammation. *Circulation* **2005**, *111*, 1448–1454. [CrossRef] [PubMed]

77. Haegele, A.D.; Gillette, C.; O'Neill, C.; Wolfe, P.; Heimendinger, J.; Sedlacek, S.; Thompson, H.J. Plasma xanthophyll carotenoids correlate inversely with indices of oxidative DNA damage and lipid peroxidation. *Cancer Epidemiol. Biomark. Prev.* **2000**, *9*, 421–425.

78. Alavez, S.; Vantipalli, M.C.; Zucker, D.J.; Klang, I.M.; Lithgow, G.J. Amyloid binding compounds maintain protein homeostasis during ageing and extend lifespan. *Nature* **2011**, *472*, 226–229. [CrossRef] [PubMed]

79. Regitz, C.; Dußling, L.M.; Wenzel, U. Amyloid-beta (Aβ(1-42))-induced paralysis in *Caenorhabditis elegans* is inhibited by the polyphenol quercetin through activation of protein degradation pathways. *Mol. Nutr. Food Res.* **2014**, *58*, 1931–1940. [CrossRef] [PubMed]

nutrients

Article

Reduced Carotenoid and Retinoid Concentrations and Altered Lycopene Isomer Ratio in Plasma of Atopic Dermatitis Patients

Renata Lucas [1], Johanna Mihály [2], Gordon M. Lowe [3], Daniel L. Graham [3,4], Monika Szklenar [5], Andrea Szegedi [1], Daniel Töröcsik [1] and Ralph Rühl [2,5,*]

[1] Department of Dermatology, Faculty of Medicine, University of Debrecen, 4032 Debrecen, Hungary; renata.lucas@rocketmail.com (R.L.); aszegedi@med.unideb.hu (A.S.); dtorocsik@gmail.com (D.T.)
[2] Department of Biochemistry and Molecular Biology, University of Debrecen, 4032 Debrecen, Hungary; johanna@med.unideb.hu
[3] School of Pharmacy and Biomolecular Sciences, Liverpool John Moores University, Byrom Street, Liverpool L3 3AF, UK; g.m.lowe@ljmu.ac.uk (G.M.L.); D.L.Graham@ljmu.ac.uk (D.L.G.)
[4] Faculty of Science, Liverpool John Moores University, Byrom Street, Liverpool L3 3AF, UK
[5] Paprika Bioanalytics BT, 4002 Debrecen, Hungary; monikaszklenar1@gmail.com
* Correspondence: ralphruehl@web.de; Tel.: +36-30-2330-501

Received: 23 July 2018; Accepted: 22 September 2018; Published: 1 October 2018

Abstract: Carotenoids and retinoids are known to alter the allergic response with important physiological roles in the skin and the immune system. In the human organism various carotenoids are present, some of which are retinoid precursors. The bioactive derivatives of these retinoids are the retinoic acids, which can potently activate nuclear hormone receptors such as the retinoic acid receptor and the retinoid X receptor. In this study, we aimed to assess how plasma carotenoid and retinoid concentrations along with the ratio of their isomers are altered in atopic dermatitis (AD) patients ($n = 20$) compared to healthy volunteers (HV, $n = 20$). The study indicated that plasma levels of the carotenoids lutein (HV 198 ± 14 ng/mL, AD 158 ± 12 ng/mL, $p = 0.02$; all values in mean $\pm$ SEM), zeaxanthin (HV 349 ± 30 ng/mL, AD 236 ± 18 ng/mL, $p \leq 0.01$), as well as the retinoids retinol (HV 216 ± 20 ng/mL, AD 167 ± 17 ng/mL, $p = 0.04$) and all-*trans*-retinoic acid (HV 1.1 ± 0.1 ng/mL, AD 0.7 ± 0.1 ng/mL, $p = 0.04$) were significantly lower in the AD-patients, while lycopene isomers, α-carotene, and β-carotene levels were comparable to that determined in the healthy volunteers. In addition, the ratios of 13-*cis*- vs. all-*trans*-lycopene (HV 0.31 ± 0.01, AD 0.45 ± 0.07, $p = 0.03$) as well as 13-*cis*- vs. all-*trans*-retinoic acid (HV 1.4 ± 0.2, AD 2.6 ± 0.6, $p = 0.03$) were increased in the plasma of AD-patients indicating an AD-specific 13-*cis*-isomerisation. A positive correlation with SCORAD was calculated with 13-*cis*- vs. all-*trans*-lycopene ratio ($r = 0.40$, $p = 0.01$), while a negative correlation was observed with zeaxanthin plasma levels ($r = -0.42$, $p = 0.01$). Based on our results, we conclude that in the plasma of AD-patients various carotenoids and retinoids are present at lower concentrations, while the ratio of selected lycopene isomers also differed in the AD-patient group. An increase in plasma isomers of both lycopene and retinoic acid may cause an altered activation of nuclear hormone receptor signaling pathways and thus may be partly responsible for the AD-phenotype.

Keywords: lycopene; carotene; retinoic acid; retinoid; vitamin A; RAR; RXR

1. Introduction

Carotenoids and retinoids are considered to have beneficial effects in the prevention of many major diseases and they play a crucial role in skin physiology and allergic responses [1–11]. Retinoids regulate a variety of physiological processes, such as proliferation, differentiation, immune regulation, and epidermal barrier function [12]. The term retinoids include both natural forms of vitamin A, retinaldehyde, and retinoic acid, as well as synthetic retinol analogs. A high vitamin A or high pro-vitamin A containing diet resulted in increased plasma levels of all-*trans* retinoic acid thus providing an important association between nutritional factors and retinoic acid signaling mediated pathways [13]. Several skin diseases have been associated with alterations of the retinoid metabolism and signaling [14].

Vitamin A and retinoid derivatives play a pivotal role in cutaneous physiology, and various skin diseases have been associated with altered retinoid metabolism and signaling, such as atopic dermatitis (AD), which is a common chronic inflammatory skin disease, showing structural abnormalities of the epidermal barrier, and is characterized by increased IgE secretion and Th2 response [9,14–18]. Bioactive retinoic acids activating RXR and RAR mediated signalling are also affecting various aspects concerning an allergic skin inflammation like a systemic Th1/Th2 shift [9,17,18] as well as a topical allergic skin inflammation [2,3,19,20].

Carotenoids, a family of more than 600 compounds, are important micronutrients in the human diet and are also present in the human plasma [10,21,22]. The major and most studied of the dietary carotenoids are β-carotene, lycopene, lutein, and zeaxanthin. β-Carotene is known to have the greatest pro-vitamin A activity [23]. Lycopene, a red pigment mainly originating from tomatoes and tomato products, is found as all-*trans* isomer in the majority of the food sources [24]. *Cis*-isomers of lycopene are also found in biological systems with the 5-*cis*, 9-*cis*, 13-*cis*, and 15-*cis* isomers being the most predominant forms [25]. Lycopene is also present in human and animal tissues, but it is found mainly as *cis*-isomers, of which the 5-*cis* isomer is the most predominant form [26,27]. Unfortunately, except for β-carotene and partly lycopene functioning as known or potential precursors for RAR and RXR ligands [28–30], no clearly defined mechanisms of action has been found, except the controversial discussed and partly non nutritional and physiological relevant antioxidant potential [31].

Supporting the importance of retinoic acid in AD (patho)-physiology, in our previous study, we showed that retinoic acid levels are lower in the skin of AD-patients in comparison to healthy volunteers. Based on the observed alterations in retinoid transport, synthesis and plasma concentrations we concluded that retinoid signaling pathways might contribute to AD pathogenesis [2,3,11,32], which is also confirmed by numerous studies in rodents [1,6,33]. In this study, we aimed to assess whether the concentrations of carotenoids and retinoids, especially lycopene and its isomers along with retinoic acid and its isomers, which are potential indirect or direct activators of the RAR- and RXR-mediated signalling pathways, differ in the plasma of healthy volunteers and AD-patients.

2. Materials and Methods

Study population: After informed consent and the approval of the local Ethics Committee of the University of Debrecen, Hungary, Medical, and Health Science Centre, peripheral blood was collected from 20 AD-patients (8 male, 12 female; mean age 20 years, range 15–32 years). A group of 20 healthy age-matched volunteers (six males, 14 females, mean age 21 years, range 19–24 years) served as controls in this study. All AD patients fulfilled the diagnostic criteria established by Hanifin and Rajka [34].

The severity and activity of the disease was determined by the SCORAD (SCORe Atopic Dermatitis) index [35] and in our AD-patients the mean SCORAD was 35.2 (range 13–64). Patients were additionally tested for plasma total IgE by ELISA (ADALTIS Italia S.p.A., Casalecchio di Reno, Italy) according to the manufacturer's instructions. The white cell count along with the absolute count of eosinophils in whole blood was determined while using an Advia 120 haematology analyzer (Siemens, München, Germany).

The plasma samples that were used in this study originate of a larger pool of plasma that was obtained from previous published studies and where different lipid profiles were analysed [36–38]. In eight of the 20 patients the disease started in the first year of life, in two patients between ages 3–4, in seven patients between 6–18 years, and in three patients in adulthood (>18 years). In patients' history, 10 patients of 20 had rhinitis, three had asthma, and three had both rhinitis and asthma (Table 1). Patients, have not been treated with oral glucocorticosteroids, non-steroidal anti-inflammatory drugs or other systemic immunomodulatory agents for at least four weeks, also did not receive antihistamins and topical corticosteroids for at least five days prior to blood sampling.

Plasma sample preparation: Peripheral blood was collected into EDTA-containing BD vacutainer blood collection tubes (Becton-Dickinson, BD Diagnostics, Le Pont de Claix, France) and transferred to a 15 mL falcon tube (Sigma Aldrich, Budapest, Hungary) immediately after collection and then centrifuged under dimmed yellow light at room temperature, 2500 rpm for 15 min. Plasma was removed after centrifugation and kept on −80 °C.

HPLC analysis for carotenoids: The plasma samples were obtained in Debrecen and transported on dry ice and exclusion of light to Liverpool John Moores University, for analysis. Upon arrival the samples were stored at −80 °C until they were processed.

Carotenoids were extracted from plasma samples using the following method: 1.0 mL aliquot of patient's plasma was added to a glass flip-top squat vial and 1.0 mL of ethanol added. The sample was vortexed immediately for 2 s, prior to the addition of 1.5 mL diethyl ether. The sample was vortexed again for 2 s prior to the addition of 1.5 mL hexane. The sample was then vortexed a last time for 2 s and then allowed to stand for partition. The top layer of the sample was then removed with a glass Pasteur pipette and dried down in a 4 mL amber screw-top vial under oxygen free nitrogen conditions. The dry sample was further resuspended in 100 μL tetrahydrofuran and 400 μL methanol and then transferred to an amber 2 mL HPLC vial prior to analysis.

All HPLC was performed on an Agilent 1100 series fully automated HPLC (Agilent Technologies UK Ltd., Berkshire, UK) with diode array detection. All samples were analysed using either a C18 or C30 column (VWR International Ltd., Lutterworth, UK). C18 column was used for estimating the concentration of carotenoids (β-carotene, lycopene, β-cryptoxanthin, lutein) [39] and the C30 was used for determining the concentration of isomers for β-carotene and lycopene. Concentrations were estimated via comparison with purified external standard samples under identical HPLC conditions [40,41].

The precision of carotenoid analysis in plasma samples was determined ahead of the analysis of our examined samples. The *intra*-day precision was determined based on $n = 12$ consecutive analysis of the same human plasma sample under the same separate extraction and analysis conditions and it resulted in %-coefficient of variation (%CV) = 7.7 for all-*trans* lycopene (determined concentration of 0.65 ± 0.05 μmol/L) and %CV = 7.5 for β-carotene (determined concentration of 0.53 ± 0.04 μmol/L).

The *inter*-day precision of the carotenoid analysis was determined using the same pooled samples which were previously aliquoted and individually stored at −80 °C. Each individual consecutive day ($n = 10$) ahead of the analysis these aliqots were defrosted, freshly extracted and analyzed each day ($n = 10$). We determined a %CV = 8.8 for all-*trans* lycopene and %CV = 9.1 for β-carotene.

Analysis using a C18 reverse phase column: From our plasma aliquot, a sample volume of 50 μL was injected into the HPLC using an auto-sampler. The solvent system used was 66/22/10 acetonitrile/tetrahydrofuran/methanol (0.005% *w/v* ammonium acetate). The solvent mixture was delivered at a rate of 0.8 mL/min. The column was held at a temperature of 22 °C. The column was protected using a C18 guard column supplied by Phenomenex (Macclesfield, UK). Lycopene and its components were separated using a 5 μm Gemini (Phenomenex, Macclesfield, UK) C18 reverse phase column (4.6 × 250 mm).

Analysis using a C30 column: A sample volume of 50 μL was injected into the HPLC while using an auto-sampler. The solvents system comprised of 50/40/10 methyl-tert-butylether/methanol/ethyl acetate. This was delivered isocratically at a rate of 0.45 mL/min. The column was a YMC (VWR, Lutterworth, UK) C30 (5 μm 4.6 × 250 mm), and kept at a temperature of 40 °C. The column was protected using a C30 guard column that was supplied by YMC (VWR, Lutterworth, UK).

The run times for the C18 column was 20 min, whilst for the C30 column, it was 35 min. The detection of the eluted compounds was by diode array screening between 300–600 nm and integration of each peak was performed using the Chemstation software (v10A) (Agilent Technologies UK Ltd., Berkshire, UK).

HPLC MS-MS analysis for retinoids: Concentrations of 13CRA, ATRA and retinol were determined in human plasma samples by our high performance liquid chromatography mass spectrometry—mass spectrometry (HPLC MS-MS) method as described previously [42]. In draft, high performance liquid chromatography mass spectrometry (2695XE separation module; Waters, Waters, Budapest, Hungary)—mass spectrometry (Micromass Quattro Ultima PT; Waters, Budapest, Hungary), analyses were performed under dark yellow/amber light while using previously validated protocol. For sample preparation, 100 μL plasma was diluted with a threefold volume of isopropanol vortexed for 10 s, put in an ultra-sonic bath for 5 min, shaken for 6 min, and centrifuged at 13,000 rpm in a Heraeus BIOFUGE Fresco at +4 °C. After centrifugation, the supernatants were dried in an Eppendorf concentrator 5301 (Eppendorf, D) at 30 °C. The dried extracts were resuspended with 60 μL of methanol, diluted with 40 μL of 60 mM aqueous ammonium acetate solution and transferred into the autosampler and subsequently analyzed. Quantification was performed like previously described in [42].

Statistics: The data comparing healthy volunteers and AD-patients are shown as mean and standard error mean (SEM) based on 2 x *n* = 20 samples using a paired student's *t*-test analysis while considering a *p* value of less than 0.05 significant. Statistical analysis was performed using Graph pad Prism (version 7.04 for windows; GraphPad Software, Inc., LaJolla, CA, USA).

For the correlation analysis displayed in Figure 1 and Figure S1 data from *n* = 40 individuals, healthy volunteers and AD-patients were combined, and Spearman correlation (*r*- and *p*-value) were calculated in "R" 3.3.2. version to determine the relationship between the carotenoid and retinoid levels as well as calculated percentile amounts and ratios with the three clinical AD-markers (IgE, %-EOS and SCORAD). The correlation analysis and heap map creation was also performed in "R" 3.3.2. version via the "heatmap.2" function [43]. Significance indicated by a *p*-values > 0.05 were marked with a black frame in Figure 1.

3. Results

3.1. Characterisation of the Study Cohort

Characterisation of the study cohort: The same patient cohort was used as in our previous studies [36–38]. AD-patients and healthy volunteers did not differ in age and gender. Clinical markers of atopic dermatitis (AD) like SCORAD [35] and ongoing discussion about validity as an AD marker [44], total IgE, and percental amount of eosinophils from peripheral blood mononuclear cells (PBMCs) were significantly increased (Table 1).

Table 1. Clinical and basic demographic data from healthy volunteers and atopic dermatitis (AD)-patients, mean ± SEM.

	Healthy Volunteers	**AD-Patients**	*p*-Value
	n = 20	*n* = 20	*n* = 20
Age in years	21 ± 0.3	20 ± 1.2	0.48
Gender	70% female	60% female	-
SCORAD	0 ± 0	35.2 ± 4.3	**<0.01**
Total IgE (KU/L)	32 ± 0.0	2941 ± 1134	**0.01**
EOS %	2.5 ± 0	7.3 ± 1.1	**<0.01**

Total IgE—plasma total IgE levels in kilounits (KU)/L; SCORAD units [35], EOS %—number of eosinophils as percentage from PBMCs, AD-atopic dermatitis. This table is adapted from Mihaly et al., 2013 [38]. Numbers in bold letters indicate significance.

3.2. Carotenoid Concentrations and Lycopene Isomers

Reduced concentrations of lutein and zeaxanthin in AD-patients: Lutein and zeaxanthin plasma levels were significantly decreased in the plasma of AD patients compared to healthy volunteers (from 198 ± 14 ng/mL to 158 ± 12 ng/mL, *p* = 0.02, respectively, from 349 ± 30 ng/mL to 236 ± 18 ng/mL, *p* > 0.01). α-Carotene and β-carotene levels were comparable in healthy volunteers and AD-patients (Table 2A).

Altered lycopene isomer concentrations in healthy volunteers and AD-patients: Total lycopene levels display a non-significantly trend to be lower in the plasma of AD-patients (from 281 ± 30 ng/mL to 248 ± 160 ng/mL). Individual concentrations of lycopene isomers, like all-*trans* (from 126 ± 15 ng/mL to 107 ± 17 ng/mL), 9-*cis* (from 25 ± 3 ng/mL to 21 ± 3 ng/mL), and 5-*cis* (from 94 ± 10 ng/mL to 80 ± 12 ng/mL), isomers also showed a non-significant trend of lower levels in the plasma of AD-patients, while the concentration of 13-*cis*-lycopene displayed a non-significant trend of lower levels when compared to healthy volunteers from 36 ± 3 ng/mL to 40 ± 3 ng/mL (Table 2B).

Table 2. Carotenoid concentrations (**A**), total sum and individual concentrations of lycopene isomers (**B**), calculated %-amounts of individual lycopene-isomers from the sum of lycopene isomers (**C**) and calculated ratios of selected lycopene isomers all calculated based on determined plasma concentrations from healthy volunteers and AD-patients.

(A) Carotenoid Concentrations in Plasma of Healthy Volunteers and AD-Patients.			
	Healthy Volunteers	**AD-Patients**	*p*-Value
	n = 20 (ng/mL)	*n* = 20 (ng/mL)	
lutein	198 ± 14	158 ± 12	**0.02**
zeaxanthin	349 ± 30	236 ± 18	**<0.01**
α-carotene	171 ± 21	149 ± 24	0.24
β-carotene	492 ± 77	394 ± 65	0.17
(B) Total Sums and Concentration of Lycopene Isomers in Plasma of Healthy Volunteers and AD-Patients.			
lycopene (sum)	281 ± 30	248 ± 35	0.24
lycopene (all-*trans*-)	126 ± 15	107 ± 17	0.20
lycopene (13-*cis*-)	36 ± 3	40 ± 5	0.29
lycopene (9-*cis*-)	25 ± 3	21 ± 3	0.17
lycopene (5-*cis*-)	94 ± 10	80 ± 12	0.19
(C) Calculated %-Amounts of Lycopene Isomers in Plasma of Healthy Volunteers and AD-Patients.			
	n = 20 (in%)	*n* = 20 (in%)	
lycopene (all-*trans*-)	44.1 ± 0.9	41.8 ± 1.3	0.08
lycopene (13-*cis*-)	13.3 ± 0.4	17.2 ± 1.7	**0.01**
lycopene (9-*cis*-)	8.8 ± 0.2	8.6 ± 0.3	0.24
lycopene (5-*cis*-)	33.8 ± 0.8	32.4 ± 1.2	0.16
(D) Calculated Ratios of Selected Lycopene Isomers in Plasma of Healthy Volunteers and AD-Patients.			
	n = 20 (Ratio)	*n* = 20 (Ratio)	
13-*cis*-/all-*trans*-lycopene	0.31 ± 0.01	0.45 ± 0.07	0.03
5-*cis*-/all-*trans*-lycopene	0.78 ± 0.03	0.79 ± 0.04	0.40

13-*cis*-lycopene%-amounts of lycopene isomers show significant alterations in AD-patients compared to healthy volunteers: The 13-*cis* lycopene ratio has been significantly increased (from 13.3 ± 0.4% to 17.2 ± 1.7%). All-*trans*-, 9-*cis*- and 5-*cis*-lycopene did not show significant alterations in their %-amounts (Table 2C).

Calculated ratios of selected lycopene isomers in healthy volunteers and AD-patients: 13-*cis*-/all-*trans*-lycopene ratios has been significantly increased in AD patients as compared to healthy volunteers from 0.31 ± 0.01 to 0.45 ± 0.07 (*p* = 0.03), while no alteration could be observed in the 5-*cis*-/all-*trans*-lycopene ratio (from 0.78 ± 0.03 to 0.79 ± 0.04) (Table 2D).

3.3. Retinoid Concentration and Ratios of Retinoic Acid Isomerization

Reduced retinoic acid and retinol concentrations in AD-patients compared to healthy volunteers: All-*trans*-retinoic acid concentrations were significantly lower in the plasma of AD-patients (0.7 ± 0.1 ng/mL) compared to healthy volunteers (1.1 ± 0.1 ng/mL, $p = 0.04$), while 13-*cis*-retinoic acid concentrations just display a non-significant trend of lower levels (from 1.2 ± 0.1 ng/mL to 1.0 ± 0.1 ng/mL). Retinol concentrations were also significantly decreased in plasma of AD-patients when compared to healthy volunteers (from 216 ± 20 ng/mL to 167 ± 17 ng/mL, $p = 0.04$). Our results showed that both ATRA and retinol were present in a lower concentration in the plasma of atopic individuals (Table 3A).

Ratio of plasma levels of retinoic acid isomers 13CRA/ATRA: The 13CRA/ATRA ratio was significantly increased in the plasma of AD-patients from 1.4 ± 0.2 to 2.6 ± 0.6 ($p = 0.03$) (Table 3B).

Table 3. (**A**) retinoic acid and retinol concentrations and (**B**) a calculated ratio of the plasma levels of retinoic acid isomers 13CRA/ATRA determined, based on plasma concentrations originating from healthy volunteers as well as AD-patients.

(A) Retinoic Acid and Retinol Concentrations in Human Plasma from Healthy Volunteers as well as AD-Patients.			
	Healthy Volunteers	**AD-Patients**	*p*-Value
	n = 20 (ng/mL)	*n* = 20 (ng/mL)	
ATRA	1.1 ± 0.1	0.7 ± 0.1	**0.04**
13CRA	1.2 ± 0.1	1.0 ± 0.1	0.17
ROL	216 ± 20	167 ± 17	**0.04**
(B) Ratio of the Plasma Levels of Retinoic Acid Isomers 13CRA/ATRA.			
	n = 20 (Ratio)	*n* = 20 (Ratio)	
13CRA/ATRA	1.4 ± 0.2	2.6 ± 2.6	**0.03**

Data are shown as mean and SEM based on *n* = 20 samples. ATRA—all-*trans*-retinoic acid, 13CRA—13-*cis*-retinoic acid, ROL—retinol. Numbers in bold letters indicate significance.

3.4. Correlation Analysis

Zeaxanthin levels negatively and 13-*cis*-/all-*trans*-lycopene ratios positively correlate to clinical AD-markers: The correlation analysis between plasma values and calculations of percentile amounts and ratios of selected retinoids and carotenoids originating from *n* = 40 individuals (healthy volunteers and AD-patients) determined a significant positive correlation of plasma 13-*cis*-/all-*trans*-lycopene ratios ($r = 0.40$, $p = 0.01$) with SCORAD. A negative correlation of plasma lutein levels with SCORAD ($r = -0.36$, $p = 0.02$) and IgE ($r = -0.33$, $p = 0.04$), a negative correlation with plasma zeaxanthin levels with %-EOS ($r = -0.41$, $p = 0.01$), IgE ($r = -0.45$, $p < 0.01$), and SCORAD ($r = -0.42$, $p = 0.01$).

Further positive or negative correlation with plasma levels of ACAR with %-EOS ($r = -0.35$, $p = 0.02$), %-13CLYC with %-EOS ($r = 0.33$, $p = 0.04$) and SCORAD ($r = 0.34$, $p = 0.03$), and 9CLYC with IgE ($r = -0.34$, $p = 0.03$) were calculated. No further significant correlations were found with the clinical AD-markers number of eosinophils as percentage from PBMCs (%-EOS) or plasma total IgE levels in kilounits per liter (IgE) (outlined in Figure 1).

In addition, we did not observe any significant correlation between plasma ATBC and ATRA levels of *n* = 40 individuals (healthy volunteers and AD-patients) (Figure S1).

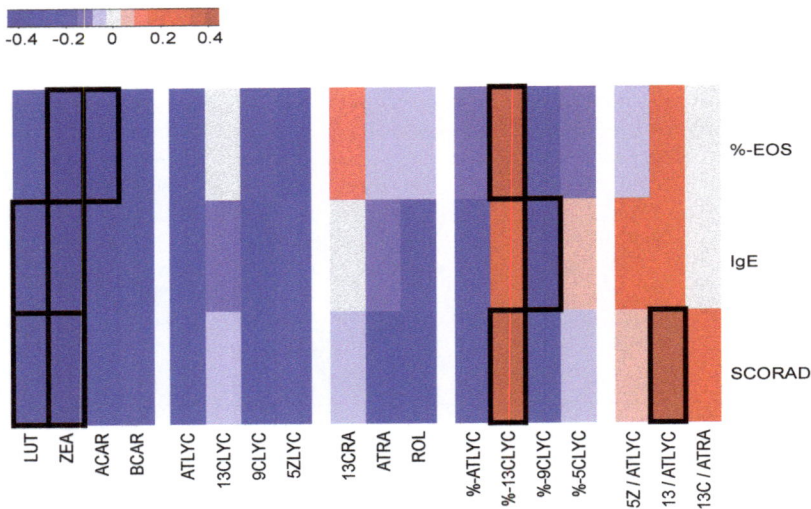

Figure 1. In this clustered image map analysis individual correlations between two values are displayed based on the determined concentrations and calculated ratios and percentile amounts originating from *n* = 40 individuals (healthy volunteers and AD-patients). A specific correlation value is indicated by a two-dimensional colored image (red to blue), where each entry of the matrix is colored on the basis of its correlation value indicated by the displayed color legend in a distance matrix analysis from plasma levels of the carotenoids zeaxanthin (ZEA), lutein (LUT), β-carotene (BCAR), α-carotene (ACAR), all-*trans*-lycopene (ATLYC), 13-*cis*-lycopene (13CLYC), 9-*cis*-lycopene (9CLYC), 5-*cis*-lycopene (9CLYC) and the retinoids 13-*cis*-retinoic acid (13CRA), all-*trans*-retinoic acid (ATRA), retinol (ROL) levels in addition to calculated ratios and percentile amounts of selected carotenoids or retinoids with the clinical AD-markers, like number of eosinophils as percentage from PBMCs (%-EOS), plasma total IgE levels in kilounits per liter (IgE), and SCORAD units [45,46]. Significant correlations are indicated by a *p* > 0.05 and were additionally marked by a black frame.

4. Discussion

Based on analysis of retinoids and carotenoids in human plasma, as well as calculating relevant ratios, we observed that specific carotenoids, like zeaxanthin and lutein, as well as the retinoids retinol and all-*trans*-retinoic acid (ATRA) are lower in plasma of AD-patients vs healthy individuals. Additionally, we observed that the %-amount of 13-*cis*-isomers based on calculated ratios of lycopene and retinoic acid isomers were higher in AD-patients. If these found alterations are a cause or a consequence of AD needs to be discussed and further examined.

Naturally occurring forms of vitamin A and other synthetic retinoid analogues are mainly present in the all-*trans*-configuration form, but *cis*-isomers also have relevant biological roles, maintaining essential physiological processes in the human organism, such as vision, cellular growth and differentiation, reproduction, normal growth and development, healthy immune system, and healthy skin and barrier functions [45,46]. Increased ATRA concentrations have been shown to increase retinoic acid-mediated signaling via RAR- and RXR-mediated signalling pathways. Surprisingly, in this study calculating the individual correlation between the plasma values of ATRA with plasma levels of the main pro-vitamin A all-*trans*-β-carotene (ATBC, Figure S1) displayed no significant correlation determined in all individuals (*n* = 40, healthy adults and AD-patients). Therefore, the connection between increased carotenoid intake with following increased plasma ATBC levels and further increased RAR- and RXR-mediated signalling seems to be questionable, like discussed and co-found in recent reports [4,23].

The retinoic acid exists in three major stereoisomeric forms: (ATRA), 9-*cis*-retinoic acid (9CRA), and 13-*cis*-retinoic acid (13CRA, also known as isotretinoin) [47]. ATRA binds only to retinoic acid receptors (RARs), while 9CRA can bind to both RARs and retinoid X receptor (RXRs). Recently the real endogenous RXR ligand 9-*cis*-13,14-dihydroretinoic acid was identified by our group [48,49]. The 13-*cis*-isomer of retinoic acids does not bind specifically to RXRs and it has a lower affinity to RARs then ATRA or 9CRA [50]. In general, 13CRA is considered to be a non active form of the biologically active ATRA and it is generated endogenously, non-enzymatically by spontaneous isomerization from ATRA [51], or enzymatically by means of a novel identified enzyme 13-*cis* specific isomerohydrolase, which generates exclusively 13-*cis*-retinol, a precursor of 13CRA [52]. A reduced, non-enzymatic or enzymatic, isomerization back to the all-*trans* configuration maybe an alternative reason of this altered isomer-distribution occurring in serum of atopic dermatitis patients [53,54]. This retinoid- and carotenoid-isomerization is still a highly controversial topic and multiple mechanisms might occur.

Lycopene isomers were analyzed as well as %-amounts of lycopene isomers all-*trans*-, 13-*cis*-, 9-*cis*- and 5-*cis*-lycopene were calculated and we found significantly increased %-amounts of 13-*cis*-lycopene in plasma of AD-patients as well as an increased ratio of 13-*cis*-/all-*trans*-lycopene compared to healthy volunteers. We conclude that a specific 13-*cis* isomerization of retinoic acid and of lycopene might likely be a consequence of the AD-phenotype. The 13-*cis*-/all-*trans*-lycopene ratio positively correlated with the clinical AD-marker SCORAD. This indicates that neither the systemic AD-markers IgE and %-EOS, but more likely the skin with the specific AD-marker SCORAD, might be co-involved with that specific 13-*cis*-isomerization. It stays still elusive, which cells, which specific enzymes and which alternative specific 13-*cis*-isomerization inducing mechanisms are behind this phenomenon.

In our examined AD-patients, lutein and zeaxanthin concentrations were also significantly lower in the plasma compared to healthy volunteers and both negatively correlated with the AD-markers SCORAD, IgE and %-EOS (excluding lutein) as visualized in Figure 1. This let us postulate potential positive effects of food rich in zeaxanthin on AD. As serum levels of lutein and zeaxanthin are increased by dietary intake of food rich in fruits and vegetables [55,56] as well as supplements rich in zeaxanthin and lutein [57] a potential connection between a healthy diet and AD can be postulated [58]. Recently, positive effects of lutein and zeaxanthin supplementation and local administration of these carotenoids have been shown to be beneficial for the human skin [59,60]. The precise mechanism of action and a mechanistic explanation of this found phenomenon remain elusive. In addition, the reduced serum zeaxanthin levels may also be used as a biomarker and an indicator for AD.

Other carotenoids, like α-carotene, β-carotene, and lycopene were not altered in AD-patients, which is partly in agreement with a previous study in atopic children [7,61]. Unfortunately carotenoid levels in humans have a high variability due to altered individual nutrition, chronic inflammatory background, and genetic background [7,62–64]. A direct functional comparison of found alterations of carotenoid plasma levels of data originating from adults, examined this study, and children [7] cannot be done because the levels found in children were determined to be much lower. If these reduced levels of carotenoids in AD-patients or atopic children are based on higher chronic inflammation and a further pro-oxidant stress, a targeted selected downregulation of carotenoid absorption and binding, a different occurrence of carotenoid-levels specific polymorphisms or reduced intake of healthy food [21,23], remains speculative. One possible mechanism might be that lutein/zeaxanthin are known to negatively interfere with ATRA-RAR-mediated signalling [65–67] which was associated with pro-allergic Th2 immune responses [1,6,9,17,19].

Much larger cohorts examining carotenoid and retinoid levels, as well as genetic background, transcriptomic based regulation of these lipid pathways in children and adults also for multiple AD-relevant organs like the immune cells and healthy and affected skin should be planed and performed to get further knowledge about this phenomenon.

A special focus should be put on examining carotene oxygenases, which may cleave lycopene to yield retinoid-like derivatives [28–30], as still non-identified bioactive lycopene-metabolites. As retinoids have been shown to play important roles in skin homeostasis and signaling, alterations of retinoid signaling are related to several skin diseases and malignancies mediated by RARα, RARγ, RXR, and PPARδ-mediated signaling [2,3,68–70]. Altered retinoid concentrations, also based on potential novel lycopene-metabolites in the skin, might lead to altered nuclear hormone receptor pathway activation resulting in skin abnormalities and a further AD-specific phenotype.

We postulate that specific differences in retinoid or carotenoid isomerization via an enzymatic or non-enzymatic specific 13-*cis*-isomerisation in AD might be a consequence of chronic skin inflammation present in AD. These 13-*cis*/all-*trans* isomer ratios may have a still unknown biological meaning and may serve as a biological relevant priming factor for the AD-phenotype. Markers of 13-*cis*-isomerisation like the 13-*cis*-/all-*trans*-lycopene ratio positively correlates well with SCORAD and might also serve as a plasma biomarker for AD.

The limitations of the study are that the number of involved individuals was relatively low ($n = 20$), no food intake diaries were taken to evaluate differences in the dietary pattern of healthy volunteers and diseased patients in addition to skin/PBMC transcriptomics, retinoic/carotenoid profiling and following systems biology analysis, which would have resulted in a deeper advanced knowledge with a focus on the influence and connection of individual retinoids/carotenoids on AD.

5. Conclusions

Our study let us conclude that in the plasma of AD-patients various carotenoids and retinoids are present at lower levels, while the ratio of lycopene- and retinoic acid-isomers was also altered. These alterations might be a consequence of chronic skin inflammation that is present in AD and might cause an altered activation of nuclear hormone receptor signaling pathways that could be partly responsible for the AD-phenotype.

Supplementary Materials: The following are available online at http://www.mdpi.com/2072-6643/10/10/1390/s1, Figure S1: Displayed is an MS-Excel calculated graphical direct correlation of plasma levels of all-*trans*-β-carotene (ATBC) and all-*trans*-retinoic acid (ATRA) plasma levels from $n = 40$ individuals (healthy volunteers and AD-patients). R- and p-values were previously calculated using "R" and were displayed graphically in Figure 1 and indicate no significant correlation between these two variables.

Author Contributions: Investigation, A.S., R.L., J.M., G.L., D.G. and M.S.; Writing—original draft, R.R. and D.T.

Funding: This project was funded by the OTKA project OTKA K109362 (R.R.), NKFIH K 128250 (A.S.) and GINOP-2.3.2-15-2016-00005 (D.T.). D.T. is recipient of the János Bolyai research scholarship of the Hungarian Academy of Sciences.

Conflicts of Interest: The authors declare no conflict of interest.

References

1. Rühl, R.; Garcia, A.; Schweigert, F.J.; Worm, M. Modulation of cytokine production by low and high retinoid diets in ovalbumin-sensitized mice. *Int. J. Vitam. Nutr. Res.* **2004**, *74*, 279–284. [CrossRef] [PubMed]

2. Gericke, J.; Ittensohn, J.; Mihaly, J.; Alvarez, S.; Alvarez, R.; Törőcsik, D.; de Lera, A.R.; Rühl, R. Regulation of retinoid-mediated signaling involved in skin homeostasis by RAR and RXR agonists/antagonists in mouse skin. *PLoS ONE* **2013**, *8*, e642–e643. [CrossRef] [PubMed]

3. Gericke, J.; Ittensohn, J.; Mihaly, J.; Dubrac, S.; Rühl, R. Allergen-induced dermatitis causes alterations in cutaneous retinoid-mediated signaling in mice. *PLoS ONE* **2013**, *8*, e71244. [CrossRef] [PubMed]

4. Bohn, T.; Planchon, S.; Leclercq, C.C.; Renaut, J.; Mihaly, J.; Beke, G.; Rühl, R. Proteomic responses of carotenoid and retinol administration to Mongolian gerbils. *Food Funct.* **2018**, *9*, 3835–3844. [CrossRef] [PubMed]

5. Rühl, R. Non-pro-vitamin A and pro-vitamin A carotenoids in atopy development. *Int. Arch. Allergy Appl. Immunol.* **2013**, *161*, 99–115. [CrossRef] [PubMed]

6. Rühl, R.; Hanel, A.; Garcia, A.L.; Dahten, A.; Herz, U.; Schweigert, F.J.; Worm, M. Role of vitamin A elimination or supplementation diets during postnatal development on the allergic sensitisation in mice. *Mol. Nutr. Food Res.* **2007**, *51*, 1173–1181. [CrossRef] [PubMed]

7. Rühl, R.; Taner, C.; Schweigert, F.J.; Wahn, U.; Gruber, C. Serum carotenoids and atopy among children of different ethnic origin living in Germany. *Pediatr. Allergy Immunol.* **2010**, *21*, 1072–1075. [CrossRef] [PubMed]

8. Stephensen, C.B.; Jiang, X.; Freytag, T. Vitamin A deficiency increases the in vivo development of IL-10-positive Th2 cells and decreases development of Th1 cells in mice. *J. Nutr.* **2004**, *134*, 2660–2666. [CrossRef] [PubMed]

9. Stephensen, C.B.; Rasooly, R.; Jiang, X.; Ceddia, M.A.; Weaver, C.T.; Chandraratna, R.A.; Bucy, R.P. Vitamin A enhances in vitro Th2 development via retinoid X receptor pathway. *J. Immunol.* **2002**, *168*, 4495–4503. [CrossRef] [PubMed]

10. Rühl, R. Effects of dietary retinoids and carotenoids on immune development. *Proc. Nutr. Soc.* **2007**, *66*, 458–469. [CrossRef] [PubMed]

11. Mihaly, J.; Gamlieli, A.; Worm, M.; Rühl, R. Decreased retinoid concentration and retinoid signalling pathways in human atopic dermatitis. *Exp. Dermatol.* **2011**, *20*, 326–330. [CrossRef] [PubMed]

12. Elias, P.M. The skin barrier as an innate immune element. *Semin. Immunopathol.* **2007**, *29*, 3–14. [CrossRef] [PubMed]

13. Rühl, R.; Bub, A.; Watzl, B. Modulation of plasma all-trans retinoic acid concentrations by the consumption of carotenoid-rich vegetables. *Nutrition* **2008**, *24*, 1224–1226. [CrossRef] [PubMed]

14. Roos, T.C.; Jugert, F.K.; Merk, H.F.; Bickers, D.R. Retinoid metabolism in the skin. *Pharmacol. Rev.* **1998**, *50*, 315–333. [PubMed]

15. Leung, D.Y. New insights into atopic dermatitis: Role of skin barrier and immune dysregulation. *Allergol. Int.* **2013**, *62*, 151–161. [CrossRef] [PubMed]

16. Heise, R.; Mey, J.; Neis, M.M.; Marquardt, Y.; Joussen, S.; Ott, H.; Wiederholt, T.; Kurschat, P.; Megahed, M.; Bickers, D.R.; et al. Skin retinoid concentrations are modulated by CYP26AI expression restricted to basal keratinocytes in normal human skin and differentiated 3D skin models. *J. Investig. Dermatol.* **2006**, *126*, 2473–2480. [CrossRef] [PubMed]

17. Iwata, M.; Eshima, Y.; Kagechika, H. Retinoic acids exert direct effects on T cells to suppress Th1 development and enhance Th2 development via retinoic acid receptors. *Int. Immunol.* **2003**, *15*, 1017–1025. [CrossRef] [PubMed]

18. Spilianakis, C.G.; Lee, G.R.; Flavell, R.A. Twisting the Th1/Th2 immune response via the retinoid X receptor: Lessons from a genetic approach. *Eur. J. Immunol.* **2005**, *35*, 3400–3404. [CrossRef] [PubMed]

19. Mihaly, J.; Gericke, J.; Lucas, R.; de Lera, A.R.; Alvarez, S.; Torocsik, D.; Rühl, R. TSLP expression in the skin is mediated via RARgamma-RXR pathways. *Immunobiology* **2016**, *221*, 161–165. [CrossRef] [PubMed]

20. Mihaly, J.; Gericke, J.; Aydemir, G.; Weiss, K.; Carlsen, H.; Blomhoff, R.; Garcia, J.; Rühl, R. Reduced retinoid signaling in the skin after systemic retinoid-X receptor ligand treatment in mice with potential relevance for skin disorders. *Dermatology* **2013**, *225*, 304–311. [CrossRef] [PubMed]

21. Bohn, T.; Desmarchelier, C.; Dragsted, L.O.; Nielsen, C.S.; Stahl, W.; Ruhl, R.; Keijer, J.; Borel, P. Host-related factors explaining interindividual variability of carotenoid bioavailability and tissue concentrations in humans. *Mol. Nutr. Food Res.* **2017**, *61*, 1600685. [CrossRef] [PubMed]

22. Krinsky, N.I. Possible biologic mechanisms for a protective role of xanthophylls. *J. Nutr.* **2002**, *132*, 540S–542S. [CrossRef] [PubMed]

23. Bohn, T.; Desmarchelier, C.; El, S.N.; Keijer, J.; van Schothorst, E.M.; Rühl, R.; Borel, P. B-Carotene in the Human body—Metabolic acitivation pathways—From Digestion to Tissue Distribution. *Proc. Nutr. Soc.* **2018**, in press.

24. Boileau, T.W.; Boileau, A.C.; Erdman, J.W.J. Bioavailability of all-trans and cis-isomers of lycopene. *Exp. Biol. Med.* **2002**, *227*, 914–919. [CrossRef]

25. Boileau, T.W.; Clinton, S.K.; Erdman, J.W.J. Tissue lycopene concentrations and isomer patterns are affected by androgen status and dietary lycopene concentration in male F344 rats. *J. Nutr.* **2000**, *130*, 1613–1618. [CrossRef] [PubMed]

26. Stahl, W.; Schwarz, W.; Sundquist, A.R.; Sies, H. Cis-trans isomers of lycopene and beta-carotene in human serum and tissues. *Arch. Biochem. Biophys.* **1992**, *294*, 173–177. [CrossRef]

27. Erdman, J.W.J. How do nutritional and hormonal status modify the bioavailability, uptake, and distribution of different isomers of lycopene? *J. Nutr.* **2005**, *135*, 2046S–2047S. [CrossRef] [PubMed]

28. Aydemir, G.; Kasiri, Y.; Birta, E.; Beke, G.; Garcia, A.L.; Bartok, E.M.; Rühl, R. Lycopene-derived bioactive retinoic acid receptors/retinoid-X receptors-activating metabolites may be relevant for lycopene's anti-cancer potential. *Mol. Nutr. Food Res.* **2013**, *57*, 739–747. [CrossRef] [PubMed]

29. Aydemir, G.; Carlsen, H.; Blomhoff, R.; Rühl, R. Lycopene induces Retinoic Acid Receptor transcriptional activation in mice. *Mol. Nutr. Food Res.* **2012**, *56*, 702–712. [CrossRef] [PubMed]

30. Aydemir, G.; Kasiri, Y.; Bartok, E.M.; Birta, E.; Frohlich, K.; Bohm, V.; Mihaly, J.; Rühl, R. Lycopene supplementation restores vitamin A deficiency in mice and possesses thereby partial pro-vitamin A activity transmitted via RAR-signaling. *Mol. Nutr. Food Res.* **2016**, *60*, 203–211. [CrossRef] [PubMed]

31. Erdman, J.W.J.; Ford, N.A.; Lindshield, B.L. Are the health attributes of lycopene related to its antioxidant function? *Arch. Biochem. Biophys.* **2009**, *483*, 229–235. [CrossRef] [PubMed]

32. Biswas, R.; Chakraborti, G.; Mukherjee, K.; Bhattacharjee, D.; Mallick, S.; Biswas, T. Retinol Levels in Serum and Chronic Skin Lesions of Atopic Dermatitis. *Indian J. Dermatol.* **2018**, *63*, 251–254. [PubMed]

33. Stephensen, C.B.; Borowsky, A.D.; Lloyd, K.C. Disruption of Rxra gene in thymocytes and T lymphocytes modestly alters lymphocyte frequencies, proliferation, survival and T helper type 1/type 2 balance. *Immunology* **2007**, *121*, 484–498. [CrossRef] [PubMed]

34. Hanifin, J.; Rajka, G. Diagnostic features of atopic dermatitis. *Acta. Derm. Venereol.* **1980**, *92*, 44–47.

35. Stalder, J.F.; Taïeb, A.; Atherton, D.J.; Bieber, P.; Bonifazi, E.; Broberg, A.; Calza, A.; Coleman, R.; De Prost, Y.; Stalder, J.F. Severity scoring of atopic dermatitis: The SCORAD index. Consensus Report of the European Task Force on Atopic Dermatitis. *Dermatology* **1993**, *186*, 23–31.

36. Mihaly, J.; Sonntag, D.; Krebiehl, G.; Szegedi, A.; Torocsik, D.; Rühl, R. Steroid concentrations in patients with atopic dermatitis: Reduced plasma dehydroepiandrosterone sulfate and increased cortisone levels. *Br. J. Dermatol.* **2015**, *172*, 285–288. [CrossRef] [PubMed]

37. Mihaly, J.; Marosvolgyi, T.; Szegedi, A.; Koroskenyi, K.; Lucas, R.; Torocsik, D.; Garcia, A.L.; Decsi, T.; Rühl, R. Increased FADS2-Derived n-6 PUFAs and Reduced n-3 PUFAs in Plasma of Atopic Dermatitis Patients. *Skin Pharmacol. Physiol.* **2014**, *27*, 242–248. [CrossRef] [PubMed]

38. Mihaly, J.; Gericke, J.; Torocsik, D.; Gaspar, K.; Szegedi, A.; Rühl, R. Reduced lipoxygenase and cyclooxygenase mediated signaling in PBMC of atopic dermatitis patients. *Prostaglandins Other Lipid Mediat.* **2013**, *107*, 35–42. [CrossRef] [PubMed]

39. Lowe, G.M.; Bilton, R.F.; Davies, I.G.; Ford, T.C.; Billington, D.; Young, A.J. Carotenoid composition and antioxidant potential in subfractions of human low-density lipoprotein. *Ann. Clin. Biochem.* **1999**, *36*, 323–332. [CrossRef] [PubMed]

40. Graham, D.L.; Carail, M.; Caris-Veyrat, C.; Lowe, G.M. (13Z)- and (9Z)-lycopene isomers are major intermediates in the oxidative degradation of lycopene by cigarette smoke and Sin-1. *Free Radic. Res.* **2012**, *46*, 891–902. [CrossRef] [PubMed]

41. Graham, D.L.; Carail, M.; Caris-Veyrat, C.; Lowe, G.M. Cigarette smoke and human plasma lycopene depletion. *Food Chem. Toxicol.* **2004**, *48*, 2413–2420. [CrossRef] [PubMed]

42. Rühl, R. Method to determine 4-oxo-retinoic acids, retinoic acids and retinol in serum and cell extracts by liquid chromatography/diode-array detection atmospheric pressure chemical ionisation tandem mass spectrometry. *Rapid Commun. Mass Spectrom.* **2006**, *20*, 2497–2504. [CrossRef] [PubMed]

43. The R Development Core Team. R: A Language and Environment for Statistical Computing. Foundation for Statistical Computing. Available online: http://softlibre.unizar.es/manuales/aplicaciones/r/fullrefman.pdf (accessed on 20 July 2018).

44. Hurault, G.; Schram, M.E.; Roekevisch, E.; Spuls, P.I.; Tanaka, R.J. Relationship and probabilistic stratification of EASI and oSCORAD severity scores for atopic dermatitis. *Br. J. Dermatol.* **2018**, in press. [CrossRef] [PubMed]

45. Vahlquist, A. What are natural retinoids? *Dermatology* **1999**, *199*, 3–11. [CrossRef] [PubMed]

46. D'Ambrosio, D.N.; Clugston, R.D.; Blaner, W.S. Vitamin A metabolism: An update. *Nutrients* **2011**, *3*, 63–103. [CrossRef] [PubMed]

47. Blomhoff, R.; Blomhoff, H.K. Overview of retinoid metabolism and function. *J. Neurobiol.* **2006**, *66*, 606–630. [CrossRef] [PubMed]

48. Rühl, R.; Krzyzosiak, A.; Niewiadomska-Cimicka, A.; Rochel, N.; Szeles, L.; Vaz, B.; Wietrzych-Schindler, M.; Alvarez, S.; Szklenar, M.; Nagy, L.; et al. 9-cis-13,14-dihydroretinoic acid is an endogenous retinoid acting as RXR ligand in mice. *PLoS Genet.* **2015**, *11*, e1005213. [CrossRef] [PubMed]

49. de Lera, A.R.; Krezel, W.; Rühl, R. An Endogenous Mammalian Retinoid X Receptor Ligand, At Last! *Chem. Med. Chem.* **2016**, *11*, 1027–1037. [CrossRef] [PubMed]

50. Idres, N.; Marill, J.; Flexor, M.A.; Chabot, G.G. Activation of retinoic acid receptor-dependent transcription by all-trans-retinoic acid metabolites and isomers. *J. Biol. Chem.* **2002**, *277*, 31491–31498. [CrossRef] [PubMed]

51. Kane, M.A.; Folias, A.E.; Wang, C.; Napoli, J.L. Quantitative profiling of endogenous retinoic acid in vivo and in vitro by tandem mass spectrometry. *Anal. Chem.* **2008**, *80*, 1702–1708. [CrossRef] [PubMed]

52. Takahashi, Y.; Moiseyev, G.; Nikolaeva, O.; Ma, J.X. Identification of the key residues determining the product specificity of isomerohydrolase. *Biochemistry* **2012**, *51*, 4217–4225. [CrossRef] [PubMed]

53. McBee, J.K.; Van Hooser, J.P.; Jang, G.F.; Palczewski, K. Isomerization of 11-cis-retinoids to all-trans-retinoids in vitro and in vivo. *J. Biol. Chem.* **2001**, *276*, 48483–48493. [CrossRef] [PubMed]

54. Redmond, T.M.; Poliakov, E.; Kuo, S.; Chander, P.; Gentleman, S. RPE65, visual cycle retinol isomerase, is not inherently 11-cis-specific: Support for a carbocation mechanism of retinol isomerization. *J. Biol. Chem.* **2010**, *285*, 1919–1927. [CrossRef] [PubMed]

55. Muller, H.; Bub, A.; Watzl, B.; Rechkemmer, G. Plasma concentrations of carotenoids in healthy volunteers after intervention with carotenoid-rich foods. *Eur. J. Nutr.* **1999**, *38*, 35–44. [CrossRef] [PubMed]

56. Brevik, A.; Andersen, L.F.; Karlsen, A.; Trygg, K.U.; Blomhoff, R.; Drevon, C.A. Six carotenoids in plasma used to assess recommended intake of fruits and vegetables in a controlled feeding study. *Eur. J. Clin. Nutr.* **2004**, *58*, 1166–1173. [CrossRef] [PubMed]

57. Conrady, C.D.; Bell, J.P.; Besch, B.M.; Gorusupudi, A.; Farnsworth, K.; Ermakov, I.; Sharifzadeh, M.; Ermakova, M.; Gellermann, W.; Bernstein, P.S. Correlations Between Macular, Skin, and Serum Carotenoids. *Investig. Ophthalmol. Vis. Sci.* **2017**, *58*, 3616–3627. [CrossRef] [PubMed]

58. Devereux, G.; Seaton, A. Diet as a risk factor for atopy and asthma. *J. Allergy Clin. Immunol.* **2005**, *115*, 1109–1117. [CrossRef] [PubMed]

59. Shegokar, R.; Mitri, K. Carotenoid lutein: A promising candidate for pharmaceutical and nutraceutical applications. *J. Diet. Suppl.* **2012**, *9*, 183–210. [CrossRef] [PubMed]

60. Schwartz, S.; Frank, E.; Gierhart, D.; Simpson, P.; Frumento, R. Zeaxanthin-based dietary supplement and topical serum improve hydration and reduce wrinkle count in female subjects. *J. Cosmet. Dermatol.* **2016**, *15*, e13–e20. [CrossRef] [PubMed]

61. Grüber, C.; Taner, C.; Mihaly, J.; Matricardi, P.M.; Wahn, U.; Rühl, R. Serum retinoic acid and atopy among children of different ethnic origin living in Germany. *J. Pediatr. Gastroenterol. Nutr.* **2012**, *54*, 558–560. [CrossRef] [PubMed]

62. Wood, L.G.; Garg, M.L.; Blake, R.J.; Garcia-Caraballo, S.; Gibson, P.G. Airway and circulating levels of carotenoids in asthma and healthy controls. *J. Am. Coll. Nutr.* **2005**, *24*, 448–455. [CrossRef] [PubMed]

63. Canfield, L.M.; Clandinin, M.T.; Davies, D.P.; Fernandez, M.C.; Jackson, J.; Hawkes, J.; Goldman, W.J.; Pramuk, K.; Reyes, H.; Sablan, B.; et al. Multinational study of major breast milk carotenoids of healthy mothers. *Eur. J. Nutr.* **2003**, *42*, 133–141. [PubMed]

64. Borel, P. Genetic variations involved in interindividual variability in carotenoid status. *Mol. Nutr. Food Res.* **2011**, *56*, 228–240. [CrossRef] [PubMed]

65. van Vliet, T.; van Vlissingen, M.F.; van Schaik, F.; van den Berg, H. beta-Carotene absorption and cleavage in rats is affected by the vitamin A concentration of the diet. *J. Nutr.* **1996**, *126*, 499–508. [CrossRef] [PubMed]

66. van Het Hof, K.H.; West, C.E.; Weststrate, J.A.; Hautvast, J.G. Dietary factors that affect the bioavailability of carotenoids. *J. Nutr.* **2000**, *130*, 503–506. [CrossRef] [PubMed]

67. Grolier, P.; Duszka, C.; Borel, P.; Alexandre-Gouabau, M.C.; Azais-Braesco, V. In vitro and in vivo inhibition of beta-carotene dioxygenase activity by canthaxanthin in rat intestine. *Arch. Biochem. Biophys.* **1997**, *348*, 233–238. [CrossRef] [PubMed]

68. Shaw, N.; Elholm, M.; Noy, N. Retinoic acid is a high affinity selective ligand for the peroxisome proliferator-activated receptor beta/delta. *J. Biol. Chem.* **2003**, *278*, 41589–41592. [CrossRef] [PubMed]

69. Romanowska, M.; al Yacoub, N.; Seidel, H.; Donandt, S.; Gerken, H.; Phillip, S.; Haritonova, N.; Artuc, M.; Schweiger, S.; Sterry, W.; et al. PPARdelta enhances keratinocyte proliferation in psoriasis and induces heparin-binding EGF-like growth factor. *J. Investig. Dermatol.* **2008**, *128*, 110–124. [CrossRef] [PubMed]

70. Romanowska, M.; Reilly, L.; Palmer, C.N.; Gustafsson, M.C.; Foerster, J. Activation of PPARbeta/delta causes a psoriasis-like skin disease in vivo. *PLoS ONE* **2010**, *5*, e9701. [CrossRef] [PubMed]

nutrients

MDPI

Communication

Serum Carotenoids Reveal Poor Fruit and Vegetable Intake among Schoolchildren in Burkina Faso

Jean Fidèle Bationo [1], Augustin N. Zeba [2], Souheila Abbeddou [3], Nadine D. Coulibaly [2], Olivier O. Sombier [2], Jesse Sheftel [4], Imael Henri Nestor Bassole [5], Nicolas Barro [5], Jean Bosco Ouedraogo [2] and Sherry A. Tanumihardjo [4],*

[1] Centre Muraz, Post Office Box 390, Bobo Dioulasso 01, Burkina Faso; jeanfidelebationo@gmail.com
[2] Institute de Recherche en Sciences de la Santé, Post Office Box 545, Bobo Dioulasso 01, Burkina Faso; nawidzeba@gmail.com (A.N.Z.); nadineyd@yahoo.fr (N.D.C.); sombieolivier@yahoo.fr (O.O.S.); jbouedraogo@gmail.com (J.B.O.)
[3] University of Ghent, Department Public Health; 9000 Ghent, Belgium; Souheila.Abbeddou@ugent.be
[4] University of Wisconsin-Madison, Nutritional Sciences Department; Madison, WI 53706, USA; jsheftel@wisc.edu
[5] Université Ouaga 1 Joseph Ki-Zerbo, Ouagadougou 03, Burkina Faso; ismael.bassole@gmail.com (I.H.N.B.); barronicolas@yahoo.fr (N.B.)
*   Correspondence: sherry@nutrisci.wisc.edu

Received: 27 August 2018; Accepted: 27 September 2018; Published: 4 October 2018

Abstract: The health benefits of fruits and vegetables are well-documented. Those rich in provitamin A carotenoids are good sources of vitamin A. This cross-sectional study indirectly assessed fruit and vegetable intakes using serum carotenoids in 193 schoolchildren aged 7 to 12 years in the Western part of Burkina Faso. The mean total serum carotenoid concentration was 0.23 ± 0.29 µmol/L, which included α- and β-carotene, lutein, and β-cryptoxanthin, and determined with serum retinol concentrations in a single analysis with high performance liquid chromatography. Serum retinol concentration was 0.80 ± 0.35 µmol/L with 46% of children ($n = 88$) having low values <0.7 µmol/L. Total serum carotene (the sum of α- and β-carotene) concentration was 0.13 ± 0.24 µmol/L, well below the reference range of 0.9–3.7 µmol carotene/L used to assess habitual intake of fruits and vegetables. Individual carotenoid concentrations were determined for α-carotene (0.01 ± 0.05 µmol/L), β-carotene (0.17 ± 0.24 µmol/L), β-cryptoxanthin (0.07 ± 0.06 µmol/L), and lutein (0.06 ± 0.05 µmol/L). These results confirm the previously measured high prevalence of low serum vitamin A concentrations and adds information about low serum carotenoids among schoolchildren suggesting that they have low intakes of provitamin A-rich fruits and vegetables.

Keywords: α-carotene; β-carotene; β-cryptoxanthin; carotenoids; lutein; provitamin A; retinol; vitamin A

1. Introduction

Micronutrient deficiencies among children under 5 years old in low- and middle-income countries are common. Iron, vitamin A (VA), and zinc are specifically targeted for improvement by the World Health Organization (WHO) because their deficiencies are prevalent and lead to increased mortality and morbidity [1]. Community randomized controlled trials have shown that administering preformed VA solely or in combination with zinc to children in regions with a high prevalence of malaria, reduced morbidity [2,3]. Targeting preschool children is of specific interest because of long-term detrimental effects of undernutrition on cognitive development and adulthood work productivity [1]. VA deficiency (VAD) can lead to anemia, stunting, weakened resistance to infection, blindness, and death [1]. A random-effects meta-analysis of several VA trials showed reduction of mortality rates by 24% among children 6–59 months of age [4]. Current strategies to alleviate VAD are targeted high–dose VA supplementation to 6–59 months old children, fortification of foods that have high population coverage with preformed VA as retinyl palmitate and promoting the consumption of foods rich in VA and provitamin A carotenoids [1,5–7].

Despite a biannual campaign distributing high-dose VA capsules to children <59 months, VAD remains a public health problem in Burkina Faso [8]. Schoolchildren are assumed to be at lower risk but likely also suffer from micronutrient deficiencies in areas with low dietary diversity. Worldwide, 190 million preschool children are estimated to have VAD [1]. In Kaya ($n = 214$, age 8.5 ± 1.6 years) and Bogandé ($n = 337$, age 6.2–10.3 years), Burkina Faso, 47.2% and 37.1% of schoolchildren, respectively, had low serum retinol concentrations [9]. In Ouagadougou, 38.7% of 7–14 years old children attending private or public school had low serum retinol, which was more prevalent among children in public schools [10]. Low serum retinol concentration is an indicator of VAD and is currently recommended by WHO when used along with other biomarkers or surveys [11]. A prevalence of 20% of a given population group with serum retinol concentrations <0.7 μmol/L, is used to define VAD as a severe public health problem [11].

Micronutrient deficiencies in schoolchildren represent an additional concern because of increased nutritional needs for development, health, and academic performance [12]; this period can be as important as early infancy due to catch-up growth. In low- and middle-income countries where intake of animal-based foods is low, dietary provitamin A carotenoids are the main source of VA [13]. Due to economic reasons, limited knowledge and restricted access, the frequency and quantity of fruits and vegetables consumed by Burkinabe children is inadequate [7]. Increased access by children allowing repeat exposure showed an association with enhanced taste preferences towards increased fruit and vegetable intake [14]. Green leafy vegetables, orange fruits, and yellow-colored vegetables are rich in carotenoids and important for optimal health [13], and serum carotenoid concentrations reflect consumption of these fruits and vegetables [15,16]. The present study assessed serum retinol and carotenoid concentrations among schoolchildren (aged 7–12 years) in Burkina Faso to compare carotenoid-containing fruit and vegetable intake with other countries to determine relative consumption.

2. Materials and Methods

2.1. Study Area and Subjects

This study was conducted according to the guidelines laid down in the Declaration of Helsinki and all procedures involving human subjects/patients were approved by the Ethical Review Committee of Center MURAZ (001-2014/CE-CM). Written informed consent was obtained from parents or guardians of all included subjects. The trial was registered at Pan African Clinical Trials Registry (PACTR201702001947398). This cross-sectional study was conducted in the Western part of Burkina Faso, in primary school "A" of Kou's Valley in Bama's village in the Dande health district, between March and April 2014 during the dry season. Children attending the school were considered eligible if they were 7–12 years old living in the area; and their caregiver signed an informed consent form.

Exclusion criteria included clinical VAD, i.e., physiological ocular symptoms; severe anemia defined as hemoglobin (Hb) concentration <70 g/L [17]; serious illness (based on a medical examination) including tuberculosis and symptomatic human immunodeficiency virus infection. Children were enrolled based on the list of classes as a sampling frame. Recruitment covered the first four years of school (preparatory and elementary levels). We used z-score to assess anthropometric parameters based on WHO growth standards. Underweight was defined as WAZ more than 2 standard deviations (SDs) below the WHO median [18].

2.2. Data and Sample Collection

Venous blood (7 mL) was collected in EDTA tubes by trained nurses of Bama's clinic and stored immediately on ice before transportation to the laboratories of the Institut de Recherche en Sciences de la Santé (IRSS, Bobo-Dioulasso). Vials were protected from light with aluminum foil to avoid photo-degradation of the carotenoids and retinol. Blood samples were centrifuged the same day at

3000 rpm for 10-min with a Universal 320R centrifuge (Hettich Zentrifugen, D-78532, Tuttlingen, Germany). Serum was transferred into brown 1.5 mL microcentrifuge tubes (Eppendorf, Hamburg, Germany) and stored at −20 °C until shipment to the University of Wisconsin–Madison, USA (UW) for serum retinol and carotenoid analyses. Malaria parasites were counted on thick blood smears prepared in the field from capillary blood. Hb concentration was measured on capillary blood with the 301+ Hemocue system (Angelholm, Sweden). Height to the nearest 0.1 cm and weight to the nearest 10 g were measured.

2.3. Serum Retinol and Carotenoid Extraction and Analys

Serum retinol and carotenoid analyses were performed at UW in the same analysis as described [15,16] with modifications. To 200–300 µL serum, ethanol (1.5 X v) containing 0.1% butylated hydroxytoluene (MP Biomedicals, Solon, OH, USA) as an antioxidant, and 25 µL C23-β-apo-carotenol as an internal standard were added followed by 3 extractions with hexanes (2.5 X v) (Fisher Scientific, Pittsburgh, PA, USA). Pooled hexanes were dried under nitrogen and reconstituted in 100 µL 50:50 (v/v) methanol:dichloroethane; 50 µL was injected into a Waters HPLC [16] equipped with a Sunfire C18 (5 µm, 4.6×250 mm) analytic column (Waters, Inc., Milford, MA, USA) and a guard column. Samples were run at 1 mL/min using 70:30 (v/v) methanol:water (solvent A) and 80:20 (v/v) methanol:dichloroethane (solvent B), both with 10 mM ammonium acetate as a modifier, with the following gradient: 10-min linear gradient from 50% A:50% B to 100% B, followed by a 20-min hold, 3-min linear gradient to 50% A:50% B, and 6-min equilibration. Chromatograms were evaluated at 450 nm for carotenoids and 325 nm for retinol using purified external standards.

2.4. Statistical Methods

All analyses were conducted using the Statistical Analysis System (version 9.4; SAS Institute, Cary, NC, USA). Descriptive statistics were used to summarize nutritional characteristics of the participants as well as the main outcomes. z-scores were determined using WHO Anthro. The difference between Body Mass Index (BMI)-for-age z-scores of the children and WHO standards (z-score = 0) was analyzed by one-sample Student's *t*-test. Data were reported as mean ± SD.

3. Results

3.1. Subject Characteristics

All children (n = 193) met inclusion criteria and none of them had ocular signs of VAD, had blood samples drawn, and had anthropometric parameters measured (Table 1). Comparing BMI-for-age z-score (BAZ) of the children to WHO growth standards determined that mean BAZ was −1.24 ± 1.05 (different from 0, $p < 0.0001$) with 20% (CI = 95%) low BAZ (Figure 1). The children had high prevalence of underweight (17.6% with CI = 95%), stunting (25.2% with CI = 95%), asymptomatic malaria (39.5%), and prevalent anemia (23% with Hb concentration < 110 g/L).

Table 1. Subject characteristics.

Parameter	Value (mean ± SD)
Age, year	9.3 ± 1.48
Height, cm	129.1 ± 12.4
Weight, kg	24.8 ± 5.37
Height-for-age z-score	−0.73 ± 1.25
Weight-for-age z-score	−1.16 ± 1.01
BMI-for-age z-score	−1.24 ± 1.05
Hemoglobin, g/L	122.4 ± 10.9
Positive malaria blood smear, % (average parasitemia)	39.5 (1655 parasites)

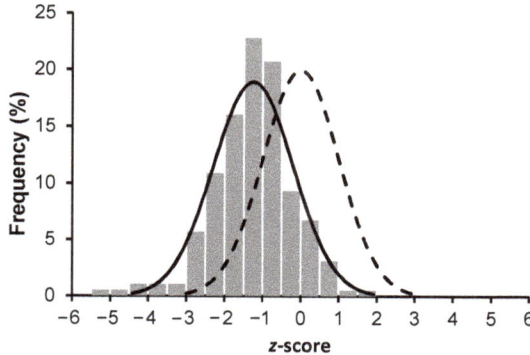

Figure 1. BMI-for-age z-score distribution in children aged 7–12 years (n = 193) in Burkina Faso. The normal distribution curves are given for these children (solid line, mean = −1.24 ± 1.05) compared with the WHO child growth standards for children 61 months–19 years of age (dashed line, mean = 0 ± 1).

3.2. Serum Retinol and Carotenoid Concentrations

Mean serum retinol concentration was 0.80 ± 0.35 μmol/L with 46% (88 children out of 193) <0.7 μmol/L (the cut-off for VAD) [11], although the utility of this measurement is limited as markers of chronic or acute inflammation (e.g., α_1-acid glycoprotein and C-reactive protein) were not taken to correct for the impact of infection in these children, which is known to lower serum retinol independent of VA status [11]. Concentrations of lutein, α- and β-carotene, and β-cryptoxanthin were determined and compared with other studies in children (Table 2). Total serum carotenoids ranged from 0 to 2.82 μmol/L with a mean of 0.23 ± 0.29 μmol/L. Serum carotenes had a low mean concentration (0.13 ± 0.24 μmol/L) well under the reference range (0.9–3.7 μmol/L) [16], and only two children were within this reference range. Serum retinol concentrations were not related to serum provitamin A carotenoid concentrations $p > 0.05$).

Table 2. Serum carotenoids concentrations (±SD) or range in apparently healthy children from various regions of the world.

Country	Age Years	n	Retinol μmol/L	β-Carotene μmol/L	α-Carotene μmol/L	β-Cryptoxanthin μmol/L	Lutein μmol/L	Reference
Burkina Faso	7–12	193	0.80 ± 0.35	0.12 ± 0.21	0.01 ± 0.04	0.05 ± 0.06	0.05 ± 0.06	Present study
(range)			(0.17–1.92)	(0.004–2.15)	(0–0.4)	(0–0.3)	(0–0.2)	
Zambia	5–7	123	0.98	0.76	0.62	0.10	0.86	Mondloch et al. [16]
Senegal	2–4	281	-	0.16	0.030	0.020	0.46	Rankins et al. [19]
India	2–11	50	1.10	0.31	0.035	0.12	0.42	Das et al. [20]
Philippines	9–12	27	0.87	0.23	0.03	0.07	0.23	Ribaya-Mercado et al. [21]
USA	6–7	839	-	0.34	0.075	0.21	0.34 [a]	Ford et al. [22]

[a] This value includes the xanthophyll zeaxanthin.

4. Discussion

The high prevalence of low serum retinol concentrations is consistent with research in other regions from Burkina Faso [9,10]. One-quarter of the children were anemic, with almost 40% of those presenting with asymptomatic malaria. Serum concentrations of the four measured carotenoids were lower than those found in other populations, including young preschool children and schoolchildren from other low-income countries (Table 2) [16,19–22]. Low fruit and vegetable consumption has been recognized as a key contributor to inadequate micronutrient intake and deficiencies affecting optimal health [13]. Intake of provitamin A-source foods was likely quite low as demonstrated by low serum retinol and carotenoid concentrations. Previous studies in Bolivia, Burkina Faso, and the Philippines showed that the diets are mainly based on cereals and tubers with white flesh and poor in carotenoids [23,24].

Staple foods are traditionally consumed with vegetable sauces in Burkina Faso, which could be good carotenoid sources, but data from previous studies showed high intake of vegetables that are poor in carotenoids, such as dried okra. The Institute of Medicine has defined bioefficacy of dietary β-carotene, α-carotene, and β-cryptoxanthin as 12, 24, and 24 μg to 1 μg retinol activity equivalents (RAE), respectively [25]. Dried okra, which has <5 μg RAE/100 g [26], is reportedly consumed daily in rural areas of Burkina Faso [24]. On the other hand, sorrel (*Rumex acetosa* L., a small green herb), which contains 75–140 μg RAE/100 g [26,27], is reportedly consumed in only one meal in six days (generally market day) [26,27] while Weber and Grune showed that carotenoids make up 35% of total VA intake in industrialized countries [28].

Dietary data collected previously showed that intake of fruits and vegetables is generally poor, due to low frequency and small quantities [26,27]. An intervention was able to increase intake of sorrel leaves and provide an extra 35–49 μg RAE to schoolchildren (Table 3) [27]. A previous 4-month trial in Burkina Faso showed that red palm oil (RPO), naturally rich in carotenoids, can cover daily VA needs for children <6 years of age [10]. Rural populations cannot afford refined VA-fortified oil, so promotion of RPO may be appropriate and add VA-value to vegetables poor in RAE, such as okra [26,27]. Although, the study village produced agricultural products, dietary habits, and cultural restrictions limited food diversification. Most people do not consume papaya for cultural reasons [26]. Other fruits and vegetables constitute an income for mothers who sell the products instead of using them for household consumption [26].

Table 3. Retinol activity equivalents (RAE) from sauce intake in Burkinabe children and coverage of the child's vitamin A requirements in relationship to the Estimated Average Requirements and the Recommended Daily Allowances.

Characteristic	Normal Diet [a]		Added Green Leafy Vegetables [b]	
Age group	6–9 years	≥10 years	6–9 years	≥10 years
Number of meals/day	2	2	2	2
Amount of sauce ingested/day (g)	184	254	184	254
Mean retinol activity equivalents (μg RAE/100 g)	36	36	55	55
Mean intakes of retinol activity (μg RAE/day)	66	91	101	140
Estimated average requirements [25] (μg RAE/day)				
Boys (6–13 years)	275–445	445	275–445	445
Girls (6–13 years)	275–420	420	275–420	420
Recommended Daily Allowance [25] (g RAE/day)	400–600	600	400–600	600
% of estimated average requirements met/day for boys and girls aged 6–13 years	15–24	20–22	23–37	31–33

Data based on information in reference [26]; RAE, Retinol Activity Equivalent. [a] The normal diet included 60 households in a rural area of Burkina Faso that received no intervention. [b] Green leafy vegetables were provided to 30 households from the same area that received locally-produced sorrel leaves at intervals corresponding to market day frequencies (every 3 days) for two months.

Serum carotenoid concentrations are influenced by bioavailability and overall bioefficacy [13,25]. Bioconversion of provitamin A carotenoids to retinol is affected by VA status and host-related facts, such as genetic polymorphisms [25]. Because different bioefficacy factors are used for provitamin A carotenoids, food composition data tables should report food VA content in mass of each carotenoid whenever possible [13]. We recognize some shortcomings of the study. The lack of dietary data on intake of fruits, vegetables, and other sources of VA and carotenoids is an important limitation of our study. Furthermore, biomarkers of inflammation like C-reactive protein and α_1-acid glycoprotein, which impact serum retinol, were not measured [11]. Malaria in Burkina Faso is endemic, which has been associated with low serum retinol concentrations in Thailand [29]. Serum retinol also suffers from sensitivity and specificity issues [11]. Determination of confounding factors and the causes of low serum retinol concentrations in this community would allow us to form more concrete associations among variables.

This study biochemically suggested poor fruit and vegetable intake among schoolchildren in Burkina Faso and demonstrated that low serum retinol concentrations are still prevalent, adding to the body of evidence that large-scale promotion of naturally rich RPO or food fortification may be needed among the population. Vegetable intervention studies should be pursued to evaluate whether locally available vegetables could be efficacious. Additionally, it shows the importance of behavior change education to increase dietary diversity. Dietary validation studies of self-reported fruit and vegetable intake should ideally include measurement of serum biomarkers of intake. Fruits and vegetables should be available and accessible both at home and at school. To date, there are a limited number of studies evaluating school-based environments to increase availability and accessibility of fruits and vegetables for children.

Author Contributions: Writing—original draft preparation, J.F.B., S.A. and J.S.; conceptualization, A.N.Z., J.B.O. and S.A.T.; methodology, J.F.B. and J.S.; validation, S.A. and S.A.T.; investigation, J.F.B., A.N.Z., N.D.C. and O.O.S.; academic oversight for J.F.B., I.H.N.B. and N.B.; resources, J.B.O. and S.A.T.; data curation, J.F.B. and J.S.; writing—review and editing, S.A., J.S. and S.A.T.; supervision, A.N.Z., J.B.O. and S.A.T.; project administration, A.N.Z. and J.B.O.; funding acquisition, A.N.Z., J.B.O. and S.A.T.

Funding: This work was supported by the International Atomic Energy Agency and Global Health funds at UW-Madison during sample analysis and manuscript preparation.

Conflicts of Interest: The authors declare no conflict of interest. The funders had no role in the design of the study; in the collection, analyses, or interpretation of data; in the writing of the manuscript, and in the decision to publish the results.

References

1. *Essential Nutrition Actions: Improving Maternal, Newborn, Infant and Young Child Health and Nutrition*; World Health Organization: Geneva, Switzerland, 2013.
2. Zeba, A.N.; Sorgho, H.; Rouamba, N.; Zongo, I.; Rouamba, J.; Guiguemdë, R.T.; Hamer, D.H.; Mokhtar, N.; Ouedraogo, J.B. Major reduction of malaria morbidity with combined vitamin A and zinc supplementation in young children in Burkina Faso: A randomized double blind trial. *Nutr. J.* **2008**, *7*, 7. [CrossRef] [PubMed]
3. Shankar, A.H.; Genton, B.; Semba, R.D.; Baisor, M.; Paino, J.; Tamja, S.; Adiguma, T.; Wu, L.; Rare, L.; Tielsch, J.M.; et al. Effect of vitamin A supplementation on morbidity due to Plasmodium falciparum in young children in Papua New Guinea: A randomised trial. *Lancet* **1999**, *354*, 203–209. [CrossRef]
4. Imdad, A.; Mayo-Wilson, E.; Herzer, K.; Bhutta, Z.A. Vitamin A supplementation for preventing morbidity and mortality in children from six months to five years of age. *Cochrane Database Syst. Rev.* **2017**, *3*, CD008524. [CrossRef] [PubMed]
5. Krause, V.M.; Delisle, H.; Solomons, N.W. Fortified foods contribute one half of recommended vitamin A intake in poor urban Guatemalan toddlers. *J. Nutr.* **1998**, *128*, 860–864. [CrossRef] [PubMed]
6. Engle-Stone, R.; Nankap, M.; Ndjebayi, A.O.; Brown, K.H. Simulations based on representative 24-h recall data predict region-specific differences in adequacy of vitamin A intake among Cameroonian women and young children following large-scale fortification of vegetable oil and other potential food vehicles. *J. Nutr.* **2014**, *144*, 1826–1834. [CrossRef] [PubMed]
7. Nana, C.P.; Brouwer, I.D.; Zagré, N.M.; Kok, F.J.; Traoré, A.S. Impact of promotion of mango and liver as sources of vitamin A for young children: A pilot study in Burkina Faso. *Public Health Nutr.* **2016**, *9*, 808–813. [CrossRef]
8. *Plan National d'Action Pour La Nutrition, Version Révisée*; Ministère de la Santé: Ouagadougou, Burkina Faso, 2001.
9. Daboné, C.; Delisle, H.F.; Receveur, O. Poor nutritional status of schoolchildren in urban and peri-urban areas of Ouagadougou (Burkina Faso). *Nutr. J.* **2011**, *10*, 34. [CrossRef] [PubMed]
10. Zeba, A.N.; Prével, Y.M.; Somé, I.T.; Delisle, H.F. The positive impact of red palm oil in school meals on vitamin A status: Study in Burkina Faso. *Nutr. J.* **2006**, *5*, 17. [CrossRef] [PubMed]
11. Tanumihardjo, S.A.; Russell, R.M.; Stephensen, C.B.; Gannon, B.M.; Craft, N.E.; Haskell, M.J.; Lietz, G.; Schulze, K.; Raiten, D.J. Biomarkers of nutrition for development (BOND)-Vitamin A review. *J. Nutr.* **2016**, *146*, 1816S–1848S. [CrossRef] [PubMed]
12. Schoenthaler, S.J.; Bier, I.D.; Young, K.; Nichols, D.; Jansenns, S. The effect of vitamin-mineral supplementation on the intelligence of American schoolchildren: A randomized, double-blind placebo-controlled trial. *J. Altern. Complement. Med.* **2000**, *6*, 19–29. [CrossRef] [PubMed]
13. Tanumihardjo, S.A.; Palacios, N.; Pixley, K.V. Provitamin A carotenoid bioavailability: What really matters? *Int. J. Vitam. Nutr. Res.* **2010**, *80*, 336–350. [CrossRef] [PubMed]

14. Blanchette, L.; Brugg, J. Determinants of fruit and vegetable consumption among 6-12-year-old children and effective interventions to increase consumption. *J. Hum. Nutr. Diet.* **2005**, *18*, 431–443. [CrossRef] [PubMed]

15. Yang, Z.; Zhang, Z.; Penniston, K.L.; Binkley, N.; Tanumihardjo, S.A. Serum carotenoid concentrations in postmenopausal women from the United States with and without osteoporosis. *Int. J. Vit. Nutr. Res.* **2008**, *78*, 105–111. [CrossRef] [PubMed]

16. Mondloch, S.; Gannon, B.M.; Davis, C.R.; Chileshe, J.; Kaliwile, C.; Masi, C.; Rios-Avila, L.; Gregory, J.F., 3rd; Tanumihardjo, S.A. High provitamin A carotenoid serum concentrations, elevated retinyl esters, and saturated retinol-binding protein in Zambian preschool children are consistent with the presence of high liver vitamin A stores. *Am. J. Clin. Nutr.* **2015**, *102*, 497–504. [CrossRef] [PubMed]

17. *Vitamin and Mineral Nutrition Information System—Haemoglobin Concentrations for the Diagnosis of Anaemia and Assessment of Severity*; World Health Organization: Geneva, Switzerland, 2011.

18. *Comparison of the World Health Organization (WHO) Child Growth Standards and the National Center for Health Statistics/WHO International Growth Reference: Implications for Child Health Programmes*; World Health Organization: Geneva, Switzerland, 2011.

19. Rankins, J.; Green, N.R.; Tremper, W.; Stacewitcz-Sapuntzakis, M.; Bowen, P.; Ndiaye, M. Undernutrition and vitamin A deficiency in the Department of Linguère, Louga Region of Sénégal. *Am. J. Clin. Nutr.* **1993**, *58*, 91–97. [CrossRef] [PubMed]

20. Das, B.S.; Thurnham, D.I.; Das, D.B. Plasma α-tocopherol, retinol and carotenoids in children with falciparum malaria. *Am. J. Clin. Nutr.* **1996**, *64*, 94–100. [CrossRef] [PubMed]

21. Ribaya-Mercado, J.D.; Maramag, C.C.; Tengco, L.W.; Blumberg, J.B.; Solon, F.S. Relationships of body mass index with serum carotenoids, tocopherols and retinol at steady-state and in response to a carotenoid rich vegetable diet intervention in Filipino schoolchildren. *Biosci. Rep.* **2008**, *28*, 97–106. [CrossRef] [PubMed]

22. Ford, E.S.; Gillespie, C.; Ballew, C.; Sowell, A.; Mannino, D.M. Serum carotenoid concentrations in US children and adolescents. *Am. J. Clin. Nutr.* **2002**, *76*, 818–827. [CrossRef] [PubMed]

23. Melgar-Quinonez, H.R.; Zubieta, A.C.; MkNelly, B.; Nteziyaremye, A.; Gerardo, M.F.; Dunford, C. Household food insecurity and food expenditure in Bolivia, Burkina Faso and the Philippines. *J. Nutr.* **2006**, *136*, 1431S–1437S. [CrossRef] [PubMed]

24. Belesova, K.; Gasparrini, A.; Sié, A.; Sauerborn, R.; Wilkinson, P. Household cereal crop harvest and children's nutritional status in rural Burkina Faso. *Environ. Health* **2017**, *16*, 65. [CrossRef] [PubMed]

25. Institute of Medicine, Food and Nutrition Board. *Dietary Reference Intakes for Vitamin A, Vitamin K, Arsenic, Boron, Chromium, Copper, Iodine, Iron, Manganese, Molybdenum, Nickel, Silicon, Vanadium, and Zinc*; National Academy Press: Washington, DC, USA, 2001; pp. 65–126.

26. Nana, C.P.; Brouwer, I.D.; Zagré, N.M.; Kok, F.J.; Traoré, A.S. Community assessment of availability, consumption, and cultural acceptability of food sources of (pro)vitamin A: Toward the development of a dietary intervention among preschool children in rural Burkina Faso. *Food Nutr. Bull.* **2005**, *26*, 356–365. [CrossRef] [PubMed]

27. Ayassou, K.; Ouedraogo, M.; Mathieu-Daudé, C.; Alain, B.; Chevalier, P. Amélioration de L'alimentation Burkinabè Avec Des Aliments Riches en Caroténoïdes. Available online: http://horizon.documentation.ird.fr/exl-doc/pleins_textes/divers11-08/010036323.pdf (accessed on 13 July 2018).

28. Weber, D.; Grune, T. The contribution of β-carotene to vitamin A supply of humans. *Mol. Nutr. Food Res.* **2012**, *56*, 251–258. [CrossRef] [PubMed]

29. Stuetz, W.; McGready, R.; Cho, T.; Prapamontol, T.; Biesalski, H.K.; Stepniewska, K.; Nosten, F. Relation of DDT residues to plasma retinol, alpha-tocopherol, and beta-carotene during pregnancy and malaria infection: A case-control study in Karen women in northern Thailand. *Sci. Total Environ.* **2006**, *363*, 78–86. [CrossRef] [PubMed]

MDPI

St. Alban-Anlage 66

4052 Basel

Switzerland

Tel. +41 61 683 77 34

Fax +41 61 302 89 18

www.mdpi.com

Nutrients Editorial Office

E-mail: nutrients@mdpi.com

www.mdpi.com/journal/nutrients

www.ingramcontent.com/pod-product-compliance
Lightning Source LLC
Chambersburg PA
CBHW051716210326
41597CB00032B/5502